On Their Own Terms

On Their Own Terms

Science in China, 1550–1900

Benjamin A. Elman

Harvard University Press

Cambridge, Massachusetts, and London, England | 2005

Library of Congress Cataloging-in-Publication Data

Elman, Benjamin A., 1946–
    On their own terms : science in China, 1550–1900 / Benjamin A. Elman
        p. cm.
    Includes bibliographical references and index.
    ISBN 0-674-01685-8 (alk. paper)
    1. Science—China—History—16th century. 2. Science—China—
History—17th century. 3. Science—China—History—18th century.
4. Science—China—History—19th century. I. Title.
Q127.C5E38 2005
509'.51—dc22   2004059654

For Susan Naquin and Nathan Sivin
and in memory of my mother, Rachel Elman

Contents

List of Maps, Illustrations, and Tables

Tables

Chinese Dynasties

Shang: 16th–11th Century B.C.E.

Zhou: 11th Century–221 B.C.E.

Qin: 221–207 B.C.E.

Han: 206 B.C.E.–220 C.E.

  (Former Han: 206 B.C.E.–28 C.E.

  Later Han: 25–220)

Wei: 220–265

Western Jin: 265–316

Nan-Bei Chao: 420–581

  (Southern and Northern Dynasties)

  (Lui-Song: 420–479

  Northern Wei: 386–534

  Northern Zhou: 557–581)

Sui: 581–618

Tang: 618–907

Liao: 916–1125

Song: 960–1280

  (Northern Song: 960–1127

  Southern Song: 1127–1280)

Jin: 1115–1234

Yuan: 1280–1368

Ming: 1368–1644

Qing: 1644–1911

Physiocratic Macroregions of Agrarian China in Relation to Major Rivers

Source: Reprinted from *The City in Late Imperial China,* edited by G. Wm. Skinner, with the permission of the publishers, Stanford University Press. Copyright 1977 by the Board of Trustees of the Leland Stanford Jr. University.

China Administrative Map

Abbreviations

BJMS   *Baijia mingshu* (Famous works of all the scholastic sects). Hu Wenhuan (fl. ca. 1596), comp. Late Ming Wanli edition, circa 1603. The Library of Congress edition contains 156 works. The edition in the Oriental Library of the Institute of Humanistic Studies, Kyoto University, contains ninety-two works.

CRZ   *Chouren zhuan* (Biographies of mathematical astronomers). Ruan Yuan (1764–1849), comp. Taibei: Shijie shuju, 1962.

DMB   *Dictionary of Ming Biography*. L. C. Goodrich, et al., eds. 2 vols. New York: Columbia University Press, 1976.

DSB   *Dictionary of Scientific Biography*. Charles Gillispie et al. N.Y.: Scribner's Sons, 1970–78.

ECCP   *Eminent Chinese of the Ch'ing Period*. Arthur Hummel, ed. Reprint. Taibei: Chengwen Bookstore, 1972.

GJTSJC   *Gujin tushu jicheng* (Synthesis of books and illustrations past and present). Chen Menglei et al. Taibei: Dingwen shuju, 1965.

GRC   *Gregorian Reform of the Calendar: Proceedings of the Vatican Conference to Commemorate its 400th Anniversary 1582–1982*. G. V. Coyne, S. J., et al., eds. Vatican City: Specola Vaticana, 1982.

GZCS   *Gezhi congshu* (Collectanea for investigating things and extending knowledge). Hu Wenhuan (fl. ca. 1596), comp. Taibei: National Library Rare Books Collection edition with forty-six works, ca. 1592–97.

GZHB   *Gezhi huibian* (The Chinese Scientific Magazine; also called *The Chinese Scientific and Industrial Magazine*). Reprint in 6 vols. Nanjing: Guji shudian, 1992. The dating of articles in this reprint are defective, and some sections are disordered. Reference to the original editions, though limited in number, is recommended.

GZJY   *Gezhi jingyuan* (Mirror of origins based on the investigation of things and extending knowledge). Chen Yuanlong (1652–1736),

comp. Original edition published in 1735. Included in the
SKQS, vols. 1031–1032.

GZSYKY *Gezhi shuyuan keyi* (China Prize Essay Contest). Annual volumes.
Shanghai: Shanghai Polytechnic, 1886–1893. From the Fryer
Private Library, University of California, Berkeley, in 2 volumes,
which include eight traditional stringed volumes (*ce*).

GZQM *Gezhi qimeng* (Primers for science education). Young J. Allen et
al., trans. Shanghai: Jiangnan Arsenal Publication, 1879–1880.
Gezhi qimeng (Primers for science education). Joseph Edkins,
trans. Beijing: Imperial Customs Office Publication, 1886.

HJAS *Harvard Journal of Asiatic Studies.*

HQJJ *Huang Qing jingjie* (Qing exegesis of the Classics). Ruan Yuan et
al., comp. 20 vols. Taibei, Fuxing Reprint, 1961.

JAS *Journal of Asian Studies.*

JIK *Jingyi kao* (Critique of classical studies). Zhu Yizun. *Sibu beiyao*
edition.

LHCT *Liuhe congtan* (Shanghae Serial), Published January 1857 to June
1858 by Mohais huguan (London Missionary Society).

LMS London Missionary Society.

MS *Mingshi* (Ming history). Taibei: Dingwen Bookstore, 1982.

SBCK *Sibu congkan* photolithograph editions. Shanghai: Commercial
Press, 1920–1922.

SCC *Science and Civilisation in China.* Joseph Needham and others.
Multi-volumes. Cambridge: Cambridge University Press, 1954–.

SKQS *Siku quanshu* (Complete collection of the four treasuries).
Reprint. Taipei: Commercial Press, 1983–1986.

SKQSZM *Siku quanshu zongmu* (Catalog of the complete collection of the
four treasuries). Ji Yun et al., comp. Taibei: Yiwen Press reprint,
1974.

TPYL *Taiping yulan* (Materials of the Taiping Xing Guo [976–983] era
read by the emperor). Reprint of the 1199 edition. 7 vols. Taibei:
Shangwu yinshu guan, 1974.

ZGJD *Zhongguo jindai xuezhi shiliao* (Historical sources of modern
Chinese educational institutions), Vol. 1, Part 1. Shanghai:
Huadong shifan daxue chuban she, 1983.

ZGKX *Zhongguo kexue jishu dianji tonghui* (Comprehensive
compendium of materials and sources on Chinese science and
technology). Multi-volumes. Shanghai: He'nan jiaoyu chubanshe, 1993, 1996, etc.

Preface

Conventions

This work is not a straightforward historical survey of the Chinese sciences that approaches that history topically. Fortunately, Joseph Needham's *Science and Civilisation in China* series provides the best sourcebook for that. Instead, I focus principally on Chinese natural studies and the literati mastery of European natural learning from 1550 to 1900. In particular, I show how Manchu rulers and Chinese scholars broadened late imperial studies of astronomy, geography, mathematics, and medicine through court and literati contacts with Jesuits and Protestants, who presented European natural philosophy as part of their religious inheritance. Other areas such as physics, chemistry, and technology are covered when pertinent.

In this volume, the term *literati* refers to select members of the land-holding gentry who maintained their status as cultural elites primarily through classical scholarship, knowledge of lineage ritual, and literary publications. The term *gentry* refers to those before 1900 who wielded local power as landlords or provincial and empire-wide power as government officials. The cultural status of both the gentry at large and the literati in their midst correlated with their rank on the civil service examinations. In addition, during the late empire, gentry and merchants intermingled, with the latter becoming part of the gentry elite.

Chinese characters have not been included for Chinese historical figures. Most are readily available in biographical collections such as *Dictionary of Ming Biography* (see DMB), *Eminent Chinese of the Ch'ing Period* (ECCP), and Howard Boorman and Richard Howard, *Biographical Dictionary of Republican China* (5 vols. New York: Columbia University Press, 1967–1971). A romanized bibliography of Chinese and Japanese sources cited in the text is presented at the end.

Terminology such as the "West" refers to early modern Europe (1500–1800) in Parts I–III and grows to include the United States and modern Europe (1800–1900) in Parts IV–V. Chinese sources typically refer to the "West" or "Western." When appropriate, I narrow the focus to Western Europe in Chapters 1–7, when Europe is the focus rather than the West. Normally, I present the "West" as a cultural construct used by Jesuits, Protestants, and Chinese to refer to themselves or to the "other." "European" and "Euro-American" will designate a regionalized view with more historical and chronological precision.

The term "Learning of the Way" (*Daoxue*) will refer to the schools of literati thought that developed in the Song, Yuan, and Ming dynasties. Unfortunately, there has been a tendency to use "Neo-Confucianism" (*xin ruxue*) to cover all classical intellectual currents from the Tang dynasty (618–907) until the fall of the last dynasty in 1911. Readers should guard against this tendency to assimilate Neo-Confucianism to the broader amorphous designations that have been attached to it. The definition and content of "Learning of the Way" were as debatable during the eleventh and twelfth centuries as they were in the seventeenth and eighteenth. The term "Cheng-Zhu orthodoxy" will refer specifically to the imperially sanctioned Cheng Yi (1033–1107) and Zhu Xi (1130–1200) school of interpretations for the official canon. These were officially accepted in civil examinations beginning in 1313 and remained the basis of the imperial examinations until 1905.

Historiographical Issues

During the Sino-Japanese War from 1894 to 1895, the Japanese army and navy decisively defeated the armed forces of the Manchu Qing dynasty (1644–1911). Since then, Chinese and Japanese patriots and scholars have assumed that Meiji Japan (1868–1911) was vastly superior to Qing China in modern science prior to 1894. Actually, prior to the war many contemporary observers thought the Qing army and navy were superior, even if only in sheer numbers. After 1895, each side rewrote their histories to validate triumphant Japan or lament the defeated Qing. For Chinese and Manchus, the Sino-Japanese War turned the Qing era of Self-Strengthening reforms from 1865 to 1895 into an alleged scientific and technological catastrophe.[1]

Naval Wars in Chinese History

One of the ironies of the Qing misfortune in 1895 was that since the Song dynasty (960–1280) China had at times supported a substantial navy, which the Mongols used to invade Japan in 1274 and 1281 and attack Java in 1293.

Subsequently, the Ming dynasty (1368–1644) navy carried out several enormous excursions into Southeast Asia and the Indian Ocean from 1403 to 1434, which ended when the court scrapped the navy in the 1460s to prepare for possible land wars in the northwest against the Oirat Mongols.

A coastal navy equipped with cannon and firearms had defended the China coast from Japanese pirates in the mid-sixteenth century, initially in vain but with eventual success. Chinese naval power further revived when the Ming helped Korea to halt Toyotomi Hideyoshi's (1536–1598) invasions of 1592 and 1598. Ming loyalists early on defeated the Qing dynasty in major naval and land battles along the Fujian coast. The naval revival lasted only until the 1680s when Qing naval forces annexed Taiwan. Thereafter, the Chinese still developed new types of sailing vessels, such as the Zhejiang junks first built in 1699. Chinese merchants used them in northern waters for the Ningbo-Nagasaki trade between Japan and China, which lasted into the eighteenth century despite Japan's alleged closed-door policies.[2]

The Qing court in the 1860s and we today might have heralded the revival of the Qing navy after the Opium wars as a return to the brighter days of the early fifteenth, mid-sixteenth, and seventeenth centuries. Instead, the late Qing navy was ridiculed after 1895. Anti-Manchu patriots appealed to the Zheng He fleets as sign of China's past greatness and current Manchu weakness. Moreover, the superiority of Japan in modern technology and science was assumed after the Sino-Japanese War. The Japanese navy dominated Pacific waters until 1945.[3]

After 1895 each side read their fates differently: back to the early Meiji period (later, even further back to Dutch Learning—but not back to Hideyoshi's defeats), in the case of triumphant Japan; or back to the failures of the self-strengthening movement after 1865 (later back to all classical learning and institutions), in the case of the defeated Qing. The possible continuity between the military strength of the Ming and early Qing naval fleets and the late Qing navy became inconceivable.[4]

Scholars revisited the first Western impact in East Asia that began in the sixteenth century and concluded that, unlike the more open-minded Japanese who absorbed Dutch scientific learning, the Ming Chinese—like their Qing successors—had also rejected the new world of scientific learning the Jesuits offered them. This perspective, for instance, still informs the discussion in the recently revised edition of *Sources of Chinese Tradition*, Volume Two, which uncritically describes "the general disinterest of the Chinese in Western science," which "had been tendered at the hands of gentle missionaries."[5]

No amount of sophistry can turn the Qing debacle in 1895 into a victory, even a hollow or moral one. Past accounts of China's failures in science and dynastic losses on modern military battlefields are instructive, but their rhetoric about that violence has usurped the actual destructive events—long since past—which placed China at the mercy of the West and Meiji Japan from the 1890s onward. Beneath the cultural narrative of scientific, technological, and military failure, which many Euro-Americans, Japanese, and Chinese still attach to Chinese history after 1895, lies another story. It will never replace the triumphal story of the march of Western and Japanese imperialism via science, technology, and empire-building, nor should it. Nevertheless, this book tells a quieter story of longstanding Chinese interests in the natural world, medicine, the arts and crafts, and commerce that set the stage for the interaction with European science, technology, and medicine from 1600 to 1900.

The impatient perspectives of China's efforts to westernize after 1865, which are still prominent among historians today, have unfortunately predisposed us to underestimate the crucial role the missionary translations of science, the industrialization and techno-science in the Qing arsenals, and the new government schools played in the emergence of modern science and technology. We should deal with the late nineteenth-century arsenals, factories, and translation schools by also considering them as a harbinger of things to come and not simply as a prelude to the end of the Qing dynasty and imperial China. China's defeat in 1895 was not due to a lack of science or technology. Late Qing Manchu and Chinese leaders inherited a legacy of regionalism and decentralization from the Taiping Rebellion (1850–1867), which appreciably contributed to corruption and the mismanagement of the war effort. Pamela Crossley has noted that the military reforms were "overwhelmed by the inability to integrate development in such a way that innovative institutions might reinforce each other rather than impede each other."[6]

Premodern and Modern Science

When Europeans reached China during the age of exploration, the highest learning (*scientia*) of their men of culture did not connote natural science. For Thomas Aquinas (1225–1274), natural philosophy, not natural science, was a field of higher learning. *Science* was a medieval French term, which was synonymous with accurate and systematized knowledge. When Latinized the word became *scientia* and represented among scholastics and early modern elites the specialized branches of Aristotelian moral and natural philosophy.[7]

Philip Melanchthon's (1497–1560) *On the Soul* (*De Anima*, 1553), upheld the scholastic distinction between *scientia* and *opinio*. The former represented certain knowledge, while the latter referred to plausible or commonly accepted beliefs. Included in the Scholastic regime for learning were the seven sciences of medieval learning: grammar, logic, rhetoric, arithmetic, music, geometry, and astronomy. Comparable to the classical ideal of the six arts in ancient China (rites, music, archery, charioteering, calligraphy, and mathematics), these seven liberal arts served in Roman education as preparation for more specialized training in philosophy, medicine, or law.[8]

In medieval times, Boethius's (c. 475–524) pioneering translations of Aristotle into Latin had prioritized four mathematical disciplines for elementary education: arithmetic, geometry, music, and astronomy. They represented the four roads to wisdom (*quadrivium*), which balanced the three disciplines of logic (*trivium*), namely, grammar, dialectics, and rhetoric. After Thomas Aquinas in Paris and others elsewhere reconciled the Muslim harvest of classical learning with Christian theology, the preferred order of Aristotelian learning for the Renaissance scholars and bookmen became: (1) logic; (2) mathematics; (3) natural science; (4) moral philosophy; and (5) metaphysics. This regime was reproduced in the *Ratio Studiorum* at Jesuit colleges in Western Europe. The Jesuits transmitted Aristotelian cosmology, physics, and meteorology through their Scholastic theory of knowledge.[9]

Like contemporary Europeans and Islamic scholars, late imperial Chinese also prioritized mathematical studies for their premodern exact sciences, which informed Chinese astronomy, geography, cartography, and alchemy in different ways. Literati also applied the naturalistic concepts of yin and yang and the five evolutive phases (*wuxing*) to elucidate the spontaneous (*ziran*) changes in the "stuff of the world" (*qi*). Rational and abstract explanations of natural things and phenomena characterized the premodern sciences worldwide, particularly Chinese elite traditions of natural studies.[10]

I will discuss popular understandings of natural events and the popularization of the exact sciences in late imperial China, but readers interested in the popular conceptions of natural phenomena among nonelites should refer to the forthcoming work of Nathan Sivin on the relationship between science, medicine, and popular culture in light of gods, ghosts, and spirits as intermediaries between humans and the cosmos. Works on Chinese Buddhism and Daoism by scholars who address popular religion rather than esoteric religion with a hereditary priesthood also advance our understanding of the history of medicine.[11]

If premodern science in China, Europe, and Islam often denoted a rational and abstract understanding of the natural world, the rise of modern science in the eighteenth and nineteenth centuries melded the exact sciences with machine-driven technologies that surpassed the rich artisanal traditions of the early modern world. In addition, historians of modern science have stressed the application of mathematical hypotheses to nature, the use of the experimental method, the geometricization of space, and the mechanical model of reality to explain the scientific revolution in the West. My study will be agnostic about these latter claims, which philosophers of science frequently use to explain why China or the Islamic world failed to develop the rigorous mental mind-set of modern science.[12]

Instead, I will explore Chinese interests in natural studies as they articulated and practiced them on their own terms rather than speculate about why they did not accomplish what the Europeans did. Whenever appropriate, I will contextualize Chinese natural studies by comparing the lingering vitality of the premodern exact sciences in China with the decisive turn toward Newtonian mechanics and industrial revolution in eighteenth-century France and England. I will also compare the rapid industrialization of Europe in the nineteenth century to the slower but equally extraordinary rise of modern Chinese machine-driven industry.

Traduttore, Traditore: Christianity, Science, and Translation in China

I should note at the outset the surplus of studies on the Jesuits and science in late imperial China and the relative surfeit of such studies for the Protestants and science. As a result, until recently most accounts have underestimated the religious activities of the former and overestimated them for the latter. I will add another layer of materials and interpretation to earlier accounts of science and the Jesuits in China. I will also emphasize the relationship between Protestants and modern science in China, which was repressed by both Chinese and Christians after China's debacle in the Sino-Japanese War in 1895.

Scholars and churchmen since 1600 have lauded and damned the role of the Jesuits in China. I will describe their many sides, but my focus on science will highlight their anti-Copernican and anti-Newtonian positions in the seventeenth and eighteenth centuries. The Jesuits should not be "black-boxed," however, as a religious enemy of science. Many were "closet" Copernicans or Newtonians at a time when they could not publicly disown the ecclesiastical positions the Pope and the Jesuit order took against Copernicus (1473–1543) and Galileo (1564–1642). In the end, as Nathan Sivin has explained, "European cosmology had been discredited by its own unexpected lack of internal coherence." In the seventeenth century, Jesuits first presented Copernicus as a follower of Ptolemy (and geocentrism) and then of Tycho Brahe (and geoheliocentrism). When Michel Benoist (1715–1774) finally presented Copernican heliocentrism in the mid-eighteenth century, Chinese literati thought the presentations too incoherent to take seriously.[13]

Similarly, my account of the Protestants will highlight their opposition to Darwin's theory of evolution, particularly the notion of the survival of the fittest through natural selection. Lacking a unified church such as Catholicism under the Pope, the Methodists, Presbyterians, and other Protestants who came to China after the Opium War (1839–1842) nevertheless were surprisingly even more unanimous in opposing Darwin than the Jesuits had been contra Copernicus. Few Protestants in China were "closet" Darwinians; almost all favored Christianized versions of evolution. Still, the Protestants were not unrelenting enemies of science either, even though they delayed the translation of modern biology in China by three decades.

What united the Jesuit and Protestant eras in Chinese history was the religious agenda they brought as foreigners to China. Their primary project of religious conversion required that they translate Christianity into Chinese, orally and textually, as well as in ritual practice. Hence, Christians worked with Chinese literati converts to present Catholic and Protestant beliefs orally and in readable classical Chinese, the lingua franca of educated Chinese until the twentieth century. The majority of works produced by this workmanlike partnership were religious works. A significant minority dealt with the sciences, early modern for the Jesuits, and modern for the Protestants.[14]

How the Jesuits and Protestants worked with their native Chinese informants to legitimate the West after 1600 is one of the keys to understanding the translation enterprise, which packaged Christianity and European science together. Jesuits and Protestants tainted their translation project with their own conscious and unconscious predispositions by massaging the literal meaning of original European texts. They often chose to capture the spirit of the original through a wide-ranging engagement with the classical resources of the target language and audience. At other times the missionaries and their Chinese partners created new terminology to present Christian religious and scientific views. Usually, however, both sides could find acceptable terms in the vernacular and classical lexicon of Chinese to accommodate the meaning of the Western source. Accommodation was as difficult for the Jesuits during the Rites Controversy as for the Protestants. By 1850, the new disciplines of chemistry and the calculus, for example, increasingly required new concepts—not only new words. Such concepts shaped new realities, which empowered many Chinese to grasp new scientific discourses.[15]

In this volume, I bracket the religious texts that the Catholics and Protestants spent most of their time translating into Chinese. In addition, I spotlight the early modern and modern scientific texts translated jointly by Christian missionaries and Chinese literati. These science translations were not simply innocent by-products of the missionary enterprise, however. The science texts the missionaries successfully translated into classical Chinese were encoded with Christian messages and religiously induced silences. Hence, I do not focus on translation as a futile exercise in philosophical incommensurability. Instead, I demonstrate the willful infiltration of Christian beliefs in the scientific textbooks translated into Chinese.[16]

My analysis of scientific translations by missionaries and their converts will reveal the unspoken ideological predispositions that Catholics, Protestants, and Chinese all encoded. This lack of transparency has been underestimated in previous accounts of translation in China and Japan as high-minded efforts at achieving verisimilitude. Jesuit and Protestant translators did not just fail to achieve a seamless word-by-word correspondence from Latin or English to Chinese. Nor were they simply idealistic failures because they unsuccessfully sought to transmit the spirit of the source through more adventuresome literary tactics. The translation enterprise in Japan, for example, followed a different cultural trajectory altogether because the Japanese barred the Jesuits early on and stressed their own translations of science in the nineteenth century, rather than relying on Protestant middlemen.[17]

The Jesuits and Protestants in China rejected the European original when the native source betrayed their religious sensibilities. The result was not a failure to communicate. Competent in vernacular Chinese, the Jesuits and Protestants communicated well enough to their Chinese partners who produced the classical translations. Rather, the enterprise of translation became a mindful, Christian effort not to translate Copernican heliocentricity or Darwinian evolution in the original to the target audience. We see in this act of dissembling about the sciences a sense of religious commitment that we should not underestimate. Jesuit and Protestant dissembling tells us that we must further problematize the act of translation ideologically and recognize that the Chinese were not privy to the untranslated passages that we are today. What they did not know mattered.

Science, Medicine, and Social Status in Late Imperial China

Modern scholars have stressed the social distance between those in imperial China who represented the classically trained scholar-gentry elite and those who worked as physicians or professed mathematical astronomy. I have tried to account for all by presenting men such as the physician-herbalist Li Shizhen (1518–1593) and the mathematical astronomer Mei Wending (1633–1721) as part of the community of literati scholars. Many physicians and mathematicians were proficient in the triad of astronomy, nonjudicial astrology, and divination. Not until 1900 would a scientific community emerge in China that would fully encompass the doctor and the mathematician, but both were rooted in the same elite social formations since 1100 that undergirded premodern study of natural studies.[18]

Specialists will readily understand why I select Li Shizhen and his *Systematic Materia Medica* (*Bencao gangmu*) to represent in part late Ming Chinese views of natural phenomena, but they will wonder why I also choose the relatively unknown (by today's standards) Hu Wenhuan (fl. ca. 1600) and his much-maligned *Collectanea for Investigating Things and Extending Knowledge* (*Gezhi congshu*). Although it contained many rogue editions printed in Hu's Hangzhou and Nanjing print shops, the wide dissemination of the *Collectanea* in Ming-Qing China and Japan marks it as a popular work. Like Li's *Systematic Materia Medica*, it was influential among elites, although many belittled its populist audience. Hu also incorporated orthodox classical lexicons and natural histories that were representative overall of literati learning on the eve of the Jesuits in China.

Although lacking the scholarly pedigree of the more famous genre of jotted notes (*biji*), the lower-brow status of Hu's collection of natural studies grants us a unique window onto the quotidian world of learning, which existed beyond the reach of Wang Yangming's (1472–1528) revival of ancient classical values, his influential schoolmen, and other late Ming higher-brow literati intellectual currents. As my Prologue makes clear, most local literati were reading popular novels and encyclopedias whether or not they appreciated Wang Yangming's classical discourses. Hu's *Collectanea* is also uncontaminated by later accretions that mark many materials reproduced after the Jesuit presence.[19]

The subsequent rise in social status of literati who mastered mathematics, astronomy, and Western studies was appreciable after 1700. Just as mathematics enhanced the Jesuit view of themselves in Europe and China as urbane masters of *scientia,* so too when Ming and Qing literati mastered mathematical astronomy, they enhanced their social status and political value when the calendar crisis loomed over the Ming and Manchu courts. While the cognitive framework of science in the making was certainly seminal in late imperial China, as in modern Europe, its institutionalization also depended on the heightened social prestige and political patronage that experts in the exact sciences increasingly displayed in the eighteenth century.

Things as Phenomena, Affairs, and Events

Chinese interest in the natural world focused on things (*wu*). In the Chinese context, however, things meant more than simply material objects. The Chinese term referred to objects, events, mental and physical phenomena, the unknown, and the anomalous—often simultaneously but frequently delimited depending on context. Ming literati and artisans applied other terms, such as tool or implement (*qi*); controlling mechanism (*ji*); and device, contrivance, or weapon (*xie*) to utensils and cannon. Johann Schreck's (Terrenz, 1576–1630) *Diagrams and Explanations of the Marvelous Devices of the Far West* (*Yuanxi qiqi tushuo luzui*, 1627), for instance, presented traditional European craftsmanship in light of Chinese artisanship. Later, Chinese and Manchus assimilated the terms to industrial machinery in the nineteenth century.[20]

Some leading classical scholars such as Zhu Xi (1130–1200) tried to subdue the range of things to limits more to their liking. Particularly since 1000, literati who favored rational explanations of moral principles in the perceptible world sometimes distanced the unknown or the anomalous from their discourses about the investigation of things (*gewu*). They were never successful in eliminating such alleged excesses, but their efforts were influential in creating a highbrow image of a typically agnostic Confucian stereotype that modern historians have reproduced uncritically.

In this work, "things" serve as shorthand for a broader and more amorphous understanding of objects, phenomena, and events that could usually be named (*mingwu*). In encyclopedias and popular compendia, for instance in Hu Wenhuan's lower-brow editions, the rational commingled with the ineffable, the fantastic, and the magical. Late imperial references to things were multivalent and fluid, and the reader should avoid the more objectified notions of things that might at first seem relevant.

Taboos

I should speak to my efforts in this book to prioritize the history of Chinese science for a fuller understanding of premodern Chinese intellectual history. The average reader will not have heard of the "Awesome Taboo" that overcomes specialists who study Chinese art, literature, history, humanities, and social science. Nathan Sivin describes the taboo as follows:

> This stupendous prohibition rules out, under penalties so hideous that no one has even imagined them, reading, under any circumstances whatever, any primary source devoted to science, technology, or medicine.[21]

The cultural history of the calculus in China has not mattered to humanists fluent in Chinese, who tend to fall all over themselves in the presence of traditional Chinese philosophy and religion. Nor have a generation of sinologically trained social scientists or institutional specialists included the history of science and technology in their discussions of the political, social, and economic transformation of China in the nineteenth and twentieth centuries.

Unfortunately, a lesser "Science Taboo" is also common among those who speak for the sciences. While overtly welcoming studies of the sciences by humanists, they frequently assume the superior white robes of the scientist when they dismiss or refuse to read accounts by those whose careers are not primarily in the sciences. While training in science is a marvelous benefit for the historian of science, it is not a substitute for historical precision. Nor does scientific training provide the full relief of historical context. We will supersede Thomas Kuhn's analytically still-useful division of the internal versus the external histories of science only when our humanistic and scientific taboos are identified and rendered psychologically benevolent.

Late imperial rulers and literati were frequently more interested in the sciences than contemporary humanists like myself. Those afflicted by the "Awesome Taboo" or the "Science Taboo" will perhaps remain so. Hopefully, younger scholars less vulnerable to such taboos will quickly supersede my book since it is just the first intensive effort to link the Jesuit and Protestant eras to explain the rise of science in China. I especially want to thank those who have helped me cross the divide between the sciences and the humanities to produce this book.

An Overview of the Argument

This study is premised on the need for a unified narrative to describe the Chinese development of modern science, medicine, and technology since 1550. It ties together the Jesuit impact on late imperial China, circa 1600–1800, and the Protestant era in early modern China from the 1840s to 1900. Without such a synthesis, which is currently unavailable, a coherent account of the emergence of modern science in China lacks a wider historical significance for both specialists of China and students of modern science.

Arguably, by 1600 Europe was ahead of Asia in producing basic machines such as clocks, screws, levers, and pulleys that would be applied increasingly to the mechanization of agricultural and industrial production. In the seventeenth and eighteenth centuries, however, Europeans still sought the technological secrets for silk production, textile weaving, porcelain making, and large-scale tea production from the Chinese. Chinese literati in turn, before 1800, borrowed from Europe new algebraic notations (of Hindu-Arabic origins), Tychonic cosmology, Euclidean geometry, spherical trigonometry, and arithmetic and trigonometric logarithms.

If there has been one constant in China since the middle of the nineteenth century, it is that imperial reformers, early Republicans, Guomindang party cadres, and Chinese Communists have all prioritized science and technology. Yet, we have undervalued the place of science in modern Chinese history. Indeed, our understanding of Chinese natural studies before 1900 has remained muddled, despite the important contributions made by Nathan Sivin and Joseph Needham and his collaborators. This volume surveys the evolution of the native Chinese sciences from the sixteenth to the early twentieth century under the influence of the Jesuits and Protestant missionaries. In the end, however, we will discover that the Chinese produced their own science.

Parts I and II of this volume will challenge allegations that Chinese literati were not curious about European science in the seventeenth and eighteenth centuries.[22] We will describe how the Jesuits in China devised an accommodation approach focusing on mathematics and astronomy, which they generally did not employ in Japan, India, Persia, or Southeast Asia, not to mention the New World. To gain the trust of the secretive dragon throne and its literati elites, Matteo Ricci (1552–1610) and his followers prioritized natural studies and mathematical astronomy during the late Ming and early Qing dynasties precisely because they recognized that literati and emperors were interested in such fields. They realized that such interest would improve the cultural environment for converting Chinese to Christianity.[23]

The inverse of modern claims about Chinese obduracy toward modern science is the assertion that Christianity and science only had marginal influence on mainstream Chinese thought before the nineteenth century. Many historians cannot see past the key internal issues in the classical debates of Ming-Qing scholars. In the round, this claim has many merits. We will see in Part III, however, that literati interests in natural studies and Western learning continued after the Rites Controversy and actually peaked in the late eighteenth century.

The failure of the Jesuit mission and other Europeans to transmit scientific and mathematical knowledge during and after the Kangxi reign was not due to Chinese disinterest. We will contextualize the Chinese absence of knowledge about eighteenth-century scientific developments in Europe by describing the break in scientific transmission that the demise of the Jesuits and their schools in Europe during the eighteenth century caused, which deprived the Chinese of information about new trends there. The Jesuit demise delayed information from Europe about the role of the calculus as the engineer's tool kit and mechanics as the physicist's building blocks for almost a century.[24]

Earlier historical accounts have generally overvalued or undervalued the role of the Jesuits in Ming-Qing intellectual life. In many cases the Jesuits were less relevant in the ongoing changes in literati learning. In the medical field, for example, few Qing dynasty physicians before the nineteenth century took early modern European Galenic medicine seriously as a threat or complement to native remedies. On the other hand, the Kangxi revival of interest in mathematics was closely tied to the introduction of Jesuit algebra, trigonometry, and logarithms.

The technical competence of the Jesuits in the China mission during the eighteenth century ranged from surveying methods to cannon making. They also introduced pulley systems, sundials, telescopes, water pumps, musical instruments, clocks, and other mechanical devices. Their European enemies accused the Jesuits of making themselves useful to local rulers for their personal advantage rather than in the name of Christianity. In addition, emperors, their courts, and literati families welcomed Western goods and manufactures, which reverberated in the material culture of Qing novels such as *Dream of the Red Chamber*.[25]

Initially, the Chinese required a higher degree of astronomical expertise than Ricci could provide, because Ming needs focused on eclipse prediction based on cyclical time, not the determination of the linear date for Easter. Gregorian reformers in the 1570s had not worried about eclipses. Because of their preference for better cosmology, the Jesuits in China went beyond the Ptolemaic world after Ricci's death and mastered the new Tychonic world system. Once this was accomplished, however, the work of Jesuits in the Astro-calendric Bureau became regulatory rather than explorative.

Armed with the intellectual and instrumental tools provided by Brahe and his followers, the Jesuits solved the problems they were hired to undertake. They did not keep up with newer scientific developments in Europe, which more and more were products of northern European Protestants outside the Church. Consequently, the Tychonic system was still used to train astronomers in Qing China during the late nineteenth century. The Kangxi emperor (r. 1662–1722) reproduced the institutional models for translation that Xu Guangqi (1562–1633) had created with the help of Matteo Ricci and Li Zhizao (1565–1630). Later, on an even larger scale, the Protestant missionaries collaborated with Chinese to translate post–Industrial Revolution science and technology into Chinese after 1850.[26]

Parts III and IV assess the growing interest among elites in classical medicine, mathematics, and astronomy during the eighteenth century. When Pope Pius VII in 1814, after the defeat of Napoleon, restored the Jesuits as a religious order, it was in the context of the Congress of Vienna and its efforts to reinstate the ancient regime across Europe. In this climate, the Jesuits became even more conservative, and its members generally opposed the progressive-liberal ideologies of the Protestant missionary organizations sent to China after 1840.

Whether due to an unnecessary controversy, or because of a Jansenist conspiracy, the breakdown of the Jesuit consensus in the eighteenth century coincided with an increasing Chinese self-reliance in mathematical training and acceptance of European learning as rooted in China's ancient Classics. Under imperial patronage, literati upgraded mathematical studies from an insignificant skill in 1700 to an important domain of knowledge that complemented classical learning by 1800. One irony of the failure of the French Jesuits to keep up in mathematics and science was that although French Jesuits had corresponded with Leibniz, the inventor of the notational forms for the calculus that engineers employed during the eighteenth century in Europe, the Jesuits—not the Chinese—failed to see beyond the Figurist mysteries they read into the binary mathematics in the order of the sixty-four hexagrams of the *Change Classic* (*Yijing*), a divination text of possible imminent transformations that took its final form as one of the Five Classics during the Han dynasty.

The chapters in Part IV recount how Protestant missionaries and their Chinese assistants finally translated Newton's laws of physics in the 1687 *Principia* and modern mathematical analysis into Chinese when the missionaries arrived in China's newly opened treaty ports after the Opium War. The leaders of the 1793 Macartney mission had defined their historical role in relation to their Jesuit predecessors by presenting Great Britain, and thus themselves, to the Manchu court and Chinese literati as the manufacturing leaders of Europe and as enthusiastic teachers of their new scientific knowledge.

Lord Macartney (1737–1806) believed that the gifts that he had brought, chief among them a solar planetarium synchronized by "the most ingenious mechanism that had ever been constructed in Europe," were more sophisticated than the geocentric armillary spheres, mechanical clocks, and telescopes previously introduced by the Jesuits. He also thought that such gifts would convince the Qianlong emperor of Britain's dominance in science and technology.[27]

Neither Macartney nor his ship's mechanic and mathematician cum astronomer James Dinwiddie (1746–1815) thought much about the fact that the British East India Company had purchased the German-made planetarium at an auction and coated it in oriental style for the Chinese market. Only when he visited the lavish Qing imperial gardens filled "with spheres, orreries, clocks, and musical automatons of such exquisite workmanship" did Macartney stop to consider the limits of his scientific apparatus. Savants in London had already ridiculed the intended gifts for China as a banal effort to redress Britain's large deficits in the China trade. When someone suggested that the Chinese might be more interested in British machinery, that was set aside out of fear that the clever Chinese would quickly learn how to copy export machinery, as American machinists mischievously had.[28]

Indeed, Macartney never presented the pulleys, air pump, chemical and electrical contrivances, or steam engine models that he had on board. Nor did the mission present the chronometer Macartney also brought as a possible gift. As a new means to determine longitude, the chronometer would have been more efficient than the Jesuit method for surveying that the Manchus used to appraise their domains. Instead the apparatuses were returned to the British East India Company or given to Dinwiddie, who lectured on them and presented some experiments in Guangzhou to the English Factory, which was attended by Chinese merchants. Macartney remarkably noted, "Had Dinwiddie remained at Canton and continued his courses, I dare say he might have soon realized a very considerable sum of money from his Chinese pupils alone."[29]

Qing China did not yet require the specific "ingenious objects" or "manufactures" from England that Macartney actually presented, any more than India before the Manchester textile factories exploited the prodigious Indian cottage clothing industry. Such material needs in China followed on the heels of an illegal opium trade that created a commodity that millions of Chinese could not do without. Nevertheless, Chinese merchant interests in Dinwiddie's experiments and contrivances in 1793 suggests that we should not be surprised that many Chinese quickly took notice of the engineering fruits of the industrial revolution after the Opium War.[30]

Because of Britain's role in defeating Catholic France, her Protestant missionaries increasingly traveled to India, Singapore, and China. In the late eighteenth century, Protestant churches became influential arbiters of public attitudes toward the social issues of the day. Evangelism, the belief in salvation by faith alone, was formed through the auspices of Protestant churches, particularly the Methodists and low-Church Anglicans, and informed Victorian public values. Great Britain's scientific ethos based on Newtonian mechanics and machine-driven industry was transmitted worldwide by enlightened missionaries and Church evangelicals.[31]

Resurgence of emigration in the mid-eighteenth century paralleled growth of the British empire. The middle classes joined in the nation's civilizing mission at home and abroad. The 1780s and 1790s were a time when missionary philanthropy for spiritual conversion found widespread support. Eighteenth-century missions abroad focused on keeping European settlers in the church and were supported by a small circle of wealthy aristocratic sponsors associated with English monopolies. In contrast, the modern missionary movement in the "second" British Empire, like its free traders, depended on the financial support of large numbers of small contributors. The missionary enterprise turned from saving lives through aid of the poor at home to reforming manners and saving souls abroad.

Protestant initiatives to introduce European and American political and cultural institutions to China in the 1830s built on and surpassed earlier Jesuit geographies of the world and descriptions of European countries. Chinese scholars based in Guangzhou soon gained a more precise understanding of the new missionaries and the various countries they came from. The Qing dynasty's collection of strategic information to deal with the opium menace was prioritized by those most closely involved in the maritime crisis. Both Lin Zexu (1785–1850) and Wei Yuan (1794–1856) realized that China's knowledge of the West gained during interactions with the Jesuits in the seventeenth and eighteenth centuries was obsolete. Indeed, Wei hoped that the Russians on land and the newly established United States at sea might become allies of China in future wars with the British.[32]

Stress on science for dynastic self-strengthening (*ziqiang*) attracted Chinese literati interest, particularly after the debacle of the Taiping Rebellion. The Chinese terminology for modern science in the nineteenth century was less and less compatible with the Ming-Qing traditions of natural studies. The Jesuit use of this traditional terminology had not implied that European countries were economically and politically superior to the Ming or Qing Empire. The linking of state power to science had not occurred to the Jesuits because Christianity still took precedence. They linked science to the Church and its religious mission.[33]

From 1850 to 1870, a core group of missionaries and Chinese co-workers in Guangzhou, Ningbo, Beijing and Shanghai translated many works on astronomy, mathematics, medicine and related fields of botany, geography, geology, mechanics, and navigation. Parallel to the arsenals and official schools, private initiatives also popularized modern science (*gezhi xue*) in the treaty ports and among Qing officials and literati. The second half of the nineteenth century began China's modernization, which was initially perceived as westernization (*Xihua*).[34]

In Part V, we reconsider the new arsenals, shipyards, technical schools, and translation bureaus, which are usually undervalued in earlier failure narratives about China's early industrialization. In light of the increased training in military technology and education in Western science that was available in Qing China after 1865, we will see that—a decade before similar efforts in Meiji Japan—Qing reformers forged a union of scientific knowledge and experimental practice among literati and artisans, first and most notably in Shanghai at the Jiangnan Arsenal and at the Fuzhou Navy Yard. Japan's Iwakura Embassy noted both industrial sites when it visited Shanghai on the way home during its worldwide fact-finding mission from 1871 to 1873.[35]

We also scrutinize longstanding claims made by contemporaries of the Sino-Japanese War that China's defeat demonstrated the failure of the Foreign Affairs Movement to introduce Western science and technology successfully. Readers will be surprised to learn that, based on their comparable levels of science and technology in the 1890s, Qing China might have defeated Meiji Japan in their great war over Korea, just as the late Ming had helped Koreans defeat Hideyoshi's invading forces in the 1590s. When we look more carefully at the total picture of the self-strengthening period from 1865 to 1895, the view that Qing China was irrevocably weak and backward, in contrast to a powerful Europe and a rapidly industrializing Japan, is a relic of the impact of the 1894–1895 war on international and domestic opinion.[36]

Changes in elite and popular opinion after 1895 forced literati to reconsider the suitability of the Manchu-Chinese domestication of Euro-American science and technology using the interpretive lens of the investigation of things. Westernizing radicals such as Yan Fu concluded that the accommodation between Chinese ways and Western institutions, which had informed the self-strengthening movement since the 1860s, had failed. The Sino-Japanese War altered the frame of reference for the post-1895 period for both the new Chinese and the new Japanese intelligentsia that was emerging from Qing and Tokugawa scholarly elites. For late Qing overseas Chinese students of modern science and medicine, Japan replaced the Jesuits and Protestants as the messenger of science.

In Part V, we also trace the beginnings of the failure narrative for Chinese science (why China did not produce science or technology?), which paralleled the story of political decline (why no democracy?) and economic deterioration (why no capitalism?). After 1900, Chinese elites, particularly reformers and revolutionaries, increasingly demeaned their traditional sciences as incompatible with the universal findings of modern science. They disparaged Chinese natural studies and medicine. This negative rationale allowed them to transfer support to a positive set of revolutionary institutional changes that required modern science and medicine based on Western models as mediated by Meiji Japan.[37]

The elimination of traditional terms for science empowered reformers and radicals to elide the scientific texts that had been translated since 1850. After 1895, reformers no longer considered the arsenals, navy yards, and factories, where the manufacture of armaments and ships had begun, as Western-style enterprises. Instead, they dismissed them as backward Qing dynasty military factories. Single-minded promotion of Western learning caused a cultural rupture in attitudes. Radicals separated modern science from the preservation of the Qing China.[38]

Because civil examination credentials no longer confirmed elite status after 1905, everyone—not just civil examination failures as before—turned to other avenues of learning and careers outside officialdom. They traveled to treaty ports such as Shanghai or abroad to seek their fortunes as members of a new gentry-based Chinese intelligentsia that would be the seeds for modern Chinese intellectuals, scientists, doctors, and engineers. The failure of the 1898 reforms enhanced the mediation of science in China via Japan. The new texts translated from Japanese made modern science available to a wider audience and raised the level of knowledge among students and teachers.

Chinese students educated abroad at Western universities such as Cornell or sponsored by the Rockefeller Foundation after 1914 for medical study in the United States, as well as those trained in the sciences locally at higher-level missionary schools and the new universities, regarded modern science in light of the Japanese version of Western science, rather than traditional frameworks. They believed terms from the latter, derived from the language of the discredited Chinese past, were inappropriate for universal, modern science.

After 1915, the teleology of a universal and progressive science first invented in Europe replaced the Chinese notion that Western natural studies had their origins in ancient China. The dismantling of the traditions of Chinese natural studies (*gezhi*) and natural history (*bowu*), among many other categories, which had linked the premodern sciences and medicine to classical learning from 1370 to 1905, climaxed during the New Culture Movement from 1915 to 1919. When their opposition to classical learning and its traditions of natural studies peaked, New Culture advocates helped replace the imperial tradition of natural studies and classical medicine with modern science and medicine.[39]

If we now have seen through the pretensions of an earlier narrative in the twentieth century that adored the technological superiority of Western Europe over the non-West, then we will not welcome replacing that smug narrative with a self-satisfied story about the rise of modern science in China that a Chinese patriot might write in the twenty-first century. I prepare the narrative of science in China in this volume as a qualified success story, but I would not subscribe to any nationalistic claim that the march of science in contemporary China should appropriate imperial and technological pretensions vis-à-vis others. Hopefully, the fateful accommodation between techno-science (*keji*) and Chinese cultural history in modern times will be more measured, more hybrid, and less exuberant than the European "measure of man" was circa 1914, when English, French, and German politicians and generals authorized an unprecedented technological catastrophe during the Great War until 1918.[40]

I

Introduction

Prologue

In a famous 1966 "Preface," Michel Foucault recounted his laughter while reading a passage by Jorge Luis Borges that revealed to him "the exotic charm of another system of thought." Borges quoted from "a certain Chinese encyclopedia" titled *Celestial Emporium of Benevolent Knowledge,* in which the Chinese had classified animals into a series of bizarre divisions whose order was impossible to fathom, such as "those that have broken a flower vase" or "those that resemble flies from a distance." Borges attributed the title and content of the encyclopedia to Dr. Franz Kuhn, a German translator of Chinese novels. Foucault drew the lesson that Chinese exoticisms revealed the limitations in our system of thought. Foucault noted, for instance, that Western use of an alphabetical series separated all entries from their natural affinities, linking them instead arbitrarily.[1]

Borges's fable about enumerations from China challenges our efforts to establish an unproblematic order among things. Borges's caricature enabled Foucault to reframe fruitfully the early modern European act of classifying new curiosities in the eighteenth century, which, arguably, gave the sciences of life a new precision and scope. By comparison, the Chinese versions of natural history, more accurately, "an historical array of entries about things" (*bowu*), still resembled natural history in the classical period in the West, when a dividing line between living and nonliving things was not decisive.

Like their classicist counterparts in sixteenth-century Europe, the Chinese authors of natural histories sought to identify and classify natural phenomena through a correct language of words. For them, all phenomena originated from the stuff of the world (*qi*) formed through the spontaneous shaping of all things (*zaohua*) by some sort of ultimate shaping force (*zaohua zhu*) or internal power (*zaowu zhu*). They mediated efforts to make sense of the world of things through their prevailing social, political, and cultural hierarchies.[2]

Finding the Correct Conceptual Grid

Before the arrival of the Jesuits, the classical method of investigating things (*gewu*) and extending knowledge (*zhizhi*) was a key feature of Chinese literati efforts to set the limits of the natural and establish the boundaries for the known and unknown. Finding the "correct conceptual grid" had also required "investigating meanings, conceptions, and ideas" (*geyi*) when foreigners introduced new religions such as Buddhism to China after A.D. 200. To this end, Chinese scholars compared foreign terms in light of native doctrines to determine any systematic correspondences between them.[3]

After 1600 the Jesuits linked the Chinese "investigation of things" (*gewu*) to European higher learning (*scientia*, i.e., philosophical, theological, and natural studies). Natural studies in premodern China had long implied that things (*wu*) included visible objects, physical or mental phenomena, and historical events. Since the Song dynasty (960–1280), literati contended that things and affairs were synonymous. Each revealed universal principles. Chinese classifications thereafter reflected the cultural priorities of literati elites who saw principles in all things.[4]

Things also extended to the unnamed or unknown (*wuming*), that is, the supernormal and otherworldly realms of human experience that were never completely elided by literati uncomfortable with the unseen and unfathomable. Natural wonders were also considered natural objects, such as luminescent amber (*hubo*), which gave off yellow light; magnetic lodestones (*cishi*, lit., "loving stone") with marvelous qualities; and exotic plants and animals. Discovery of magnetism, for instance, led to the invention of the magnetic compass circa 1080, which was used for both navigation at sea and geomantic siting on land. As in Renaissance Europe, the Chinese believed the lodestone possessed healing properties.

Things that had medicinal uses, such as herbal medicines, or valued as mirabilia, were also commodities to be bartered locally, purchased in the marketplace, sold to traders, collected at home, or more literally consumed. Since the Ming, Chinese cooks, for example, sometimes laced exotic foods with imported opium. Writers textually deployed things chronologically or topically in premodern Chinese encyclopedias with accompanying glosses so their significance could be widely fathomed. As a result, natural studies became a field for Chinese scholars who were fascinated with the etymologies of the words for natural and marvelous things, phenomena, and events. Accordingly, Chinese classical scholars normally presented the social lives of things textually.[5]

Unlike early modern European scientific culture, where natural history was increasingly encompassed in a museum, the historical array of entries about things, phenomena, events, and affairs (*bowu*, lit., "broad learning of things") represented a nuanced account of natural phenomena as words in a text that needed to be decoded primarily through the analysis of language by connoisseurs. Each historical or natural event, natural object or man-made implement, and mental or physical phenomena could be described chronologically in terms of a teleology of its usefulness or value to humans. In this chapter we will emphasize the textual lives of things as a clue to the cultural framework within which late imperial Chinese scholars classified them. We will also describe the monetarization of the late Ming dynasty (1368–1644) economy, which unleashed the commoditization of all things and turned things into objects of wealth rather than moral cultivation.[6]

How literati presented things in this volatile time will enable us to understand better why the investigation of things became the Chinese window on Jesuit natural philosophy—and the Jesuit window on Chinese learning—after 1600. Late Ming efforts to naturalize anomalies and supernormal events by appealing to the operation of yin and yang polarities through the five evolutive phases of *qi* (the stuff of the world) present us with an interesting parallel to the early modern European transformation of prodigies, such as supernatural miracles and monsters, into natural phenomena.[7]

What Should Be the Literati Theory of Knowledge?

During the Ming dynasty, literati increasingly relied on the annotations of Cheng Yi (1032–1085) and Zhu Xi (1130–1200) to interpret the core philosophical interpretations of the classical canon. Gainsaying the then-influential Mahayana Buddhist claim that all things were ephemeral and thus empty of abiding reality (*sunya, kong*), they argued that "investigating things and extending knowledge" (*gewu zhizhi*) presupposed that there was a "principle for all things" (*wanwu zhi li*) in a real, not illusory, world. They also denigrated the petty belles lettres of their time as vacuous classical learning and thereby prioritized for literati the search in the Classics for universal principles of all things, events, and phenomena (*qiongli*), which they labeled "Learning of the Way" (*Daoxue*).

Zhu Xi also discarded Shao Yong's (1011–1077) cosmological speculations, which Shao had forcefully enunciated in his influential Northern Song (960–1126) *Treatise on the Ultimate Axis for Managing the World* (*Huangji jingshi shu*). In the latter, Shao contended that human beings should not pick out things, phenomena, and affairs from a human standpoint. A sage, Shao argued, could internalize perception so that "things were perceived as things." In this way, one could exhaustively fathom the principles of all things (*qiongli; wu zhi li*). Shao favored a naturalistic epistemology that prioritized apprehending unmediated principles directly in things.[8]

Cheng Yi and Zhu Xi feared that without the classical unity provided by investigating things for moral cultivation, Shao's misplaced stress on "perceiving things" (*guanwu*) would succumb to triviality. Thereafter, the investigation of things rather than "broad learning" (*boxue*), literary flair, or alleged Buddhist nihilism became the key to wisdom for literati versed in the Classics, Dynastic Histories, and natural studies. By mastering orthodox classical learning, Ming literati hoped to pass the rigorous civil service examinations and gain a prestigious office in the dynastic bureaucracy.[9]

Zhu Xi concluded, "One must in three or four cases out of ten seek principles outside oneself." In six to seven cases out of ten, however, moral principles should be sought as part of one's moral cultivation. Although the *Documents Classic,* one of the Five Classics, warned against overly delighting in things, Zhu Xi still sought to accommodate human feelings to them.[10] Zhu Xi proposed a uniform methodology to accumulate knowledge and wisdom in both the cultural-moral and natural-political realms. He did not, however, delineate a precise classificatory system within which the investigation of things (*gewu*) represented a particular kind of knowledge, i.e., astronomical or mathematical, social or political, cultural or religious.[11]

Since the Southern Song (1127–1280), leading literati who disagreed with Zhu Xi's prioritizing the investigation of things attacked Zhu's use of the Great Learning as one of the Four Books to establish the epistemological boundaries for literati learning.[12] A "Learning of the Way" master, Zhu had elevated his own commentary on the investigation of things by reading it into the Great Learning from antiquity as if Zhu's words were the authentic commentarial voice of one of Confucius's direct disciples, Master Zeng (Zengzi). Zhu Xi's classical conflation of his own commentary with the allegedly original ten commentarial chapters was controversial and widely disputed by his contemporaries. During the Ming, however, Zhu's commentary became the orthodox standard in the civil examination curriculum. Sixteenth-century debates surrounding other versions of the Great Learning, however, increasingly challenged Zhu Xi and reopened literati efforts to reconsider the appropriate meaning of the investigation of things as the literati theory of knowledge.[13]

The mid-Ming scholar-official and statesman-general Wang Yangming (1472–1528), for instance, preferred the Old Text (*guben*) version of the Great Learning that had survived in the *Record of Rites* over the redacted version that Zhu Xi and his followers championed. This scholarly tactic enabled Wang to contradict Zhu Xi's externalist views of the investigation of things in the Four Books and claim his own classical views as orthodox. In particular, Wang Yangming gainsaid Zhu Xi's emphasis on the investigation of things and extension of knowledge (*gezhi*) ahead of morality (*chengyi*). For Wang and his late Ming followers the investigation of things and the extension of knowledge took a back seat to making one's will sincere. Literati morality should take precedence over classical erudition. Moreover, Wang equated Zhu's approach with an "amusement in things that ruins one's aims" (*wanwu sangzhi*), which the "Hounds of Lü" chapter in the *Documents Classic* had condemned.[14]

One of the stories associated with Wang Yangming related that in 1492, when he was in Beijing to take the spring 1493 metropolitan examination, he discussed with a colleague Zhu Xi's version of the investigation of all things as the means to attain sagehood. Wang suggested that his friend should begin investigating for principles by looking at the bamboo in front of their pavilion. Cheng Yi earlier had contended that "every blade of grass and every tree possesses a principle that should be examined." When his friend grew exhausted, Wang took up the task, but he also failed to fathom the principles of bamboo.[15]

Because he had concentrated so hard and had become ill, Wang Yangming now considered Cheng Yi's views on the investigation of things naive. He also pointed to the elusiveness in Zhu Xi's investigating things to fathom principles. The official account in Wang's *Instructions for Practical Living* (*Chuanxi lu*), recorded by students between 1521 and 1527 while Yangming was in retirement after a roller-coaster military career, described the episode:

> From morning till night I was unable to find the principles of the bamboo. On the seventh day I also became sick because I thought too hard. In consequence [my friend and I] sighed to each other and said it was impossible to be a sage or a worthy, for we do not have the tremendous energy to investigate things [for the principles] that they have. After I had lived among [non-Han] peoples for three years [in Jiangxi], I understood what all this meant and realized that there is really nothing in the things in the world to investigate, that the effort to investigate things is only to be carried out in and with reference to one's body and mind, and that if one firmly believes that everyone can become a sage, one will naturally be able to take up the task.[16]

Wang's alternate agenda for investigating things, which he clarified in later years, dismissed the tedious search for principles in things out there, such as bamboo, in favor of an innate knowledge of the moral consciousness (*liangzhi*). The latter was essential for moral cultivation and the inner repository of universal principles. Wang demonstrated to his followers that the moral principles of things were not easily grasped by following Zhu Xi's methods because such principles were in the mind, not in things themselves. In fact, searching for such principles outside the self was enervating, Wang contended, because piecemeal knowledge never yielded a singular moral vision unifying knowledge and action (*zhixing heyi*).

One of Wang's students noted that things (*wu*) were the same as events (*shi*), which Song scholars such as Cheng Yi had already claimed, but the student extended the point to clarify that "both referred to the mind," where their principles resided. Wang concluded from the exchange that investigating things meant preserving the clear character of the mind. The principles of things were not in the world out there, which were only things in themselves, but rather in human consciousness. Such views also reflected the Yangming schoolmen's interest in the Buddhist stress on the emptiness of things in the world and the centrality of the mind's consciousness as a reservoir of innate knowledge.[17]

Subsequently, the late Ming appearance of an ancient stone-inscribed version of the Great Learning, which was later determined to be a forgery, reopened for many literati Wang Yangming's influential claim that Zhu Xi had in Song times manipulated the original text of the key passage on the investigation of things to make his personal interpretation canonical. Others sympathized with Wang's ideas but went beyond his position on the priority of the Old Text version of the Great Learning from the *Record of Rites*. To get back to the most ancient meaning of the Great Learning, they uncritically accepted a stone version allegedly rediscovered among manuscripts and recently made public.[18]

That an object such as a rubbing from a spurious stone inscription of the Great Learning could set off a classical brouhaha during the late Ming was not unprecedented when compared to earlier examples of charges of classical forgery. In the past, however, forgeries and fakes had not threatened the aura of the original. Indeed, through reproductions Song, Yuan (1280–1368), and early Ming literati had approximated and appreciated originals of calligraphy and painting long since lost. Northern Song antiquarians, for instance, also appreciated antiquities such as bronzes and revived the classical canon such as the Great Learning as an omen of cultural and political legitimacy. Wang Yangming continued this high-minded trend even though he pointedly gainsaid Zhu Xi.[19]

Because the Southern Song (1127–1280) government valued bronze currency, officials criminalized making new vessels and confiscated ancient bronze vessels to melt down for currency. The substantial challenge an archaic literary fake posed for literati classical learning tells us that, even more so than the Song and Yuan, the late Ming intellectual marketplace was increasingly concerned with texts—not just antiquities—as cultural commodities. In an age when printed versions of the Classics were no longer the restricted prerogative of learned elites, literati increasingly demonstrated their exclusive mastery of the canon by appealing to the authenticity of their unique editions and manuscripts. Such textual trends also allowed Ming commentators in the south to repossess the northern drama scripts from the Yuan, which they later competed over bitterly to edit and reprint. These exclusive claims were in turn challenged by many other connoisseurs and experts who possessed—or claimed to possess—the requisite cultural resources.[20]

Late Ming Classicism in the Context of Commercial Expansion

Both officials and literati concerned themselves with the technical facets of maintaining the late Ming agrarian economy. The latter drew its strength from the productivity of an integrated river-canal-lake system and land-commodity-labor taxes collected from private farms in over 1,300 counties where about 90 percent of China's population of approximately 150 million people lived in 1600. Since 1381, the government had classified the entire population into social and economic categories to determine taxes and measure access to the civil and military examinations. Revised in 1391, this massive undertaking aimed at measuring the economic resources under Ming control, equalizing the distribution of the land tax (paid in kind), and obtaining fair labor services from all households.

Echoing the ancient classical models in the *Rituals of Zhou,* a text that imperial reformers since antiquity appealed to for contemporary guidance, these classifications, such as households of farmers, commoners, military men, artisans, and merchants, reflected the initial status of each family in early Ming society and how much labor service they had to provide. The government assigned each household category with a specific labor service it had to perform for the bureaucracy, and these tasks were organized according to village-family units of 110 households (*lijia*) in each community.[21]

A merchant household was expected to supply merchandise or goods on demand; a military family had to provide at least two soldiers for service; an artisan household provided one worker for imperial workshops. The land registers were supposed to be revised every ten years, and each family was required to perform its labor service in perpetuity. The wide gap between the theory and practice of Ming tax collection, however, greatly diminished government control of the economy by the sixteenth century. When regional markets gradually turned to a silver currency for large transactions out of the direct control of the government and for payment of land and labor taxes, this confirmed the dynasty's weakened hold over its agrarian tax resources.[22]

Geared to a village commodity economy circa 1400, the Ming tax system became increasingly obsolete as population rose from 65 to 150 million and the economy became more commercialized. The Ming economy was further transformed by an agrarian revolution in which cotton displaced rice production in southern coastal provinces and the influx of Japanese and New World silver monetized the sixteenth-century economy in unprecedented ways. Ming Chinese unwittingly faced a global marketplace in contrast to Song regional concerns. By the 1570s, the Ming government had bowed to the inevitable and through the Single Whip Reforms commuted the land tax and service labor systems into a single monetary payment in silver.[23]

Wang Yangming's claim that the principles of things existed only in the mind, accordingly occurred at a time when literati views of the economy, commodities, and objects and their significance were changing. As China's population grew between 1450 and 1600, the reach of the relatively static imperial bureaucracy declined. Similarly, anxious Ming literati wondered if the Cheng-Zhu classical orthodoxy could still represent universal principles of knowledge at a time when domestic goods were financially converted into objects of wealth paid for with imported silver. Ming literati such as Yuan Huang (1533–1606) worked out the tensions between morality and affluence when he created a new moral calculus for measuring private wealth by keeping track of good and bad deeds in ledgers of merit and demerit.[24]

Although literati after Wang Yangming still placed human understanding within a classical theory of knowledge, the quantity and exchange velocity of goods in the marketplace had multiplied exponentially. Ming elites were living through a decisive shift away from the traditional ideals of sagehood, morality, and frugality. Within an interregional market economy of exceptional scope and magnitude, gentry and merchant elites transmuted the impartial investigation of things for moral cultivation into the consumption of objects for emotional health and satisfaction. Ming painters presented the contemporary fondness for and connoisseurship of antiquities, for example, as a genre known as "Broadly Examining Antiquities" (*Bogu tu*). The paintings valorized the literatus as a collector of exquisite things (see Figure P.1).[25]

Late Ming antiquarianism in particular drew its strength from the economic prosperity that pervaded the Yangzi delta. There and elsewhere merchants and literati used their increased financial resources to compete for status through conspicuous consumption. Merchants and literati on their travels searched for ancient works of art, early manuscripts, rare editions, and magnificent ceramics. They paid extravagant sums when they found what they wanted. The rise in value of ancient arts and crafts also touched off increased production of imitations, fakes, and forgeries of ancient bronzes, jades, and ceramics. Late Ming antiquarians with their fixation on possessing things challenged the principled ideals of both Cheng-Zhu learning and Yangming revisionism. For the latter, the former had focused on "things too much."[26]

Wang Yangming rejected the Song Cheng-Zhu theory of knowledge because he thought its epistemology that all things were knowable in light of

Figure P.1. A Ming painting of "Broadly Examining Antiquities."

Source: Permission from National Palace Museum, Taipei, Taiwan, Republic of China.

principles was naive. He rerouted the Cheng-Zhu agenda and reduced all things in the perceptible world to the unified field of the mind's awareness, where all principles ultimately resided. Wang noted, "Seeking principles in myriad affairs and things is like saying that one should seek the principle of filial piety in one's parents."[27]

Things in themselves were banal for Wang and his followers, who demoted the value of things out there, precisely when man-made commodities of value proliferated in the marketplace. Wang Yangming's turn from things to the mind in part refuted the inroads made by connoisseurship and commoditization among Ming literati. Purists like Wang still sought enlightenment, but they no longer located true principles in the vulgar connoisseur's world of objects and wealth.[28]

Late Ming commercial expansion built on the dramatic monetarization of the Chinese economy during the Silver Age of 1550–1650 and unleashed the commoditization of things into objects of desire and affection. After the Ming state commutated village and town labor tax services into cash levies, for example, the imperial court and its bureaucracy lost control of its land and labor resources. In effect, during Wang Yangming's time a decisive shift from a predominantly subsistence livelihood based on a huge agrarian economy to a steadily expanding market economy occurred, which was linked to internal and external networks of provincial, regional, and international trade.[29]

The increase in internal (between town and village) and external (between provinces) trading links stimulated an escalation in commerce, especially merchant travels and resources. Since late medieval times, the imperial state had provided a shipping infrastructure via the Grand Canal, bridges, and roads for grain tax purposes, which had linked north China and the Yellow River to the Yangzi delta, the granary of the empire since medieval times. The delta was then linked via the Yangzi to its vast middle and upper reaches.[30]

This infrastructure fueled a revolution in domestic cotton production and clothing after 1400 whereby almost all commoners in China by 1600 wore winter garments made from cotton rather than the hemp or flax linens of Song times. Diversification of crops and stress on sericulture and cotton paralleled sophisticated rice transplantation techniques in the middle Yangzi, which increasingly replaced the lower Yangzi region, Wang Yangming's home region, as the rice granary of the empire. Sugar and other cash crops, which traded as commodities in exchange for rice from new areas that now had rice surpluses, made up for the rice deficits in the more handicraft-oriented Yangzi delta.[31]

Specialized towns emerged in which the cultivation of commercial crops such as cotton and silk replaced rice land. In Shandong and Henan, hired northern laborers also grew cotton that was shipped to the Yangzi delta for weaving. Dual cropping of summer rice and winter wheat had long been common in the south. In this commercialized environment the cultivation and manufacture of cotton and silk using multiple-spindle spinning machines tended to become separate operations with an accompanying division of labor. Local commodity production in the Yangzi delta, for instance, shifted from traditional household handicrafts in the early Ming to merchant-oriented production in family workshops.[32]

Silk, cotton, and rice markets furthered the commercialization of the rural village economy and spurred trading links with towns and cities. Improved seeds, changing crop rotations, and new cash crops such as maize, peanuts, and sweet potatoes from the New World produced a doubling of grain yields as a complement to the extension of cultivated acreage from 1500 to 1800. Commercialized handicraft production meant that changes in the rural economy produced corresponding changes in the social order for both men and women. Differentiation between urban centers and rural production in village households made peasant producers dependent on market forces and merchant middlemen.[33]

Until the Ming dynasty, it was generally true that "men till while women weave." Silk production, that is, spinning, weaving, and raising silkworms, was handled by women, who used family looms. By the sixteenth century, however, when taxes were increasingly monetarized, men and women began contributing equally to rural labor in south China. The shift in sericulture from a local, household industry to a new interregional product changed the longstanding gender division of labor. Men now worked at the loom in urban or suburban family workshops, while rural women still produced the cocoons for thread.[34]

For elite consumers, Ming cotton and silk production translated into commoditized fashion, which Tim Brook perceptively notes "traveled through the social structure just as it did through the marketing structure." Fashionable women, whether wives or courtesans, preferred Suzhou cotton embroideries or Huzhou silk brocades, which simultaneously affirmed their modesty (by covering the body) but also enhanced their stylishness (through design). Maids and concubines quickly emulated their masters' tastes. Rather than affirm the occasional warning about luxurious living or Wang Yangming's renewed call for moral cultivation, most Ming elites became agents for the transmission of extravagance through style. Purists criticized the affinity they perceived between literati collectors of antiquities and the predilection of their wives and families for silks and furs.[35]

Retreat of the dynasty from direct involvement in village affairs also magnified the role of gentrified elites as landlords in late imperial politics and society. Under the umbrella of the central government, gentry and merchants in the Yangzi delta and elsewhere diversified their hold on local power via profiteering based on land rent and commercial enterprises. As state influence lessened, local public health matters evolved under the umbrella of gentry philanthropy and local literati physicians. Such elites also monopolized positions in the imperial bureaucracy by translating their economic and social power into cultural and educational advantages that enabled mainly the sons of gentry and merchants to pass the empirewide civil examinations.[36]

Expansion of the internal economy matched growth in foreign trade. The spice trade with Southeast Asia, for instance, doubled in the sixteenth century, and Ming China increasingly imported hardwoods from Southeast Asia for furniture, palaces and temples. The Ming and Qing dynasties exported teas to Central Asia in exchange for horses until the eighteenth century. In addition to large profits from cotton goods in the domestic market, the Chinese became the world's largest exporter of manufactured goods, tea, silks, and ceramics. Production costs were kept low as a result of low overhead and a surplus of labor. Efficient agriculture also kept food prices low.[37]

Early Ming porcelain was manufactured primarily for the domestic market, principally at kilns in Zhejiang (Longquan), Jiangsu (Yixing), Fujian (Dehua), and Jiangxi (Jingdezhen) provinces (see Figure P.2). Merchants linked the pottery kilns to their imperial and literati consumers. Traders translated the external demands of the market to the local producers. Later in the dynasty the Ming exported porcelain to Japan and southeast and south Asia. The Dutch East India Company handled some six million pieces in the seventeenth century, but this number represented only about 16 percent of Ming ceramic exports. Yixing's and Jingdezhen's landlocked factories were linked via lakes and rivers in the lower and middle Yangzi region to southeastern ports such as Xiamen, Fuzhou, and Guangzhou and from there to the Indian Ocean trade and Islamic markets. The *blanc-de-Chine* styles developed at Dehua were for a time extremely popular in seventeenth century Europe and were exported in large quantities at the nearby port of Quanzhou.[38]

The largest pottery factories at Jingdezhen, for example, followed the usual imperial pattern for operating such enterprises, which involved state supervision of merchant activities (*guandu shangban*). The enterprises were based on the labor of hundreds of artisans, who produced the "Mohammedan blue" and polychrome ware in the "five colors" for which the Ming became famous. Imperial taste and literati connoisseurship deflected the technical discourses of the producers into a sublime discussion of porcelain aesthetics for consumers. With the fall of the Ming, imperial purchases declined, but Jingdezhen revived and remained the major domestic and international producer of porcelain in China until 1800.[39]

After 1750, Europe gained access to the technical secrets that had made China the leader in pottery making and ceramic ware for centuries. Many as-

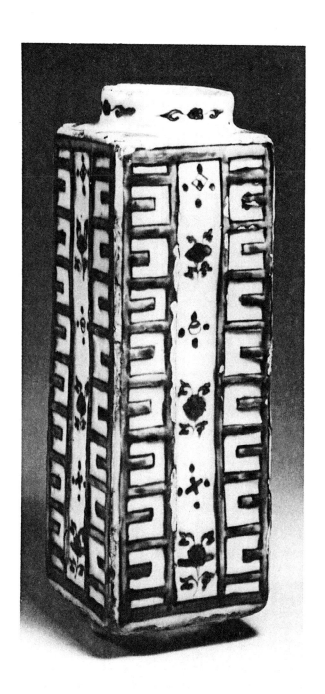

Figure P.2. Late Ming pottery—vase in the shape of a ritual jade blue-and-white porcelain.

Source: Permission from China Institute Gallery.

pects of Ming technologies, including ceramic ware, were included by Song Yingxing (1587–1666?) in his late Ming *Heavenly Crafts Revealing the Uses of Things* (*Tiangong kaiwu*), for example. Because Song described processes that were government monopolies, however, his work was not widely available during the Qing dynasty (1644–1911), even though it now constitutes our major source for Ming and Qing technologies and handicrafts.[40]

Printing Technology and Publishing

Woodblock printing developed during the Song and Ming periods into a sophisticated art that by the late Ming included multicolor printing, woodblock illustration, the use of copper movable type, and woodcut facsimiles of earlier editions. Government printing during the Song dynasty had encouraged paper production (invented during the Later Han dynasty), and an explosive expansion of printing in the mid-Ming yielded a wider circulation of erudite and practical knowledge empirewide. By 1800, scholarship, book production, and libraries were at the heart of China's cultural fabric. Highbrow printing of classical works for literati elites in the Yangzi delta was paralleled by printing of popular literature, vernacular novels, almanacs, encyclopedias, and literacy primers for a wider audience of commoners in the Yangzi delta, Fujian in the southeast, and Sichuan in the southwest.[41]

Classical learning revived during the Song dynasties in part due to the increased circulation of books and the diffusion of classical texts brought on by the spread of printing, and this development was surpassed in the late Ming. Widespread woodblock printing of classical texts created repositories for books used by scholars and students. Such publishing enterprises in turn increased interest in scholarship and education. Editions were now more readily available, and prices were low enough to make book collecting a possibility for most scholars.

Printing in the early Ming capital at Nanjing surpassed Hangzhou, which had been a publishing center since the Southern Song. Merchants from Huizhou, who could access the cheaper wood in their home area for woodblocks, and book traders from elsewhere congregated in Nanjing. Color editions of books became a prominent feature of Nanjing editions. Woodblock printing reached its peak in technical sophistication in the mid-sixteenth century with the rise of scholar-printers in Suzhou, Nanjing, Hangzhou, Huizhou, and Yangzhou in the Yangzi delta and more commercially oriented printers in Jianyang, Fujian. During the late Ming period, Suzhou became the center for quality printing, and outstanding xylographers staffed the printing shops. Fujian, the center for commercial publishing, produced a larger quantity of novels, dramas, and popular manuals (including medical handbooks) than elsewhere.

Family businesses predominated in Fujian publishing, for example, particularly in the western mountains where the local lineages established successful low-cost enterprises through an intricate network of bookstores and traveling merchants. The kinship links among local publishers, distributors, and sellers instantiated a nested geographical hierarchy for producing and selling books for elite and popular audiences locally and regionally. As family businesses the local print shops fulfilled their lineage's cultural aspirations by catering to the local civil examination market for books on the Classics and classical primers. In this manner, merchant families that profited from the book trade also invested in education for their kin in the hope that their social status would improve through an association with books and the classical scholarship of elites.[42]

When successful, this strategy allowed the more prominent family print businesses to become highly cultured literati enterprises. We will see this change more clearly when we look closely at the Hu family print shop in Hangzhou in Chapter 1. By the late Ming, Hangzhou printers such as Hu Wenhuan (fl. ca. 1596) operated shops in several cities in the Yangzi delta to disseminate their publications. Encyclopedias such as Hu's *Collectanea for Investigating Things and Extending Knowledge* (*Gezhi congshu*), because they were printed in so many different combinations, were widely available in China and Japan.

Although Chinese printers experimented with movable type, xylography was generally more economical when publishing many copies of a particular work. Woodblocks were easily stored and, with reasonable care, preserved for frequent reuse. In fact, printers often used the same woodblocks for collectanea of different titles and compilers. Books set with movable type accommodated misprints and errors more easily, whereas a woodblock, once proofread, was more permanent.

Moreover, breaking down type matrices after a printing rendered later editions very expensive. A previous woodblock, on the other hand, could be used for cheap later printings. When relatively few copies were needed (for imperial projects with limited circulation), then it was feasible to employ movable type (see Figure P.3). Woodblocks were especially economical for books and manuals produced by lowbrow print shops in Fujian and Sichuan.[43]

The proliferation of books and manuals during the late Ming led to the printing of numerous encyclopedias (*leishu*, lit., "classified digests"). Encyclopedias functioned as repositories and manuals of popular knowledge during the late Ming, in addition to serving as scholarly compendiums for students preparing for the imperial examinations. From this environment of readily available reference books, practical manuals, and popular compendiums of knowledge emerged a book-oriented atmosphere conducive to the development of scholarship and the practical arts.[44]

Figure P.3. Setting moveable type in the Qianlong Imperial Printing Office.
Source: Qinding Wuying dian juzhen ban chengshi (Beijing, 1776).

For instance, the "Street of the Glazed Tile and Glass Factory" (*Liuli chang*), located in the southern, Han Chinese city inside Beijing, originally a factory site, by the eighteenth century was the major book emporium and center for antiques in Qing China. The Street Factory reached its height as a book market during the Qianlong era, 1736–1795. Because it was located close to the Hanlin Academy, the emporium was a gathering spot for intellectuals, scholars, and degree candidates who came to Beijing. Its cultural atmosphere stressed the value of rare works and ancient artifacts, promoted the exchange of books, and stimulated scholarship during the eighteenth century. Books and manuscripts of all kinds moved freely between Beijing and the main book markets in the Yangzi delta and southeast China.

In addition, the printing of "daily use" encyclopedias (*riyong leishu*) in the 1590s was emblematic of a widening publishing world that appealed to the lesser lights of late Ming society, namely, merchants, artisans, and licentiates (*shengyuan*, i.e., those only licensed to take higher examinations). Presented as repositories of useful information for daily life, popular encyclopedias provided nonelites with a wide choice of subjects dealing with medical prescriptions, divination formulas, ancient lore, astrology, geomantic almanacs, calligraphy, etc. Unlike reference books for elites that focused primarily on the civil examinations, elite family ritual, and classical learning, many late Ming encyclopedias included information on travel and lodging useful to merchants. Such attention to the practical needs of nonelites in provinces such as Fujian meant that compilers and printers were no longer limited to orthodox topics. They could present the material aspects of normal life in rich detail for a broad audience of new readers.[45]

Building on a new realism that also informed late Ming fiction, ribald novels such as *The Plum in the Golden Vase* (*Jin Ping Mei*) presented protagonists who owned drugstores, for example. Composed by an author who was writing as daily use encyclopedias proliferated, the book, circulating as a manuscript in the late 1590s, presented an inventory of goods, money, objects, collectables, events, and skills—from medical potions for enhancing sexual prowess to elaborating food at banquets, drinking games, and popular jokes. The fictional contents of the novel enlivened but also mirrored the categories and contents of the narrative-free encyclopedias. The latter was the source where authors obtained their detailed information about popular songs and daily life experiences, which they emplotted in the new realism of the time. In this publishing environment, novels and encyclopedias represented different aspects of a burgeoning commercial environment that was reaching nontraditional audiences.[46]

While the late Ming novel often made officials the villains, the more mundane practical encyclopedia leveled the field to include lowbrow interests alongside elite tastes and conventions. Such trends challenge our image of Ming learning. Was Wang Yangming learning really representative of late Ming literati? In Chapter 1 we address the appearance of many practical compendia on things, affairs, and phenomena, which flowered into a profusion of daily use encyclopedias in the 1590s and which contrast sharply with the high-minded claims of the Yangming schoolmen that the principles of all things were already in the mind.

Yangming idealism was more a reaction to than an obstacle to the collection and investigation of things of cultural cum financial value. Wang's highbrow idealism represented a classical rejoinder to the widespread lowbrow commodification of things during the late Ming. Many other works predated, led up to, or paralleled the late Ming encyclopedias. They also enunciated the problem of knowledge in light of the investigation of things as a textual inventory of objects.[47]

The increasing market for published works during the late Ming represented a time of expanding classical and popular literacy. Late Ming followers of Cheng-Zhu learning in turn superficially blamed the crisis of knowledge on Wang Yangming's misplaced idealism rather than rampant connoisseurship. They reacted by stressing even more the concrete aspects of Zhu Xi's search for informing principles (*li*) in the stuff of the world (*qi*). Literati authors also embodied the realm of *qi* in late Ming literature and poetry by instantiating it in human emotions (*qing*). At the same time, Wang Yangming radically prioritized the principles of things in the mind at the expense of the emotions.[48]

Naturalization of Anomalies in Ming China and Early Modern Europe

Changing attitudes towards anomalies can serve as a barometer for different outlooks concerning natural, otherworldly, and supernatural causes. The description of monsters and attitudes toward them in Europe, for example, evolved in the sixteenth and seventeenth centuries while their religious significance gradually diminished. In contrast to the monism of *qi* that most Chinese affirmed as the psycho-physical stuff of the world, the Jesuits presented to the Chinese a numinous world of a personal God, angels, and the eternal souls of the dead.[49]

Initially perceiving monsters as divine prodigies contrary to nature, early modern European elites transformed them into natural wonders. Subsequently, in the seventeenth century Francis Bacon (1561–1626) included them in natural history in his *Novum Organon*. Prodigies when treated as natural anomalies allowed Bacon to explain the secrets of nature. Such phenomena eventually lost their autonomy as prodigies, and French specialists in the Parisian Academy of Science integrated them into the medical disciplines of comparative anatomy and embryology. Monsters thereby became counter-examples to normal development, and thus examples of medical pathology, in which nature's uniformity contrasted with the exceptionalism of monsters.[50]

Similarly, with regard to the treatment of miracles in Europe, Lorraine Daston describes how "the debate over the evidence *of* miracles became a debate over the evidence *for* miracles." In moving from signs to facts, miracles as preternatural (what rarely happens) or supernatural (God's unmediated actions) eventually lost much of their religious meaning. By focusing on the natural causes for deviating instances, Bacon and others made the anomaly and the miraculous pertinent to scientific discourse. The goal in seventeenth-century naturalism was to explain the causes of marvels and prodigies in light of natural philosophy and not as pure miracles.[51]

Ming Chinese literati were also setting limits in official circles to the use of anomalies as political or religious signs. In Europe among learned and powerful elites, however, the evidentiary requirements there implied less confidence in the authenticity of supernatural events. Both Ming and European elites saw any emotionally charged enthusiasm for the marvelous as a possible act of deception or ignorance. For both, the dangers of false prodigies among the intellectually heterodox or politicized rabble-rousers were very real.

Among late imperial Chinese elites, and the Jesuits in China, however, the otherworldly as a counterpart to the natural world never diminished as much as among secular Protestant elites in northern Europe. Late Ming civil examinations also demonstrate that the naturalization of unusual phenomena among Ming literati, as for the Catholic Jesuit missionaries who were tied to a new Counter-Reformation vision of the natural world, remained incomplete. By way of contrast, among Protestant men of science the boundary between the natural and the artificial dissolved into the notion of God as a great artisan, who could be mimicked by the great artist or man of science. The mechanization of nature was the result. Nevertheless, even in Protestant Europe the nonnatural persisted as a legal category governing abnormal familial and sexual relationships.[52]

Moreover, early modern European mechanical philosophy quickly became in the hands of the Protestant faithful a significant means to head off the development of an agnostic or atheistic naturalism. This goal paralleled the Jesuit concern with pagan atheism and materialism, which they perceived in the Song dynasty cosmology still prominent in the late Ming. The new mechanistic view in Europe infused dead matter with the supernatural activity of God. Jesuits would similarly contend in Ming China that the psycho-physical states of *qi* could not produce a world infused with spiritual vision and human intelligence. Heaven and earth were inanimate things and could not produce all things in the world. For the Jesuits, entities could reproduce themselves, but only God created ex nihilo.[53]

In early modern Europe, unlike Ming China, the distinction between natural and supernatural events was effectively obliterated by the role of divine activity in mechanical philosophy. The heretic Isaac Newton (1642–1727), for example, thought the observed stability of the solar system was maintained by divine intervention and that gravity could not be accounted for by the laws of mechanics alone. For Newton, as for the Jesuits, mere matter and motion could not explain cohesion or gravity. Ming literati never saw gravity, but they explained motion in terms of dense (= falling) versus light (= rising) qi informed by the Supreme Ultimate's (taiji) patterning of principles (li) through the efficacy of yin-yang and the five phases.[54]

Theological considerations remained an integral part of early Newtonianism, which represented, surprisingly, Protestant cultural echoes of the role of divine activity in Scholastic philosophy. Hence, both Thomism and Newtonianism represented a mitigated naturalism, which unlike Aristotlelianism did not exclude divine providence from natural explanations. The French theologian and reformer John Calvin (1509–1564) had rejected occultism, for instance, because for him inert matter required the spiritual excitation of natural impulses to move and change. Later, the Cambridge Platonists who opposed Calvinism still affirmed the principle of immanent divine causation.

Because inert matter could not be accounted for by mechanics alone, Protestants preferred Newton's theory of gravity, as it brought God into the continuous activity of creation. For many, the natural required divine intervention. Similarly, Cartesianism affirmed but did not depend upon God's role as the guarantor of truth. We cannot assess the scientific revolution in early modern Europe as a purely naturalistic movement. The new divine right of kingship was likewise enmeshed in a similar question of God's immanent participation in running the mundane human world.[55]

The reinterpretation of nature, that is, the naturalization of anomalies and miracles, still required divine props—a physico-theology—for its authority in Europe. Until nature was universalized as neutral and amoral, or gravity could be explained by the laws of mechanics alone, portents and prodigies could not be reduced to nonissues.

As long as Ming-Qing literati believed that a rational and orderly cosmos informed the microcosmic patterns of differentiation and organization in the creation and evolution of all things in the world, the naturalization of anomalies in Ming and Qing times still required the cultivation of moral perfection of learned scholars and officials. Only the sage could truly investigate things and extend knowledge. In England only gentlemanly enthusiasts, such as the early Fellows of the Royal Society, possessed the circumstances, education, cultural heritage, and moral equipment to rightfully engage in the new practice of empirical and experimental science.[56]

Later in 1699, even the Jesuits were excluded from the French Academy of Science because they were no longer regarded as contributing members of the emerging scientific community in Paris. Jesuits in France had successfully moved out of their traditional classroom milieu into fields of scientific activity earlier in the seventeenth century. Many typified how early modern Catholics became members of the European scientific community.

After academy membership in Paris became more exclusive, however, the lack of Jesuit scientific work in China, for example, was clear. In the eighteenth century, French Jesuits had no official affiliation with the Parisian Academy, unlike their late seventeenth-century predecessors. The Jesuit decline in leadership in European science and mathematics after 1700 delayed the transmission of modern mathematics, such as the calculus, and modern mechanics to China until the middle of the nineteenth century.[57]

Ming China in 1600 was not awaiting the Jesuits' arrival. It was a dynamic if troubled dynasty of puzzling but powerful rulers, who had sent armies and navies to roll back a major Japanese invasion of Korea. Concerned literati, free-spending elites, and a vast peasantry tried to understand the present and future in light of past ideals. Scholar-printers during the late Ming produced encyclopedias and collectanea that presented the natural world in light of past and present knowledge, both classical and popular, and added new layers of concern for daily existence, fascination with material life, and spiritual concern for bodily health.

In 1579 on the eve of the Jesuit arrival, for instance, the Henan provincial examiners prepared a policy question (*celun*) asking several thousand civil service candidates to discuss astrology, mathematical astronomy, and the calendar, which was then in crisis. In particular, they asked candidates to "discuss the issues in detail so that the examiners could ascertain their learning based on the investigation of things [*gewu zhi xue*]." The best essay obliged by emphasizing the "natural regularities of the heavens" (*ziran zhi shu*) and the "fundamental principles" underlying them (*benran zhi li*). In Chapter 1, we will explore how mid- and late Ming compendia recycled popular lore while simultaneously reorganizing classical knowledge and natural learning into more accessible repositories of ancient wisdom.[58]

1

Ming Classification on the Eve of Jesuit Contact

By the time the Jesuits arrived in Ming China, many Han Chinese literati were debating an appropriate theory of knowledge. The debate often took the form of claims that morality (*zun dexing*) took precedence over formal knowledge (*dao wenxue*) or vice versa. The focus of the debate in Ming China was on the investigation of things and the extension of knowledge (*gewu zhizhi*). Ming literati invoked a sense of urgency in their encyclopedic efforts to reconstruct the textual lives of things at a time when the meaning and human significance of natural and man-made objects as commodities for the many betrayed the ideals of moral cultivation for the few.[1]

The Jesuits tried to reshape this research agenda by mediating between what they thought was China and their West (*Taixi*, i.e., early modern western Europe). They would add precision to the Chinese notion of investigating things and extending knowledge by exposing Ming literati to European classification schemes, forms of argument, and the organizational principles for all specialized knowledge, i.e., *scientia*. They never grasped, however, that what was happening in Ming China, namely the commoditization of things into objects of material value, was also sweeping through western Europe. Mexican silver dollars coming to China were the first steps of disenchantment about the moral investigation of things for the Chinese.[2]

Ordering Things through Names

Under Mongol rule during the Yuan dynasty, medicine as a profession became increasingly attractive when physician families were exempted from taxes and corvée labor as an occupational group. Some literati-physicians also linked investigating things and extending knowledge to medicine as a classical term to denote the technical learning of the latter. As the only eminent Yuan physician initiated in the direct line of Cheng-Zhu (Cheng Yi and Zhu

Xi) teachings, Zhu Zhenheng (1282–1358), for example, associated medical learning with classical studies. In his most famous work, titled *Discussions Based on Investigating Things and Extending Knowledge* (*Gezhi yulun*), Zhu also opposed Song standardized government medical prescriptions that had been drawn from an empirewide collection of Northern Song (960–1126) medical texts for the Imperial Pharmacy.[3]

In the *Discussions,* Zhu asked Yuan literati to acknowledge medical learning as one of the key fields of study that complemented moral and theoretical teachings by demonstrating the practical uses of the latter. Zhu's work had drawn on Liu Wansu's (1120?–1200) *Direct Investigations of Cold Damage Illness* (*Shanghan zhige*), which had also appealed to the ideal of investigating things (*gewu*). "Cold damage" disorders (*shanghan*) included influenza, common colds, and typhus or typhoid fever. In the late eighteenth century, the editors of the Qianlong emperor's (r. 1736–1795) Imperial Library favorably cited Zhu Zhenheng's claim that medicine was one of the concrete fields that informed the "investigation of things and extension of knowledge."[4]

Overlap between medicine and the investigation of things was often mentioned in late Ming writings. For example, the literati physician Yu Chang, a secondary provincial graduate of 1630, prepared a medical case book. In his 1643 preface, Yu presented the methodology of the Great Learning as the epistemological basis for his efforts to gain an impartial understanding of the root causes of a particular medical problem: "The achievement of sincerity resides in the investigation of things and the advancement of knowledge. Drawing distinctions [in this sphere] requires more rigor than the [legal] dichotomy of fraud versus honesty." According to Yu, investigating a case of illness was like investigating things. The key was to advance one's knowledge one item at a time until the proper diagnosis became clear.[5]

When the Catholic convert Wang Honghan learned about the Western anatomy introduced by the Jesuits in the late Ming, he equated Catholicism with ancient literati classical learning. In addition, Wang likened the "great physician" with the "great literatus," and stressed that study of the two traditions necessitated "exhaustively fathoming principles and investigating things." To "classify studies and clarify principles" (*gexue mingli*) was the way of the physician and the literatus. Although Wang accepted the Galenic tradition of anatomy as part of the Song-Ming investigation of things, he also turned to the Chinese therapeutic tradition for lessons in how to cure illnesses.[6]

Ordering Antiquities and New Findings

The investigation of things was conceptually also applied to the collection, study, and classification of antiquities, as in Cao Zhao's (fl. 1387–1399) *Key*

Issues in the Investigation of Antiquities (*Gegu yaolun*), which was published in the early Ming and enlarged several times. The work originally appeared circa 1387–1388 with important accounts of ceramics and lacquer, as well as traditional subjects such as calligraphy, painting, zithers, stones, bronzes, and ink slabs. It became an exemplar for late Ming antiquarians.[7]

The 1462 edition prepared by Wang Zuo (palace graduate of 1427) was considerably enlarged and included findings prepared by several members of the official Ming dynasty naval expeditions led by Admiral Zheng He (1371–1433) taken to Southeast Asia and the Indian Ocean from 1405 to 1433. Ma Huan's (fl. 1413–1451) *Captivating Views of the Ocean's Shores* (*Yingyai shenglan*, 1433), for example, had described the twenty countries the fleet visited and included detailed accounts of Yemeni towns such as Dhufar and Aden in southern Arabia.[8]

In addition to such descriptions, Wang Zuo was particularly interested in ancient bronzes, calligraphic specimens, and native curiosities. He also added native imperial seals, iron tallies, official costumes, and palace architecture to the collection. In his preface, Wang added, "Whenever you see an object, you must read all about it in the repertories, study its provenance, classify its quality, and judge its authenticity." Archaic-looking fakes produced by clever craftsmen for the Ming market of cultural commodities posed significant financial dangers for literati.[9]

Unlike the impact sixteenth-century oceanic discoveries allegedly had in early modern Europe, the new information the Zheng He fleets brought back to Ming China from southeast and south Asia in the fifteenth century did not challenge the existing frameworks of orthodox knowledge. Donald Lach has argued, for instance, that the early modern European world "underwent a transformation in the sixteenth century which produced in observers a sense of mild shock, wary fascination, or deep wonderment." Lach acknowledges, however, that many scholars "remained oblivious to the rents in the curtain obscuring the East." Those who were alert to the new findings realized that neither classical nor Christian learning in Europe could encompass the latest information, unusual artifacts, and geographical discoveries or duplicate the more advanced technical arts of India and China in textile manufacture and porcelain production.[10]

More recently, however, Michael Ryan, following Lucien Febvre, has argued that the newly discovered lands and new peoples registered little impact on the values, beliefs, and traditions of sixteenth- and seventeenth-century Europeans. We might add that the overwhelming intellectual influence in sixteenth-century Europe came from classical Greek manuscripts sent to Europe from Constantinople in 1453, and not from oceanic voyages. Similarly, Ryan rejects Lach's view that new forms of cultural relativism emerged in Europe. Instead,

Ryan has examined how these new worlds were incorporated into a European lexicon by asking how European contemporaries interpreted their world. Their use of categories, such as ancient paganism as a transhistorical framework to classify the cultures of the new worlds, enabled them to domesticate exotic peoples within the frame of Graeco-Roman pagan antiquity.[11]

Europeans understood other peoples in light of familiar genealogies, which minimized the impact of new worlds by conceptualizing new worlds within the terms of the old one. In 1669, John Webb (1611–1672) contended that after the biblical deluge Noah's son Sem and his people had entered China. Athanasius Kircher (1601–1680), a Jesuit scholar who thought China a derivative of Egypt, thought that Cham was the better choice. The Jesuits maintained that all people in the world descended from Noah's three sons (Sem, Cham, and Japheth). Europeans usually regarded Sem as the ancestor of Asian peoples. When Chinese converts translated the Jesuit argument that the earliest Chinese were of foreign origin, however, this provoked an attack by critics of the Jesuit in Qing court circles in 1664 and led to the martyrdom of the converts.[12]

Ryan has contended that the new worlds were discovered by the Europeans, which implies some sort of ownership of their discovery. Later, however, literati in Ming China also discovered and incorporated the world of Europe introduced by the Jesuits. If the discovery of the new worlds in early modern Europe coincided with the recovery of the ancient pagan world, then an alternative, a parallel, and a contemporary assimilative process occurred that we can call the Chinese discovery of the West (*Taixi*), not to mention the European discovery of China. Moreover, after the Jesuit arrival in Ming China, literati who welcomed them prepared parallel but opposite narratives to place "Western learning" within the boundaries of China's classical antiquity.

A century prior to the Jesuit arrival, Ming literati had widely applied their paradigm for investigating things and extending knowledge. The approach had enough authority to allow the compilers of the *Key Issues in the Investigation of Antiquities,* for instance, to domesticate the new materials brought by the Ming navy from the Indian Ocean within a traditional focus on encyclopedias and their already established range of classifications. The fifteenth and sixteenth centuries in China were certainly not "centuries of wonder." Leaving out the "discovered" and their reception of their "discoverers," however, produces only a one-sided historical narrative. Moreover, the Chinese had been learning about the Old World via Islam in central Asia and Persia all along and had never experienced isolation as the Europeans had.

Any claim that most Ming literati, when compared to their European contemporaries, engaged in a subjectivist and idealist discourse about things is off the mark. Indeed, Mark Elvin has misread his sources to contend that

Wang Yangming and his sixteenth-century followers led most Ming literati away from the precocious intellectual promise of objectivist science and natural studies in Song times. In Elvin's view, the pervasive influence of Wang Yangming was one of the three key factors (the others were filling in of the south China frontier and the closed-door policy of the dynasty), that had doomed Ming China to failure in its global competition with early modern Europe.[13]

Moreover, the popular encyclopedias outlined below make it clear that Ming compilers of encyclopedias never took literally Wang Yangming's efforts to find the principles of bamboo through meditative techniques. We need to address the roots of these practical compendia of things, affairs, and phenomena, which were printed as a wide variety of "daily use" compendia of the 1590s. They contrast sharply with the high-minded claims of the Yangming schoolmen that the principles of all things were innate in the mind.[14]

Natural Studies in Yuan-Ming Encyclopedias

During the Yuan and Ming dynasties, compilers constantly added material to Southern Song compilations, but the military dislocations of the period from 1350 to 1450 yielded a dramatic downturn in the numbers of such publications at the same time that the Chinese population declined significantly. The *Expanded Records of a Forest of Matters* (*Shilin guangji*), for example, was a Southern Song compilation that scholar-printers added to during the Yuan dynasty. By 1330–1332, the edition had forty-three divisions, which included heavenly correspondences, the calendar and time, topography, plants and fruits, bamboo and trees, musical harmonics, Daoism, Chan Buddhism, mathematics, tea, wines, liquors, etc. In contrast, the original *Expanded Records* had only seventeen divisions, which generally focused on government and essay questions from the civil examinations.[15]

The Yuan encyclopedia, *Complete Collection of Classified Affairs Essential for Those Living at Home* (*Jujia biyong shilei quanji*), was a precursor to late Ming encyclopedias that focused on customs, practical learning, home rituals, sericulture, health, etc. The 1301 preface to the *Complete Collection,* for instance, called it a "guide to local clerks' learning," and like Ming daily use compendia (*riyong leishu*) it focused more on the audience of lower clerks who served in yamens in cities and the countryside.[16]

The production of such daily use compendia spread more widely from the coast to the hinterlands. Late Ming publications for a popular audience centered in Fujian province, especially in the interior county of Jianyang. A noticeable increase occurred in printed works on agriculture, health, medicine, and mathematics for social groups in traditional Chinese society who were classically or functionally literate, that is, literati, artisans, and merchants. This publishing upsurge in the sixteenth century should be understood in light of a similar expansion during the Southern Song. In other words, the publishing of popular encyclopedias from the Song to late Ming increased, although there was a significant dip in the early Ming.[17]

Jianyang editions were criticized during the Wanli reign period (1573–1619) for their poor paper and for their use of softwood to carve woodblocks in order to sell editions quickly. They remained widely popular, however, as part of an expanding print culture, which took advantage of a decline in printing costs and book publication expenses. Such savings were due in part to a more efficient division of labor in the production of woodblocks for printed books and because the sea route for Jianyang editions from Fujian in the southeast to other regions, particularly the prosperous Yangzi delta, proved cheaper.[18]

The classification system for daily use compendia published in the late Ming typically ranged in content from twenty to over forty major categories. Those that appeared most frequently in late Ming practical handbooks were astrology, topography, medicine, mathematics, dream interpretation, nourishing life, sericulture, writing skills, legal matters, and plants. The encyclopedias published in Jianyang encompassed a wide variety of books, including household manuals, quotation dictionaries, and collections of anecdotes, but their common feature was their topical arrangement as "household handbooks" and manuals for everyday life. We can identify seven types of late Ming daily-use compendia: (1) general works; (2) works oriented to the civil service; (3) reference use (phrase dictionaries, letter writing, etc.); (4) literary collections; (5) sources for looking up names and people; (6) collections of stories and anecdotes; and (7) primers for children.[19]

Based on this synopsis of the classifications and bibliographic locations for natural studies used since medieval times in the Dynastic Histories, Imperial Library, and in official and private encyclopedias, we cannot identify a single, unified field of traditional natural studies before the Jesuits arrived in China. Although the four classifications (*sibu*) were usually in effect, they were often superseded by Song-Yuan-Ming popular compendia. Nonetheless, we will see that among late Ming and early Qing elites both the "wide learning of things" (*bowu*) and "investigating things and extending knowledge" (*gezhi*) were becoming a common epistemological frame for the accumulation of classical and practical knowledge.[20]

Ordering Pharmacology through Names

Biographies of Li Shizhen (1518–1593) have noted that the pharmacologist-physician had at the age of fourteen *sui* (Chinese added one year after the first new year) passed the preliminary county civil examinations held at his home province of Hubei. Assuming approximately 1,000 candidates per county for local qualifying examinations, Li was one of some 1.2 million literate local candidates. After he failed three times in the provincial examinations, like 95 percent of the candidates empirewide, Li never advanced further. As he came from a family whose patriarchs had been medical practitioners and - studied pharmacognosy for several generations, Li Shizhen turned to medicine for a livelihood. His father served in the Imperial Medical Academy and had written treatises on diagnosis, smallpox, the pulse, ginseng, and the local mugwort.[21]

In terms of classical literacy, i.e., mastery of the orthodox canon, dynastic histories, and encyclopedias, Li was by 1535 one out of some 35,820 licentiates within an approximate population of 100 million. Beginning in the fifteenth century, each stage of the Ming civil service selection process eliminated the vast majority of candidates, most, like Li, at the local and provincial competitions. A vast constituency of failures, like Li Shizhen, sought alternative careers in teaching, medicine, preparing legal plaints for others, etc., where they could use their classical skills as scholars and writers.[22]

An unexpected consequence of the empirewide civil examinations was the large number of classically literate men who as a writing elite also formed a significant reading public. In addition, literate women in elite lineages were by the late Ming exchanging poetry and educating their male and female children with equal facility. Such large numbers of classically literate men and women created an audience of some 5 million readers and their families, that an author of a popular Ming work, such as a practical digest, pharmacopoeia, or novel, could potentially reach.[23]

Li Shizhen was not directly influenced by the 1590s daily use compendia discussed above, but he was aware of the dubious quality of such earlier works. His magnum opus, the *Systematic Materia Medica* (*Bencao gangmu*, 1596), and the effusion of encyclopedias and classified pharmacopoeia in the late Ming represented improvements. An ample number of classically literate men and women, and a large audience for practical manuals, compendia, and popular fiction, created the market that late Ming printers in south China sought to profit from by selling a wide assortment of works by men such as Li Shizhen in the late Ming. Chŏson Korea, Tokugawa Japan, Le Vietnam, and the overseas Chinese communities in southeast Asia were additional markets for Ming and Qing editions of such compendia.[24]

Li Shizhen completed the final version of his *Systematic Materia Medica* in 1587. By 1590 he lined up a printer in Nanjing for the complete project. The first printed version appeared posthumously in 1596. Subsequently, the huge work sold out rapidly enough that eight reprints were issued in the seventeenth century alone, three of them by 1602. The *Systematic Materia Medica* was quickly transmitted to Tokugawa Japan via European traders. The first Japanese edition came out in 1637. Scholars in Europe began to take notice of Li Shizhen's work in the middle of the seventeenth century, and the French Jesuit academic Jean-Baptiste du Halde (1674–1743) cited it extensively in his 1735 geographical and historical account of the Manchu empire. In the late eighteenth century, the 1655 edition of the *Systematic Materia Medica* was included in the Qing Imperial Library.[25]

If we place Li Shizhen's *Systematic Materia Medica* in its medical and pharmaceutical context, we find that Li worked tirelessly on a revised materia medica that would correct the errors of identification, classification, and evaluation that had accrued in the standard Song dynasty series. In particular, Li's generation of physicians, like their Jin and Yuan predecessors, were dissatisfied with Northern Song materia medica.[26]

Unlike the Ming, the Song Bureau of Medicine, for example, had collected thousands of medical formulas that led it to publish and distribute formularies beginning in 1107–1110. The latter were based on compound medicines packaged for sale in the government dispensaries. Revised many times, these formularies were reprinted for wide use. Liu Wansu and Zhu Zhenheng, among others, had been critical of these recommendations because they found the system rigid and not effective in treating illnesses in the thirteenth and fourteenth centuries.[27]

From 1556, Li traveled widely in major drug-producing provinces, gathering comprehensive information on mineralogy, metallurgy, botany, zoology, and drug formulas. Over the span of three decades from 1552 to 1578, Li took into account some 40 pharmacopoeias by various authors, 361 medical works, and 591 classical, historical, literary, and encyclopedic works. The medical works were further divided into some 84 cited in earlier pharmacopoeia and the 277 Li cited for the first time.

The nonmedical works were from historical, canonical, technical, and literary sources, as well as from dictionaries and encyclopedias. Of these, 151 texts had been used in earlier accounts; 440 were cited by Li for the first time. Among the 932 titles, including the 792 described above, which Li cited as sources in his detailed bibliography, 36 percent were medical and pharmacological works, while 76 percent were quoted for the first time in the materia medica literature.

Hence, the first innovative aspect of Li's work was his addition to the materia medica literature of much data about drugs and diseases new since the Song dynasties. He incorporated, for example, the medicinal uses of maize and the sweet potato, New World crops introduced from the Spanish Philippines in the middle of the sixteenth century. He also reported on the emergence of a new disease, syphilis, which he traced back to the late fifteenth century in Guangdong province, but which may have had a New World origin.

Li Shizhen argued in his "Conventions" to the *Systematic Materia Medica* that the ancient headings in earlier canonical works on materia medica had been mixed up and that their true meanings were lost. Because Li saw his efforts in terms of a revisionist, tripartite research agenda that included first research and correction, then analytic procedures, and finally synthesizing procedures, recent historical accounts have described Li Shizhen as not merely a pharmacist-doctor but also a botanist-naturalist.[28]

The structure of previous works on materia medica had been based on the ancient classifications in the *Canonical Materia Medica of Emperor Shennong* (*Shennong bencao jing*), which became a classic after it was compiled circa the first century B.C. or A.D. Although the work was subsequently lost, its three-part distinction between minerals, plants, and animals was appropriated in later pharmacopoeias because of its canonical status. Drugs, for example, were classified from a therapeutic point of view into three grades. The *Canonical Pharmacopoeia of Emperor Shennong with Collected Commentaries* (*Shen bencao jing jizhu*) by Tao Hongjing (456–536) introduced a second partition in the materia medica literature. Based on the three natural grades of minerals, plants, and animals, Tao distinguished each of these categories by subdividing them into several subgroups with no attempt at rigorous taxonomy.

Completed in 1496, a century before Li Shizhen's work was published, Wang Lun's (fl. ca. 1456–1496) *Essentials of Materia Medica* (*Bencao jiyao*) distinguished between ten different sub-groups: (1) herbs, (2) trees, (3) vegetables, (4) fruits, (5) grains, (6) minerals, (7) quadrupeds, (8) birds, (9) insects and fish, and (10) humans. Chen Jiamo's (b. 1496) *Enlightenment on the Materia Medica* (*Bencao mengquan*), which was published in Li Shizhen's youth, had followed Wang Lun's model with a few modifications in the order of Wang's ten subgroups. For example, Chen made plants first (herbs, trees, vegetables, fruits) and humans second in the order.[29]

By the time Li Shizhen turned to materia medica issues, there was a clear sense that past classifications, like Song official prescriptions, had become problematic, both because of accumulated confusions in the historical record and because of a good deal of new information that had to be added based on experiences with illnesses more common in south China. In addition, Li Shizhen's organic pharmacology continued the Song-Yuan-Ming condemnation of ingesting poisonous substances, which had been associated with the traditions of external alchemy in medieval times, favoring natural herbs and less harmful minerals.[30]

In his "Conventions," Li further declared the need to reorganize the existing knowledge of drug products. He did this by hierarchically ordering the material into 16 sections, 60 categories, and 1,895 kinds. Li's use of categories as an independent structuring device was somewhat less innovative, but he did improve on earlier efforts to develop a classificatory scheme. Li Shizhen paid attention to the ecology of living things from which drugs were derived. His taxonomic order combined rubrics that were indifferent as to habitat, physical characteristics, and use.

Items were organized by their morphology and their qualitatively defined physical configurations of *qi*. To the latter, Li applied the five evolutive phases as divisions for dynamic processes or static configurations. The fire phase, in particular, was reconceptualized in Li's *Systematic Materia Medica* as an independent medical category, which represented one of the configurations of *qi* affecting the body. In this regard, Li was building on Song and Jin conceptual developments that prioritized the focus on fire within and without the body as an important heat factor in the maintenance of health and the prolonging of life (*yangsheng*).[31]

Li's sorting of plants was sometimes determined by the following categories: the habitat of plants; their morphological aspects; whether they were wild or cultivated plants; their therapeutic aspects; culinary and medicinal aspects; and the agricultural categories for grains. His taxonomy was, however, often still based on subjective judgments, medically overdetermined, and not on botany. Hence, the criteria for his categories were not exclusively naturalistic.[32]

Moreover, Li Shizhen reclassified the entire materia medica record, which revealed Li's concern with the "investigation of things." His approach also stressed the problem of nomenclature (lit., "rectification of names"), as the main entry in his classifications. Another subentry explained the names of things themselves. The structure of Li's entries in the *Systematic Materia Medica* were presented as explanations, corrections, and clarifications in the "Conventions." They provided late Ming physicians and scholars with a broad reservoir of practical medical knowledge.[33]

Accordingly, Li Shizhen's materia medica enterprise was also a scholarly project: "Although this is what doctors have called medicine, in examining and explicating their principles I have actually practiced what we literati scholars call the 'study of the investigation of things.' This can fill the gaps in the *Progress Toward Elegance* [*Erya* dictionary] and those in the commentaries on the *Poetry Classic*." Appealing to the Ming tradition of investigating things, Li contradicted in practice Wang Yangming's claims about what constituted careful research concerning the external world. Li Shizhen had little patience with any effort to employ meditative techniques to fathom the principles of a plant such as bamboo. He was too busy concerning himself with its natural history, nomenclature, and medicinal aspects.[34]

Li's materia medica and the many daily use compendia of the late Ming are a better guide than the work of Yangming schoolmen to the evolution of the investigation of things as an epistemological principle for precise scholarship before the influx of Western learning via the Jesuits in the seventeenth century. Earlier accounts that have paid attention primarily to the moral philosophy of Wang Yangming have misrepresented the intellectual complexity of pre-Jesuit natural studies in Ming China.

Like Li Shizhen's *Systematic Materia Medica*, late Ming digests cited a wide variety of books, manuals, dictionaries, and literary collections, although they did not strive for Li's standards of scholarly rigor. Their common feature was their increasingly complex topical arrangements, which were accompanied by new knowledge and more information about things. Such works were widely printed in south China by printers when lower printing costs made cheaper editions accessible beyond the classically literate elite.

Collecting the Collectors

The Ming scholar-merchant and Hangzhou bookseller Hu Wenhuan (fl. ca. 1596), like Li Shizhen, prefigured the Sino-Jesuit dialogue concerning the investigation of things and European *scientia* in the early seventeenth century when he compiled and published his widely circulated *Collectanea for Investigating Things and Extending Knowledge* (*Gezhi congshu*) in the 1590s. This collection of books, published as a set, embodied a repository of classical, historical, institutional, medical, and technical works from antiquity to the present in China. The collectanea also contributed to the growing late Ming literature of material culture. Its wide dissemination in Ming-Qing China and Japan, in many editions from Hu Wenhuan's Hangzhou and Nanjing print shops, marks it as a very influential and thus representative work. It did not have the scholarly pedigree of the *Systematic Materia Medica*, but Hu's *Collectanea* affords us a unique window onto more common divisions of knowledge. Hu had successfully commoditized classical learning in an age of Ming connoisseurship.[35]

Although provincial informants mentioned two hundred sets of the *Collectanea* in the 1780s, the compilers of the Qianlong Imperial Library catalog criticized its unevenness, looked down on Hu's profit-oriented marketing of several editions, and only summarized its content in the official catalog. It was not included in the library. Nevertheless, its extensive circulation in many published forms allows us to access a representative world of pre-Jesuit natural knowledge and lore. Analyzing its subject matter allows us to go beyond Borges's and Foucault's ahistorical musings about quaint Chinese encyclopedias. The *Collectanea for Investigating Things and Extending Knowledge* presented a cumulative account via a collection of books of all areas of knowledge important to a literati and nouveau riche audience before the Jesuits made their presence felt in Ming literati circles in South China after 1611.[36]

Because Hu Wenhuan also had wide-ranging interests in medicine and popular religion, some versions of the collectanea contained a broad scope of illumination texts and esoteric writings, which I will discuss. In addition to addressing the collectanea in light of its many variant editions, I will also analyze its initial, pre-Jesuit focus on early lexicons and natural histories, which overlapped with classical learning and natural studies during the late Ming. The smaller, more orthodox version of the collectanea thus parallels the distancing of the queer and supranormal in some Song encyclopedias of the tenth century.[37]

For example, the Northern Song *Materials of the Taiping Xing Guo Era [976–983] for the Emperor to Read (Taiping yulan)* did not include unusual phenomena or paranormal novelties from medieval times. Moreover, the compilers declined to present grotesques; strange plants, animals, and minerals, and odd countries that appeared in the *Expanded Records of the Taiping Xingguo Era (Taiping guangji)*, which focused on popular religion.

The taxonomy informing the *Materials of the Taiping Era* was more representative of highbrow literati tastes, which were reproduced in the Song genre of jottings (*biji*). The latter were generally about fictional, historical, or textual material that were preserved by their authors as odd notes. Earlier interests encompassing natural phenomena and supernormal topics in jottings had waned. Thereafter, most jottings were about recollections of court events, celebrated fellow officials, and experiences in the civil service.[38]

On the other hand, even though the *Expanded Records* was less consonant with orthodox encyclopedias, its contents were representative of lower-brow literati and echoed the nourishing of life (*yangsheng*) traditions that permeated popular religion and medical discussions on prolonging life and achieving immortality. This dual track of Song encyclopedias was never mutually exclusive, however, and the creative tension between them continued in later Ming collections of books published as a set, such as Hu Wenhuan's *Collectanea*.[39]

The Life of a Late Ming Scholar-Printer-Collector

We glean from Hu's many prefaces to the works he collated a sense of his life and work as a scholar-printer. His grandfather and father were both Hangzhou collectors and printers, and many of Hu Wenhuan's reprints came from his family's cultural traditions. Between 1592 and 1597, Hu wrote some forty-nine prefaces for works he compiled and published. Hu and his staff assembled some five hundred works in his Hangzhou and Nanjing print shops, which made their way, selectively, into the enlarged versions of the *Collectanea for Investigating Things and Extending Knowledge* and other collectanea that Hu printed.

Hu Wenhuan had several colleagues who shared his bookman's interests and helped compile his works. In addition to collation, they provided prefaces and encomia in Hu Wenhuan's series. One of them, Zhang Lun, was also the teacher in the Hu family school, which indicates that he was well versed in classical learning and that like Hu he was a licentiate who had not advanced further on the examination ladder. Another colleague, Chen Bangtai, helped in printing the books. Hu's family printing shop in Hangzhou became known as the "Hu-Chen Great Print Shop," which suggests Chen's importance to the enterprise.[40]

Hu himself was a Nanjing imperial school student, who had likely purchased his licentiate status to enable him to compete in local qualifying examinations. Like Li Shizhen, then, Hu Wenhuan had attained a high level of classical literacy and literary ability, but also never attained a higher provincial or metropolitan degree. By the late Ming, of the 50,000 candidates empirewide competing triennially for some 1,200 provincial degrees, fewer than 3 percent would succeed. Because most like Hu Wenhuan, never became provincial graduates, the Ming dynasty required licentiates to keep taking biennial renewal examinations to maintain their special legal status. Like most such students, Hu did not attend his assigned school. The rolls of local official schools were increasingly filled with candidates who had repeatedly failed higher examinations and had nowhere else to go.[41]

Hu used two studio names, one known as the "Hall of Writings Brought Together" (*Wenhui tang*) in Hangzhou and the other called the "Office of Thoughts of Retirement" (*Sichun guan*) in Nanjing. His other sobriquets, such as "Penetrating the Arcane" (*Dongxuan*), also suggest his sympathies with the esotericism associated with popular religion. As the southern capital of the Ming, Nanjing was an important publishing center in the Yangzi delta, along with Hangzhou and Suzhou. Consequently, Hu also relied on the Nanjing book market for many of his editions, although they were mainly published in Hangzhou. Hu's range of focus, from orthodox classical texts to esoteric medical writings on nourishing life (*yangsheng*), were tied to his examination studies and his weak health as a youth.[42]

In addition to books and texts, Hu Wenhuan also collected antiques and musical instruments. He was interested in tea as a cultural phenomenon, as well as wines and perfumes. The financial benefits from his printing enterprise, based on selling many different series of his printed works or individual volumes from the collection, enabled Hu to maintain the lifestyle of a literati scholar with wide cultural interests, even though he had failed to gain an official appointment. By the late Ming, merchant and literati collectors like the Hu family grew and diversified. Hu Wenhuan finally received an appointment in 1613 and served as a low-level county official in Hunan province, first as a magistrate's aide and then as an administrative clerk. While Wenhuan was away, the Hu publishing enterprise diminished.[43]

In sum, Hu was a merchant-scholar of wide-ranging literati interests, and the large-scale collectanea that he compiled, collated, and printed before 1613 preserved many rare texts for his Ming contemporaries, though his editions were later criticized for their poor quality.[44]

A Highbrow Version of the Collectanea

The contents of the more classical versions of the *Collectanea for Investigating Things and Extending Knowledge* ranged from aspects of heaven and earth to classifications of birds, animals, insects, fish, grasses, foodstuffs, architecture, and tools. A classicist frame of reference informed Hu's systematic collection of data from native sources concerning China's natural resources, the arts, and manufactures (see also Chapter 6). The portions dealing with these high-minded themes reveal Hu's efforts to republish and cumulatively build on previous works that focused on natural phenomena, names and their referents, and affairs and things.

The earliest printing of Hu's more orthodox versions included forty-six works that we will focus on first. Later, enlarged editions of 180–200 or more works were compiled by Hu in order to sell more sets of the collectanea for profit. These editions somewhat resembled the Northern Song *Expanded Records of the Taiping Era* as a repository of popular and religious lore, although the latter was not just a collection of texts available for purchase. Starting with annotations of the *Progress Toward Elegance* dictionary and the *Explication of Names* (*Shiming*), the initial edition of forty-six works emphasized a comprehensive account of etymologies and word definitions that would shed light on the golden age of antiquity, a period of enlightened governance. The "rectification of names" (*zhengming*), a perennial literati concern, was a passionate aim.

Overall, the collectanea emphasized a broad learning of phenomena (*bowu*), one of Hu's main classification categories, which encompassed natural and textual studies within a humanist, esoteric, and institutional agenda. Missing, however, was the sense of methodological unity to the investigation of things and the extension of knowledge that Zhu Xi and his Song-Ming followers had prioritized. Hu was content to introduce the lexicographers together in one place without any guidelines for synthesis or analysis.[45]

Hu's *Collectanea* represented one of a large number of late Ming collections of books published as a set that were aimed at elite and popular audiences. Their contents incorporated information on natural history and lexicography to fill the gaps in meaning that had opened between visual things and written words in classical learning. Scholars working on pharmacopoeia and medicine, such as Li Shizhen, were also concerned that the language used to designate things-phenomena-affairs had by the sixteenth century become increasingly muddled. Historical accounts of phenomena were presented as a genealogy of glosses, a sort of textualized natural history, but the order had accrued chaotically. Names for plants, animals, etc., in terms of kind, species, attributes, and uses needed reordering, i.e., rectification.

Although it did not pursue the methodological rigor that informed the Li Shizhen's *Systematic Materia Medica,* Hu's *Collectanea,* nonetheless, contained a repository of works from ancient to contemporary times, which in its initial printing presented the classical ordering of knowledge and its later recapitulations. The collection embodied the problems rather than the solutions. Overall, late Ming collectanea and encyclopedias sought to restore all the words that had been lost or muddled since antiquity through a meticulous reexamination of "things." Composed in the classical language of elites but accessible to low-level licentiates and clerks, Hu's collection of the collectors allowed him to describe and order things, phenomena, and affairs.

The extant genealogy of classical glosses was the first step in capturing the common affinity of things through the language in which they were historically represented. The late Ming collectanea became a textual site for refashioning and reproducing the findings of past sages, worthies, scholars, seekers of immortality, monks, masters, and commoners. It served as a genre for "collecting the collectors" through texts and diagrams, which drew on the intellectual lineage of things, phenomena, affairs, and esoterica dating from the ancients.[46]

Just as Li Shizhen engaged in a reformative agenda of observing, analyzing, recognizing, and naming plants for medicinal purposes, others reviewed the classical search for why things were the way they were (*suoyi ran*). By delimiting and filtering the visible structure of things drawn from a wide variety of sources, Hu's *Collectanea* derived its classical respectability from the traditions of orthodox classical learning. When the Jesuits presented an alternative natural history in an Aristotelian conceptual language (= *scientia,* i.e., "specialized knowledge") for the structure of the world, their description of the visible world was translated by their Chinese collaborators in the language of the late Ming theory of knowledge, namely investigating things and exhaustively mastering principles.

Lexicons in the Collectanea for Investigating Things and
Extending Knowledge

Available since the Han dynasty, the *Progress Toward Elegance* dictionary contained glosses for the names of things in early classical texts. The compiler was not the observer, but the secondhand entries of cryptic glosses identified anything from grammatical terms in language, labels for kinship relations, technical language for palace architecture, and names of utensils and tools to the names of musical instruments. Since antiquity, elegance in language had corresponded to elegance in material life.

In addition, the *Progress Toward Elegance* extended its classifying vision to the natural world. One of its nineteen subjects explicated astronomical, calendrical, and climatic terms. Four sections included glosses for geographical, topographical, and geological items. Finally, two sections covered the names of plants (i.e., grasses, herbs, vegetables, trees, shrubs), and five sections presented words for insects, fish, fowl, and wild, domestic, and legendary animals.[47]

As a classical dictionary, the *Progress Toward Elegance* was not written by a botanist or paleontologist. Yet its search for orderliness through understanding ancient names was a serious effort to know about things through an assessment of words. When the medieval commentator Guo Pu (276–324) annotated the dictionary, for instance, he noted in his "Preface" that it was a comprehensive compilation of theses. He added that it was the "unconfused basis for a broad learning of things."

The *Progress Toward Elegance*, Guo concluded, was the handiest contemporary source for names (= "definitions") of birds, animals, and plants. It was also a valuable source for names of curiosities and words in different dialects, based on "reliable and verified sources." The compilers of commentaries and follow-ups to the *Progress,* which Hu Wenhuan also included in his *Collectanea,* addressed the discourse of names, things, phenomena, and affairs. Mediated through the classical language, an amorphous natural history emerged. The allegedly well-constructed classical language for reality gave the reader access to the verbal structure of the visible world.[48]

The stature of the *Progress Toward Elegance* was enhanced during the Wei dynasty (220–264) when it was attributed to the sagely Duke of Zhou by the *Expanded Elegance* (*Guangya*), which was the third work included in the initial printing of Hu's *Collectanea.* Later classicists often accepted this attribution and contended that Confucius had said the *Progress Toward Elegance* was a window on ancient Zhou governance. Because of such attributions, the *Progress Toward Elegance* rose in the ninth century to the status of a Classic. Consequently, from the late Tang to the end of the Song, scholars treated it as a canonical text even though few accepted the claim that the Duke of Zhou was its author. Nevertheless, when the Qing Imperial Library was completed in the late eighteenth century, the *Progress Toward Elegance* was classified under the "classics" and given eminence of place as the paradigmatic work in etymology.[49]

Other Song classical works such as Lu Dian's *Adding to Elegance* (*Piya*) linked the *Progress Toward Elegance* as a classical collection of glosses to the *Poetry Classic*. Lu noted in his "Preface" that the civil examination essay was not used before 1056 and that prior to that candidates for public office had relied on poetry questions in the civil examinations to gain a name for themselves. As a by-product of such preparation candidates had a greater knowledge of the flora and fauna enumerated in the *Poetry Classic*.[50]

Lu Dian was paraphrasing Confucius' famous injunction to his students in the *Analects* that through study of the *Poetry Classic* they could become "more acquainted with the names of birds, beasts, and plants." Lu described how after 1077, when the examination essay took precedence over poetry, everyone mastered "classical techniques" while traditions of belles lettres declined. As a result, knowledge of the names of natural things diminished. Lu Dian clearly pointed to the importance of the *Poetry Classic* for natural knowledge.[51]

In his observations added to the heavens section of the *Progress Toward Elegance*, Lu Dian included sections on thunder and lightning. Citing ancient and medieval sources on how rain produced thunder, Lu Dian explained thunder in terms of the concatenation of yin and yang: "Thunder is the yang part of yin." He then cited the medieval work by Yang Quan *On the Principles of Things* (*Wuli lun*) to the effect that "an accretion of wind produced thunder." From this perspective, the fundamental cause of lightning was thunder and not vice versa as in modern accounts. Hence for lightning, Lu Dian noted that when "yin and yang intercept each other and dazzle, this is the same [configuration of] *qi* as thunder, which when released produced bright light."[52]

Based on traditions that thunder emanated from heavenly *qi*, while lightning came from earth's *qi*, Lu noted that thunder belonged to the fire phase, and that the bright light was the lightning while the sound was thunder. In conclusion, Lu cautioned that although related in terms of yin-yang and the fire phases, one should still see thunder and lightning as separate phenomena. Lu added that when the ruler decided court cases, such cases were seen as "counterparts of heaven" and that if there were occurrences of lightning and thunder, these were taken into account.[53]

Hu Wenhuan included his own work on the *Poetry Classic*, in which he repeated that knowledge of the names of birds, animals, herbs, and trees was important for understanding the *Poetry Classic*. Hu's views reflected what was in late Ming commonplace in civil examination policy questions on "broad learning" (*bowu*). In a 1597 provincial examination, for instance, the Shuntian examiners asked several thousand candidates for civil office to address why Confucius had focused students on "names and things" and not just "principles and nature."[54]

In a 1593 preface to one of his works, Hu replied to critiques that his research was redundant or too nitpicking: "If one says that Master Zhu [Xi] brought together [all knowledge about the *Poetry Classic* in his "Collection of Commentaries"] and that I have split it all up, then what good is erudition?" In effect, Hu revived focus on the "things-phenomena-affairs" mentioned in the *Poetry Classic,* which displaced Zhu Xi's efforts to bowdlerize the more sensuous love poems in the classic. The charge that Hu was splitting up knowledge, however, was an indication that his printing of works dealing with things, affairs, and phenomena no longer sustained an overarching methodological unity, which had informed the Song appeal to universal knowledge. In his scholarship on the *Poetry,* Hu collected items of information. He was not searching for universal principles.[55]

Perhaps because it was already widely available, Hu Wenhuan did not include the second century *Analysis of Simple Graphs as an Explanation of Complex Characters* (*Shuowen jiezi*) in the many printings of his collectanea. A paleographic dictionary of ancient writing forms, the *Analysis of Simple Graphs* frequently informed the lexicographical content of Hu's collectanea, however. Nor did Hu's first collection of books published as a set embrace important Tang and Song works on phonology, although he included Ming works on poetic rhymes, including two by himself. Overall, the lexicographical focus in this version of the collection stressed etymology over both phonology or paleography as the means to recover ancient glosses. This endeavor reflected late Ming trends in classical studies that stressed etymology and paleography, which were later superseded by Qing dynasty phonology.[56]

Early Natural Histories in Hu Wenhuan's *Collectanea*

Chinese naturalists represented the physical world through an historical array of entries about things, phenomena, and affairs (*bowu*). Several works were included in Hu Wenhuan's *Collectanea* that resembled natural history in the classical period in the West, when a dividing line between animate and non-animate objects was not yet decisive. A biology did not yet exist among the Chinese or the Jesuits, although we can identify aspects of a vitalism linked to *qi* and nurturing life (*yangsheng*) in Ming-Qing medical discourse, which was bounded by a classical concern with the structure of the visible world expressed as a theory of words about things.[57]

Like their classicist counterparts in Europe, the authors of Chinese "natural histories," whose works were included in the first printing of Hu's *Collectanea,* sought to identify and classify natural phenomena through language. The first such natural history that Hu Wenhuan included was the hoary *Classic of Mountains and Streams* (*Shanhai jing*), annotated by Guo Pu. Guo was a "master of esoterica" (*fangshi*) who had annotated several archaic texts included in Hu's *Collectanea* that dealt with anomalies and natural phenomena. Guo's official biography in the *History of the Jin Dynasty* noted that he was a student of classical techniques, whose broad studies (*boxue*) included the five phases, astrology, techniques for prognostication, warding off calamities, and turning back misfortunes.

The *Classic of Mountains and Streams* was a composite compiled over several centuries, which can be described as a *descriptio mundi*. Ascribed by some to the sage-king Yu the Great, it described the earth in terms of concentric rectangles with a central territory, four seas encompassing the central lands, and a great wilderness. Because it also contained valuable information about mountain spirits, popular medicine, divination and other mantic arts, often incorporated in later texts, it was at times classified simply as a handbook of anomalies and prodigies, in addition to a historical geography. Han dynasty scholars regarded it as a window on ancient portents and foreign customs. The Ming bibliophile Hu Yinglin (1551–1602), a Zhejiang contemporary of Hu Wenhuan, described it as the ancestor of oddities in Chinese literature. This view foreshadowed its derisive inclusion under the "idle chatter and gossip" (*xiaoshuo*) subcategory in the late eighteenth-century Imperial Library catalog.[58]

Guo Pu's preface to the work opened by noting that most scholars before him had seen the text as a "strange and unconventional collection," that everyone had doubts about. For Guo the key issue in interpreting the work was first to be clear about what an anomaly was:

> With respect to a correspondence phenomenon, when people called it an anomaly, they did not know why it is anomalous. When they did not call it an anomaly, they did not know why it is not. Why? Something is not anomalous of itself. It must await our [assessment] before it can be anomalous. Anomaly ultimately resides in us, not the thing.

Because an anomaly was apparently odd to us but not in and of itself, an assessment of a thing, phenomenon, or affair, according to Guo Pu, was based on our prejudiced judgments drawn from our unbalanced experience: "We observe based on what we habitually see, and we regard as strange what we have rarely heard about. This is the usual screen for human feelings." Moreover, because oddities reflected our assessment of them more than anything inherent in phenomena, any strange things recorded in the text of the *Classic of Mountains and Streams* were difficult to explain away simply as anomalies.

Guo concluded that such records probably reflected ancient, everyday references that no longer corresponded to today's habitual designations for things. In effect, Guo Pu was protecting the phenomena in the *Classic of Mountains and Streams* from thoughtless derision. However odd such records appeared to us today, they had to be taken seriously as ancient accounts of actual things, phenomena, and affairs. Guo later prepared a work containing over 250 appraisals of the famous illustrations that by the third century A.D. were included in the work but later lost.[59]

Next in the collectanea, Hu Wenhuan included two works on birds and beasts. First came Zhang Hua's (232–300) annotation of *Shi Kuang's Broad Classic of Aviary* (*Shi Kuang qinjing*). The work allegedly dated from the Zhou dynasty (ca. 1045–221 B.C.) and opened with 360 categories of birds. Priority was given to the male phoenix as the chief example of feathered types of animals. Like Guo Pu, Zhang Hua was a medieval master of esoterica, famous for his knowledge of underground and remote realms, skilled as a diviner, and informed about spirits and exorcism. The *Classic of Quadrapeds* (*Shoujing*), which followed, included mention of auspicious animals. Neither work was a zoology per se, though each contained descriptive accounts. Rather, they were repositories of curiosities, prodigies, and wonders and thus outside the realm of normal explanation.[60]

Two later works in the initial printing of the *Collectanea* were "records of anomalies" (*zhiguai*) that dated from the Han through the Song. The *Classic of Spirit Marvels* (*Shenyi jing*) of the Later Han dynasty was completed circa the late second century, and a later Jin commentary attributed to Zhang Hua was added. The religious tenor of the work, including the veneration of sacred texts, linked it to the increasingly porous cultural and religious boundaries of orthodoxy and esoterica, which Chinese literati could comfortably straddle up to the Song dynasty. The *Record of Accounts of Strange Things* (*Shuyi ji*) was allegedly prepared during the Liang dynasty (502–556), although the compilers of the Qing Imperial Library catalog saw it as a Tang forgery because it contained anachronisms.[61]

A preface to the *Record of Accounts of Strange Things* dating from circa A.D. 503 indicated that the author had accessed a library of 30,000 chapters when he prepared his work, which allowed him to include a great deal of stories about anomalies. In addition, his work was informed by an interest in a broad learning about things. The "Afterword" composed in 1044 by Ye Fang defended the *Record* and defined oddities as "matters that were unheard of." It linked the focus in the work to the "broad learning of the scholar," thus making writings about anomalies perfectly respectable, a claim that was contentious among Song literati, who distanced themselves from the records of anomalies genre.[62]

Perhaps the most representative texts of early natural history included in Hu's *Collectanea* were those that focused on "broad learning of things" (*bowu*). Zhang Hua's *Treatise on Broad Learning of Things* (*Bowu zhi*) and Li Shi's Song dynasty *Continuation* (*Xu bowu zhi,* ca. 1150) were included. Other such Song dynasty works in Hu's collectanea were:

- *Record of the Origins of Things and Affairs* (*Shiwu jiyuan*), compiled by Gao Cheng circa 1078–1085
- *Discussions on the Formation of Heaven and Earth and the Myriad Things* (*Tiandi wanwu zaohua lun*) by Wang Bo (1197–1274)

Several works from the early and middle Ming were also printed:

- *Origins of Things* (*Wuyuan*)
- *Origins of Things Old and New* (*Gujin yuanshi*)
- *Examination of Ancient and Contemporary Things and Affairs* (*Gujin shiwu kao*)

Zhang Hua had prepared a table of contents with thirty-nine topical headings for his *Treatise* (see Table 1.1).

Of particular note in this classification list are the highlighted items #17, "nature of things" (*wuxing*), #18, "principles of things" (*wuli*), and #19, "categories of things" (*wulei*), which used terminology that would later be refracted first by Jesuit and then by Protestant missionaries—with the help of their Chinese collaborators—to refer to late medieval and early modern scientific fields.[63]

The Jesuit translation for physics into Chinese (= *cha xingli zhi dao,* lit., "observing the nature of principles of the Way) was used in Giulio Aleni's (1582–1649) *A Summary of Western Learning* (*Xixue fan,* 1623). *Physica* or *physiologia* (= *xingxing xue*) appeared in the *De Logica* (*Mingli tan,* lit., "Chats on the principles of names," 1631), which was compiled by Francisco Furtado (1587–1653) and Li Zhizao (1565–1630). Similarly, the modern translation for physics (= *wuli*), despite its late nineteenth-century Meiji Japan lineage, could also trace its genealogy as a technical term back to Zhang Hua's *Treatise.*[64]

In general these works stressed each human event, object-implement, or natural phenomenon in terms of a teleology of their usefulness to humans and presented a genealogy of discovery that traced each item back to the appropriate sage, ruler, or scholar in antiquity. In his preface to the *Record of the Origins of Things and Affairs,* dated 1448, Yan Jing linked all myriad things and affairs to their principles, which could be investigated by studying their origins:

> The myriad things, which fill heaven and earth and extend from antiquity to the present, always have matters changing within them. Things

Table 1.1. Table of contents for Zhang Hua's *Treatise on Broad Learning of Things* (*Bowu zhi*).

1. Geography	14. Unusual insects	27. Debates on *fangshi*
2. Earth	15. Unusual fish	28. Corrections of names
3. Mountains	16. Unusual plants	29. Corrections of texts
4. Waters	17. Nature of things	30. Corrections of geography
5. General topography	18. Principles of things	31. Corrections of ceremonial rites
6. Peoples in five directions	19. Categories of things	32. Corrections of music
7. Product	20. Medicines	33. Corrections of sumptuary display
8. Foreign countries	21. Medical discussions	34. Correcting names of utensils
9. Strange peoples	22. Food taboos	35. Correcting names of things
10. Unusual customs	23. Pharmaceutical arts	36. Information about anomalies
11. Unusual products	24. Stagecraft	37. Historical supplements
12. Unusual beasts	25. *Fangshi*	38. Miscellaneous words, part 1
13. Unusual birds	26. Medicinal foods	39. Miscellaneous words, part 2[a]

a. *Bowu zhi, juan* 4–5, in GZCS. See also Campany, *Strange Writing*, pp. 49–52, Roger Greatrex, trans., "Bowu Zhi: An Annotated Translation" (Stockholm University Ph.D. diss. in Oriental Studies, 1987), and Hervouet, ed., *A Sung Bibliography*, p. 343.

have myriad variations, affairs have myriad transformations. No matter or thing is without a principle and an origin. If one does not fathom the principle [involved], then there is no way to complete the knowledge in our minds. If we do not research the origins [of a thing or matter], then how can we unravel and fathom its principle? Therefore, sagely studies give priority to investigating things and extending knowledge. Literary scholars value broad inquiry and wide learning. But a scholar [*Ru*] would be embarrassed by failing to know even one thing.[65]

The meaning of the title for the *Record of the Origins of Things and Affairs* assumed an overlap of natural things, phenomena, and human affairs with a stress on the investigation of things and exhaustively mastering principles. Its scope ranged from large (heaven, earth, mountains, streams) to small (birds, beasts, herbs, trees), to subtle (the arcana of yin and yang), and to the manifest (the institutions that use rituals and music). This spectrum moved from natural to social phenomena. The sources for each item were cited in chronological sequence to elaborate a particular item and set its origins in human terms. Further discussion was otiose.[66]

Luo Qi's 1474 preface for his *Origins of Things* opened with discussion of the sages as the creators of implements. He was critical of earlier works for not excluding patently absurd items. Within his arrangement of 18 categories, Luo included 239 total items. The afterword explained that the aim of the *Origins* was to explain the things of heaven and earth in relation to changes in human matters.[67]

The *Examination of Affairs and Things from Antiquity to the Present* by Wang Sanpin included an institutional account of historical geography (*dili*), as well as general information required for classical literacy and civil examination preparation. In addition, Wang's brief history of "numerical correspondences" (*shuxue*), later a term for mathematics in the nineteenth century, was linked to the *Change Classic* (*Yijing*) and the naturalistic process of the formation of things based on yin-yang and *qi*. In his 1538 preface, Wang Sanpin noted that he had added material to Song accounts.

Wang presented naturalistic explanations for the phenomena of wind and lightning based on yin-yang and *qi* that added some precision to Lu Dian's earlier analysis in the *Adding to Elegance*. Lu Dian had cited *On the Principles of Things* to the effect that "an accumulation of wind formed lightning." Wang Sanpin, on the other hand, explained this by contending that "the yang that is within cannot escape, and its resounding blows produce lightning. . . . When yang is outside," Wang continued, "and cannot get inside, then it swirls around without stopping, forming wind." Similarly, Wang cited Song dynasty accounts on rain: "When earthly *qi* rise and heavenly *qi* fall, they form the vapors of steam and form rain."[68]

A Theory of Cosmology and Natural Change

Although Wang Bo's work was presented before the *Origins of Things* and other such compendia on "things" and "origins," we can conclude this discussion of early natural histories with a work that dealt with the theoretical issues inherent in this genre of entries about things, phenomena, and affairs, namely, Wang's ambitiously titled *Discussions on the Formation of Heaven and Earth and the Myriad Things*.

Wang Bo had opposed Zhu Xi's interpretation of the passage in the Great Learning on the investigation of things and extension of knowledge. Nevertheless, he was in substantial agreement with the spirit of Zhu Xi's investigation of things to fathom principles. Wang, like Zhu, articulated a holistic, structural account of natural processes, which enabled literati to explain in theory the production of all things. In his preface to Wang's *Discussions,* Zhou Yong summarized this all-encompassing view of the cosmos and its naturalistic evolution:

The Great Ultimate is principle [*li*]. Yin-yang is *qi*. Heaven and earth are forms. Talking about them in aggregate, when a form is endowed with a certain *qi*, principle is complete within that *qi*. Talking about them disaggregated, then what is above form [i.e., principle = Great Ultimate] and what is below form [i.e., *qi* = yin-yang] are bound to differ. The Great Ultimate originally contains yin-yang. What moves becomes yang; what remains quiescent becomes yin. If something is alive, then all of it [i.e., yin and yang] is alive at once. One cannot speak [of this simultaneity] as something occurring before and something else after.[69]

Zhou placed Wang Bo's account of the "shaping of all things" (*zaohua*) within the theoretical framework of what in Song times was referred to as "studies of principles" (*lixue*) associated with Zhu Xi and his followers. Wang Bo, according to Zhou, had "pushed back to the origins of all things and illuminated the mysteries of the shaping of things, and was truly adequate to elaborate studies of principles and provided wings [i.e., commentary] for 'this culture' [i.e., the classical teachings preserved by Confucius]." In Zhou's view, Zhu Xi's "Learning of the Way" was not simply a moral vision of philosophical principles. It was also a theory of the principles of natural change.[70]

Wang Bo's work represented an extended essay on the cosmology of phenomenal beginnings that created all things in the world through the interaction of *qi*, form (*xing*), and materiality (*zhi*). The shaping of all things presupposed, however, the more paramount status (*ling*) of humans in the cosmos. Wang was not unusual in this claim that humans were the rulers of birds and beasts, and that, therefore, things in the world were classified, encyclopedically and otherwise, as secondary to humans and subject to their vital needs.[71]

Wang Bo's own account in the *Discussions* stated the pre-Jesuit theoretical vision informing natural history in late imperial China, which Europeans would challenge in the late Ming:

Originally, before the beginnings of distinctions, there was the Great Change, the Great Origin, the Great Beginning, and the Great Simplicity. The Great Change occurred before *qi* was manifest. The Great Origin was the beginning of *qi*. The Great Beginning was the beginning of forms. The Great Simplicity was the beginning of material. When *qi*, forms, and materiality were not yet mutually separated, then we call this chaos [i.e., undifferentiated]. Chaos refers to the myriad changes [involved], chaotic without a beginning of separation. When chaos was already separated out, then heaven and earth opened out. When heaven and earth were distinguished, then they produced two models [i.e., yin and yang]. The light and clear floated up and became heaven; the heavy and turbid congealed and became earth.[72]

Without citing his source, perhaps to disguise its unorthodox status, Wang Bo was paraphrasing classical Daoist cosmology in the *Book of Master Lie* (*Liezi*), a work that he thought was written much earlier but today would be dated circa third century A.D. The *Book of Master Lie* included the following passage under the heading of "Heaven's Fortune" (*Tianrui*):

In the past, the sages conformed to yin-yang to unify heaven and earth. If what took form grew out of the formless, then from what did heaven and earth grow? Therefore it is said: 'There was the Great Change, the Great Origin, the Great Beginning, and the Great Simplicity. The Great Change occurred before *qi* was manifest. The Great Origin was the beginning of *qi*. The Great Beginning was the beginning of forms. The Great Simplicity was the beginning of material. When *qi*, forms, and material were not yet mutually separated, therefore we call this chaos [i.e., undifferentiated]. Chaos refers to the myriad things when there was mutual chaos without a beginning of separation. It was invisible, inaudible, and unobtainable. Therefore it was called the Change.'[73]

Borrowing from a classical Daoist text was still acceptable during the declining years of the Southern Song, even when appeals to orthodoxy became more strident. Moreover, the overlap between the orthodox cosmology of the Learning of the Way and that of Master Lie indicates how closely tied classical learning and natural speculations had become. Wang Bo used this classical if unorthodox theory of the "shaping of things" to explain, precisely, natural phenomena (*ziran*, lit., "so of itself"). At times, Wang correlated phenomena based on numerological regularities, but more often his cosmological concepts were used to explain natural events.[74]

We have noted that the medieval *Adding to Elegance* and Ming *Examination of Ancient and Contemporary Things and Affairs* explained lightning and thunder by the concatenations of yin and yang. Zhu Xi had explained lightning in conversations with his students by criticizing the popular notion of a thunder spirit. Like Lu Dian and Wang Bo, Zhu also stressed the mutual rubbing and grinding of yang versus yin *qi* that produced lightning and thunder when *qi* was condensed and exploded. Nevertheless, Zhu also considered the anomalistic aspects of *qi* in light of thunder as auspicious or ominous. For Zhu, during a thunder and lightning strike the *qi* condensed into a sort of thunder ax that possessed spiritual powers. Zhu tied that thunder and lightning to the appearance of spirits and the manifestation of devils (*guishen*).[75]

For the same phenomena, Wang Bo explained that when yin and yang rubbed and ground against each other, their interaction exploded into sparks of lightning. Wang Bo's Southern Song account added more materiality to the swirling winds than earlier accounts and left out the popular understanding of the thunder ax. Nevertheless, the common denominator of these explanations, whose dates range across almost a millennium, was that rain, clouds, thunder, and lightning resulted from interactions between yin *qi* and yang *qi*.[76]

From Orthodoxy to Esoterica

In addition to his interests in classical lexicography and natural history, Hu Wenhuan also compiled and printed a wide range of works based on his broader interests in Chinese religion, which extended to popular Daoism, Buddhism, and medicine. Nourishing life (*yangsheng*) was an important theme in Chinese classical medicine and internal alchemy. The goal was to achieve the ancient ideal of bodily regeneration and longevity through internal moral discipline. Likewise, the focus in Ming publications emphasized "multiplying descendants" and "nurturing life." By linking meditative exercises, sexual discretion, and ingestion guidelines, Ming scholar-printers found a rich vein of readers when they tapped into teachings that historically were concerned with the sexual arts of the bedroom (*fangshu*) and stressed love, fertility, and health in light of the body's potential for producing offspring.[77]

This aspect of Chinese interest in the natural world grew out of the need to maintain the harmonious functioning of the human body, avoiding both excesses and wasting of the body's natural vitality. The *Basic Questions* (*Suwen*), one of the two best-known works in the medical corpus known as the *Inner Canon of the Yellow Emperor* (*Huangdi neijing*), for example, had emphasized the goal of human longevity. There the ruler was advised to cultivate serenity, detachment and nonstriving in order to "guard one's essence and spirit within." Such medical regimes were paralleled by early claims in the classical literature that an imperishable metal (gold) could serve as an elixir and that the attainment of earthly immortality was also possible using herbs of immortality. Such recipes for nourishing life (*yangsheng*) became a major focus in late Ming collectanea and encyclopedias.[78]

The goal of various techniques of breath control was physical immortality in an imperishable body, which was nurtured like an embryo through yogic disciplines. As an inner alchemy (*neidan*), such bodily transformation was comparable to how an alchemist brought an elixir to maturity in a matrix of lead (*waidan*). Older alchemical writings were later reinterpreted in light of the physiologic procedures associated with popular breath control and the meditation techniques introduced by Buddhists in medieval China.[79]

Alchemy from the start had medical overtones because the art of making elixirs of immortality represented a natural process for creating perfected substances that could be ingested or internalized. Alchemical treatises accordingly presented chemical operations as metaphors for cosmic and spiritual processes. The extraction of silver from lead (*waidan*), for example, reflected the emergence of a perfected self (*neidan*). Internal and external alchemy were complementary and regularly practiced together until the eleventh century, when adepts largely abandoned external alchemy.

Events, like things, fit into a cyclical pattern of natural processes. The life cycle of birth, growth, maturity, decay, and death paralleled the cycles of day and night, the seasonal changes, and the calendrical cycles of cosmic change. Yin and yang as dualistic, dynamic concepts were applied to spatial relations that were changing or in equilibrium. The five evolutive phases provided a parallel division of cycles or configurations of change for organizing natural phenomena. The eight trigrams and hexagrams from the *Change Classic* (*Yijing*) were also associated with the production of elixirs.[80]

In Hu Wenhuan's time, the unconventional literatus Yuan Huang (1533–1606) drew on two chief views of fate in Ming public and private life: (1) a system of merit and demerit that human agents could influence; (2) a framework of meaning beyond human influence. Ming literati such as Yuan increasingly regarded merit accumulation in the moral realm as homologous with cultural prestige and worldly material benefit. Later a Cheng-Zhu follower, Lu Shiyi (1611–1672), initially used Yuan's personal *Ledger of Merit and Demerit* (*Gongguo ge*) and titled his own work as *Chapters on Investigating Things and Extending Knowledge* (*Gezhi pian*).

The late Ming debate between moral purists (typically followers of the Learning of the Way), who thought fate was beyond human influence, and those like Yuan Huang, who favored merit accumulation among elites for success in civil examinations and among commoners for personal self-improvement in daily life, had become important enough to contest in official circles such as the civil examinations. Elaboration of a moral vision for those in late Ming society such as Hu Wenhuan, who aspired to social mobility and material success, was not so easily controlled in an era of social and moral flexibility, however.[81]

Yuan Huang also reinvigorated the traditions of inner alchemy by drawing on the psycho-physical model of the body, which informed traditions of outer alchemy. In such representations, the body became the crucible where the gross material of physical life could be refined and purified through human discipline. Yuan recommended the know-how to ensure that a fertile body was properly disciplined by sexual and spiritual exercises and by diet and pharmacy. This flawless body manifested the powers and pleasures of sexual function and the possibility of sagely transcendence and bodily immortality achieved through the procedures of inner alchemy.

By mastering the techniques of sexual hygiene, meditation, breath control, and bodily visualization, an adept could produce an elixir out of his or her own body. Yuan Huang emphasized the moral management of the seminal essence and breath through the three primary psycho-spiritual-material vitalities of the body, *qi,* and psyche (*jing qi shen,* lit., "essence, psycho-physical configuration, and psyche"). An adept could link the refinement of the body's essence, purification of one's *qi,* and the self-transcendence of the psyche to the inner cultivation of an immortal embryo. In a perfect state of emptiness, the circulated vitalities of the body would assume an increasingly concentrated and ethereal form.

This life-affirming overlap of sex, cultivation, and male fertility for nourishing life appealed to elite and popular audiences of men and women. Charlotte Furth has described how the Ming ideal of inner alchemy required moral cultivation and enlisted the sexual drive for the inward goals of classical sagehood, a short step, we might add, from a concept of the sublimation of the affects into spiritual ideals. Male sexual powers in particular were associated with the reproductive function. Consequently, late Ming versions of nourishing life that appeared in popular daily use compendia published separately or in collections focused more on human health and longevity than on the utopian quest for immortality. Inner alchemy was popularized as the moralization of health in Ming cultural life, and the goal of individual fertility displaced human longevity as the goal.[82]

Other Editions of Natural Lore: Expanded Versions of Hu Wenhuan's Collectanea

The 1609 version of the *Collectanea for Investigating Things and Extending Knowledge* was likely the most authoritative edition because it contained the only datable preface by Hu Wenhuan and a general table of contents of 140 works in the collection. Hu's preface presented fourteen categories of works that he and his staff had collated from the early and later masters and the hundred lineages of classical learning:

1. etymology
2. philology·

3. poetic ability
4. literary criticism
5. astrology
6. geography/topography
7. astronomy/harmonics
8. penal laws
9. low-ranking officials
10. medicine and prognostication
11. Buddhism/Daoism
12. border peoples/foreigners
13. fauna
14. flora

Except for the lexicons and natural histories, which had appeared as a separate edition, Hu's division of categories for the 1609 edition of his collectanea generally reflected those that we have previously described in the daily use encyclopedias of the 1590s.[83]

Like many Ming scholar-printers, Hu mixed and matched the contents of his editions to sell more copies of his printed works. He also changed several original works to conform with his own printing needs. The Qianlong editors of the Imperial Library project found this aspect of Hu's editions especially egregious as they castigated him for his blatant profit-seeking. The editors made clear in their 1780s critique of Hu's massive *Collectanea* that they also disapproved of his inclusion of works on natural history (*bowu*) because such accounts represented writings previously classified as fictional accounts.

In addition, the editors noted that the version of Hu's collectanea that they reviewed contained 181 works divided into 12 classifications. At least four of these dealt with natural events:

1. classics/commentaries: fifteen works
2. unofficial histories: twenty-one works
3. officialdom: twelve works
4. legalists: twelve works
5. lessons/admonishments: fourteen works
6. masters and others: eight works
7. respect for life: eighteen works
8. seasons/agriculture: eight works
9. crafts: ten works
10. virtuous/praiseworthy: seventeen works
11. fictional works: eleven works
12. arts collections: thrity-five works[84]

What is interesting here is that Hu Wenhuan's publications on poetry, literature, the arts, nourishing life, etc., were added to the earlier, smaller versions of the collectanea, such as the version with forty-six classical works summarized above.

Later, many were also included in Hu Wenhuan's other major collectanea known as the *Famous Works of All the Scholastic Lineages* (*Baijia mingshu*), which can be dated to 1603. The preface by the Hanlin academician Zhu Zhifan (b. 1564) made it clear that Hu's *Famous Works* was also a comprehensive library of works stressing poetry, astrology, astronomy, flora and fauna, medicine and the arts of longevity, in addition to the more classical themes of philology and technical glosses on things and affairs. In his preface, Zhu Zhifan used the term for investigating things and extending knowledge (*gezhi*) to suggest the overarching unity Hu Wenhuan applied to such works, which linked the *Famous Works* collection to the *Collectanea for Investigating Things and Extending Knowledge*.[85]

Each collectanea contained many of the same works that Hu Wenhuan compiled or wrote himself. In the late sixteenth and early seventeenth century, Hu Wenhuan compiled and published more than five hundred works in Hangzhou and Nanjing. Moreover, the various editions of the *Collectanea for Investigating Things and Extending Knowledge* and *Famous Works of All the Scholastic Lineages* that have circulated over time are almost all different in content.[86]

Using works from earlier editions of the *Collectanea*, Hu Wenhuan published the *Famous Works* as an alternative way to present his collection of books. In effect, he had combined in his collectanea Ming works that were similar to two types of Song encyclopedias—the *Materials of the Taiping Xing Guo Era for the Emperor to Read* and *Expanded Records of the Taiping Xing Guo Era*. Other, smaller collectanea that Hu compiled also included many works from his two major collectanea (see Table 1.2). The *Collectanea for Prolonging Life* (*Shouyang congshu*), for example, contained a series of Ming works in which we see a nascent vitalism, or a philosophy for nourishing life, which was representative of similar currents in late Ming digests and collectanea.[87]

Late Ming Statecraft, Mathematics, and Christianity

Most of the collectanea and practical compendia common in the late Ming were produced by lower-level gentry-literati or scholar-printers. In contrast to Li Shizhen's *Systematic Materia Medica* and Hu Wenhuan's *Collectanea*, the scholar-official Feng Yingjing (1555–1606) while in jail compiled an important encyclopedia titled the *Compendium of Practical Statecraft in the*

Table 1.2. Other collectanea compiled by Hu Wenhuan.

- *Secret Works for Investigating Things and Extending Knowledge* (*Gezhi mishu*), with thirty-seven of the forty-six works on lexicography and natural history in the original edition of the *Collectanea of Works Investigating Things and Extending Knowledge*.
- *Collectanea for Prolonging Life* (*Shouyang congshu*), which dealt with some sixteen to twenty-three medical texts associated with "nourishing life," many of which were included in the larger versions of the *Gezhi congshu* and *Baijia mingshu*.
- *Broad Review of Original Roots* (*Yuanzong bolan*), with thirty-one works mainly dealing with "nourishing life," which were also included in the *Gezhi congshu* and *Baijia mingshu*.
- *Comprehensive Collection of Maps and Works on the Great Ming Dynasty* (*Da-Ming yitong tushu*), with nine works that were included in the larger versions of the *Gezhi congshu* and *Baijia mingshu*.
- *Origins of Things Old and New* (*Gujin yuanshi*), with three works, all of them natural histories that were included in the original version of lexicographies in the *Gezhi congshu*.
- *Four Authorities Authors on Music, Calligraphy, and the Arts* (*Youyi sijia*), including four works.[a]

a. See Wang Baoping, "Zhongguo Hu Wenhuan congshu jingyan lu," pp. 19–25.

August Ming Dynasty (*Huang Ming jingshi shiyong bian*), which was published in 1604 with a 1603 preface by Feng. A palace examination graduate in 1592, Feng was jailed for four years beginning in 1601, while serving as an assistant surveillance commissioner, for refusing to allow a eunuch tax collector in Huguang free rein in accumulating revenues in the name of the emperor.[88]

Also jailed with Feng was He Dongru (fl. ca. 1589), who was a prefectural judge in Huguang and had also resisted eunuch tax demands on miners in the area. Together they discussed the political chaos of their time, and each resolved to compile works while in jail that would stress statecraft policies to help ameliorate the dangers of their time. They advocated a return to the policies of the first Ming emperor, who had unified the empire after the anti-Mongol conflagrations in the middle and late fourteenth century. Feng Yingjing's *Compendium* thus returned to the repertoire of statecraft techniques and moral policies derived from the *Rituals of Zhou* that had informed politics in the early Ming.[89]

In addition, Feng Yingjing had befriended Matteo Ricci and taken an avid interest in the Christian religion and the Western studies that were associated with it. It is said that Feng would have been baptized had he not died prematurely, just two years after his release from jail. The Christian impact on

Feng's *Compendium of Practical Statecraft in the August Ming Dynasty* was made implicit in the opening chapter, which comprised the first Ming emperor's own official writings. Under the cover of an annotation to the emperor's orthodox evocation of classical doctrines, namely that "the sages and worthies had the mind-heart of a gentleman," Feng interpolated a discussion of an "enlightened antiquity" when "the spirituality of a unified mind served the spirituality of the minds of ten thousand people."[90]

Feng then turned to a discussion of the "Lord on high" (*shangdi*) and his rectifying power to discern right from wrong, a comforting lesson for a man jailed by a corrupt eunuch. The affinity of human spirituality with such a power, Feng contended, carried over to the founding emperor, who had been endowed with the requisite spirituality to sustain his reign. This reference to the Lord on high, a term that Ricci and the early Jesuits had settled on as a translation for *Deus* because of its ancient provenance in the *Documents Classic*, was permissible in Feng's notes because the first Ming emperor had himself used the term in his discussion of palace sacrifices. Feng regarded this reference as the "origins of the emperor's methods of the mind." Building on this imperial precedent, Feng could then claim that "we are all the Lord on high's people." Feng's rhetorical appeal to spiritual salvation was manifest: "How much heaven must love me as a person to enable me to suffer the pain of the world."[91]

Feng's *Compendium* was composed as a statecraft digest and not as a Christian tract, however. Its focus on imperial governance incorporated sections on officialdom, ritual studies, musical studies, archery, calligraphy, and arithmetic, in addition to sections on orthodox studies. Feng's inclusion of arithmetic and mathematics in the compendium was intended to correct excessive stress on belles lettres, which had emerged during the Ming, particularly since the eight-legged examination essay became a literary rage in the late fifteenth century. Here, too, we discover Feng's contact with Ricci and European studies because the arithmetic that Feng detailed in his digest was drawn from the *Systematic Treatise on Computational Methods* (*Suanfa tongzong*) by Cheng Dawei (1533–1606), which had been published in 1592–1593 and quickly emulated as *wasan* (Japanese mathematics) for practical computations in Japan.[92]

Cheng Dawei's *Systematic Treatise* comprised a collection of 595 difficult arithmetic problems. Designed as a general and practical arithmetic using the abacus, rather than more cumbersome counting rods (*chousuan*) used to solve higher-degree equations during the Song and Yuan dynasties, Cheng's work was widely disseminated before the Jesuits encouraged the Kangxi emperor's interest in it later in the seventeenth century. Cheng's *Treatise* reflected the arithmetic language of Ming merchants, who were the key guardians of the *Computational Methods in Nine Chapters* (*Jiuzhang suanshu*) tradition.

One of the "Ten Mathematical Classics," the *Computational Methods* was compiled between 200 B.C. and A.D. 300, with later commentaries added to it. Some regarded it as the most important work in the mathematics canon. It was also the model for mathematical language and the pattern for computation since medieval times, when the official canon was used at the School for Dynastic Students during the Sui and Tang dynasties to teach mathematics.[93]

The oldest known edition of the *Computational Methods in Nine Chapters* dated from the thirteenth century, however. Because the latter was not fully reconstituted and reprinted until the eighteenth century, Cheng Dawei's *Systematic Treatise on Computational Methods* represented the best available late Ming work in the *Nine Chapters* tradition for arithmetic problems and their numerical solutions. Some of Cheng's problems were later adapted into a more classical style for literati in Ricci's arithmetic, which was based on European and Chinese sources.[94]

Matteo Ricci's 1614 *Translations of Guidelines for Practical Arithmetic* (*Tongwen suanzhi*), compiled with Li Zhizao's (1565–1630) help, adapted its arithmetic from Clavius' *Epitome Arithmeticae Practicae*. It was also based on arithmetic from the Yuan and Ming periods. Hence, Ricci's treatise on European arithmetical knowledge added problems from the *Systematic Treatise on Computational Methods* by following the Chinese arithmetical tradition. Although Ricci introduced the Western pen-and-paper approach to arithmetical calculations, overall his *Translations* did not go beyond the limits of the Chinese traditional mathematics.[95]

By including portions of Cheng Dawei's *Systematic Treatise,* Feng Yingjing's statecraft digest incorporated a definitive arithmetic in which problems, rules, and solutions were composed in verse with rhyming formulas for teaching arithmetical terms, computation with the abacus, and multiplication tables. Feng added that although the Duke of Zhou had in antiquity developed the *Computational Methods in Nine Chapters,* its arithmetic had not developed the relational features for computing the sides of a right triangle (*gougu*, lit., "base and altitude") that were later found in the *Sea Island Computational Canon* (*Haidao suanjing*) by Liu Hui at the end of third century A.D. Feng added that Westerners such as Matteo Ricci had mastered the computational procedures, i.e., geometry and trigonometry, necessary to measure long distances in fields by mastering similar right triangle procedures.[96]

Feng's interest in more advanced mathematics, which went beyond the simple arithmetic of Cheng Dawei's *Systematic Treatise,* indicates that the Jesuits were dealing with literati already concerned about the need to develop more expertise in mathematics. This is shown by Feng's inclusion of earlier Ming right triangle studies in his section on mathematics. A generation earlier, for example, the literatus Tang Shunzhi (1507–1560) prepared a series of essays on triangles and segments of a circle that were essential in geometric calculations, which Tang precociously called "studies of measured amounts" (*jihe xue*, i.e., magnitude or quantity), a term later selected by Ricci and his Chinese collaborator for the late Ming translation of Euclid's geometry.

Before the arrival of the Jesuits, Tang Shunzhi had distinguished between new and old computational methods, a division that would reappear in later discussions of Western learning. His essays preceded by several decades the earliest Jesuit transmission of medieval and Renaissance mathematics to China. The latter were essential to integrate ancient classical theory with contemporary technical expertise in order to reform the Ming calendar.[97]

Feng Yingjing's stress on practical problems in mensuration involving right triangle calculations in his digest also occurred before Ricci's translation of Euclid's *Elements of Geometry* was completed. Moreover, problems in late Ming calendar reform that we will address in Chapter 2 increasingly contributed to the Ming reception of early modern European mathematics and computational astronomy. Consequently, Feng Yingjing's *Compendium* foreshadowed late Ming and early Qing literati and imperial court interest in European trigonometry, trigonometric tables, and logarithms. It also prefigured statecraft collections that were compiled by convert-officials such as Xu Guangqi.

Collecting Things in Texts

The proliferation of late Ming daily use compendia—many of which Hu Wenhuan's *Collectanea* prominently contained—reflected a widening audience for information about things, phenomena, and affairs of all sorts. The accruing knowledge of things and affairs among Ming scholars was still subsumed within the moral and philosophical frameworks that informed the orthodox literati classification of the natural world and drew on the classical repertoires of knowledge outlined above. These repertoires notably included medieval masters of esoterica such as Zhang Hua, who were central to the late imperial definition of a cumulative knowledge of things and phenomena. Although the classical lexicons and natural histories were the beginning points, the knowledge in ancient canonical texts and their commentaries were insufficient.[98]

Literati deployed things in digests and the collectanea by presenting chronological or topical presentations of past glosses about them. Things, events, and anomalies were displayed textually and sometimes pictorially. In time, words as glosses, that is, the textual lives of things, took precedence over any analysis of the things signified. As a result, natural studies became a venue for Chinese textual scholars who were fascinated with the etymologies of the words. Sages had created such words to encompass phenomena. Hence, they were also important as a genealogy of items in the classics. Unlike early modern European scientific culture, where natural history was increasingly displayed as concrete items in a museum, the array of entries about things included in Hu Wenhuan's collectanea of early lexicons and digests converted natural phenomena into words in a text that needed to be decoded primarily through the analysis of language.[99]

Paula Findlen has described the new attitudes toward nature as a collectable entity and new techniques of investigation that informed natural history in early modern European scientific culture. The first science museums were repositories of technology, curiosities, and wonders that built on Pliny's encyclopedic definition of nature in his *Natural History* as everything that was worthy of memory. Europeans coped with the empirical explosion of materials that the wider dissemination of texts, increased travel, voyages of discovery, and more systematic forms of communication had made possible by establishing private museums, which became state-sponsored institutions from the eighteenth century onward. Such museum collections became symbols of prestige and power, and collectors entertained the image of knowledge without end more widely in the seventeenth century.[100]

Paula Findlen focuses on the linguistic, philosophical, and social matrices that gave museums a precise intellectual and spatial configuration. Museums became a venue to experience nature. From the sociology of collecting and its cultural logic, we can see that in late imperial China this sort of collecting and classifying knowledge about things occurred within the pages of collectanea and encyclopedias. Just as the museum was firmly set in the premodern European encyclopedic tradition of catalogs and the vocabulary of collecting, so the daily use compendia of the late Ming were sites of classically derived knowledge where individuals of privilege and learning earned the right to collect and classify the world. Others could participate in such collecting by reading about things in the texts that they purchased or borrowed.

Hu Wenhuan's merchant-scholar-printer status in Hangzhou and Nanjing publishing circles allowed him to participate in highbrow activities at the same time that he published works for profit aimed at popular audiences. The first increased the classical prestige of his collecting and reprinting information from lexicons and natural histories. Hu elevated his own curiosity about things and phenomena to a virtue that was entirely appropriate for gentlemanly behavior. Likewise, Li Shizhen's much more analytic work on pharmacopoeia approximated the role of the collector of natural objects—or information about them—as parallel to the Learning of the Way scholar who investigated things, or someone who collected objects.[101]

If the museum became a site of encyclopedic dreams and humanist sociability in sixteenth-century Europe, collecting information about things in the late Ming was not yet a prelude to display (in museums) or manipulation (in laboratories). Li Shizhen certainly shared the naturalist's agenda. Moreover, his pilgrimages to collect medicines and herbs were done through fieldwork and perusing texts. His natural studies remained focused on remedies. Secondarily, he was fascinated with the etymologies of terms for living things, which could then be applied via the investigation of things to classify appropriate medicines. Ming collectors of encyclopedias (such as Hu Wenhuan) never expressed a penchant for purely experiential knowledge obtained in the laboratory, although medical men continued to produce empirewide a rich plethora of medicines and accessories for traditional Chinese healing. Li Shizhen did so as well.[102]

In early modern Europe, gentlemen enriched their collecting experiences by increasing the presence of animal skeletons and fossils in the museum. The new culture of experiential demonstration transformed the museum into a site of medical knowledge, within which competition over the control of knowledge between apothecaries and physicians-professors ensued. Nevertheless, the modern category of natural science had no formal meaning for the European naturalist in the sixteenth century any more than it did in the late Ming. To compare the historical context for early modern scientific culture in Europe and natural knowledge in late Ming collectanea and digests, we need to problematize the historical teleologies that turn the past purely into prologue.[103]

Collecting as a research agenda rearranged the boundaries of natural studies in Europe, but the coexistence of the old and new, the occult and the demonstrable, in early natural history was as prevalent in Europe as in the late Ming. Even after the Jesuits arrived in Ming China, no one there or in Europe singled out and privileged natural science. Later in the seventeenth century, via Bacon, European scholar-gentlemen dismantled older forms of natural philosophy in favor of a new empiricism. Galileo, Descartes, and Newton transformed such high-minded empiricism into the concrete beginnings of physical science in Europe after the Jesuits arrived in China.[104]

Efforts to normalize the marvelous in Ming China turned the collection of information about things into a form of classical knowledge gained through encyclopedic research, which defended itself using the rhetoric of orthodox moral cultivation, i.e., to investigate things and extend knowledge. Rather than microcosms of nature, collections of early lexicons and Ming digest created textual museums for their theater of marvels. Hu Wenhuan's efforts in the 1590s to collect the collectors within a single collection of books published as a set was not unique. His economic resources allowed him to produce and publish several collectanea he deemed appropriate to place under the general heading of investigating things and extending knowledge (*gezhi*). His initial reconstruction of the ancient lexical texts allowed him to use them as an orthodox base for enlarged editions of his *Collectanea,* which included domains of knowledge that exceeded the boundaries of the official canon.

When the military dust settled after the final Manchu victories in the 1660s, Ming loyalist scholars from the Yangzi delta such as Gu Yanwu (1613–1682) and Huang Zongxi (1610–1695) took their predecessors to task for their naiveté concerning classical controversies such as the Old Text and stone versions of the Great Learning. Chen Que (1604–1677) in particular singled out this controversy and criticized late Ming scholars for their credulity and lack of the textual expertise needed to recognize the dubious stone version as an obvious forgery. In general, the focus on the investigation of things in the early Qing shifted from a pathway to sagehood to a more rigorous methodology for extending all knowledge, whether moral, textual, or worldly. The Jesuit appeal to *scientia* in China was received by literati in the midst of a sea change in classical learning from the Ming to Qing dynasty.[105]

Seventeenth-century Chinese collections and digests drew on Zhu Xi's scholarly eminence, when investigating things and extending knowledge (*gezhi*) became a popular phrase among literati and the Jesuits in their discussions of the proper approach to the diversity of knowledge. In the late nineteenth century, literati and their Protestant missionary collaborators chose Zhu Xi's terminology to translate modern science into Chinese as the study of investigating things and extending knowledge (*gezhi xue*). In opening remarks for the inaugural spring 1876 issue of the *Chinese Scientific Magazine* published in Shanghai, Xu Shou (1818–1882) as cofounder commemorated Zhu Xi's gloss for the investigation of things as an eloquent statement of the universal search for the knowledge of all "principles underlying things and affairs" (*shiwu zhi li*). The naming of science in this way, Xu Shou declared, would empower Chinese to master the practical fields of modern science.[106]

II

Natural Studies and the Jesuits

2

The Late Ming Calendar Crisis and Gregorian Reform

For Chinese emperors and their officials, the mandate of heaven since antiquity measured their cosmological, political, and historical worth. The emperor as the "Son of Heaven" was ritually responsible to establish a system of computational astronomy, which made an accurate calendar a realistic goal for his government. Relying on an accurate astronomical system for a mathematized cosmos, the emperor could disseminate an official calendar.[1] The first record of an articulated astronomy bureau (hereafter, "Astro-calendric Bureau") with clearly delineated duties occurred in the *History of the Later Han* (A.D. 25–220; *Hou Hanshu*).[2]

The *Documents Classic* (*Shangshu*), one of the original Five Classics, contained a famous injunction by the legendary Emperor Yao to regulate the calendar so that celestial events could be predicted. To order the empire, literally, "all under Heaven," Chinese and non-Chinese dynasties required technical mastery of heavenly and atmospheric events such as nonperiodic stellar novae, solar halos, and meteors and meteorites, in addition to periodic lunar and solar eclipses. Celestial events that imperial officials could not predict, as well as earthquakes, famines, etc., were portents that potentially pointed to the emperor's lack of virtue.[3]

In addition, an accurate calendar organized specific economic, political, and religious rituals to demonstrate authority over the agrarian cycle, affirm the cosmic order, and normalize relations with other rulers. The calendar's astrological and ideological significance made its imperial compilation and promulgation a critical ceremonial duty. The symbolic power of the calendar to order daily life via timekeeping affirmed the effectiveness of the dynasty in place, especially a conquest dynasty such as the Yuan under the Mongols and later the Qing under the Manchus. The calendar first and foremost mattered politically, and its impact on the agrarian cycle of seasons was secondary.

Mathematical astronomy (*lifa*), never an abstract study for its own sake, was the key to preparing calendars and horoscopes. Once a computational system was in place—that is, most of the time—the annual work of calendar making involved the lowest-ranking officials, computists with minimal skills. The bulk of the effort of bureau officials went into mathematical astronomy and the observation and interpretation of omens.[4]

Astronomical prediction, astrological interpretation, and the auspicious performance of daily tasks and ceremonies in court and at home presumed an accurate lunisolar calendar. Both the dynasty in power and the people at large were concerned with the dates of events, actions, and rituals. It was fortuitous for the Jesuits who began arriving in Ming China in the 1580s that their order had been involved in resolving one of the major calendrical problems bedeviling the Church in Europe, namely the controversy over the date for the movable feast of Easter.

Pope Gregory XIII (1502–1585) had appointed the German Jesuit Christoph Clavius (1538–1612), a leading mathematician and churchman, to the Gregorian reform commission. Clavius ascertained the proper date for Easter, originally derived from the Jewish Passover, with the promulgation of the Gregorian calendar in 1582. The Gregorian reforms occurred before the Copernican heliocentric controversy climaxed in 1616, but astronomers such as Clavius were already aware of Copernicus's challenge to the geocentric model of the solar system. The earth was a planet in Copernicus's system, and it orbited around the sun with the other planets, as well as rotating daily on its own axis. This brouhaha between the Church and astronomers later affected the Jesuit's role as transmitters of European astronomy to Ming and Qing China.[5]

In Ming China, the remarkable astronomical advances of the Yuan dynasty's Season Granting astronomy system (*Shoushi li*) of 1280, which was repromulgated by the Ming dynasty after 1368, was a major factor in the stability of the calendar until the 1580s. The very success of the Yuan calendar in predicting celestial events, conversely, produced an entrenched complacency in the Astro-calendric Bureau. Fortuitously, the Jesuits, whose main base of activity was in Japan, arrived in south China. Michele Ruggieri (1543–1607), for example, entered Macao in 1579 and initiated a Chinese mission in 1582. Versed in Renaissance mathematics, Ruggieri introduced European mathematics, mechanical clocks, and prisms to gain favor among local elites near Guangzhou. He was joined in the Jesuit residence by Matteo Ricci (1552–1610) in 1583.

Ricci had studied law in Rome and mathematical astronomy under Clavius at the Collegio Romano. Since 1581, one year before the Gregorian calendar was promulgated, he had been in Macao. Many of Ruggieri's and Ricci's initial conversions were "via the door of Mathematics." Moreover, when Ricci arrived in Beijing for the first time in 1601, he brought with him a number of books on mathematics, and he remained in contact with Clavius while in China. Later, when he realized the seriousness of the Ming calendar problem, Ricci hoped the court would employ Jesuits to correct the calendar. Equally important, he realized that the Astro-calendric Bureau offered a place for foreigners in the court bureaucracy, with access to the top.

Ricci's goal was to curry favor with the court and literati elites to gain their tolerance of Jesuit efforts to convert both elites and commoners. The success of some Jesuits such as Ricci in Beijing allowed Jesuits in the field to gain Christian converts openly. Less transparent, however, were their attempts to inscribe precise mathematical predictions of the periodicity of repeating celestial events, such as eclipses, with a vision of time that dated events linearly from creation.[6]

Consequently, when after three hundred years of use the discrepancies between the Yuan-Ming lunisolar calendar and the solar year became astronomically and politically significant, the Jesuits were well equipped to participate in calls for calendar reform during the Wanli reign (1573–1619) because of the legacy of training in mathematics with emphasis on astronomy, that Clavius had established in Rome. The curriculum at the Collegio Romano was copied in Jesuit colleges all over Europe, including the Portuguese College of Coimbra, where the Jesuits who were sent to China received their final training. They were worthy technical adversaries for the hereditary Muslim and Han Chinese officials entrenched in the Astro-calendric Bureau. The battle over the correct time was more than a technical rivalry, however.[7]

Development of the Ming Astro-calendric Bureau

Because the official production of the calendar provided the institutional venue for the development of Chinese astronomy, monographs on astronomy and astrology were regularly included in the dynastic histories. Afterward such information was also added to some medieval and late imperial encyclopedias and gazetteers. Technical knowledge of the heavens mattered for the government, which is why the calendrical advantages and astronomical models demonstrated by foreigners, such as the Indians, Muslims, and Jesuits who were hired by the court, were incorporated into the Tang, Yuan, and Ming-Qing calendars. Moral cultivation was rhetorically preeminent for literati, but understanding the calendar was an important aspect of the classical tradition and thus pertinent to them.[8]

Although rural life was unaffected by the inevitable calendrical errors that ensued from the anomalistic motions of the sun and moon, the official calendar in China was always linked to celestial phenomena. The goal of those who designed computational systems was to represent accurately the periodicity of heavenly bodies. The astrological interpretation of heavenly patterns (*tian-wen*) was based first on accurate observations and secondly on the contextualized sociopolitical interpretation of observable celestial events.

The luni-solar calendar in place combined solar years and lunar months. Since twelve lunar months amounted to only 354 or 355 days, it was essential periodically to add leap months, or intercalary months, to maintain an average of a little over 365 days in a tropical year. Officials were needed, in addition to their more demanding tasks, for calendar maintenance and occasionally for minor modifications. The designers of computational systems marked the divisions of the tropical year, projected the locations of the five visible planets among the stars, and predicted solar and lunar eclipses. Such expertise enabled others in the bureau to base ceremonial dates and interpretations of anomalies on accurate information.[9]

The bureau was a midlevel agency in the Ming government. Under the courts of Imperial Sacrifices and Imperial Entertainments in the Ministry of Rites, it was essentially a service organization. Dates for important events and ceremonies performed by the court or high officials required an accurate calendar. Otherwise, the political consequences for the court, and hence for the Bureau, could be volatile. Located between the court and the bureaucracy, the Astro-calendric Bureau was on par with the Imperial Academy of Medicine, the Office of Transmission for Official Documents, and the Directorate of Imperial Parks. It had less prestige than the Hanlin Academy, whose academicians often served dual roles as grand secretaries within the imperial household and as high officials within the Ministries of Rites and Personnel.[10]

The first Ming emperor simply appropriated the Yuan dynasty's Season Granting system—calling it the Grand Concordance astronomy system (*Datong li*)—and similarly took over its Astro-calendric Bureau intact to assert imperial legitimacy. Now called the "Directorate of Astronomy" (*Qintian jian*), the Astro-calendric Bureau continued in Nanjing with segregated Islamic and Han Chinese sections. When the third Ming emperor moved the metropolitan government to Beijing in 1414–1415, which he made his official capital in 1421, an auxiliary Astro-calendric Bureau remained in Nanjing until the end of the dynasty. Although complete with accurate versions of Yuan instruments, its importance declined.

A separate Islamic office was eliminated in 1398, but its Muslim personnel were incorporated into the main bureau to provide an accurate Islamic calendar for the many Muslims in China, many of whom had been invited by the Mongols to serve the Yuan dynasty.[11] The Ming bureau maintained the four divisions of the Jin-Yuan bureau: astrological interpretations, timekeeping,

Islamic methods, and mathematical astronomy (*lifa*). The Islamic division and its Muslim calendar in China echoed the Nine Governors System of the Tang, when Indian calendrical officials had served in the bureau.[12]

Expertise in the Ming Astronomical Bureau

The staff in the Ming Astro-calendric Bureau was responsible for general observations and specific predictions that were divided into five general duties: anomaly observation, eclipse prediction, calendar compilation, auspicious time and place selection, and timekeeping. The equipment and computational tables they employed were similar to the sophisticated observational tools that had been employed by the Yuan bureau under Kublai Khan after 1280 and by the Mongol Ilkhans at the Marâgha Observatory in Persia until 1305:

- An armillary sphere comprised of three layers of uniformly revolving rings representing the fundamental celestial circles or spheres. A sighting tube in the plane of this ring pivoted around the center to make observations and measure distances on the spheres.
- The "simplified instrument" served the same purpose but with fewer parts. It consisted of two rings, and the siting tube was fixed to the center to allow it to align with any celestial object. Its simplicity, based on equatorial coordinates until the Jesuits introduced ecliptic coordinates, made it a mainstay of Yuan-Ming astronomy.[13]
- Yuan dynasty gnomon (shadow-casting instrument) to determine the solstices when the sun's shadow produced the longest or shortest times of daylight.
- A water clock for timekeeping, namely, the three-chamber inflow type clepsydra.[14]

Although the Mongols used Islamic methods for planetary computations, Kublai also used other scholars skilled in Islamic star studies, such as the Persian Jamal al-Din, in his Astro-calendric Bureau. Jamal al-Din and others came to the Mongol court in 1267 with diagrams of an armillary sphere, a parallactic ruler, a sundial-gnomon, an astrolabe, a terrestrial globe, and a celestial sphere. Initially, the Yuan Astro-calendric Bureau was divided into three parts, with Jurchen, Islamic, and Chinese offices, although there were Han Chinese in each office. Kublai also adopted the traditional Chinese ban on the private ownership of astrological texts and private practice of astrology but as usual made no attempt to enforce it.[15]

The astronomical instruments that Jamal al-Din described to the Mongol court in Beijing strongly suggests that the astronomers at the Marâgha Observatory, which was founded about 1260 near Tabriz in Persia for the Mongol Ilkhan Hūlâgû (r. 1256–1265), Kublai's (r. 1260–1294) brother, were in touch with their Beijing counterparts. The plans for the astronomical instruments that Jamal al-Din described at the Yuan capital were the most advanced instruments of the time for both Islamic and Chinese mathematical astronomy, as were the Ilkhani computational tables completed in 1271.[16]

Yuan-Ilkhanid relations were also important factors in the physical and decorative similarities of ceramic types in both regions. These shared forms in the crafting of ceramics combined the geometric characteristics of the Near East with the floral designs of China. The Islamic experts who filled Kublai's Astro-calendric Bureau in 1280 in Beijing were paralleled by similar developments in 1259 in the Marâgha Observatory during the Mongol era of conquests in Asia.[17]

The Yuan establishment of a separate Muslim Astro-calendric bureau in 1271, which in principle used the Ptolemaic geocentric world system, segregated Mongol experts from their Chinese counterparts in the bureau, however. The claim that the Islamic Bureau held an Arabic or Persian version of Euclid's *Elements* and Ptolemy's *Almagest* remains speculative, however.[18] Nevertheless, the Islamic influence in the design of Guo Shoujing's (1231–1316) "Simplified Instrument" (*jianyi*), a simplified armillary sphere or astrolabe, and large masonry gnomon towers was clear. The reforms associated with the Yuan calendar introduced many innovations and represented a new astronomical synthesis in the long history of the Astro-calendric Bureau.[19]

In particular, the Season Granting astronomy system promulgated in 1281 applied new computational techniques of a prototrigonometric nature to convert the equatorial coordinates observed on the celestial spheres into numerical equations solvable using counting rods (*chousuan*). The compilers of the new calendar also recognized the computational differences in the procedures for positioning lunar and solar movements. This led to a gradual understanding of the lunar nodes (where the moon's path intersects the ecliptic) and thus of the conditions in which eclipses take place.[20]

By 1478, however, a memorial from the Astro-calendric Bureau acknowledged that its observational instruments were in disrepair. A 1612 memorial from the Ministry of Rites painted a darker picture of instrument dilapidation in the bureau, which has suggested to many scholars that before the Jesuits arrived the bureau was unable to use the astronomical instruments bequeathed from the Yuan dynasty. Although they were excellent by thirteenth-century standards, because the instruments were never very precise, reports on their deficiencies may have been inflated later by the Jesuits and their supporters. The mandated responsibilities of the bureau did not require precise observation until Jesuits challenged the competence of official astronomers.

Even then, Jesuit instruments quickly became obsolete as European standards rose.

For instance, the Ming bureau could not predict the location of an eclipse, only its time, duration, and magnitude. As the accuracy of calendrical computations improved, the standards for rationalizing errors in the calendar also heightened. Accuracy in predicting eclipses had gradually improved since antiquity, and what constituted error also changed. In the Han period, eclipses were predicted to the nearest day. During the Tang such predictions pointed to the nearest hour and gradually improved. By the Yuan, precision was possible to the nearest quarter of an hour. The Jesuits went further by predicting a solar eclipse to the nearest minute. Hence, by 1630 the Ming bureau was considered deficient because it was unable to achieve this new level of accuracy.[21]

Because such determinations occurred within the Observatory of the Imperial Palace, imperial complicity with bureau staff was unavoidable, particularly when the political stakes of an astrological interpretation were high, as when the Jesuits threatened to take over the Astro-calendrical Bureau. Such collusion usually took place, however, after an astronomical observation had been made by the staff, and they had turned it over to the personnel for astrological interpretations.[22]

Accordingly, the need for accurate observations inside the bureau usually outweighed the potential for the fabrication of celestial events based on politically biased motives. In addition, the Islamic calendric system was treated as a second opinion. Thatcher Deane has studied official Ming notices of the planet Mercury (known as the "Water Star") as a case example. He has found that the record of observations of Mercury in the Ming *Veritable Records* were in close accord with the planet's known position based on modern calculations. In one case, however, after the third Ming emperor came to the throne following a civil war, the mistaken record of a celestial anomaly in 1404, whereby Mercury intruded on a star, may have been fabricated to legitimize the new emperor.[23]

Even though routine observations were separate from astrologic interpretations, the manipulation of recorded anomalies by those charged with astrological interpretation was not uncommon. Because of discrepancies in the prediction of lunar and solar eclipses, astrological interpretation could impact the observation of celestial events. A formal "Relief Ceremony," for example, was enacted for solar or lunar eclipses when the predicted eclipse was visible or not. This curious ritual was more frequently performed for solar eclipses, which were the most serious anomaly. When an unpredicted solar eclipse occurred, an imperial apology on the part of the ruler and his government was obligatory. In a parallel way, when a predicted eclipse was not observed, then officials were expected to express their congratulations to the ruler for his governing achievements.[24]

Procedures for Computing Eclipses

In the Chinese lunisolar calendar, each month began on the day of a sun-moon conjunction, when no moon was visible. Total solar eclipses happened only on the day of a new moon. The coincidence of New Year's day and a solar eclipse, for instance, was improbable. A solar eclipse on any day other than the first of the month indicated an error of prediction, giving the event astrological significance. After three hundred years of use, however, the Yuan-Ming calendar in 1580 was off by one full day. This calendrical error mattered because a solar eclipse could now occur on the last or second day of the month. In addition, numerous ceremonies and events were tied to the first day of the new month. Because of the superiority of the Islamic Ptolemaic system in computing lunar-solar conjunctions, the court authorized its use in 1584.[25]

Half of the Ming bureau's solar eclipse predictions were spurious. This tendency toward overprediction was a way to protect the bureau from unpredicted anomalies and minimize the chances of missing a solar eclipse. Unverified but predicted eclipses were interpreted as Heaven's approval of imperial virtue. Failure to predict an actual solar eclipse, however, was a serious offense for the Bureau. A consistently correct prediction of local solar eclipses was not possible using the tables of periodicities in the Ming calendar until more accurate European geometric techniques arrived. Accordingly, the astronomical constraints within the bureau combined with ideological issues in the court to promote a tolerable level of overprediction.

An unpredicted lunar eclipse that occurred, however, was not treated as a culpable error. There were also fewer spurious overpredictions of lunar eclipses. Those that were predicted but spurious were regarded as failures, again the reverse of spurious solar eclipses. Because lunar eclipses were easier to predict than solar eclipses, higher standards were applied in such cases. A predicted but spurious lunar eclipse, however, stood out as a major error because of the higher expectation of accuracy. In other words, the Ming government treated solar and lunar eclipses differently because of the varying difficulties in computation. The court cushioned itself with ritual responses to unexpected solar eclipses because they were truly uncharacteristic, while lunar eclipses were not.[26]

Timekeeping, the Calendar, and the Selection of Auspicious Dates

To overcome the limits of sundials to mark time during darkness, the Chinese created water clocks. In 1090, a Northern Song government official, Su Song

(1020–1101), constructed a forty-foot-high hydro-mechanical clock that he called the "Cosmic Engine." Its interior mechanism of hooks, pins, interlocking rods, coupling devices, and locks served as a mechanical escapement that alternately engaged and disengaged to control the movement of a revolving wheel. Su's demonstrational water clock included a star globe and an armillary sphere. Each was connected by chains and turned by gears. What remained of Su Song's clock mechanism after the Song dynasty was conquered, however, finally vanished in the fourteenth century, when Ming forces drove the Mongols from Beijing, and was forgotten. Hence, when Jesuit missionaries brought European clocks to China in the late Ming, such mechanical devices were welcomed by the Chinese as an ingenious invention and an improvement over their simple water clocks.[27]

Nevertheless, regular operation of Ming water clocks continued in large cities with drum and watchtowers. There the daily changing of placards for the twelve double hours of every day were routine tasks, which urban dwellers could see in addition to hearing the drum rolls. Also routine was the compilation of the annual calendar for empirewide distribution. The dating was based on the governmental assumption that calendricizing the world of time would bring order and predictability to the sequence of celestial phenomena, which would in turn order social, political, and cultural events within the empire.

The bureau was responsible for selecting auspicious days and hours for various court and public activities such as imperial funerals. The Ming government distributed approximately 2.7 million printed calendars, or one for every seven households, for officials and commoners to follow these determinations. Although these calendars were not doled out casually, they were available beyond most administrative units and profusely reprinted by private publishers.

Neighborhood specialists trained in local yin-yang schools complemented the Astro-calendric Bureau. The government established prefectural and county schools of yin-yang divination to train functionaries to divine auspicious days and related matters of geomancy based on topography and directional orientations. Such efforts were symbolic and ineffective. A pool of hereditary families provided such personnel, but often, as in 1458, the Ming bureau received authorization from the Ministry of Rites to initiate an empirewide call for officials who were expert in astrological interpretation, mathematical astronomy, geomancy, and fortune-telling.

The politically sensitive selection of auspicious times and days by the bureau was tied to geomantic considerations in naming and siting venues for imperial ceremonies, selecting appropriate tombs for the dead, and bestowing honorific titles on the deceased. Geomantic service was a bureau duty in siting tombs, for instance. We will see in the next chapter that this became a serious bone of contention in 1658 when the Jesuits in charge of the bureau allegedly chose an inauspicious date for the funeral of an emperor's favorite son.[28]

Career Patterns in the Ming Astronomical Bureau

The first Ming emperor's 1373 regulations for the Astro-calendric bureau specifically excluded the descendants of bureau personnel from other positions in the government. Although this interdiction was lifted in 1458, the long-term effects of this policy lingered on. A career there was therefore likely a dead end. Given the politically sensitive nature of the bureau's work, moreover, it was in the interests of the imperial court and the top levels of the bureaucracy to confine astronomical and astrological expertise to the bureau.

To make such positions more attractive, the bureau granted its personnel special privileges, in addition to allowing the descendants of calendrical officials to serve in other bureaucratic appointments. Like regular officials they were exempted from corvée duties, and likewise they faced a reduced scale of punishments when involved in civil litigation. These privileges, however, were granted to all civil examination licentiates—some 500,000 by 1600—and were not comparable to the benefits and prestige accorded the highest degree holders who passed the triennial metropolitan and palace examinations in the capital.

With less career mobility, fewer degree-holders among their midst, and less governmental prestige when compared even to physicians in the Imperial Academy of Medicine, the staff of the bureau, like doctors, were at best middle-level personnel in a system that guaranteed the dominance of those proficient in classical learning. Directors of the Five Offices in the bureau typically served ten to fifteen years. They usually spent ten years first as students or yin-yang specialists.[29]

Outsiders were sometimes brought in to supervise the bureau, particularly in periods of badly needed reform. In such times, the Astro-calendric bureau became a venue for struggles over the mathematical astronomy that that spilled over into the court. Below we will see that the Astro-calendric bureau was penetrated by Jesuits beginning in the late Ming. Like Indians under the Tang and Muslims under the Yuan, the Catholic Jesuits correctly perceived mathematical astronomy as a convenient means to influence the imperial court and the emperor. The Jesuits, unlike their Indian and Muslim precursors, eventually gained control of the bureau during the Qing and were not easily relegated to an auxiliary role.

Evolution of the Late Ming Calendar Crisis

As early as 1525 officials recognized that the Ming calendar and the Astro-calendric Bureau were failing, but the prescribed remedy, as in the past, was a call for new personnel and for better management of the bureau's offices. The reform of the calendrical system itself was not yet considered necessary. Moreover, the bureau, because of the low prestige of its middle level personnel, was not in a strategic position to impact calendrical debate in the court. Additionally, Ming mathematics, in general, was linked more to commerce, such as wide use of the abacus, than to further development of its rich crafts of algebra based on calculation techniques such as counting rods, which had peaked during the Song-Yuan period and which the abacus did not support.[30]

Calendar reform, that is, the replacement of the computational system, remained an imperial prerogative only insofar as it received criticisms and recommendations from outside the bureau. Until officials associated with the Jesuits appealed to the emperor and his court for a substantial reform of the Grand Concordance astronomy system in the 1630s, the government was content to rely on special supervising officials or to recruit new talent for the bureau. This policy, however, opened the door for the Ming and Qing dynasties to accept Jesuits as calendrical experts, just as Tang and Yuan rulers had accepted Indians and Muslim specialists.[31]

The usual pattern was for an official to send a proposal for calendar reform in a memorial to the emperor, which the Ministry of Rites would adjudicate as the chief supervising agency of the Astro-calendric bureau. In theory, if the proposal had merit, then an official trial to compare the new proposal with the old methods would be scheduled. Often, however, the results of a debate over a new proposal were never clear-cut enough to justify a trial run. In other words, bureaucratic inertia prevailed because the Yuan-Ming calendar had achieved unprecedented success, unlike in earlier dynasties where numerous reforms had been required.

In 1481 the Ming court received a peculiar recommendation by Yu Zhengji, an instructor in a dynastic school, to scrap all the calendrical systems since the Han dynasty and return to the ancient nineteen-year Rule Cycle. Han dynasty astronomers had thought that at the beginning of the world cycle there had been a general conjunction of planets and that at the end of that cycle this would recur. From this fixed point in a cyclical cosmos, future reckoning of seasonal solstices or lunar-solar conjunctions and eclipses could begin. Han astronomers also thought that every 138,240 years all the planets would repeat their motions. This cycle was combined with another cycle to yield a Great Polarity Superior Epoch of 23,639,040 years.[32]

Nathan Sivin has described the Rule Cycle of 6939¾ days, adopted at the end of 105 B.C., which had been used for intercalation before the Grand Inception reform. This cycle was based on 235 lunations of 29.53085 days, which roughly equaled 19 solar years of 365¼ days. Because this Rule applied ancient values for the year and lunations, its lack of rigor perpetuated obsolete solar year and lunation values, and thus necessitated continual calendar reform.

Since nineteen lunar years of twelve months comprised only 228 lunations, the Rule Cycle required adding seven intercalary months every nineteen years to equal the 235 months in nineteen solar years. Nineteen years was the smallest interval in which the winter solstice and new moon (or any other combination of solar and lunar events) would recur on the same day (although not at the same hour). A cycle of intercalations that repeated after as few years as possible would minimize cumulative errors. After nineteen years, the errors of adding an intercalary month seven times at an interval of approximately once every 2.71 years would cancel out because the number of months and years that had passed were both integral.[33]

Yu Zhengji's claim that the placement of the intercalary months in the Ming calendar was incorrect was taken seriously, even though his views would have ended eclipse predictions. The proposal was debated but not accepted. This bit of arcane literati fundamentalism concerning the calendar may have been a caricature of Ming efforts to return to antiquity, but the nineteen-year cycle, which China had abandoned, was still important in early modern Europe as a calendrical tradition before the 1582 Gregorian reform.[34]

The Metonic Cycle

The Metonic cycle of nineteen years had been used in ancient Greece and Babylon to correlate the periods of the sun and moon for calendrical purposes. Meton had discovered circa 432 B.C. that nineteen solar years contained 235 lunations. On that basis he also computed the length of solar year as 365¼ days. The Metonic cycle accepted 6,940 days or 235 synodic months between observed new moons over nineteen solar years. Because nineteen years of 12 lunar months equaled a total of only 228 months, the Metonic cycle also required an intercalation of 7 months in the cycle, the same as the Rule cycle in the Han calendar that Yu Zhengji wished the Ming to reauthorize.[35]

The problems that both ancient Chinese and European calendrical scholars faced were the incommensurable periodicities of the sun and moon. One mean tropical year equaled 365.2422 mean solar days, which matched the average period of changing seasons in the solar year. On the other hand, one mean synodic month equaled 29.53059 mean solar days.[36] Neither the mean tropical year nor one synodic month was an integral number of days, however, and thus neither could provide exactitude in the calendar. The incommensurability of the two figures meant that the tropical year did not contain an integral number of months but, rather, an average of 365.2422 days that when divided by 29.53059 days in a month equaled 12.368266 lunar months in a solar year. This discrepancy was the raison d'être for intercalary months in both China and the ancient world.[37]

Discrepancies in Easter computations continued in Eastern and Western Christendom until Victorius discovered the first true Easter cycle, known as the 532-year Paschal cycle of twenty-eight Metonic cycles (28 × 19). In A.D. 457, Victorius used Alexandrian computations based on the nineteen-year cycle to improve on past calculations and accepted March 21 as the date for the vernal equinox. In the process, he discovered that Easter dates began to repeat themselves every 532 years. Armed with this Paschal cycle, Victorius successfully dated all 430 Easters (based on Jesus' death and resurrection) up to 457 and then for 102 years into the future, although opposition kept the solution from being immediately implemented in the Latin world.[38]

The Paschal cycle remained important in early modern Europe before and after the 1582 Gregorian reform. The French Jesuits Joseph de Prémare (1666–1736) and Joachim Bouvet (1656–1730), who arrived in China in the late seventeenth century, believed for example that the Paschal eclipse, which occurred upon the death of Christ, had been a supernatural event that was allegedly recorded in Chinese records. This claim was quickly challenged by other Jesuits, such as Antoine Gaubil (1689–1759), who accepted the independent integrity of Chinese records. Similarly, the systematic identification of Biblical and Chinese personages was common among the Jesuits and their converts. Some maintained that Chinese observational records came from Jewish patriarchs before the great deluge.[39]

Such calendrical claims reveal the European chronographic mind-set then in place when the Jesuits competed with Han Chinese and Muslims for control of the Astro-calendric Bureau in the early seventeenth century. Moreover, the nineteen-year Metonic cycle slipped by one day in roughly 308 years because its 235 calendar months were slightly longer than nineteen solar years. The pre-Gregorian calendar in Europe had not surmounted this limitation and was now ten days ahead, while the Chinese had stopped depending on the Rule Cycle shortly after the fall of the Han in A.D. 220.

The Ming calendar employed a more sophisticated approach since the Han and was significantly more accurate. The astronomical systems established by the Han, Tang, and Song dynasties all followed the precedent of the Triple Concordance system of 104 B.C. The Triple Concordance system adopted a day divisor of 81, which amounted to setting the length of a lunar month as 29.53086 days, and counted accumulated days from an epoch 143,727 years in the past. It was inevitable that the small discrepancy in the constant of

29.53086 minus the modern value of 29.53059 days, which equaled an error of only one day per 310 years, would accumulate over such a great span of time so as to become appreciable.

The Yuan dynasty calendar treatise completed in 1280 by Guo Shoujing (1231–1316) improved upon the Triple Concordance system and its successors to give more precise measurements of the length of a solar year. The Mongol system adopted precise decimal constants and counted from a recent solstice rather than an ancient epoch.

Ming Recognition of a Calendar Crisis

Criticisms of the Ming calendar occurred within a climate of rising expectations of accuracy. In 1483, discrepancies in the calendar were attributed to the difference in location between the bureau's observations in Nanjing, where the instrumentation to measure celestial phenomena had originally been designed, and Beijing. In a pre-Jesuit environment, such discrepancies were usually within the margins of permissible error. Hence, errors in computation were rationalized because the bureau's predictions were still sufficiently accurate.

No consensus was yet possible about the urgency of the problem. For instance, in 1498 a secretary in the auxiliary Astro-calendric bureau in Nanjing called for selection of a high-level minister to study why failures in lunar eclipse prediction had increased. The appointment of a plenipotentiary to recruit talented scholars to produce a new calendric system would have been a major undertaking, however. The emperor followed instead the advice of the Ministry of Rites that the request was impractical.

After a series of inaccurate eclipse predictions in 1517–1518, Zhu Yu, a timekeeping official in the Beijing bureau, contended that the secular difference in the annual difference introduced by the Yuan calendar to deal with the drift forward of the calendar was the root of the problem with eclipse predictions. Because of this discrepancy, the bureau had been over-predicting solar eclipses less often but under-predicting them more often, an ominous sign for the dynasty. The debate over this proposal was argued in terms of the distinction between new methods (*xinfa*) and old methods (*gufa*).[40]

When Zhu Yu's own prediction of a lunar eclipse turned out no better, however, Zhou Lian, another official in the bureau, convinced the court that continued readjustments in computations would suffice. Despite the problems in the Grand Concordance astronomy system, no consensus emerged that new and more accurate observations were needed to compose a new system. The focus was still on trial and error adjustments of the old and new regimes for calculation. The introduction of a Western calendar by the Jesuits in the 1620s was perceived in light of these Ming categories of old and new methods.[41]

When a new emperor took the throne in 1521, Zhu Yu, still a timekeeping official, called for new instrumentation because accurate measurements were no longer possible using the observational instruments in Beijing. Zhu called on the emperor to construct a new bronze gnomon, which was needed in Beijing to adjust for its latitude. Later in 1523 an outside overseer of the bureau, Hua Xiang, proposed that officials be dispatched to Henan province to study the giant masonry gnomon erected there for the Yuan calendar. This proposal prioritized improving instrumentation for constructing a new computational system. Hua was stalemated at once, however, by another supervisory official, who contended that the Grand Concordance astronomy system should not be changed. Both the court and Ministry of Rites waffled.[42]

The prestigious legacy of the Yuan calendar promulgated in 1281 was part of the problem. Its very success for three centuries, coming after some twenty deficient calendars launched during the Song dynasties, was a policy obstacle that reformers would have to hurdle. Ming adoption of a reformed Yuan calendar created an institutional constituency in the bureau bound to oppose any major reform on the grounds that the system in place was reliable. In effect, the institutional status of the Ming calendar limited debates over its future reform.

Although no major changes in the calendar occurred, the issue was serious enough to be taken up in civil service policy questions of the time. For example, civil examiners raised the reform of the calendar for approximately three thousand candidates taking the 1525 provincial examination in Jiangxi. In one of the policy questions, the examiners asked the candidates to think over the history of the official Ming calendar and discuss its accuracy. In this way, the examiners ingeniously solicited literati opinion about the official calendar.[43]

To answer these specific questions, the candidates needed to comprehend both mathematical astronomy and its history, which was available to candidates in the Jin and Yuan Dynastic Histories and statecraft-oriented encyclopedias. The Yuan calendar developed between 1274 and 1280 gave more precise measurements of the length of a tropical year as 365.2425 days (compared to the modern value of 365.2422 mean solar days). Without the earlier sources of error, a major revision of the Ming system was unnecessary until the sixteenth century.[44]

The 1525 policy question dealt with proposals to alter the Yuan-Ming system because, like its predecessors, it was bound eventually to show accreted errors. The examiners raised the perennial issue of calendrical calculations in the Astro-calendric Bureau, namely the tension between prediction based on continual observation and extrapolation to adjust computational procedures on one hand, and, on the other, forecasts derived from rigorous determinate mathematical techniques that did not require continual infusions of new data. The examiners had opened the door to calendar reform through an official avenue for submitting opinions to the throne. It was also a time when an embattled emperor and his entourage of powerful eunuchs in the imperial household were increasingly faced with aroused literati factions inside and outside the court.[45]

The published answer to the question by one of the highest-ranked examinees focused on the chief theoretical question: "Creating an astronomical system is a matter of conforming celestial phenomena [lit., "the sky"] to find what [techniques] accord with them, and not forcing accord, so that predictions can be validated [by phenomena]." This led him naturally to his answer. "The orbits in the sky are not uniform, but astronomy [between Han and Song] was restricted to set methods, because they were not aware of 'conforming to the heavens to find what accords with it.'"[46]

Although the Yuan system was the first to master it, a succession of astronomers had worked out simple empirical corrections that minimized precessional error. The candidate listed them and accurately summarized the techniques used by each. It took considerable understanding to see that these earlier adjustments had anticipated the annual difference. The candidate had obviously digested the account in the *Yuan History* of the Season Granting system's predecessors.[47]

Writing fifty years before the arrival of the Jesuits, the candidate recommended reviving the instruments used in the Yuan reforms of 1280. His appeal to Yuan instrumentation as a corrective measure suggests that the 1525 policy question was meant to draw responses about what to do with the Ming calendar. Accordingly, the candidate argued that the dynasty should rely on instruments to reform the calendar. Rather than technical manuals, which were unavailable, the candidate cited dynastic histories as his sources of information.[48]

In addition to this official airing of the 1521–1523 debate about the problems with the Ming calendar, private scholars—many also involved in the partisan politics that became routine in late Ming governance—also voiced their opinions about the discrepancies between the calendar and celestial events. The scholar-official Tang Shunzhi, for instance, had come to prominence when he finished first on the 1529 metropolitan examination. After a turbulent political career in which he was banished twice, Tang explored his literary and statecraft interests during his forced absences from politics. Generally regarded as one of the leading scholars of mathematics in his time, his studies also encompassed Islamic traditions in astronomy.[49]

Seeking a resolution between a fixed versus a flexible system to resolve calendrical problems, Tang assessed the Yuan-Ming calendrical system in light of what he called "concrete studies" (*shixue*). Such critical assessments of the calendar, according to Tang, had lost their vitality during the Ming. Hoping to broaden literati education to include technical learning, Tang Shunzhi also sought to reintegrate mathematics and mathematical astronomy. According to Tang, calculations based on the relational features for the sides of a right triangle (*gougu*), had been separated during the Ming from the underlying principles of mathematical astronomy. Tang wrote:

> One must know both mathematical and astronomical principles. This is where I differ from literati students [today]. One must know both fixed calculations and variable calculations. This is where I differ from officials in charge of the calendar. Principles and computations are not separate things. Computations represent the concrete applications and extensions of principles. Variable and fixed calculations are not separate things [either]. Fixed computations represent the basis for variable calculations.[50]

Tang Shunzhi stressed the "six arts" (*liuyi*) of antiquity as the model for reintegrating calendrical theory with concrete computations. He then blamed the Ming lack of integration on its literati, who had denigrated the study of numbers as a "lesser technique" not worthy of their attention, thus leaving the calendar for the mediocre staff of the Astro-calendric Bureau to compute. Tang's call for a return to the more comprehensive vision of antiquity, in which the mathematical arts held a prominent place, represented a precocious effort to recapture classical antiquity in its full technical complexity.

Usually we attribute this positive perspective of mathematics to Qing dynasty literati and their response to the introduction of Western learning in the early eighteenth century, when they called for rediscovering the Chinese origins of Western learning (*Xixue Zhongyuan*). Efforts to recover antiquity during the late Ming, however, had already carried over from astronomy to mathematics before the Jesuits arrived in south China. Tang Shunzhi used traditional crafts of early trigonometry to solve simultaneous algebraic equations in mathematical astronomy.[51]

The straw that broke the camel's back came in 1592 when the Ministry of Rites charged that the Astro-calendric Bureau had erred by a full day when predicting a lunar eclipse, which meant that the Relief Ceremony had to be rescheduled. An error of a full day also affected ceremonies tied to the first day of the month, season, and new year. In 1595, a Ming prince with exceptional mathematical talent, Zhu Zaiyu (1536–1611), proposed a new calendar, which he called the "Ten Thousand Year System for Sagely Longevity."

Zhu tried to combine the Yuan and Ming calendar systems, but because his synthesis was not based on new observations, it failed to make the prediction of eclipses any more precise at a time when accuracy was expected within one mark (= 14 minutes).[52]

The Ministry of Rites, however, never tested Zhu's proposal. Zhu was later listed in the official *Ming History* as a meritorious proponent of reform who was unjustly ignored. His biographers were Qing scholars who had the advantage of hindsight. In 1596, Wang Honghui (1542–1601?) broached with Matteo Ricci the possibility that Ricci might contribute to the reform discussions. In 1597 the scholar-official Xing Yunlu (1573–1620), who had written a work comparing ancient and recent calendars, claimed that the procedures employed in the Ming calendar to determine the winter solstice were in error by one day, thus extending the error to encompass a full season. The director of the bureau dismissed its critics, however, and defended the calendar as a stable institution that should not be recklessly changed.[53]

Gregorian Reform

The European calendar aimed at simplicity and convenience for dating civil and religious events. This was essential if hundreds of sovereign states were to accept it. To this end the Gregorian calendar reform followed the Julian, which had come roughly 1,500 years earlier, in reducing the amount of observation and computation and provided an artificial system independent of astronomical tables or observations of celestial phenomena. But, as in the time of Julius Caesar, localities continued to use incompatible lunisolar calendars, some long after the Gregorian reform. Because of its focus on physical anomalies, the Ming calendar was always linked to observable celestial phenomena such as solar eclipses. The Astro-calendric Bureau recorded celestial phenomena so that the government could deal with the ominous ones and take credit for the propitious ones.[54]

Pre-Gregorian Efforts at Calendrical Reform

For both Catholics and Protestants after the Reformation, the death and resurrection of Jesus were the most important events in a linear history. Even so, there was no agreement when the year began. Some reckoned it from Christmas Day. In the civil calendar it was January 1. Another system had it in March, the time of the vernal equinox, as in pre-Julian and some old Germanic calendars. England used March 25 for the feast of the Annunciation and the old Roman vernal equinox as New Year's Day, until it accepted the Gregorian calendar in 1752.[55]

Victorius' 532-year Paschal cycle was successful because it was a self-repeating Easter cycle since God's creation of the world, but its computational procedures required technical improvements. When compared to the Julian calendar of 365¼ days (19 inaccurate Julian years = 6,939 days, 18 hours), the Metonic ratio of 19-year cycles of 235 calendar months (= 6,939 days, 16 hours, 31 minutes, 14 seconds) was too short by 1 hour and 29 minutes. Consequently, the lunar cycle allowed calendrical slips back of one full day in roughly 308 years, which necessitated a lunar correction.[56]

The length of the tropical year was computed as 365.2423154 days at the beginning of the Christian era, compared to the Chinese value of 365.2425 and the modern value of 365.2422. The date assigned to the vernal equinox was problematical. By the year 1000, the Julian calendar (one year = 365.25 days) had drifted ahead of the tropical year, which meant that the calendar was drifting forward one extra day in a little over 130.1 years. This discrepancy required a solar correction. In 1500, the drift was one day in 128.6 years.[57]

Between the time of Julius Caesar and Pope Gregory, the precession of the equinoxes had shifted by roughly fifteen days—a figure that made the Ming Chinese problems negligible by comparison. Paul of Middleburg had asked Copernicus to advise the Fifth Lateran Council in 1512 on proposed calendar reforms. Although Clavius had reservations about Copernicus's heliocentric cosmology, he recognized him as an authority on astronomical measurements and used his data in the Gregorian reforms. The Gregorian commission heeded the Prutenic tables of Erasmus Reinhold (1511–1553), which were a revised and expanded version of the tables from Copernicus's On the Revolution of the Heavenly Spheres, published in 1543. Copernicus's astronomical reforms were intended to provide an accurate prediction of the correct dates for Easter and Christmas, not just a delineation of space but also a demarcation of time.[58]

The Council of Trent met in 1562–1563 and authorized the pope to deal with calendar reform at a time when the calendar date for the vernal equinox had receded to March 11. Petrus Pitatus (Verona, 1564, written 1539) advocated that at least fourteen days should be dropped from the solar calendar to bring it back to that of Julius Caesar when the vernal equinox was March 25. Curiously, Martin Luther preferred Easter as a fixed day rather than a movable feast and thought that calendars were irrelevant to faith and only reflected worldly authority. The Gregorian reform of 1582 managed to overcome these divergent views because of its elegant simplicity, which contrasted with the contemporary paralysis in the Ming court concerning elaborate calendar proposals.[59]

The Gregorian Calendar and Its Reception

After serving at the Council of Trent, Gregory XIII quickly nominated a commission when he became pope in 1572, which met for ten years until 1582 and reviewed various proposals for calendar reform. Alosius Lilius, a lecturer in medicine at the University of Perugia, presented an ingenious plan to restore the date of March 21st for the vernal equinox through a ten-day displacement. Once Lilius's proposal was accepted by the commission in 1577, Gregory authorized preparation of a *Compendium* based on Lilius' original, which was circulated for review. When these outside views were received and evaluated, Clavius attended to a residual error of one more day in 308 years. Gregory then introduced the new calendar via an apostolic letter in February 1582. It was written in the form of a papal bull and represented ecclesiastical law.[60]

The final rules for intercalation in the Gregorian calendar were as follows:

1. A common year contained 365 days, a leap year 366, the extra day being added to the end of February.
2. Every year of the Christian Era after 1582 that is evenly divisible by 4 is a leap year,
3. Except for centennial years, which are leap years only if evenly divisible by 400 (to repair the solar cycle).[61]

In this manner, 1600 and 2000 were designated as leap years, but 1700, 1800, and 1900 were not. The Gregorian result was ninety-seven additional days in four hundred years, or three days less than the Julian calendar, which intercalated every four years without exception, i.e., one hundred days in four hundred years. One day in the lunar cycle was omitted every 312.5 years.

Calendar reform was easily accomplished in Catholic Italy, Spain, and Portugal, but it became a bone of contention in the Protestant world because the papacy presented the reform as an act of the Catholic renewal to challenge the Reformation. The Eastern Patriarch in Orthodox lands was cautious, but Rome's unilateral declaration made it impossible for him to accept. In the sixteenth century, church and state in Europe were so interwoven that civil acceptance of the new calendar was determined by the religious affiliation of respective rulers. Criticism of the Gregorian calendar was colored by different religious and intellectual positions.[62]

In this light, Protestant Germany saw the pope as an Antichrist for instituting the calendar. Others contended that Gregory had stolen ten days from a farmer's life. Protestant sovereigns did not relent until Protestant astronomers increasingly supported it. In 1724 and 1744, the celebrations for

Easter in Protestant countries were one week earlier than for Catholic celebrations. In 1775, Germany finally adopted the Gregorian calendar by the decree of the German Emperor and King of Prussia, Frederick II.[63]

In England, secular authorities were sympathetic with the Gregorian reform, but the Anglican clergy was not. The latter insisted on consultations regarding the pope's unilateral promulgation of the new calendar, and this procedural obstacle doomed the reform in England. Scotland accepted the new calendar from 1600. When England finally passed an Act of Parliament in 1752 to enact the calendar by omitting eleven days, the Anglican clergy's hatred of Rome was no longer an obstacle.[64]

The mathematical scheme used in the Gregorian reform was prudent because it was independent of physical hypotheses and emphasized instead computational mechanisms to produce the calendar. The new calendar served mainly as a puzzle solver, that is, a set of computational practices that were increasingly rule-like regardless of the nature of the cosmos. In this sense, the Gregorian reforms marked a climax of the Medieval Computus, which survived the Copernican revolution as a calendrical tool kit.[65]

The Impact of Clavius on Jesuits in China

When the Gregorian calendar reform began in 1572, Clavius was still an isolated mathematical scholar in his Roman college. By its end he was a leading mathematical astronomer. Clavius prepared a series of mathematical textbooks at the Collegio Romano, beginning with his edition of Euclid's *Elements of Geometry*, which included a voluminous collection of notes and commentaries. Matteo Ricci would later use this edition for the classical Chinese translation of Euclid that he helped to prepare in Ming China. Knowledge of geometry, especially the first four or six books of Euclid, became particularly important in sixteenth-century Europe for engineers, cartographers, architects, land surveyors, and painters, while in the late fifteenth century expertise in geometry was increasingly valued for fortifications and ballistics.[66]

Clavius also planned to prepare a work on theoretical astronomy that would revise the planetary system and provide an alternative to Copernicus's heliocentrism. He never completed that work. In his own work, Clavius still accommodated observational tables from Copernicus's work within a geocentric cosmology. Clavius's personal opposition to astrology was strengthened by the Collegio Romano's placement of mathematical astronomy separately in the faculty of arts and medicine, where astrology was also important. The tension between astronomy and astrology among Jesuits would later affect their role in the Qing Astro-calendric Bureau, where astrological interpretation based on portents remained important.[67]

We return below to the calendrical crisis in late Ming China, where the puzzle-solving mechanisms for maintaining an accurate calendar had unraveled. The Chinese calendar had since the Han dynasties been reformed through computational means, as in Europe, and, like European calendrical specialists, the staff of the Astro-calendric Bureau did not require allegiance to any particular view of the cosmos to employ their tool kit of computational puzzle solvers.[68]

Ricci and his immediate successors in China drew on the Ptolemaic geocentric framework, which had been questioned but reaffirmed by Clavius and the Gregorian reform commission. Later, after Ricci's death in 1610 and Clavius's death in 1612, Johann Adam Schall von Bell (1592–1666) and Ferdinand Verbiest (1623–1688) were increasingly influenced by Tycho Brahe's new findings, which helped them gainsay in Europe, and elide in China, the heliocentric cosmology of the Copernicans in favor of geoheliocentricity.[69] Jesuits in China quickly adapted their religious goals to the late Ming local context and accommodated their order, with some dissent, to the inescapable priorities of the Chinese imperial system, its cultural elites, and the astrological applications of the calendar to portents and hemerology (the art of determining auspicious days).[70]

Jesuits and Late Ming Calendar Reform

The Jesuits entered China during a time of heated debate and discussion about the calendar in the late Ming. Literati inside and outside the government had noted the drift forward of the Chinese date for the lunar new year by one day since the Yuan dynasty, but the Astro-calendric Bureau had not yet provided a solution. Despite some superficial similarities, the situation in which the Jesuits would find themselves in seventeenth-century China was not really comparable to Clavius's role in Rome in the 1570s. The Gregorian calendar was formulated by a Papal commission and not a sovereign state. Although there were parallels between the Papal Curia and the Ming imperial court, the former was in scope and magnitude a minor political authority, while the latter was not a religious organization. Nor was the Astro-calendric Bureau the haven for religious debates.

Moreover, the Christian concern was to date Easter accurately in light of God's creation and Jesus' resurrection as linear events. For the Ming, however, the concern was not with linear time, which since antiquity had not been stressed. Rather, the Chinese calendricists were concerned with recurring events within a cosmology that had no beginning or end. To domesticate time meant to predict the periodicity of repeating celestial events and not to date them accurately from creation.

The key issue for the Astro-calendric Bureau was to predict anomalous events, such as solar and lunar eclipses, whose periodicity affirmed an orderly political reign by a legitimate ruler. The auspiciousness or inauspiciousness of the present depended on its interpretive place in a cyclical calendar. Different versions of time were at stake in the interaction between the Jesuits in China, who had experienced the Gregorian reforms in Europe, and Chinese literati and dynastic rulers, who believed in cosmically recurring events in time.

The Calendar and Time

Jesuits such as Clavius were insiders in the Gregorian reforms, whereas Jesuit missionaries were always outsiders in Ming China, opposed by native Han Chinese in the Ming bureaucracy and additionally contested by the Muslim calendrical staff in the Astro-calendric Bureau. The social positions of the Jesuits, who first dressed as Buddhists, had to be clarified and heightened before their calendrical proposals would be taken seriously by the Chinese literati, who served the imperial court as government officials and elite partners in local governance.

The Jesuits had to deliver their message to intelligent and cultured elites who could appreciate their spiritual sincerity and technical prowess or would gainsay both as dangerous. Likewise, the Jesuits would have to master via language, dress, and experience the different discourses of political and cultural supremacy that informed the Ming calendar and made even a one-day error in the onset of the seasons, the first day of the month, and New Year's Day a potential calamity. Getting the calendar right in China meant applying a cyclical blueprint to grasp recurring events.[71]

Nevertheless, the Ming bureau and the Christian Computus still shared, functionally and historically, a nexus of social, political, cultural, and religious elements that informed the calendar. Although such elements were divided up differently and integrated uniquely in Europe and China, Clavius's followers and the later students of Brahe and Kepler could understand why an accurate and up-to-date calendar mattered to the Ming and Qing state. More importantly, Ricci, Schall, and Verbiest understood the important role the new Christian calendar played in unifying the faith in early modern Europe.[72]

The calendar crisis was not merely an opportunity for the Jesuits in China to influence literati and an emperor enshrouded by a sophisticated bureaucracy. The crisis also allowed them access to the technical fundaments of the Chinese chronological construction of cyclical time, its conceptual limits, historical interpretation, and technical determination, which were woven into the ideological fabric of the imperial dynasty. Lurking in Jesuit efforts to promulgate a new calendar for China, constructed "in accordance with new Western methods," for example, was an effort to Christianize creationless imperial time.

For instance, Jesuits claimed, based on the biblical chronicles after the Great Deluge, that the Chinese were descendants of Noah. Chinese critics saw this as an assault on the integrity of their classical antiquity. Chinese and Manchu political power stood in the way of and transmuted the Christian agenda, triumphant in the New World and for a short time in parts of Japan, into more muted forms of technical improvement in the Chinese calendar based on Chinese time. The Jesuits had to preserve the Chinese time-measurement system and the basic structure of the cyclical imperial calendar. From the Ming point of view, what needed improvement were the technical aspects of the calendar, such as the tables used in calculation. Which world system was used for cosmological prolegomena mattered neither to the non-Christian Chinese nor the anti-Copernican Jesuits, many of whom were open to change in cosmology but all of whom were prevented from teaching it.[73]

The Jesuit Novitiate and Mathematics

The Jesuit era in China coincided with the Thirty Years War that overwhelmed Catholics and Protestants in bloody conflicts from northern to central Europe. European wars took precedence over concern with non-European matters. Similarly, England was beset with social and religious battles and civil wars that led to the execution of the king in 1649. Until the Glorious Revolution four decades later, England was not a factor in Asian affairs.

The Portuguese support of the Jesuits in China, Japan, and India, and the Spanish sponsorship of the Dominican, Franciscan, and Augustinian orders in the New World and the Philippines meant that Catholics still monopolized interactions between China and western Europe from the sixteenth to the eighteenth century. Because of growing influence of Jewish converts after the Inquisition, however, the General of the Order in 1593 commanded that the applications of all new Christians should be rejected. Luis de Almeida, for example, was the first Jesuit to introduce European medicine to Japan circa 1583, but because he descended from an old Portuguese Jewish family, his reputation for this was effaced in the seventeenth century.[74]

By 1600 the Jesuit order included 8,272 members and operated 236 colleges in southern Europe and Germany, in addition to those in Spanish and Portuguese colonies. The order's growth coincided with the unification of the Spanish and Portuguese thrones from 1580 until 1640. Over fifteen thousand members in 1626 had been trained in 444 Jesuit colleges and 100 seminaries. Within a century of its founding in 1540, the Society of Jesus controlled more than 700 educational institutions and held a virtual monopoly over chairs of the arts faculties at almost every Catholic university.

While preparing to become upholders of the Church, Jesuits such as Matteo Ricci became novices for two years, which was twice as long as any other religious order. Novices were introduced to the society's rules and discipline, and they were tested for their fitness for the Jesuit way of life. For those who entered the priesthood, many but not all had several more years of academic training in grammar, dialectics, and rhetoric, typically at one of the society's several colleges, such as the Collegio Romano. Upon completion of their philosophical and theological studies, the candidates were ordained as priests but had to finish a third year of the novitiate before they were admitted into the society.

Initially the Collegio Romano was the center of the Jesuit educational system, and its evolution influenced the development of a uniform educational curriculum, the *Ratio Studiorum,* throughout other Jesuit colleges in Europe. For the study of mathematics, it was extremely important that the formation of a Jesuit mathematical program had been organized by Clavius at the Collegio Romano, which increasingly took shape during and after the Gregorian reforms in Rome. The focus on Euclid's *Elements of Geometry,* for instance, represented Clavius's imprint on Jesuit mathematical training. Clavius had a following in the 1570s, and Matteo Ricci studied mathematics at the Collegio Romano from 1573 to 1577.[75]

Clavius's role in the promulgation of the Collegio Romano's *Ratio Studiorum* in 1586 was pivotal in the important role of mathematics in Jesuit pedagogy. Among the Jesuits sent to China at this time, besides Ricci, Giulio Aleni (1582–1649) studied under Clavius, as did the physician Johann Schreck (Terrenz, 1576–1630) and Giacomo Rho (1592–1638). Terrenz also befriended Galileo when both studied at Padua. Schall came to Rome in 1608, when Clavius was probably too old to teach, but he also belonged to the Collegio Romano's Academy of Mathematics. Verbiest's formal Jesuit studies at the Jesuit college in Leuven, Belgium, included training in medicine, mathematics, and astronomy. Later, Verbiest also studied at the Collegio Romano and the Jesuit college in Seville.[76]

The largest number of Jesuits sent to China were trained in the first tier of Portuguese colleges at Coimbra and Évora. Moreover, the majority of Jesuits who went to China were Portuguese, although they have not received the attention they have deserved until the recent study by Liam Brockey. Most were in their late twenties or early thirties when they reached China, and who had by then spent some fifteen years in Jesuit schools as students or teachers prior to their departure. The majority were involved in the religious conversion of commoners after their arrival, and they benefited from the local toleration they gained when Italians such as Ricci and Germans such as Schall worked among Han Chinese elites and for the imperial court.

Emphasizing grammar and rhetoric, philosophy, and theology, the education promulgated at Coimbra and Évora was based on the *Ratio Studiorum*. The curriculum, although not always strictly followed, included Greek and Hebrew and was intended to provide intellectual training and teaching experience for those assigned at home or sent abroad. Philosophy courses were spread over three years and included logic, ethics, physics, and mathematics (which included astronomy).

Much of the latter was drawn from Greek natural philosophy, particularly the writings of Aristotle (384–322 B.C.) and Euclid (ca. 325–265 B.C.). Scholastic theology courses usually took four years. Often, however, earlier students completed or abridged training after receiving permission to sail to China. Ricci received only two years of training in theology before departing Portugal, for example. Giulio Aleni, on the other hand, taught mathematics for two years before arriving in China in 1610.[77]

The ideology of the Jesuits drew from Ignatius's vision of an elite corps dedicated to reforming the Church and protecting Catholic lands from the Reformation. It required a long and exacting internalization of the values and ideals of the society. Full members completed the probationary period and were accepted into the society. They were divided into three groups: temporal coadjutors who were lay brothers and served as cooks, secretaries, and apothecaries, etc.; spiritual co-adjutors; and those who professed the fourth vow, that is, in addition to the three traditional vows of poverty, obedience, and chastity, they also professed obedience to the pope. The last two groups were ordained priests and received advanced training in theology. Only the professed could attend the society's General Congregation or be sent on a mission.

Ruled from Rome by its superior general, the Jesuit order penetrated the Old and New Worlds through a residence system, which was the basic unit of organization and daily life for its members. In the Beijing and Hangzhou residences, for example, novices received their training from the Jesuits who lived there. Language learning took place in the residences. In Beijing, the residence also served as the venue for a sophisticated translation project that included the fathers and their converts. The "multi-functional, military-like organization" of the Jesuit residences was flexible enough to take on secular tasks, such as the translation of European mathematical astronomy into classical Chinese.[78]

Outside of Europe, a provincial head led the order. Jesuit unity of purpose was generally achieved through a detailed protocol of written communication between the residences in each province and between provinces. Disagreements among the Jesuits over their engagement in mathematical studies, however, affected this unity. In 1615, for example, the order's provincial head for Japan and China, Valentim Carvalho, sent a mandate to Macao reversing Ricci's policy in China and prohibited the teaching of mathematics and *scientia* by the Jesuits. The mandate was later withdrawn, but it shows how Jesuit unity could be complicated by differences of opinion.

When the Augustinian, Dominican, and Franciscan orders openly opposed the Jesuits in China and elsewhere, the venue for adjudicating such disagreements shifted to the pope in Rome. Similarly, to get around papal control, which was tied to the Portuguese monopoly of Jesuit missions in Asia since 1493, Louis XIV (1643–1715) sent five French Jesuits to China in 1685 as the "King's Mathematicians." The Missions Étrangères de Paris, a congregation of secular priests, who missionized as part of the Rome-based Congregation Propaganda Fide, had since the late seventeenth century challenged Portugal and the Jesuits by establishing a clerical hierarchy in China independent of Portugal. Louis XIV's representatives also refused to take oaths that would have subjugated them to the bishops in Southeast Asia and China.[79]

As a religious order, Jesuits lived and worked as missionaries for the salvation of their neighbors. Ignatius had not intended them to live a cloistered life. Thus, they were experienced and tested before being admitted to the society. Their elevated status in the Church was also based on a dedication to learning, which was combined with worldly involvement. Jesuits participated in the intellectual changes that informed the Society's textbooks and commentaries. They also defined the religious boundaries of orthodox Catholic natural philosophy, which climaxed during the debates over the heliocentric world system developed by Copernicus and championed by Galileo.[80]

Like other Jesuits sent to China, Matteo Ricci, once he had completed his training at Coimbra in 1577–1578, embodied the intellectual affinity between Jesuit religious values and Aristotelian natural studies and philosophy. His expertise in mathematics and *scientia*, drawn in part from study under Clavius, served him well in China. After Ricci, a succession of Jesuits with mathematical and astronomical abilities were sent because Valignano honored requests from Ricci in 1605 and other Jesuits in Beijing after Ricci died in 1610 that their mission required clerics with technical expertise in mathematical astronomy. They were needed to aid in the calendar reform projects then underway in the 1620s under Chinese government supervision. In effect, the Jesuits in China now served two masters, the Church and the Ming government.[81]

Translations from Europe and the Beginning of Ming Calendar Reform

In December 1610, the Astro-calendric Bureau predicted a solar eclipse that was challenged in and outside the Bureau. The prediction was two marks late (between fifteen and twenty-nine minutes) according to the observatory chief in the Bureau, or three marks (twenty-nine to forty-three minutes) late according to a high official in the Ministry of War, who had also sent in a report on the eclipse. Diego de Pantoja (1571–1618), in China since 1599, had a made a more accurate prediction, which Chinese associates of the missionaries knew about. The supervising Ministry of Rites called for imperial approval to study calendar reform, but, as before, the Wanli emperor's court did not act on the request. In mid-1611, however, the Ministry of Rites again urged the court to initiate calendrical reforms.[82]

In its report, the ministry this time noted that for the ten eclipses reported during the Wanli reign the error in prediction usually ranged from one to two marks (fourteen to twenty-nine minutes) and rarely exceeded four marks (forty-four to fifty-eight minutes). To keep such errors from getting worse, however, reforms were needed, according to the ministry. Officials should be appointed to compare alternative systems with the Ming calendar, they urged. Moreover, the ministry reported that the Astro-calendric Bureau director, Zhou Ziyu, had recommended translation of some relevant overseas works in the possession of two foreigners from the "Great Western Ocean" (*Da xiyang*): Pantoja and Sabatino de Ursis (1575–1620). The report justified this recommendation by citing a 1382 precedent for translating Muslim works.[83]

Xing Yunlu, who had called for calendar reform in 1597, and Fan Shouji, who had noted that the 1610 eclipse prediction was off by three marks, were recommended for appointments to supervise the reform. In addition, the ministry put forward the names of two officials who also wished to work on the translation of foreign works on astronomy. They were Xu Guangqi (1562–1633), by then serving in the Hanlin Academy after taking his palace degree in 1604, and Li Zhizao, a 1598 palace graduate and the former vice-director in the Nanjing Ministry of Rites, then in mourning for the death of his father.

Both Xu and Li were confidants of the Jesuits and admirers of Matteo Ricci. Xu had been baptized in 1603. Along with Li Zhizao, whom Ricci baptized in 1610, Xu had studied under Ricci continuously from 1604 to 1607. They were first impressed in 1600 with Ricci's revised map of the world and its five continents, completed after his 1582 arrival. It was published in Chinese as the *Complete Map of the Myriad Countries on the Earth* (*Kunyu wanguo quantu*) in 1602 with the help of Li Zhizao.

The group had already completed a number of translations, such as Xu's contributions to the 1607 Chinese version of the first six chapters of Euclid's

Elements of Geometry. Ricci's writings on mathematics and astronomy generally relied on his teacher Clavius's Latin editions. Ricci's translation of cosmology, printed posthumously in 1614, called the *Structure of Heaven and Earth* (*Qiankun tiyi*), was based on Clavius's commentary on the *Tractatus de sphaera* of Sacrobosco (fl. ca. 1232), an elementary textbook that defended the Ptolemaic system. Along with the *Elements,* Sacrobosco's cosmography was taught at the Jesuit colleges in Portugal and elsewhere as an aid to navigation using applied mathematics.[84]

Until the eclipse controversy of 1610, Xu Guangqi had been more interested in the new mathematics, which he linked to traditional "measurements based on the relational features for the sides of a right triangle" (*gougu celiang,* i.e., geometry), for reform of Ming agricultural practices and for water control measures. His translation and publication of the *Western Techniques of Hydraulics* (*Taixi shuifa*) was a case in point. Xu Guangqi and Xiong Mingyu (1579–1649) together worked with Sabatino de Ursis to translate the *Western Techniques* and had it published in 1612 to introduce European principles of water management (*shuili*).

The preface to the translation equated "studies of numerical correspondences between the heavens and earth" (*xiangshu zhi xue*) with studies of the investigation of things and exhaustively mastering principles, an indication of the accommodation between European and Chinese learning that the Jesuits and their converts were seeking in their translations. The *Western Techniques* was later included in Xu's magnum opus, published posthumously as the *Complete Works on Agrarian Management* (*Nongzheng quanshu*), and reveals how early modern European learning was also applied to statecraft issues and traditional agrarian crafts, pharmacopoeia, and horticulture.[85]

Although Xu Guangqi and Li Zhizao were interested in working on the translations for the calendrical reforms, the Wanli court again took no action. Up to 1619, the emperor tolerated calendrical reform proposals but did nothing about them. Rather than entrenched conservatism, it was his inaction that made such proposals moot. The Ministry of Rites, however, had clearly recognized the serious dimensions of the calendar problem. In 1613, Li Zhizao was summoned to Beijing from mourning in Hangzhou to begin work on the reforms, while Zhou Ziyu as director of the bureau ordered revision of Chinese methods used to calculate the calendar and predict eclipses.[86]

Zhou intended to use Western methods to correct the errors in the Grand Concordance astronomy system, but this was not yet done because the Jesuits only had Clavius's original works on astronomy and some minor instruments such as an astrolabe. They needed more technical treatises on mathematical astronomy before they could proceed. Ricci had written a letter on May 12, 1605, requesting that a mathematician, an astronomer, and calendrical treatises be sent to China. Ricci's successor as head of the China mission, Nicolas Longobardo, called on Rome several times to send specialized books and instruments, as well as Jesuits with mathematical training, to further the cause of Christianity in China. In 1612, the Belgian Jesuit Nicholas Trigault (1577–1628) was sent to Rome to report the order's progress in China and to collect as many mathematical and astronomical works for the mission in Beijing as possible.

To gain support from the Ming court and literati for the larger missionary enterprise, Ricci decided to present the Jesuits in China as specialists in mathematical astronomy. After Trigault's return, they mastered how these technical fields had changed and advanced since the Gregorian calendar reform in 1582. Until 1616, no papal fiat prevented them from transmitting more dangerous parts of the new learning, such as Copernicanism, although Copernicus's cosmology was generally neglected. For instance, Longobardo wrote to Rome, "If this plan is successful, two or three missionaries will professedly occupy themselves with these scientific studies, while the rest of the missionaries will calmly and safely attend the Christians in other parts of the kingdom." This plan was worked out in advance with Xu Guangqi and Li Zhizao, then high officials in the Ming government. What Longobardo did not say was that this would mean a significant portion of the Jesuit contingent in China would become subordinates of the Ming and Qing dynasties.[87]

In 1613, Li Zhizao memorialized the throne, itemizing fourteen new features of Western learning when compared to Chinese calendrical procedures. Among these, Li included the mathematical device of deferents and epicycles from Ptolemaic astronomy; the graduation of 360 degrees in a circle rather than the 365 degrees based on the tropical year of the Chinese; and the geometrical representation of celestial phenomena that provided a spatial theory of eclipses. Li also regarded the Western notion of the precession of the equinoxes as novel without seeing its parallel to the annual difference used since the Yuan to adjust for the displacement of the winter solstice. At the same time, Li recommended that the proposed translation project should move forward with the assistance of Pantoja, de Ursis, Longobardo, and Manuel Diaz the younger (1574–1659). Xu Guangqi planned to use European astronomical tables as supplements to the existing ones, just as Arabic tables were used in the past.[88]

After these early translations were authorized, however, the vice-minister of the Ministry of Rites in Nanjing, Shen Que (d. 1624), proscribed the Jesuits in 1616 for establishing illegal residences in China and "staining the minds of the people." Shen urged that the Jesuits and converts be executed for clandestine meetings similar to the subversive White Lotus sect's religious groups. He was influenced by local literati who feared the Jesuit presence in

Nanjing. Shen also opposed any Jesuit role in calendrical reform. In Nanjing, Shen jailed Alfonso Vagnoni (1566–1640) as the leader of the mission, along with other Jesuits and several of their converts. The Jesuit house in Nanjing was dismantled, and its building materials were used to restore a shrine honoring an early Ming martyr. Vagnoni and Alvaro Semedo (1585–1658) were beaten and then transferred in cages to Guangzhou for expulsion. A sympathetic prefect there freed them, fortunately, and provided for their care before they were sent on to Macao.[89]

In contrast, Yang Tingyun (1562–1627), a convert, and Li Zhizao, who again retired temporarily, provided the Jesuits from Beijing with a haven in Hangzhou. Some Jesuits had already arrived in Hangzhou in 1611 at the invitation of Li Zhizao while Li observed a period of mourning for his father. Yang Tingyun met the Jesuits at that time following his retirement to Hangzhou in 1609. Yang then collaborated in Hangzhou with Pantoja to produce the *Seven Overcomings* (*Qike*), a tract on the seven deadly sins. In 1615, the *Explicatio Sphaerae Coelestis* (*Tianwen lue*) was translated by Diaz the younger. It included an updated geocentric model of the Ptolemaic cosmos and also presented a description of the telescope and Galileo's discoveries using it. After 1616, most of the fourteen Jesuits remaining in China studied Chinese texts and language in seclusion in Hangzhou.[90]

Shen renewed the persecution of the Jesuits from 1621 to 1627 when he became a grand secretary in Beijing during the final factional debacles of the Ming. After brief returns to office, Xu Guangqi and Li Zhizao again resigned. This time of palace infighting between asymmetrical factions of eunuchs and officials, many sympathetic with the Donglin Academy partisans, coincided with wars against the Manchus in the northeast. Shen was able to take advantage of a calamity in Beijing in 1621 to continue his attacks on the Jesuits. By this time, however, they were presented in Beijing as military experts, which gained them permission to reside there. Xu ordered a shipment of European cannon, procured through the Jesuits, to be used in the anti-Manchu campaigns, but two of them exploded, causing loss of life when they were fired in Beijing.[91]

Several years later, when the Chongzhen emperor (r. 1628–1644) came to power, he recalled Xu as a court diarist. In 1629, Xu became a senior vice-minister in the Ministry of Rites. In his new position, Xu Guangqi reported on the competition for predicting the solar eclipse of June 21, 1629. Chinese and Muslim predictions were not as accurate as the Jesuit determination of the time and length of the eclipse. When calls for calendar reform continued and drew government support, Xu then announced the need for immediate calendar reform. The Astro-calendric Bureau continued in its normal duties, but in 1629 Xu Guangqi proposed establishing an independent agency in

Beijing. When approved, the agency resided at a separate venue in a former academy conveniently located next to the Catholic church established by Ricci.

Once in charge of calendrical reform, Xu presented a plan to manufacture ten new astronomical instruments. He also prioritized the key astronomical problems that required technical attention. The reforms for a Sino-Jesuit calendrical system were presented in stages between 1631 and 1635, but the final steps of implementation failed, in part due to Xu's death in 1633, and definitively at the fall of the Ming dynasty due to internal revolts and external invasion in 1644.[92]

Xu could proceed with an ambitious translation project to solve these problems because Trigault had returned in 1621 from his 1613–1614 assignment to Rome with some 7,000 new works, many scientific, published in Europe in the 1610s. Most were still Ptolemaic, i.e., geocentric, in orientation. By 1628, Li Zhizao, for instance, had used several of these works to compile the *First Collection of Celestial Studies* (*Tianxue chuhan*), which included nineteen of the books Trigault brought back. The *First Collection* included the *Translations of Guidelines for Practical Arithmetic,* which had been translated by Ricci and Li jointly, as well as Xu's and Ricci's *Elements of Geometry* (*Jihe yuanben*). Li's collection represented the last stage of Ptolemaic astronomy in China. It would be replaced there in the 1630s by the geoheliocentric system devised by Tycho Brahe at the end of the sixteenth century.[93]

Many works of Aristotelian natural philosophy were also translated into classical Chinese by the Jesuits and their collaborators. *A Summary of Western Learning* (*Xixue fan*), for example, was compiled by Aleni to introduce the Scholastic educational system to China. It was first published in Hangzhou in 1622 and later was given eminence of place in Li Zhizao's *First Collection of Celestial Studies*. It discussed the system of schools, curricula, examinations, and degrees in a Thomistic manner and was based on both the Jesuit *Ratio Studiorum* for pedagogy at the Collegio Romano and the Coimbra texts known as the *Commentarii Collegii Conimbricensis,* which represented a new synthesis of the works of Aristotle. Aleni presented the six divisions of Scholasticism and *scientia*, i.e., specialized learning.[94]

The Coimbra textbooks used to train Jesuit missionaries were famous in Catholic and Protestant countries, where they were also widely used. Li's *First Collection* marked the initial group of European teaching materials made available in Chinese. To encompass Western learning, it was divided into two parts: (1) "principles or patterns" (*li*) and (2) applications, literally, "implements or utensils" (*qi*). The first part included Aleni's summary of Western European fields of knowledge and world geography; the second included geometry, astronomy, and hydraulics.

Similarly, the *Inquiry into the Universe* (*Huanyou chuan*, 1628) by Li Zhizao and Francisco Furtado (1587–1653) represented a translation of

Aristotle's' *De coelo et mundo*. Other Aristotelian materials that were translated included *De Logica* (*Mingli tan*, 1631), again by Furtado and Li Zhizao; the *Treatise on the Composition of the Universe* (*Kongji gezhi*) by Alphonso Vagnoni (1566–1640), from the Coimbra edition of Aristotle's *Meteoriologica* in 1633; and Vagnoni's *Creation of Heaven and Earth* (*Huanyu shimo*) in 1637. Other works on medicine, mining, and mechanics were also translated, such as Agricola's (Georg Bauer, 1490–1555) *De Re Metallica* (*Kunyu gezhi*, lit., "investigation of the earth") by Schall in 1640.[95]

For the future calendrical reform, however, the most significant Jesuits who arrived in China with Trigault in 1621 were the German Johann Adam Schall von Bell and the Italian Giacomo Rho (1593–1638). Both brought with them a new orientation to the cosmos drawn from the work of Tycho Brahe, which replaced the Ptolemaic system with a geoheliocentric one in which the sun and its planets revolved around the earth. In the face of the Copernican challenge, this development represented a compromise position beyond Ptolemy's increasingly discredited geocentrism, but one not yet acknowledging Copernican heliocentrism.[96]

The Compilation of the Mathematical Astronomy of the Chongzhen Reign

The inauguration of a formal astronomical reform system in 1629 by Xu Guangqi, Li Zhizao, and their coterie of Jesuit translators led to the official production of an essential series of translations from the latest Western works, which was titled the *Mathematical Astronomy of the Chongzhen Reign* (*Chongzhen lishu*, 1635). Although the work was never published as a collectanea, i.e., a collection of books published as a set, many works from it were. Some works included had been translated before 1629, such as Li's and Schall's *On the Farseeing Optic Glasses* (*Yuanjing shuo*, i.e., the telescope) completed in 1626 with help of Li Zubo, later martyred in 1665. In the early 1630s, a telescope, the first of many, was sent from Rome.

Published in 1630, *On the Farseeing Optic Glasses* introduced some findings but no cosmology by Galileo. It also presented information on the manufacture and practical use of telescopes. It was arguably the earliest book that introduced the Tychonic system to China. The earth was immobile, i.e., non-orbiting and nonrotating; the planets rotated around the sun, and all en masse rotated around the earth. The choice of this system by Xu Guangqi as a replacement for the "spherical heavens" (*huntian*) cosmology preceded the inauguration of calendrical reform in 1629 and meant that a decision had already been made by the Jesuits not to adhere to the Ptolemaic system and by the Chinese not to adhere to "spherical heavens" cosmology.[97]

Two ancient Chinese models of the cosmos were initially influential, since the account in *The Gnomon of the Zhou Dynasty and Classic of Computations*

(*Zhoubi suanjing*) appeared in the early Han period. The "vaulted heavens" (*gaitian*) cosmology was usually interpreted as a hemispherical dome, an umbrella-like canopy, over a flat earth. Its classical alternative, beginning in the transition from the early to later Han, was called the "spherical heavens" (*huntian*) cosmology, which became influential in the period between A.D. 100 to 180, but was not further elaborated, and was seldom discussed afterward. The Chinese computus was essentially numerical, like modern computer programs. Therefore, mathematicians in the Astro-calendric Bureau after the Han did not correlate their methods with an actual mapping of the cosmos.[98]

As director in charge of the reform effort until his death in 1633, Xu Guangqi organized a calendrical team to systematically adopt European astronomical knowledge. Moreover, as the official supervisor of Jesuit translators, he allowed them to implement the Tychonic system as the theoretical and observational foundation of the calendar reform. In early 1631, Xu presented the throne with the first group of eight works in twenty-three volumes translated by his team.

To avoid the charge that European learning took precedence over Chinese learning, Xu's memorial to the emperor stressed that the translations of Western methods were the first step in preparing a comprehensive system of correspondences between European and Chinese measurements: "Melting the material and substance of Western knowledge, we will cast them into the mold of the Grand Concordance system." Such accommodative language diverted attention from what the translations actually signified, that is, a new cosmological system. Unlike Li Zhizao, who earlier in 1610 had stressed the newness of Jesuit calendrical methods, Xu in 1631 minimized novelty and stressed accuracy instead. This emphasis on unifying European and Chinese knowledge was a tactic that would continue during the Qing dynasty.[99]

Xu noted that, for the 1,350 years between the Han and Yuan dynasties, calendrical methods had developed continuously. Because no major improvements in the Yuan-Ming calendar had been achieved in the 350 years since the Mongol system, it was now time for reform. Xu added that the usual rule of thumb was that any calendrical system had to be reformed every 300 years, which was close to the slip of one day every 308 years that so concerned the Gregorian reformers.

Stressing accuracy based on improved observation, Xu contended that every calendar must be reformed to maintain accuracy. Islamic materials had been used for Yuan and early Ming reforms, and now Xu would use Western methods to provide newer and more up-to-date source materials for contemporary reform. Measurement and mathematics were the key for Xu. He regarded mathematics as the root of European expertise, which followed Ricci's view, derived from Clavius. To find a precedent, Xu appealed to the ancient

Rites of Zhou (Zhouli) for technical corroboration of the importance of mathematics for early Chinese scholar-officials.[100]

In 1632 Xu Guangqi presented a second set of seven translated works to the emperor dealing mainly with fixed stars (which brought the total to 72 volumes), and a star atlas. Falling ill in September 1633, Xu recommended Li Tianjing (1579–1659) to succeed him as head of calendrical reform. Rho and Schall continued their work on the project, and they later made a third presentation of seven more translations. In 1634, Li Tianjing presented the fourth group of translated calendrical works, which comprised twenty-nine volumes and one star atlas. A last installment was presented in 1635. The five sets of works were then combined into the *Mathematical Astronomy of the Chongzhen Reign*, totaling 137 volumes and two star atlases. The traditional goals of reform, to determine seasonal divisions and verify eclipses by observation, were satisfied by 1642.[101]

Unfortunately for the Jesuits, Li Tianjing allowed Muslim calendricists in the Astro-calendric Bureau to establish a fourth branch in 1634, which was known as the Eastern Office. The reform center was renamed the Western Office. In effect, the Ming bureau now had competing Chinese, Islamic, and European branches. After 1634, Li negotiated a compromise to compare the results of all the systems, even though the European remained the most accurate. After various predictions for the 1642 solar eclipse were evaluated, the old calendrical system was replaced with the new system, but this decision came too late for the Ming dynasty.[102]

The works included in the *Mathematical Astronomy of the Chongzhen Reign* were quickly taken by Adam Schall to the conquering Manchu court in 1644. When the Shunzhi emperor (r. 1644–1661) appointed Schall as the first European head of the Qing Astro-calendric Bureau, Schall strategically changed the title to *Mathematical Astronomy According to New Western Methods* (*Xiyang xinfa lishu*) and had it recopied in 1645 for the Manchu court in 103 volumes with minor rearrangements. The Qing dynasty's computus for the Temporal Model astronomy system (*Shixian li*) accordingly was based on the European system adapted for the Ming dynasty.

The Jesuits' change of loyalty to the conquering Qing did not please literati loyal to the fallen Ming, however. This episode confirmed their view of Schall and the Jesuits as opportunists, particularly since Schall was also eager to work with the anti-Ming rebel leader Li Zicheng (1605?–1645) when he briefly captured Beijing in 1644. Some later argued to the Shunzhi court in 1660 that Schall subordinated the Chinese empire to Europe. They accused Schall and the Jesuits of touting the superiority of Christianity through their calendar. After Schall became director of the bureau, except for a successful attack on the Jesuits from 1664 to 1669, a Jesuit was in charge of the bureau from 1645 until 1775, although the society was suppressed by the pope in 1773. The Jesuits were unable to substitute a linear calendar based on Christian notions of time for the cyclical Ming and Qing calendars.[103]

Schall also added a few new works to the *New Western Methods,* such as Aleni's *Essential Methods for Determining Quantities* (*Jihe yaofa*), and the *Methods for Using Counting Rods* (*Chousuan*), which included Napier's device (without mentioning John Napier, 1550–1617) for quickly multiplying, dividing, and taking square and cubic roots of numbers by using multiplication tables on rods. Schall also included a translation introducing Galileo's proportional compass (again without cosmology) and a work by Xu Guangqi answering his Chinese critics (often attributed to Schall).[104]

Aleni's *Essential Methods* was read more widely in the late Ming and early Qing than the now-more-famous Ricci/Xu translation of Euclid's *Elements of Geometry.* Measuring a quantity, literally, "measuring magnitudes" (*jihe*) had initially referred in Aleni's *A Summary of Western Learning* to the Aristotelian category of continuous quantity, that is, a theory of magnitudes. In the early Qing, Mei Wending (1633–1721), separating geometry from arithmetic, followed Ricci's translation of the term rather than Aleni's. Aleni's book collected applications of mathematical astronomy and was not a geometry for beginners. Just as in medieval and early modern Europe, Euclid's geometry was introduced to the Chinese in different forms, which were aimed at different audiences for different purposes.[105]

From the "Spherical Heavens" to the Tychonic World System

When Tycho Brahe obtained a higher degree of precision for his celestial observations by using new astronomical instruments, his observations opened a new era of precision in astronomy. Clavius had checked Copernican data to correct earlier positions because Brahe's findings had challenged the astronomical tradition. Clavius apparently also had his doubts about the Ptolemaic system, based on Galileo's observations. Building on such developments, Rho and Schall as the chief astronomers in Jesuit circles introduced the Tychonic system to Ming China.

Because the Copernican system replaced the Tychonic system in Protestant European astronomy, the continued use of the Tychonic system in Ming and Qing China bequeathed disadvantages to the further development of astronomy in China. These limits were set by the Jesuits, whose intellectual stratagems in this regard have been described by Nathan Sivin. The development of Chinese mathematical astronomy would have been different if the Copernican system had been introduced in a timely fashion by the Jesuits in the Astro-calendric Bureau. A few of Galileo's discoveries, for example, were noted in Ming Jesuit translations, but not his support of heliocentricity.[106]

Nevertheless, the Tychonic age in the Ming and Qing Astro-calendric Bureau meant that Chinese specialists by the 1630s had available to them a rich tool kit of new computational techniques, more accurate observations, a new view of the cosmos that modified the "spherical heavens" system, and the latest precision instruments that went beyond those available to the Gregorian reformers in Europe a generation earlier. Johannes Kepler and Christian Longomontanus (1562–1647), Brahe's assistant, became the European authorities on geoheliocentricity named in Chinese accounts. Longomontanus's *Astronomia Danica,* published in 1622, was crucial for the adoption of the Tychonic system in China.

Furtado, an Aristotelian, refused to accept the corruptibility of the heavens, a finding which had resulted after the discovery of sunspots by using the telescope. Schreck's *Brief Description of the Measurement of the Heavens* (*Cetian yueshuo*), presented to throne in 1630, had already included a description of sun spots. Opposition by older Jesuits in China suggests that the newer Jesuits were able to introduce more recent astronomy so long as they observed the proscription against Copernicus. Furtado and others, for instance, also published objections to Longomontanus's claims about the rotation of the earth in the modified version of geoheliocentricity.[107]

Rho's translation titled the *Treatise on the Motion of the Five Planets* (*Wuwei lizhi*) was subsequently revised by Schall with help of Chinese scholars and presented to the throne in 1635. It presented Longomontanus's semi-Tychonic system and included illustrations of the old Ptolemaic and new Tychonic world systems, which were integrated into the *Mathematical Astronomy of the Chongzhen Reign*. The planets, it was contended, moved freely by themselves through the clear ether of the atmosphere. Longomontanus had also denied the Aristotelian boundary between a changing sublunar world and an unchanging celestial world.

Rho gave no clear indication of what Copernicanism was when he presented its peripheral features by perhaps intentionally mixing it up with Brahe's general method. The motions of the sun, moon, and five planets all relied on the Tycho-Longomontanus interpretation. For the general model of lunar motion, however, the works in the *Mathematical Astronomy of the Chongzhen Reign* presented both the Copernican and Tychonic mapping of that motion. Both remained, of course, geocentric. Rho's mathematical explanations of the motions of the planets were presented in three categories: Ptolemaic, Copernican, and Tychonic. It marked the first transcription of epicycles in Chinese. Similarly, Rho worked out a sort of hybrid Tycho-Copernican system.[108]

The geometric understanding of eclipses also came from Rho's and Schall's reliance on Longomontanus. The influence of Kepler's optical astronomy on the geometrical theory of solar and lunar eclipses was also essential. For instance, Schall's treatise and table on eclipses, prepared under Xu Guangqi's supervision from 1632, included Kepler's chart of optical refraction. Kepler's optics had noted the important effect of atmospheric refraction. The aim of the translators was clear: to present a precise theory of eclipses, the prediction of which was so important to the Chinese. Interestingly, the materials included in the *Mathematical Astronomy of the Chongzhen Reign* portrayed Kepler as a student of Brahe. He was also seen as contributor to the development of an unexplained theory of the motion of Mars, but Kepler's elliptical laws were not introduced for another century, although he was the first to correctly explain planetary motion around the sun in 1605.

Numerical Tables, Star Maps, and Instrumentation

To enhance the accuracy of the Chinese computus, Xu Guangqi recognized how important European astronomical tables, maps of the heavens, and instruments were for the reform. All of these accompanied the treatises presented to the throne from 1631 to 1635. For his astronomical tables, Rho had used early seventeenth-century calendars. Others used the 1551 Copernican Prutenic tables by Reinhold, which the Gregorian reformers had also employed. Kepler's Rudolphine Tables of 1627 were not available in China until a full set was sent to Beijing in 1646. The Chinese created correspondence tables between European calculation tables and Chinese measurements through a synthesis (*huitong*, lit., "combining and interpenetrating").[109]

In addition, the Jesuits introduced trigonometric tables in the *Mathematical Astronomy of the Chongzhen Reign* and translated the Englishman Henry Briggs's (1556–1630) *Arithmetica Logarithmica*, which was published in Chinese in 1624. Briggs was a professor of Gresham College, the original site of the British Royal Society until 1710, and later professor of geometry at Oxford. Nikolaus Smogulecki (Jean Nicholas, 1611–1656) and his collaborator Xue Fengzuo (1600–1680) introduced trigonometric logarithms in the early Qing. In an astronomical work, circa 1656, which pioneered spherical trigonometry and logarithms in China, Smogulecki introduced the latest European method for calculating eclipses. The use of both arithmetical and trigonometric logarithms enhanced the ease of astronomical calculations. For example, when Verbiest was asked by the Kangxi emperor to prepare the groundwork for the Eternal Calendar (*Yongnian li*), which was completed in 1678, he had to calculate its accuracy for two thousand years in the future. To accomplish this task, the Astro-calendric Bureau had to go through an enormous number of calculations that were made easier using trigonometric logarithms.[110]

In addition, the Jesuits provided the Ming court with up-to-date star maps, which culminated a classical tradition in China dating back to the *Clas-*

sic of Stars (*Xingjing*) around 70 B.C. The "Catalog and Atlas of Fixed Stars" in the late Ming improved on the mapping of fixed stars earlier in China, but in this astronomical field the Jesuits had to follow the Chinese system of constellations because they were unable to correlate all the stars known to them with the Chinese star maps. Hence Chinese constellations survived, and the Jesuits could only contribute further to the Chinese tradition of star catalogs.

Xu Guangqi and Adam Schall completed a remarkable star atlas in 1631 while also presenting the throne with a treatise on fixed stars. The latter followed Brahe's methods and other works on positional astronomy. Schall had to transform the two coordinate systems, equatorial in China and ecliptic in Europe, even though Brahe, like late Europeans, chose equatorially mounted instruments and recorded in equatorial coordinates rather than ecliptic. Because trigonometric logarithms were not introduced until the early Qing by Smogulecki and Xue Fengzuo, the system of equatorial coordinates used in the Jesuit star atlas was classically Chinese and became universal in Europe only after Brahe's time.[111]

The 1631 atlas, although a sort of hybrid between Chinese and European star maps, gave the positions for 1,366 stars in the epoch 1628. The planisphere assimilated European knowledge into Chinese constellations. The 1634 "General Atlas of Fixed Stars in the Southern and Northern Hemispheres Divided by the Equator" increased the number of stars mapped to 1,812. The outside was graduated into 360 Western degrees, while the inside used 365.25 Chinese degrees. Beside these, it employed the twelve houses of the zodiac and the twenty-eight Chinese equatorial mansions in the graduated circumferential circle belt along the border.[112]

Finally, Xu was very enthusiastic about instrument making for the calendar reform. Ricci had noted the old instruments preserved since the early Ming in Nanjing and Beijing, but many new instruments were acquired by the Bureau after reforms began in earnest in 1629. These included the quadrant for measuring the altitude of celestial bodies; the parallactic ruler for measuring time; the celestial globe; the sextant; an equatorial astrolabe to observe the movement of the stars; an ecliptic-equatorial armillary sphere to measure planetary motions; and gnomons to find the declination of the sun through the year.[113]

For example, until the Jesuits introduced the ecliptic armillary sphere, which worked from coordinates based on the sun's path through the fixed stars, the Chinese had employed equatorial coordinates, i.e., the equator of the earth extrapolated to the celestial sphere. These instruments were all illustrated and explained by Brahe in his 1598 work titled *Astronomiae Instauratae Mechanica*, which Ferdinand Verbiest (1622–1688) duplicated in 105 woodcut illustrations prepared in Beijing circa 1674. They were discussed and printed in the *Astronomia Europaea* and later included in the *Syn-*

thesis of Books and Illustrations Past and Present encyclopedia. The illustrations were based on instruments produced in 1669 and the early 1670s, when Verbiest succeeded Adam Schall in 1670 as the Jesuit codirector of the Astro-calendric Bureau and reequipped the imperial observatory (see Figure 2.1).[114]

The telescope gave a special boost because it allowed study of the phases of Venus and the motions of the four Jovian moons around Jupiter in terms of the Tychonic world system. In addition, Kepler's optical astronomy and Galileo's telescopic works were very important in China and became part of

Figure 2.1. Verbiest's instruments in the Beijing Observatory.
Source: F. Verbiest, *Xinzhi yixiang tu* (Beijing, 1674).

the new astronomy. A telescope was used to observe a solar eclipse in China on October 25, 1631. Kepler also made a wooden pinhole device to measure the apparent diameters of the sun and moon and to observe solar eclipses in 1600, which was introduced into China circa 1632.[115]

Sighting devices were graduated (as they had been in China for centuries), but now used Brahe's degree subdivisions. Later, when Adam Schall was appointed to the Qing bureau in 1644, he replaced Chinese *du* with European degrees (i.e., a circle = 360 degrees rather than the Chinese 365.25) and time (one day divided in ninety-six units rather than the Chinese one hundred). Yang Guangxian (1597–1669) and other literati attacked this change in units in 1667 when Yang objected to instruments for astronomical prediction calibrated according to European methods. In the end, however, the Astro-calendric Bureau adopted European measurements.[116]

Just as timekeeping in Europe had provided a ritual regularity within society, so too furnishing standard time to the dual capitals of Beijing and Nanjing was the most routine function of the Astro-calendric Bureau. Official timekeeping in China's largest cities had depended on a water clock for quantitative astronomical observation. Timekeeping officials in the belltower and drumtower in large cities marked public time. The Chinese reception of European mechanical clocks, which Ricci and the early Jesuits first introduced in part to amaze the Chinese, should also be seen in light of the Ming and Qing dynasties' imperial horological needs, although the mechanical clock never replaced the three-chamber inflow type clepsydra then in use.[117]

The principles of mechanical clock making was part of the second-year curriculum at the Collegio Romano, for example, when Ricci studied under Clavius, along with studies of the astrolabe, planetary theory, perspective, and ecclesiastical computation. The Jesuits also became the Qing dynasty court's clock makers, and a Jesuit, usually Swiss, was placed in charge of the emperor's clock collection. Later under the Kangxi emperor, Chinese who worked under the Jesuits were expected to make native mechanical clocks. Imperial clock-making workshops were also established in Suzhou and Hangzhou. By the eighteenth century, both telescopes and mechanical clocks became common playthings in the imperial palace.[118]

The Ming calendrical crisis created an institution in the palace, where Chinese literati and Jesuit scholars interacted in translating European mathematical astronomy and instruments into Chinese texts and artifacts. Despite their Tychonic features, the instruments that Xu Guangqi and Verbiest produced for the imperial observatory represented a cultural hybrid. Allan Chapman has noted that "the European circles and technical parts were mounted upon stands contrived in the form of lions, dragons, flaming pearls, and other oriental motifs. The technology is wholly European, while the decorative features are characteristically Chinese."[119]

On the other hand, Louis Le Comte (1655–1728), one of the French Jesuits who arrived in China in 1687, noted in his 1696 memoirs that while the Beijing instruments that Verbiest produced were the finest in the world in terms of design and execution, the quality of the scale graduations was very poor by European standards. In other words, the accuracy of Chinese instruments remained deficient even after the late Ming calendar reforms. Yet, the late Ming and early Qing calendrical reforms had been successful in correcting the calendar and predicting eclipses to the minute. The implications of this paradox requires further elucidation before we turn in Chapter 3 to the further consequences of the Jesuit presence in seventeenth-century China.[120]

Chapman has described how Verbiest in his 1687 *Astronomia Europaea* illustrated a wide variety of Tychonic instruments in the fields of astronomy, ballistics, mechanics, meteorology, navigation, and optics as they were understood up to the 1650s. For instance, there were several illustrations of telescopes in the treatise, but Verbiest failed to discuss the telescope as a measuring device as well as a viewing instrument. Nor was there any awareness of astronomical instruments incorporating the telescopic sight or micrometer. In other words, Verbiest had not gone beyond Brahe's 1598 astronomical instruments.

European positional astronomy, according to Chapman, still functioned in a Tychonic context for over half a century after Brahe died in 1601. Consequently, when Verbiest left Europe for China in 1656, he sailed on the eve of a string of key advances in European instrument technology. In 1658, Huygens produced the first efficient pendulum escapement, which transformed the clock into a precise measuring instrument. Horological developments were paralleled by the invention of the micrometer in the 1660s, which enabled astronomers to read celestial angles more precisely than by using Tychonic scale graduations. Simultaneously, the telescopic sight significantly improved the accuracy of the sextant and quadrant by overcoming the physiological limits, for even Tycho Brahe, of the naked eye.[121]

These new inventions dramatically transformed the manner in which astronomical instruments in the observatories at Greenwich and Paris, for instance, were designed. Smaller, more delicate, and much more accurate than the large and unwieldy instruments in the Beijing observatory, the new astronomical instrumentation and timepieces in Europe after 1658 left the late Ming and early Qing far behind. When Verbiest arrayed his huge Tychonic instruments at the Beijing observatory in 1669–1670, he had outfitted them with obsolete equipment. Moreover, the production of more accurate timepieces in western Europe was increasingly tied to efforts to determine the thorny problem of the longitude of ships in open seas. This level of immediate accuracy went well beyond the needs of a Gregorian or Qing calendar that calculated a solar year within several minutes of accuracy. At sea, an error of just a few minutes in plotting longitude could be fatal and lead to a shipwreck.[122]

On the one hand, the Jesuits in China were certainly cut off from the latest developments in European science and technology, Schall in particular after 1656. On the other hand, just as the Jesuits made no effort to introduce the Chinese to the Copernican cosmology that eventually superseded Brahe's geoheliocentric system, so too they—especially Verbiest and his Jesuit successors in the Astro-calendric Bureau—failed to keep up with subsequent developments in astronomical instrumentation, precision timepieces, or, as we will see in Chapter 5, the revolutionary implications of analytical geometry and the differential and integral calculus for European engineers in the eighteenth century. Because Jesuit mathematicians stressed that mathematics could not explain motion, they never went beyond the focus on Euclidian geometry that Clavius had imprinted as a curriculum in the Jesuit colleges.[123]

By 1660, the Jesuits were no longer on the cutting edge of the emerging scientific community in Europe. A century earlier, under the leadership of Clavius and others in Rome, the papacy had forged a Gregorian calendar that was revolutionary in impact if not in conception. After the Church's proscription of Copernicanism in 1616 and the condemnation of Galileo in 1633, however, the Catholic Church and its leading intellectual lights, the Jesuits, rarely forged ahead in *scientia* and its new instrumentation. Even when they had just arrived in China in the late sixteenth and early seventeenth centuries, Ricci and his colleagues—because they were not specialists—were at first unable to provide the Ming court with the technical expertise needed to resolve the problems in the Ming calendar. After Ricci, the other Jesuits required a crash course based on the new materials Trigault brought back in 1621 dealing with the post-Gregorian changes in astronomy.[124]

The Jesuits were not in China to further the boundaries of *scientia* or to improve European-style astronomical instruments. They sought to make China a Catholic country. Nevertheless, a few of the French Jesuits were sent by Louis XIV to the Kangxi court as royal mathematicians and as members of the Paris Academy of Sciences founded in 1666. In addition, Clavius and his contemporaries redefined Jesuit education at the Collegio Romano to include astronomy, mathematics, and clock making. Finally, Schall and Rho in the late Ming, and Verbiest in the early Qing, could not have accomplished what they did in astronomy without the accruing astronomical literature in Latin that had made the Gregorian reforms possible.

Chapman is right, however, when he concludes, "The Jesuit observatory [in Beijing] never produced any discoveries that were of significance to Western science." The technical work the Jesuits performed for Xu Guangqi in the late Ming and for the Shunzhi and Kangxi emperors in the early Qing gave them access. They were willing minions of the state because, as their letters back to Europe indicated, this approach sanctioned their religious success in China. The Ming and Qing emperors required an accurate calendar and precise eclipse predictions. To give the Chinese what they wanted, the Jesuits provided the imperial government with a new cosmological system, mathematical tables, and instrumental techniques needed to satisfy the carefully itemized list of demands that Xu Guangqi outlined in 1629.[125]

More importantly, perhaps, a model translation laboratory was created in China from 1600 to 1635 that enabled Europeans, Chinese, and later after 1669, Manchus to work together for a common technical goal, even though many on each side had ulterior religious (the Jesuits) or political (the Ming and Qing governments) motives. This translation project far outweighed what the Jesuits achieved as individuals or the Church achieved in aggregate. The translations bequeathed an archive of books, manuscripts, maps, and instruments that would become part of the imperial library and the literati intellectual world. The intellectual impact of Jesuit works in natural studies and their assimilation by the literati, for example, belied how few Jesuits lived in China. Even the Japanese sought these works.

Despite Xu Guangqi's political and administrative success in reforming the Astro-calendric Bureau, Adam Schall's imprisonment in 1665 indicates that the Jesuit reform of the imperial calendar and Chinese time was not yet complete when the Ming dynasty buckled in 1644. Moreover, it was clear to Ricci that Ming literati were usually more interested in his knowledge of *scientia* than the teachings of Christianity he conveyed. With the help of Li Zhizao, he tried to encompass both in his use of the term "studies of Heaven" (*Tianxue*), which designated both astronomy and theology in the *First Collection of Celestial Studies*. Qing literati, with some exceptions, would also welcome the former and elide the latter. Rather than the linear date since creation for Easter, they remained focused on the prediction of cyclical celestial events in a creationless cosmos.[126]

3

Sino-Jesuit Accommodations During the Seventeenth Century

One of the ironies of the Jesuit-Chinese exchange is that Western scholars have often argued that the Chinese turned a blind eye to the European sciences introduced from Europe in the seventeenth century. To this view they usually add that the resistance of China's rulers and elites contributed to the Qing dynasty's backwardness when compared to the rise of the West. In fact, however, before 1800 Chinese literati had chosen that part of Renaissance *scientia* that they found interesting, namely studies of natural phenomena, astronomy, mathematics, and anomalies. What they usually rejected was Christian theology and the scholasticism that was even then declining in Europe.

European *Scientia* and Natural Studies in Ming-Qing China

In Ming China, when the Jesuits and their Chinese converts translated terms such as *scientia* from Latin into classical Chinese, the elite written language of literati, the translations reflected the personal views and intellectual frames for natural studies of both sixteenth-century China and Europe. Such studies were not the sciences of more modern times. Literati converts chose scholarly learning (*xuewen*) as the classical Chinese equivalent to correlate native categories of specialized studies with *scientia*. One of the most influential of the early Jesuits in late Ming China, Matteo Ricci, for instance, equated *scientia* with knowledge (*zhi*). Following Clavius, his translations of mathematics stressed geometry with proofs in the Euclidian style.[1]

In the face of the Protestant Reformation, Catholicism reaffirmed Scholasticism in the new Jesuit colleges of the sixteenth century, such as the Jesuit College of Arts at Coimbra in Lisbon and the Collegio Romano in Rome, which offered the best training in mathematics and natural studies then available in Europe. Many of the Jesuits who went to China studied at the Collegio Romano's Academy of Mathematics and at Coimbra first.[2]

The logically precise mode of presentation that marked Ricci's translations of European mathematics into Chinese differed from the inductive approach in calculation problems associated with native manuals. Ricci's *Explanation of Right Triangles* (*Gougu yi*) presented fifteen solutions for right triangles proved in the Euclidean style. Ricci and Li Zhizao compiled treatises on European arithmetical knowledge, such as the 1614 *Translations of Guidelines for Practical Arithmetic* (*Tongwen suanzhi*), which added problems and solutions they copied from Chinese mathematical texts.[3]

Ricci and Li also introduced the European pen-and-paper approach to calculations, replacing counting rods and the abacus. Although more onerous, this method facilitated checking calculations, i.e., using subtraction to check addition, etc. The traditional counting board in Yang Hui's (fl. in Southern Song) *Practical Rules of Arithmetic for Surveying* (*Tianmou bilei chengchu jiefa*) for instance, provided a matrix notation to keep track of mathematical operations performed mentally.[4]

In addition, the new method for recording fractions was introduced. Despite their logical rigor these initial mathematical translations did not supersede the practical limits of traditional Chinese mathematics. Although arithmetic had been part of an elementary education in the ancient *Record of Rites*, the early Qing call by Yan Yuan (1635–1704) and Li Gong (1659–1733) to teach children more practical learning such as arithmetic likely reflected an accommodation between European and Chinese reckoning.[5]

Jesuit Translations of Scientia in Classical Chinese

Unlike his *mappa mundi* discussed below, Ricci's *Structure of Heaven and Earth*, completed circa 1605–1608, was based on Clavius's work in Rome, as was his translation of the first six books of Euclid's *Elements*, printed in 1607, which was drawn from Clavius's 1574 annotated Latin version. Refuting skeptics of the Ptolemaic astronomical system, which he included in his commentary on Sacrobosco's work, Clavius used the terms *scientia* and *opinio* in their traditional Scholastic senses to show that Ptolemy's efforts to trace the positions and motions of heavenly bodies were consistent with the principles of physics and thus consistent with *scientia*. Because he believed that geometrical proofs were syllogistic, Clavius contended that geometry was also important for mathematics and philosophy as a window on God's divine plan for nature.[6]

Between 1592 and 1606, the College of Coimbra produced a series of Latin textbooks and commentaries for the required works of Aristotle, which the missionaries trained there brought to China. In the seventeenth century, for example, the Scholastic uses of experience were adjusted to fit novel ends, when new experiences became legitimate in historical reports of events and in citing witnesses. Reliable evidence became an empirical component of natural philosophy. For Jesuit Aristotelians, discrete experience was the focus rather than experiment as an arbiter of hypothesis (Latin *experimentum* encompassed both). In those cases where miracles seemed an implausible explanation for an anomaly, the sixteenth-century European intellectual community began to require reliable evidence *of* miracles.[7]

In their translations, the Jesuits and their converts frequently stressed the syllogism as the tool to pursue *scientia*, which was a mark of specialized learning. The initial Latin translation of the Doctrine of the Mean, one of the Four Books, was included in the *Sinarum scientia politico-moralis* (Political and moral learning of the Chinese). The translation was published partially in Canton in 1667 and partially in Goa in 1669. It was rendered into French and published in Paris in 1673 as "La science des chinois" in a collection of travel literature. The Latin translation assimilated this part of the Four Books into Aristotelian moral inquiry. Use of *scientia* as a general term by the Jesuits also implied that studies of God and humanity were unified.[8]

Likewise, the translation of the Great Learning, another of the Four Books, was initially titled the *Magna scientia* (Great learning) as part of the *Confucius Sinarum philosophus, sive Scientia Sinensis Latine exposita* (Confucius the Chinese philosopher, or Chinese learning exposited in Latin). The section in the *Confucius Sinarum philosophus* that included the translations of the Great Learning, Doctrine of the Mean, and the Analects of Confucius (the third of the Four Books) was originally titled the *Scientiae Sinicae* (lit., "Learning of China"). The final version of the *Confucius Sinarum philosophus* published in 1687 in Paris elevated Confucius's name to the main title while reducing *scientia Sinensis* (Chinese learning) to a subtitle. These translations attempted to show that the Chinese possessed a natural religion based on a rational nature.[9]

Giulio Aleni's *A Summary of Western Learning* roughly corresponded to the curriculum based on the commentaries on Aristotle at Coimbra. The curriculum prioritized the study of philosophy and reasoning, which was the beginning point for astronomy and mathematical studies. *Scientia* as systematic learning in Latin (*epistemè* in Greek) contrasted with art (*ars* in Latin; *technè* in Greek). The content of *scientia* was generated by the demonstrative syllogism from Aristotle's *Posterior Analytics*. Art and opinion (*Ars: opinio*) were generated by the probable syllogism based on dialectical reasoning, i.e., reasoning from authoritative opinions.[10]

To communicate and overcome the difficulties of translation, the Jesuits talked about their knowledge systems in the classical Chinese terms that Ming literati chose. The translated terms and transcriptions that were adopted suggested correspondences and parallels when Chinese terms were commensurable. Transcriptions were necessary for incommensurable differences in conceptions. To convey the meaning of *philosophia* in Western learning to literati, Aleni first presented a transcription: *feilusuofei*. With the help of his Chinese adepts, he then linked this transcription to Chinese terminology for philosophy (*li ke*), associated since the Song dynasty with Cheng-Zhu teachings.

Hence, in *A Summary* Aleni equated *philosophia* with exhaustively mastering principles (*qiongli*) and the investigation of things. Aleni and his Chinese informants presented *physica* (*feixijia*) as a subset of *philosophia* equivalent to the "investigation of the principles of the nature of things," again drawing on terminology associated with Song classical learning. The choices Aleni made were repeated in his *Account of Countries not Listed in the Records Office* (*Zhifang waiji*), a treatise on world geography, and later reproduced verbatim by Verbiest in his *Essential Records about the West* (*Xifang yaoji*). Similarly, Furtado and Li Zhizao compiled their *De Logica* in Hangzhou in 1631 from a text on Aristotle's dialectics prepared in 1606 at the Jesuit college in Coimbra. This view of Jesuit-Chinese collaboration challenges the standard view of the Chinese collaborator as the passive partner in the Catholic mission in China.[11]

Aleni's *A Summary of Western Learning* presented a classification of the sciences that corresponded to the general western European standards of the sixteenth century, which attributed a more important place to mathematics and technology. Li and Furtado turned *De Logica* into a version of the Aristotelian *Categories* for a potential Chinese literati audience that never materialized.

In his *Treatise on the Composition of the Universe* (*Kongji gezhi*, 1633), Vagnoni distinguished between the ancient *scientia* of the Chinese and the new *scientia* of Europe in light of the investigation of things and extension of knowledge. The new learning fathomed the inner workings of things that the ancient learning found difficult. Similarly Francisco Sambiasi (1582–1649) translated philosophy as the "study of investigating things and extending knowledge" and the investigator as "one who investigated things" (*gewu zhe*). Adam Schall later equated the inquirer of knowledge with "one who investigates things" (*gewu zhe*). For physics, the Swiss Jesuit Johann Schreck's *Diagrams and Explanations of the Marvelous Devices of the Far West* (*Yuanxi qiqi tushuo luzui*, 1627), compiled with the help of Wang Zheng (1571–1644), complemented the *Mathematical Astronomy of the Chongzhen Reign* as translations that introduced Renaissance mechanics (*zhongxue*), machine technology (*jishu*) such as clocks, and topics such as optics and heat.[12]

Between 1600 and 1773, Jesuit authors worldwide wrote more than 4,000 books, 600 journal articles (almost all after 1700), and 1,000 manuscripts dealing with *scientia*. The vast majority were by educators. They and their converts in China translated or compiled 437 works between 1584 and 1790. Thirty percent of that total (131) were in the sciences, and 57 percent (251) were on Christianity. Usually a Jesuit orally translated the Latin texts and dictated it to a Chinese, who turned it into passable literary Chinese, choosing terminology and concepts. This process was repeated after 1850 when Protestant missionaries and their collaborators translated many works of modern science primarily from English into classical Chinese.

In the late Ming, Matteo Ricci translated Clavius' 1607 *Elementorum* in this manner, as did de Ursis for the 1612 *Western Techniques of Hydraulics*. Smogulecki and his collaborators introduced European methods for calculating eclipses in an astronomical work circa 1656, which was also the first to introduce spherical trigonometry and logarithms. Later in 1672, Ferdinand Verbiest furnished further information on world geography beyond Ricci's earlier *mappa mundi*. By way of contrast, most eighteenth-century European translations in China were theological works, and the Jesuits turned instead to translating Chinese works into European languages.[13]

The late Ming literatus Fang Yizhi (1611–1671) stressed "concrete investigations" (*zhice*) in his magnum opus, titled *Notes on the Principles of Things* (*Wuli xiaozhi*). His goal was to comprehend the seminal forces underlying the patterns of natural change. Presented as a collection of observations, findings, and other sources, the *Notes* as a form of jottings resembled the late Ming encyclopedias and collectanea. Fang generally accepted European explanations of natural phenomena, such as a spherical earth, Tychonic cosmology, and human physiology, but he was critical of the Jesuits for leaving concrete investigations behind and ending in unverified religious positions, such as Schall's claim that God's existence was revealed in the world's design. Fang favored, instead, descriptive knowledge of the natural world, and he infused the investigation of things with a new view of the accumulation of knowledge, which gainsaid both the moralist focus of Zhu Xi and the introspective focus of Wang Yangming, the two most influential voices of moral philosophy and knowledge theory in China since the Song dynasties.[14]

Most Chinese literati labeled as perverse or heterodox those Jesuit works on the "Learning from Heaven" (*tianxue*, or "celestial studies") that made what they considered unacceptable claims about the world or the power of an otherworldly God (*Tianzhu*, lit., "the Lord of Heaven"). Acceptance of a portion of European *scientia* was possible in Ming and Qing China precisely because Chinese literati themselves had a longstanding and commensurable interest in natural phenomena, which had undergone a significant revival in the sixteenth century before the Jesuits, Augustinians, Dominicans, and Franciscans arrived in South China.[15]

The Accommodation Project

One of the most controversial aspects of the Jesuits' initial success in China was their monopoly, under the auspices of Portugal and Rome, over the transmission of Christianity to Asia. Until the 1630s, when with Spanish support the Dominican and Franciscan friars were finally able to overcome Jesuit opposition and join the China mission, Ricci and his followers had adopted an accommodation approach to the family rites practiced by the Chinese. The Jesuits' approach was not threatened until the middle of the seventeenth century, when the Spanish orders began preaching in south China. Rather than treating Chinese ancestor worship and veneration for Confucius as religious forms, and thus idolatrous, Ricci—despite some opposition—treated such behavior as civil rites, which he considered commensurable with the Catholic faith.

In contrast, the Dominicans and Franciscans demanded entry into China from the New World and the Philippines because they charged that the Jesuits permitted their converts to retain idolatrous observances. This brouhaha with the Jesuits simmered throughout the seventeenth century. In 1700, it burst its barriers, and the Jesuits were publicly chastised by the Jansenists in Paris for their role as "mandarin-missionaries" in preaching an impure form of Christianity in China. Domingo Fernández Navarrete (1610–1689), in particular, led the Dominican opposition in China during his twelve years there from 1655. Nevertheless, the Jesuit accommodation policy remained in place until the early eighteenth century.[16]

One aspect of the accommodation strategy, which characterized the Jesuit-Chinese interaction in the seventeenth century, was the joint use of Chinese terms such as the "investigation of things," "exhaustively mastering principles," and "knowing heaven" (*zhitian*). Such expressions indicate some of the commensurable concepts that were at the core of the intellectual encounter between China—as represented by Ming literati—and early modern Europe—as represented by the Jesuits.

Mutually commensurable terms dealing with *scientia* and natural studies suggested to both Chinese literati and the Jesuits intellectual correspondences between Catholic Europe and Ming China. Classical Chinese terms were used by literati and Jesuits to accommodate both European and Chinese views of practical studies, which included the emerging sciences of mathematics and astronomy in Europe and China. We can also see an overlap between religious and scientific work on the part of the Jesuits and their Chinese converts and sympathizers.[17]

It is significant that the term "investigating things and extending knowledge" (*gezhi*) was chosen by late Ming literati as one of the native categories of appropriate learning (*xuewen*), with the latter term equivalent for many Ming literati to European *scientia*. For example, a manuscript by Liu Junxian titled the *Essential Criteria for Specialized Learning* (*Xuewen yaolun*), which dated from the late Yuan dynasty, was finally printed in 1700. It divided classical learning into six major fields: (1) heaven and earth; (2) studies of principles and patterns; (3) statecraft; (4) morals and regulations; (5) antiquities; and (6) miscellany. As a mid-fourteenth-century classification scheme, the *Essential Criteria* focused on fields that were still considered important for literati studies in the seventeenth century.[18]

Early Jesuit translations of Aristotle's theory of the four elements in the *Treatise on the Composition of the Universe* and Agricola's *De Re Metallica* (*Kunyu gezhi*, 1640) into classical Chinese had also used investigating things and extending knowledge in light of the Latin *scientia* in their titles. Such translations suggest that orthodox doctrine and natural studies, particularly medical and mathematical astronomy, were not mutually exclusive.[19]

Commensurability was fraught with tensions that lurked just beneath the surface of well-intentioned accommodation agendas. Each side sought to efface the other by simple reduction of the other to themselves. Their actual common ground was a hybrid that assumed each side had the same agenda, but each aimed to achieve diametrically opposite results. Ricci and the Jesuits tried to efface the classical content of the investigation of things with western European natural studies, which would then enable the Chinese to know heaven and accept the Church. Chinese effaced Western learning with native traditions of investigating things and extending knowledge, which would allow them to assert that European learning originated from China and thus was assimilable.

The early Jesuits and their collaborators acquired a reputation for learning and thereby gained a role in Chinese society through mathematics and astronomy. However, furthering the Church in China remained their ultimate objective. Ming literati, on the other hand, used the debate between Zhu Xi and Wang Yangming over investigating things and exhaustively mastering principles (*gewu qiongli*) to domesticate Western learning within native traditions of natural studies.

Jesuits who came to Ming China, such as Aleni, Furtado, and Ricci, focused on promoting good relations with literati to protect the conversions of commoners by other clerics in the field, as well as to augment their accommodation project for Christianity and literati learning. Thus they generally stayed out of Ming imperial politics and received support from local Chinese scholars and officials. Ricci and the Jesuits also had strong ties to the fundamentalist position in late Ming literati learning. The accommodation approach that Ricci and his generation of Jesuits adopted allowed them to take advantage of the late Ming "return to antiquity" (*fugu*) currents of classical learning without political interference.[20]

Ricci distinguished what he considered the original teachings of Confucius and the Classics from the materialism he associated with the Learning of the Way orthodoxy of the officially prescribed Cheng-Zhu school, which he also considered atheistic. In the early Qing, such denunciations would become dangerous. Ricci favored the earlier interpretations of the Classics over those given by Zhu Xi and his Ming followers. Ricci also tried to convince Chinese scholars (perhaps after having been convinced by other Chinese) that Zhu Xi's metaphysics were not an integral part of original classical doctrine. The Jesuits thereafter contended that the ancient Classics were ultimately assimilable to Catholicism, but only when purged of Buddhist and Daoist corruption. Ricci wrote:

> The doctrine most commonly held among the Literati at present seems to me to have been taken from the sect of idols, as promulgated about five centuries ago [that is, the Song period] . . . This philosophy we endeavor to refute, not only from reason but also from the testimony of their own ancient philosophers to whom they are indebted for all the philosophy they have.[21]

The language and allusions in Ricci's translations, for example, were in classical Chinese, but his humanist tool kit drew on the tactics and arguments of western European Renaissance culture. For instance, Ricci used Epictetus's manual of Stoic moral philosophy for his Chinese work titled *Twenty-five Sayings* (*Ershiwu yan*), what Goodman and Grafton aptly call "rhetorical Chinoiserie." Updating pagan and biblical texts had preoccupied Renaissance scholars, so Ricci's use of Stoicism followed in this tradition. Justus Lipsius's presentation of Seneca as a wise pagan was another example of how Catholics had massaged pre-Christian texts to serve official Church doctrine, while meeting the needs of reform-minded Catholics in 1600. In the process, the Jesuits inscribed Graeco-Roman culture with the precepts of Christian morality.[22]

Ricci's efforts to find Chinese points of contact with Christianity paralleled humanist concerns with finding natural theology in the paganism of ancient antiquity in Greece and Rome. His use of allegorical forms for updating Stoicism to promote the European classics in China was guided by humanist tactics. Ricci's encounter with Zou Yuanbiao (1551–1624) is instructive when seen in this light. Zou argued Ricci had realized that Christian teachings were contained in the ancient Chinese Classics, an idea that later became the kernel for the Chinese notion that all Western learning was derived from China. Zou, however, suggested that the Jesuits use the *Change Classic* to elaborate their ideas further, but this idea was not taken up until the French Jesuits arrived in China after 1689.[23]

Likewise, Ricci's belief in the transforming power of ancient Greek mathematics reflected the peculiar place of Euclid's geometry in Renaissance European thought, which saw it as a sourcebook for ancient wisdom. Geometry was regarded as the basis for astronomy because it was essential in the European astronomy informing the Gregorian calendar reform. As an important participant, Clavius regarded mathematics as the mediator between metaphysics and physics, and he studied the "Book of Nature" using an Aristotelian framework.

Applying the views of his teacher to China, Ricci also thought mathematics would prepare the Chinese for the higher truths of Christianity. His use of logic was intended to convince Chinese of the necessary existence of God, as his collaborator Xu Guangqi made clear in his introduction to the Chinese translation of Euclid's *Elements of Geometry*. A clear notion of geometry as a deductive system was not yet apparent, however. Earlier we saw that the Chinese initially regarded Euclid's geometry as a theory of magnitudes, i.e., mathematics. Moreover, Fang Yizhi and Mei Wending successively suppressed the proofs in the *Elements* in favor of computational rules and theorems. Like many Europeans, they reduced Euclid to a set of fundamental, if preliminary, procedures and principles.[24]

Ricci's *True Meaning of the Lord of Heaven* (*Tianzhu shiyi*, 1603) did not focus on Christian scriptures of the Old and New Testament. Instead he emphasized a rational knowledge of God in a dialogue between a Chinese literatus and a "literatus" from the West (*Xiru*, i.e., *Ru*, "conventional scholar," often mistranslated as "Confucian"). Rather than quoting Latin scriptures, Ricci reinterpreted the Chinese Classics using Christian arguments. Ricci used the notions of "investigating things and exhausting principles" in his *True Meaning* to propagate Western learning. In the same manner Zhang Xingyao (b. 1633) referred to the Jesuits as "scholars from Western countries" versus "native scholars."[25]

The Jesuits and some of their Chinese counterparts in the late Ming believed that Christianity and classical learning shared a common heritage. Hence a literatus such as Zhang Xingyao could sympathize with Ricci when he claimed that ancient scholars had adhered to an ancient natural religion and that later scholars had departed from the original teachings in the ancient Classics. Even in the late Ming, however, Ricci's repudiation of Song literati, primarily Zhu Xi, was not widely accepted by Chinese such as Zhang, primarily because Zhang focused on the more ancient Five Classics in his accommodations, unlike Ricci, who used the more recent Four Books.

Although Ricci attacked Zhu Xi's Song commentaries for what he perceived as the philosophical materialism and atheism they contained, he showed no awareness of the Lu-Wang (Lu Xiangshan, 1139–1192, and Wang Yangming) tradition of classical learning, which earlier in the sixteenth century had been critical of the Cheng-Zhu school in favor of more ancient Chinese classical learning. Wang Yangming's heralded attack on Zhu Xi's interpretation of the Great Learning may have predisposed some late Ming literati to recognize that Ricci's critique of Cheng-Zhu learning was not unprecedented. Had Ricci been aware of Wang's critique, his own attack on Zhu Xi might have been strengthened.[26]

Later in 1711, the French Jesuit Francois Noël (1651–1729) produced the last fruits of the Jesuit translation project in Europe. His *Sinensis Imperii classici sex* (Six classical books of imperial China), which added the *Mencius*, *Classic of Filial Piety* (*Xiaojing*) and *Lesser Learning* (*Xiaoxue*) to earlier Jesuit translations, was received unenthusiastically in Europe. Noël joined the early eighteenth-century Rites Controversy to support the Kangxi emperor, but his position was out of favor among Catholics in Paris, whose Sorbonne theologians were Jansenists. In 1700 they impeached five propositions in Louis Le Comte's (1655–1728) memoirs on China. Specifically they attacked the French Jesuit Figurists, Noël and Le Comte included, for verging on heresy in their fascination with hieroglyphic symbols such as the trigrams of the *Change Classic,* which the Figurists saw as repositories of ancient celestial knowledge and associated with a sublime philosophy concealed in Chinese texts.[27]

The contributors to the *Confucius Sinarum philosophus* had initially translated the Great Learning as *Magna scientia,* but Noël changed this to *Adultorum doctrina* (learning for adults). Moreover, Noël's version did not disparage Zhu Xi and his followers as "Neoterics" as Ricci and earlier Jesuits had done. In the Kangxi reign, the Jesuits could no longer afford to attack Zhu Xi and the Song scholars as idolaters or as corrupters of the ancient Classics, although such a position in 1700 might have spared them the wrath of the Sorbonne's Jansenists.[28]

Preliminary Debates between Chinese Literati and Jesuits

Chinese literati moderately transformed their own traditions of natural studies due to their interaction with European *scientia.* The Jesuits in late Ming and early Qing China saw the investigation of things and exhaustively mastering principles as a step in conveying the experience of God to the Chinese they hoped to convert. For late Ming Chinese such as Fang Yizhi, their recovery of the "concrete studies" (*shixue*) of antiquity predisposed some literati to accept Western learning in its Jesuit version as an alternate form of the investigation of things and a confirmation of Chinese ancient learning.[29]

"Investigating Things" as an Object of Tension

According to Matteo Ricci, the Chinese entertained several absurdities concerning the nature of the world: (1) the earth is flat and the sky is a canopy; (2) there is only one celestial sphere instead of ten, and the stars move in a vacuum; (3) they do not know what air is and they only see a vacuum between the celestial spheres; (4) they count five elements instead of four; (5) for solar eclipses, the moon diminishes the sun's light when it approaches the sun; (6) during the night the sun hides under a mountain near the earth.

To gainsay such errors, half of which were misreadings, Ricci used ancient Chinese texts to confirm Christian readings and Aristotelian natural studies. He attributed the mistaken literati doctrine of "man forming one body with heaven and earth" to Buddhist influence in later times. Ricci's Scholastic tool kit enabled him to assert that ancient China had a Christian religion but that it was lost after the despotic Qin empire came to power and eliminated ancient religious traditions. Ricci's tactics facilitated Christian reinterpretations of ancient Chinese learning. Through accommodation, Ricci disqualified later Chinese learning, called Learning of the Way, by appealing to reason and logic as the basis for correct textual interpretations. The Jesuits also appropriated the Chinese stress on the innate goodness of human nature and appealed to a notion of a rational soul endowed spiritually by God to overcome the sinful material nature of humanity.[30]

Although not a Jesuit convert, Xiong Mingyu accepted Ricci's attacks on Zhu Xi and Cheng Yi in favor of ancient Chinese learning. Ricci's tool kit focused on the Chinese phrase for the Lord on high (*Shangdi*), which he contended was the ancient Chinese name for God. Xiong paralleled Ricci's tactics by focusing on investigating things and arguing that the learning of the Jesuits corrected native traditions.

For Xiong, the Jesuit conception of *scientia* corresponded to Chinese natural studies. Xiong's views carried over to Fang Yizhi, whose father had discussed Jesuit learning with Xiong in his son's presence. Xiong Mingyu together with his own son, Renlin (fl. 1637), published a two-volume work on Jesuit *scientia* titled *General Survey of Natural Studies and Geography* (*Hanyu tong*). His own volume in the work was titled *Draft for Investigating Things and Extending Knowledge* (*Gezhi cao*) and addressed natural studies, while Xiong Renlin prepared a volume on geography, which was indebted to Ricci and reproduced his *mappa mundi*. Father and son each had access to the late Ming translations included in the *Account of Countries Not Listed in the Records Office* and the *First Collection of Celestial Studies*.[31]

Xiong noted in the late Ming preface for his *Draft* that "literati took to heart the Great Learning and of course spoke of the necessity of first investigating things and extending knowledge." Unfortunately, literati failed to capture the unity of knowledge that the Doctrine of the Mean and the *Mencius* had extolled. What was required, Xiong explained, was a detailed examination of the heavens, earth, stars, constellations, the transformations of *qi,* plants and animals: "Fathoming each actual phenomenon one by one, one realized the reason why things are as they are, which illuminated the principle for why they could not be any other way."[32]

Five Phases versus Four Elements

Both sides saw an order and purpose in the cosmos and on earth, which the Jesuits linked to a physico-theology encompassing theology and geography to unify God and nature.[33] Most Chinese literati also saw the earth and heavens as a harmonious whole, but they framed arguments for the design of the cosmos around an eternally evolving Way rather than around the linear chronology of a divine providence. In place of a cosmos made up of four elements (air, fire, earth, water), the Chinese conceived of change in light of an undifferentiated Supreme Ultimate with no opposites. It spontaneously split into a completely differentiated stage of yin and yang. The interaction of yang and yin on the stuff (*qi*) of the universe then set in motion the five phases (earth, fire, metal, water, and wood) of cosmic change and yielded the concomitant production and destruction cycles of the myriad things (*wanwu*) in the world.[34]

The Jesuits' use of Aristotelian natural philosophy to refute Chinese notions of *qi* was another part of this Christianizing agenda. They regarded *qi* as air, one of four elements, within the category of matter. Neither spiritual beings nor the human soul were reducible to *qi.* Cheng-Zhu scholars had explained ghosts and spirits, for instance, through an appeal to refined but disintegrating *qi,* which was the spiritual cum material basis of both. Zhu Xi argued that although "the essence of *qi* was spiritual," the soul eventually dispersed. Ricci as a believer in the immortality of the soul saw this position as heresy and sought to illuminate the true nature of spiritual beings by refuting what he perceived as Cheng-Zhu atheism.[35]

Vagnoni's *Treatise on the Composition of the Universe* was in part a refracted presentation of the theory of the four elements from the *Conimbricenses* edition of Aristotle's works. In his translation, for example, Vagnoni vainly tried to convince the Chinese of the error of their ways for including wood and metal and excluding air as the building blocks of things in the world. His use of translation for the Jesuit encounter with Chinese literati built on Ricci's earlier efforts to expound a theory of the four elements in his *Summary of the Four Elements* (*Sixing lunlue,* 1595).[36]

Vagnoni coined the new term of a primary element (*yuanxing*) to render Aristotle's concept of element into classical Chinese and to demarcate it from the Chinese notion of five phases:

> An element is a pure substance in that when divided it does not form any other sort of thing. It can only form a composite thing made of several sorts. What is a pure substance? It is a substance of a single nature with no other composite elements. Accordingly, the myriad things in the world are distinguished by their being pure or composite. The pure elements are the four elements of earth, water, air, and fire. Composite things take on five forms such as the category of rain, dew, thunder, and lightning, the category of metals and stones, the category of plants, trees, and the five grains, the category of birds and animals, and the category of humans. All five of these forms are composites of the four elements.[37]

The key concern for Vagnoni from the outset was to refute the Chinese inclusion of metal and wood in the elements. The latter were not pure substances like air, which the Chinese had not included. Replying to a query by a Ming literatus, Vagnoni gainsaid whether there were the five phases at all:

> If one observes how the myriad things are formed, it is not from metal and wood. The many categories of humans, reptiles, birds, and animals are examples of this. Therefore, metal and wood cannot be the primary elements of the myriad things. Moreover, who does not know that metal and wood are in actuality themselves composites of water, fire, and earth. As composites they cannot be primary elements. There are many additional composites that might be called primary elements, such as plants, stones, and other things, and which could be arrayed as primary elements. Consequently, they are not limited to just five. Why only select metal and wood?[38]

In the dialogue that Vagnoni imagined with a literatus, there was an assumed agreement, however, that at least three of the Chinese five "elements," namely, water, fire, and earth, overlapped with the four elements of Aristotle. In other words, Vagnoni thought that what he meant by an element as a pure constituent of reality was the same as what the Chinese meant by the five phases. The issue was, simply put, that the Chinese had miscategorized the elements by failing to ascertain that in addition to water, fire, and earth only air was another pure substance that made up composite things such as wood and metal. Vagnoni and his Chinese informant Han Yun made no effort to conceptualize the five phases in Chinese terms, namely that the five phases were not substances but marked phases in every sequence of change, whose configurations in time were the key to the delineation of the myriad things in the world.

Of particular importance to the Jesuits in their efforts to promote Aristotle's views of "nature and principles" (*xingli*) was their belief that air was one of the four primary elements. This entailed an equation of the Chinese notion of *qi* as matter-energy with the Aristotelian element of air. Aristotle had complemented the four elements with ether as a fifth substance to account for celestial bodies that were not susceptible to change and remained in eternal, circular motion. The sublunar world was composed of the four elements and subject to finite motions of all kinds in a world of constant change and transformation.

On the other hand, the Chinese notion of *qi* worried Ricci and his colleagues because they saw it as a materialist continuum that encompassed all matter. In other words, *qi* verged on materialistic pantheism, leaving no place for the unlimited spiritual power of God. Consequently, Vagnoni tried in his dialogue to equate *qi* and air, thus demoting it to one of the four pure physical substances. The conversion of *qi* from an unlimited, physico-spiritual presence, undergirding all physical and spiritual things, into a single substance making up some but not all things was based in part on what Vagnoni perceived, incorrectly, as the physical notion of *qi*.

To prove that air was a pure element, Vagnoni appealed to the physical presence of *qi* as a sort of ether filling in the vacuum of space within the celestial spheres. Otherwise, "without *qi* the space within the heavens would be empty," and there would be no natural motion possible up or down from the earth. Moreover, without air to breathe no thing could live. Without air, humans would perish: "From this we can infer that the qualities of humans and things—whether they are wise or ignorant, beautiful or ugly, old or young, strong or weak, etc.—all are wholly related to the air that one breathes."[39]

The Jesuits were trying to delimit *qi*, rejecting both classical and popular notions of *qi* in China favoring the equation of air and *qi*. By modern standards, this interpretive effort by the Jesuits seems a mistranslation and misrepresentation. In its time, however, the interpretation of texts and doctrines was a time-worn practice for both Jesuits and Ming literati. Christian efforts to rework ancient Greek and Latin texts, including Aristotle's writings, represented an earlier phase of this willful interpretive process. The Christianization of Galen (b. A.D. 129), which eliminated his materialist conception of the soul and established instead the notion of *pneuma* as a physical instrument and agent of the immortal soul, is a clear precedent for what Vagnoni was trying to do in his *Treatise*. Aleni, for example, also adopted *qi* in this sense as both a physical and psychic *spiritus*.[40]

Rather than building blocks of the universe, Chinese literati perceived in the five phases evidence for the successive changes in all things. Rather than the lifeless element air (= ether) enunciated by Vagnoni, Chinese perceived in *qi* a more fundamental material (*zhi*), vital, and spiritual (*shen*) unity, which pervaded all things in the cosmos and undergirded all space-time change of yin and yang. While inconclusive in the seventeenth century, Vagnoni's fascinating opposition of the four elements and the five phases was replayed in the late nineteenth century when Protestant missionaries presented the modern concept of a chemical element in their translations of chemistry.[41]

Nor should we assume that the Jesuit refraction of the five phases into the four elements was completely antithetical to Ming literati. In a discussion titled "Views on the Four and Five Phases" ("Sixing wuxing shuo"), Fang Yizhi presented an ambiguous account in which he explained that even in Chinese classical learning there had not been unanimity on whether there were four or five, or six, phases. Fang noted that Shao Yong (1011–1077), for example, had spoken of water, fire, earth, and stone as four phases, thus leaving out both metal and wood. Fang also pointed to the tradition that defined *qi* as a single continuum, which was described in terms of two phases, namely yin and yang.

Fang Yizhi concluded from this survey that there was no set way to talk about reality. Four or five phases were equally plausible conventions as were many other arrangements. Fang did not give ground to the Jesuits, however, in his insistence that even when given material form, the phases were describable in light of the duality of yin and yang and not as pure elements that composed reality. Fang maintained that water and fire were the key transformative agents of *qi* that made it increase or decrease in amount. Movement was due to the action of fire, and congealing was the result of water.[42]

Other literati in China had talked about the "four greats" (*sida*) of earth, water, fire, and air, which were more comparable to the Jesuit's four elements. The Chinese had dealt with the idea of four elements this way in medieval times when Buddhism first introduced to China the *mahābhūta* (earth, water, fire, and air) as the substances of all material. In addition, Muslims had introduced the original Aristotelian four elements to China in Islamic texts translated into Chinese from the Arabic and Persian during the Mongol era.[43]

On the other hand, there were late Ming literati who thought that *qi*, when it took physical shape through congealing, made up everything between heaven and earth. *Qi* could be explained as both organized activity or energy and the material structures that embodied it. In this view, the five phases actually referred to qualities of *qi* that made up all substances—another side of their denotation as five sorts of *qi* as fundamental processes.

For Song Yingxing (b. ca. 1600), who was interested in traditional technology and agriculture, two of the five phases instead become basic ingredients of matter but not elements.[44]

Song's chief works were published in 1636–1637, and his magnum opus, *Heavenly Crafts Revealing the Uses of Things* (*Tiangong kaiwu*) was compiled as a practical compendium in the 1630s, with a preface dated to 1637. It echoed the Ming daily use compendia, which we have discussed earlier. The *Heavenly Crafts* had one important difference with the Ming compendia by Li Shizhen and Hu Wenhuan, however, namely that it was compiled in an era when Jesuits were prominent in literati and Ming court circles in Beijing and the Yangzi delta.

Song's work made no mention of European agricultural and hydraulic techniques that were introduced in works by the Jesuits and their converts, such as the *Western Techniques of Hydraulics* and Xu Guangqi's *Complete Works on Agrarian Management,* although he did make some mention of European weaponry. Song's delineation of eighteen categories of traditional agrarian and craft production documented native success in all these areas. He excluded materials he had prepared on subjects that were often included in other daily use digests, such as celestial observations and mathematical harmonics, because they were not pertinent to techniques of production.[45]

The Jesuits and Mappa Mundi in Ming China

Because of their travels, the Jesuits could also claim expertise in world geography. Their charts of the globe were based on the New World discoveries of the Portuguese and Spanish explorers they often accompanied. The *mappa mundi* that Ricci, his collaborators, and others produced for literati in the late sixteenth century caused an uproar because the Ming imperial system was based on a cosmological assertion of its geographical centrality in the world, which the Jesuit maps seemed at first to challenge. Some literati, for example, resisted the Jesuit claim that the earth was spherical, because this suggested that the "vaulted heavens" (*gaitian*) was a sphere as well. Many literati considered the earth flat and the sky a finite vault overhead.[46]

Native attempts at serious cartography and descriptive geography provided Chinese with a foundation that cumulatively helped bridge the gap between the orthodox symbolic geography popular since the Song period and the Jesuit techniques that stimulated the emergence of more rigorous cartography in the seventeenth century. Chinese achievements in geography climaxed during the Yuan and late Ming dynasties with the production of schematic grid maps, sailing charts, and relief maps. Two extant maps that demonstrate the remarkable development of mapmaking in China were carved on stone in the twelfth century. Informed by Tang models, the maps were titled the *Map of China and Foreign Areas* (*Huayi tu*) and *Map of the Tracks of Emperor Yu* (*Yuji tu*). The latter, carved in 1137 as a grid based on a scale of approximately 1: 1.5 million, is described by Needham as "the most remarkable cartographic work of its age in any culture" (see Maps 3.1 and 3.2).[47]

The medieval cartographer Pei Xiu (224–271) first used the method of indicating distances by a rectangular grid system. Since then, cartography in China was often based on the grid tradition, although many important maps still followed the *Map of China and Foreign Areas* model and were not ren-

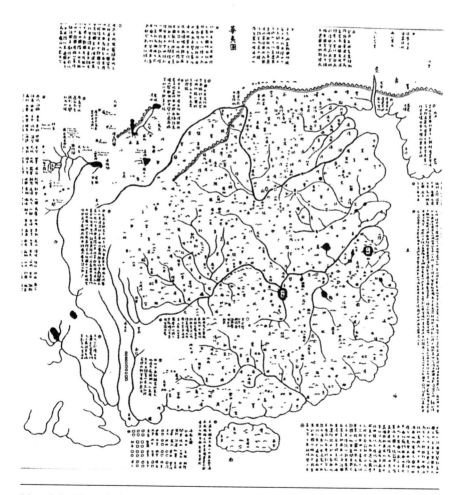

Map 3.1. Map of China and Foreign Areas (*Huayi tu*)

Source: Reprinted from *Science and Civilisation in China*, vol. 3, by Joseph Needham, with the permission of the publishers, Cambridge University Press.

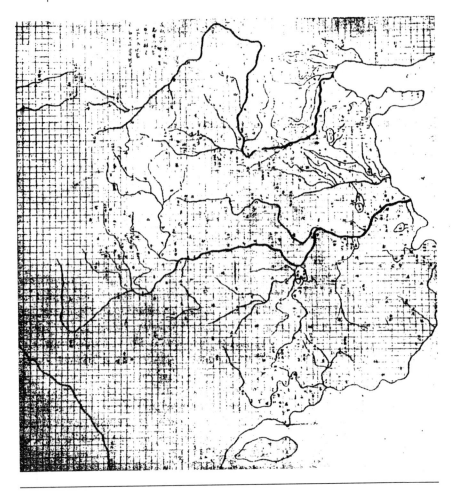

Map 3.2. Map of the Tracks of Emperor Yu (*Yuji tu*)

Source: Reprinted from *Science and Civilisation in China,* vol. 3, by Joseph Needham, with the permission of the publishers, Cambridge University Press.

dered using a grid. Jia Dan (730–805), the greatest Tang cartographer, constructed a map of Chinese and foreign regions for the emperor. The map, now lost, was thirty feet long, thirty-three feet high, and used a grid scale of one inch for one hundred Chinese miles, i.e., equal to about thirty-three English miles, the same scale used in the *Map of the Tracks of Emperor Yu.* Jia Dan's map likely depicted all of known Asia because of its huge scale.[48]

Zhu Siben (1273–1337) inherited this grid format. Perhaps the best known Chinese cartographer until the twentieth century, Zhu used grid-maps to summarize the large body of new geographical information that the

Mongol conquests in Asia had added to the earlier fund possessed by Tang and Song geographers. Zhu's map (*Yutu*), circa 1320, became known in Europe as the "Mongol Atlas of China." An early Ming map of about 1390 replicated the imperial pretensions of Mongol maps, asserting the empire's sway from Central Asia and Japan to the Atlantic via southeast Asia and the Indian Ocean world.[49]

In the sixteenth century, Luo Hongxian (1504–1564) discovered a manuscript copy of Zhu Siben's map of the known world, which had been prepared between 1311 and 1320 but never printed. Lo revised and enlarged Zhu's grid format in 1541. He also added new information, some derived from the early Ming naval explorations led by Admiral Zheng He. Luo's map (*Guang yutu,* literally, "Enlargement of a map of the earth") printed about 1555, included clear depictions of the Cape of Good Hope and southern Africa, suggesting that some of Zheng's ships had traversed from the Indian to the Atlantic Ocean and back some seventy years before Vasco da Gama (ca. 1469–1524) did it in reverse.[50]

Cartography and Ming Military Defense

As the early Mawangdui maps reveal, interest in cartography and geography in China frequently was tied to problems of border or maritime defense. Luo Hongxian became interested in geography because of Japanese pirate raids on the maritime provinces in the Yangzi delta and along the southeast coast in the sixteenth century, which had been aided by Chinese accomplices. The Ming government urgently needed maps and geographical advisors to cope with the situation. Luo spent three years collecting geographical materials for military defense and in the process discovered the manuscript of Zhu Siben's gridmap.

Zheng Ruozeng (1505–1580), a native of Suzhou prefecture, was also involved in coastal defense because of the Japanese pirate threat. In the 1540s, Zheng compiled a strategic atlas of China's coastal region extending north from the Liaodong Peninsula to southern Guangdong. Maps and strategic information were included. More importantly, Zheng Ruozeng compiled a work on coastal geography under the auspices of Zhejiang Governor Hu Zongxian (1511–1565). First published in 1562, Zheng's *A Maritime Survey: A Compilation of Plans* (*Chouhai tubian*) was modeled on Luo Hongxian's earlier work.

The *Maritime Survey* was an ambitious presentation of geographical minutiae about the China coast, containing accurate accounts of China's neighbors, Japan and Korea. Hence, it was not simply an atlas. A major contribution to Chinese historical geography, the *Maritime Survey* marked a major turning point in Chinese geographic studies. Its completion arguably provided the impetus for later literati to incorporate Jesuit information into the Chinese tradition of geography and cartography.[51]

In the past, threats to China's security had mainly come from the north and northwest, e.g., the Mongols. Hence, earlier geographers focused attention on northern frontier areas in their geographical accounts. In the sixteenth century, however, the primary military threat was along the South China Sea coastline. Zheng Ruozeng's research on the menace posed by Japanese pirates and their Chinese collaborators in the southeast thus stimulated a shift in geographically inspired research. A number of sixteenth-century geographical treatises on maritime defense were either inspired by or followed the pattern of the *Maritime Survey*.[52]

Ming Knowledge of Foreign Countries

Geographers and cartographers in Ming China were fairly knowledgeable about the southern regions (*Nanyang*, i.e., what is now called southeast Asia) and countries in the Indian Ocean and Arabian Peninsula (*Xiyang fan'guo*, i.e. "Western tributary states"). Information of this type peaked in China after the early fifteenth-century voyages of the Ming fleet. Zheng He, the eunuch commander of seven Ming expeditionary fleets launched between 1405–1433, reached at one time or another Sumatra, Java, Ceylon, Hormuz, Aden, and eastern Africa.

The expeditions overlapped with the extent of the Ming maritime tributary system around which trade and diplomatic relations were organized. Moreover, Ma Huan left a descriptive account of Zheng's fleets in 1433, which was based on his service as Muslim translator for three of the seven expeditions (1413–1415, 1421–1422, 1431–1433). Ma's *Captivating Views of the Ocean's Shores* (*Yingyai shenglan*) described the countries the fleet had visited and portrayed the customs, religions, and lifestyle in each place. He also depicted topography, geology and wildlife.[53]

Subsequently, other members of Zheng He's expeditions described their adventures. For example, Fei Xin's (1388–1436?) *Captivating Views from a Star Guided Vessel* (*Xingcha shenglan*) and Gong Zhen's *Gazetteer of the Western Tributary States* (*Xiyang fan'guo zhi*, 1434) were based on firsthand visits and help fill the gap in Indian Ocean accounts between Marco Polo's *The Travels* and Portuguese reports in the late fifteenth century. References to the expeditions and the exotic gifts brought back to China were added by 1462 to Cao Zhao's circa 1387–1388 inventory of artifacts discussed in Chapter 1.[54]

In the early eighteenth century, imperial compilers of the *Synthesis of Books and Illustrations Past and Present,* the largest encyclopedia in Chinese history, subsequently included excerpts and extracts from such Ming accounts. The Qianlong Imperial Library also included a 1520 annotation of Ma Huan's account by Zhang Sheng (1442–1517). In addition, the Imperial Library summarized the content of Gong Zhen's work and mentioned Huang Shengzeng's 1520 *Records of Western Tributaries* (*Xiyang chaogong dianlu*), although neither was copied into the collection. Huang's work drew on and cited earlier sources.[55]

Within China, Ming explorers such as Wang Shixing (1547–1598), Xie Zhaozhi (1567–1624), and Xu Hongzu (also known as Xu Xiake, 1586–1641) traveled to the most remote borders of the empire. In their travels, they prepared notebooks outlining in descriptive terms the river systems, topography, and cultural aspects of the places they visited. Xu Hongzu, for example, discovered the main source of the West River in his travels through much of Southwest China and was able to determine that the Mekong and Salween were different river systems.[56]

Matteo Ricci's *Mappa Mundi*

Jesuits added to the geographical knowledge that Ming literati already had in the late sixteenth century. Produced with the help of Chinese converts, the first edition of Matteo Ricci's *mappa mundi*, titled *Complete Map of the Earth's Mountains and Seas* (*Yudi shanhai quantu*), for instance, was printed in 1584. A flattened-sphere projection with parallel latitudes and curving longitudes, Ricci's world map went through eight editions between 1584 and 1608. The third edition was titled the *Complete Map of the Myriad Countries on the Earth* (*Kunyu wanguo quantu*) and printed in 1602 with the help of Li Zhizao (see Map 3.3). Li studied European mathematics and astronomy, in addition to geography, after meeting Ricci in Peking in 1601.[57]

Ricci's description of the form and size of the earth obliged many seventeenth-century Chinese literati to revise their views of the world. For the first time, Chinese became aware of the exact location of Europe in relation to their own country. In addition, Ricci's maps contained technical lessons for Chinese geographers: (1) Ricci taught Chinese cartographers to localize places by means of circles of latitude and longitude; (2) he invented many geographical terms and names, including Chinese terms for Europe, Asia, America, and Africa; (3) his maps transmitted to China the most recent discoveries made by European explorers; (4) he described the existence of five terrestrial continents surrounded by large oceans; (5) the maps introduced the sphericity of the earth; (6) finally, he spoke of five geographical zones and their location on the earth, i.e., the Arctic and Antarctic circles, and the temperate and tropical zones.[58]

Map 3.3. Complete Map of the Myriad Countries on the Earth (*Kunyu wanguo quantu*)

Source: http://geog.hkbu.edu.hk/GEOG1150/Chinese/Catalog/ Catalog_main_11.htm.

An indication of the initial impact of Ricci's maps, especially the 1608 edition printed in Beijing, was its inclusion in geographical works produced by late Ming literati scholars. For example, Zhang Huang (1527–1608), who had met Ricci in Jiangxi in 1595, added Ricci's *Complete Map* to his own massive illustrated collection, which Zhang called the *Compendium of Maps and Writings* (*Tushu bian,* 1613). In addition to Ricci's 1584 map of the world, Zhang also included European depictions of the northern and southern hemispheres along with traditional maps of the four seas.[59]

The only known copy of the first Chinese map of the world produced in 1593 was titled the *Comprehensive Map of Heaven and Earth and the Myriad Countries and Ancient and Modern Persons and Artifacts* (*Qiankun wanguo quantu gujin renwu shiji*), which included geographical information

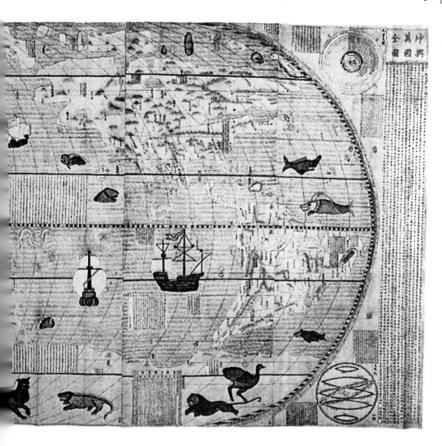

from the Jesuits. It was based on Ricci's first world map of 1584, which is now lost. The map offered a traditional representation of China with foreign lands arranged around the periphery. It served chiefly as an administrative map for officials and thus included statistical information such as population (conventionally based on families) and locally produced commodities.

Topological rather than topographical, the 1593 Chinese version fit European lands in along its edges without affecting traditional cartography. The New World was shown as a series of small islands surrounding China. Methodologically, however, the compiler stressed the "achievements of investigating things and extending knowledge," which by now served as a native trope for the accumulation of knowledge. The linkage of the world map of 1593 to such studies heralded the application of native terminology to European *scientia*.[60]

The **1602** *Complete Map of the Myriad Countries on the Earth* corresponded to one of the first issues of Ricci's third world map and is the earliest version that survives. This 1602 edition followed the *Typus Orbis Terrarum* by Abraham Ortelius (1527–1598), a Flemish scholar and geographer, which was first published in a 1570 European atlas called the *Theatrum Orbis Terrarum*, which was itself based on Gerardus Mercator's (1512–1594) prominent 1569 world map. The chief alteration for the Chinese version was that Ming China was placed at the center of the map to appeal to Chinese dynastic sensibilities. The New World was located on the eastern borders. A fourth edition of the map was prepared in 1604.[61]

Longobardo and Manuel Diaz the elder (1559–1639) continued Ricci's mapmaking, in 1623 producing a lacquered wooden globe, which updated Ricci's map and also stressed the sphericity of the earth. Aleni and Yang Tingyun quickly followed with their *Account of Countries not Listed in the Records Office*, which was later included in the 1628 *First Collection of Celestial Studies* compiled by Li Zhizao. Aleni's *Account* grew out of notes on Ricci's 1601 *mappa mundi* that Pantoja and Ursis had prepared for the Ming court.[62]

Aleni's translation represented the first detailed exposition in Chinese of a world geography that drew on Renaissance traditions of local lore and Ptolemaic cosmography to describe Asia, Europe, Africa, and the Americas. Another section focused on the oceans. In addition to the *First Collection of Celestial Studies*, other Chinese collectanea reproduced Aleni's treatise in the eighteenth century, thus making it more influential during the Qing dynasty than Ricci's maps, which were quickly forgotten by literati. In the nineteenth century, Aleni's work was reprinted three times in sixty years after world geography became de rigueur for literati trying to understand the military consequences of the Opium War (1839–1842).[63]

A world map composed by the Neapolitan Jesuit Father Francesco Sambiasi (1582–1649) in 1648 was designed as simplified version of the large maps by Ricci. It was drawn as an oval projection with China at the center. Later, Michael Piotyr Boym (1612–1659), a Polish Jesuit who had served the embattled Southern Ming in the 1650s, produced the *Map of the Middle Kingdom* (*Zhongguo tu*), circa 1652, which he took back to Rome in 1656. The original manuscript conveyed Jesuit knowledge of China to Europe. This version was superseded by Martinus Martini's (1614–1661) *Novus Atlas Sinensis* published in an Amsterdam atlas series in 1655.[64]

Although Matteo Ricci introduced the system of longitude and latitude to Ming China, the grid system employed by Luo Hongxian and his predecessors still exercised a dominant influence on Chinese cartography throughout the late Ming and Qing periods. The Chinese grid still conceived maps and texts as integral rather than independent. For example, an admired genre known as "Complete Maps of All under Heaven" (*Tianxia quantu*) was initiated by the Ming loyalist Huang Zongxi (1610–1695) in 1673 and continued by a number of talented literati scholars interested in mapmaking. Cao Junyi's ambitious "Complete Map" of 1644, for instance, mixed inaccurate classical geographical lore with precise recognition of Europe, Africa, India, and central Asia, providing longitudes and latitudes for estimating their distances from Nanjing, the Ming southern capital.[65]

In the early Qing, Verbiest, with the help of others, produced two works that dealt with world geography. His 1674 world atlas included a comprehensive map in two hemispheres with gazetteer information about each part of the globe. It was based largely on Aleni's *Account of Countries not listed in the Records Office*. Similarly, Verbiest's *Essential Records about the West* drew on the topical organization in Aleni's 1637 *Answers About the West* (*Xifang dawen*), which compared China to Europe in light of geographical lore. The compilers of the Imperial Library catalog during the Qianlong reign considered these works important enough to copy them into the collection in the 1780s.[66]

The final stage in the development of traditional Chinese mapmaking came between 1708 and 1718 when French Jesuits on behalf of the Kangxi emperor systematically surveyed the entire Manchu realm. They drew up a series of maps of the Qing empire and its border areas, which became known as the 1718 Kangxi Atlas (*Huangyu quanlan tu*). Along with succeeding maps in the Yongzheng and Qianlong periods, the Kangxi Atlas surpassed Jesuit surveys completed earlier. The Manchu court restricted access to and local reproduction of these surveys and maps. Nevertheless, when brought to Europe the atlas remained the West's chief source of China's geography until the twentieth century.[67]

Rather than switching to Jesuit cartography, late literati such as Ma Junliang sometimes presented both a traditional picture of a China-centered world and global maps that loosely duplicated Ricci's *mappa mundi*. Ma's *Capital Edition of a Complete Map Based on Astronomy* (*Jingban tianwen quantu*; see Map 3.4) circulated widely in the 1790s. It included two smaller global maps of the Euro-Asian-African and Pacific hemispheres above his detailed map of the Qing empire, which roughly corresponded to Ricci's *Map of the Myriad Countries on the Earth* rendered in a late Ming encyclopedia.[68]

Despite displaying substantial variations, a prominent feature of the "Complete Maps" was the mingling of traditional and newer techniques, such as the hybrid overlap between the grid and latitude-longitudinal approach for large-scale land maps. Indeed, Li Zhaoluo's (1769–1841) famous 1832 atlas of the Qing empire showed grid as well as latitude and longitude lines on the same map, indicating the reluctance of native geographers to give up the traditional system long after the Jesuit *mappa mundi* were introduced. This

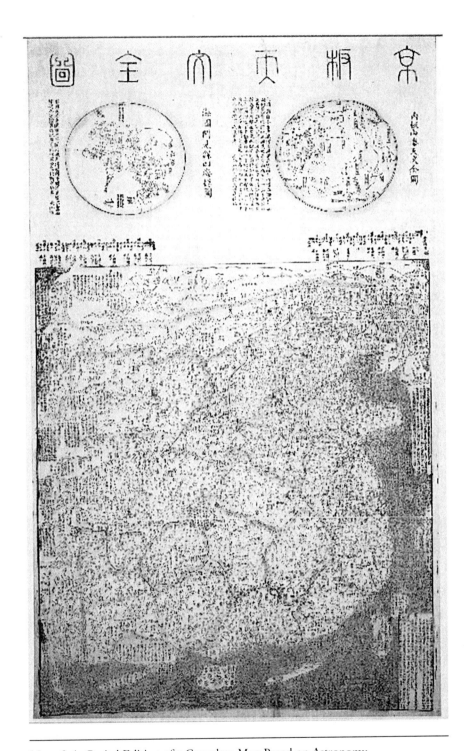

Map. 3.4. Capital Edition of a Complete Map Based on Astronomy
(*Jingban tianwen quantu*)

Source: The Woodson Research Center of the Fondren Library, Rice University.

nativist pattern for domesticating European learning was repeated in mathematical astronomy during the late eighteenth century.[69]

Literati Attacks on Calendar Reform in the Early Qing

Scholarly debates between literati and Jesuits intensified during the early Qing dynasty, particularly after 1660 when the issues of Christian creationism and biblical chronology challenged Chinese assertions of their cosmic centrality. In addition, the fundamental differences between the orthodox study of principles among literati and Jesuit views of how God had created the world and endowed humans with a eternal soul became political flashpoints. In the 1660s, the Manchu court's brief tolerance of Jesuit religious beliefs was severely tested when Chinese, Muslims, and Manchus attacked Adam Schall and his staff in the Astro-calendric Bureau as enemies of the dynasty and classical civilization.

Although the Jesuit mission survived the 1664–1667 purges, from then on the court carefully scrutinized its missionary impulses. As long as the Jesuits provided the court with needed expertise in mathematical astronomy and geographical matters, the emperor kept them on, but increasingly the needs of the Qing government, rather than the goals of the Church, defined their activities. The Jesuits could no longer protect the conversion projects of their local priests. As minions of the Qing dynasty, the Jesuits would pay dearly after 1700 when Jansenists, Dominicans, and Franciscans in Rome and Paris and the Kangxi emperor in Beijing challenged their accommodation policy. In 1704, for example, Pope Clement XI issued an ordinance prohibiting the use of the Chinese terms for "heaven" (*tian*) and "the lord on high" (*shangdi*) to designate God.[70]

Late Ming concerns about the accuracy of the official calendar forced literati to evaluate and apply specific European techniques to reform the Ming calendar. Given the interest early Manchu emperors exhibited in astronomy, we might expect that topics dealing with the calendar, as in the Ming, to carry over to the civil examinations. Because the Jesuits had penetrated the Astro-calendric Bureau, however, the throne preferred to limit this potentially volatile area of expertise to Manchu bannermen, princes, and select Chinese literati. The contemporary calendrical debates between Jesuits and officials, which challenged the Yuan-Ming calendrical system during the Ming-Qing transition, gave the court pause about adding possibly divisive calendrical questions to civil examinations. We have seen in Chapter 2 that Ming emperors had permitted such debate after 1525. In contrast, the Manchu throne proclaimed its legitimacy by resolving the Ming calendar crisis. For self-legitimation, the Qing court presented a new calendar that like its Mongol predecessor would link empire and time far into the future.[71]

Unlike Ricci and his late Ming Italian and Portuguese cohorts, German, Belgian, and French Jesuits who came to China during the rise of the Manchus to power, such as Adam Schall, Ferdinand Verbiest, and Joachim Bouvet (1656–1730) focused on the imperial courts of the Shunzhi (r. 1644–1662) and Kangxi emperors in Beijing to promote Catholic interests. They were there at a time when Cheng-Zhu orthodox learning, which Ricci and his cohort had vigorously criticized as atheistic, again flourished as orthodox under imperial ideologists such as Li Guangdi (1642–1718). Many early Qing literati now blamed Wang Yangming and his late Ming followers, not Cheng-Zhu learning, for promoting empty scholarship that had helped bring the Ming to its knees.[72]

From Wu Mingxuan to Yang Guangxian: The Chinese-Muslim Computus

Beginning in 1657, Adam Schall, director of the reconstituted Qing Astro-calendric Bureau since 1644, eliminated the Muslim section of the bureau, although he maintained the other three sections for astrological interpretations, timekeeping, and calendrical matters. His appointment meant the Jesuits and their converts would take charge of the calendar, but it also meant that they would deal with the astrological applications of the calendar to determine auspicious days and choose propitious burial sites for the imperial family. Their technical prowess now extended into thornier aspects of Chinese culture and Manchu imperial politics, which drew on the calendar as a marker of dynastic legitimacy.[73]

Functionaries in the Muslim section had long been jealous of their Jesuit counterparts, but they were doubly upset when Schall was installed as bureau director and they lost their political influence in 1645. Because their eclipse predictions appeared in some cases to be less accurate than those of Schall using recently improved Ptolemaic methods, the influence of Muslims in the bureau now waned. The elimination of the Muslim bureau, filled with hereditary astronomers since the Yuan dynasty, was intolerable to Muslims.

Wu Mingxuan, one of the Muslim hereditary functionaries, attacked Schall for allegedly making faulty predictions. When the charges proved wrong, Wu was jailed for false accusations. Upon his release from prison, Wu Mingxuan and a Han Chinese collaborator, Yang Guangxian (1597–1669), together pursued an anti-Jesuit campaign.

In the late Ming, Yang had submitted two memorials accusing officials of corruption and incompetence. One official had called for abolition of the civil service examinations, and the other was a grand secretary and an imperial favorite. Yang flamboyantly took his coffin with him to Beijing in case his charges were not sustained. Yang's life was spared even though the charges against the grand secretary proved unfounded, but the Ming court had Yang flogged and banished to the northeast. When the Ming fell in 1644, Yang was freed. By 1659 he was again in Beijing, when he joined Wu Mingxuan to dislodge the Jesuits from their position of power and influence.[74]

Beginning in 1659, Yang wrote several works attacking the Christians for their beliefs and assailing Schall for serious mistakes in the Western-inspired Qing calendar. Like many literati since Zhu Xi, Yang distrusted astronomical officials because their computational expertise was not tied to an understanding of the metaphysical workings of the cosmos. Unschooled in either European computational techniques or traditional mathematical astronomy, Yang colluded with the Muslim astronomer Wu Mingxuan to craft his charges.

Yang appropriated methods he gleaned from the Classics and early dynastic histories, to which he added Muslim computations, to counter Ming techniques for intercalations and fortnightly periods that the Jesuits in the Astro-calendric Bureau continued. Jesuit methods for computing intercalations and fortnightly periods were not new. The Jesuits had not been very attentive to the day-to-day positions of these luminaries. Nor had their Tychonic cosmology fundamentally changed the Ming bureau's computational procedures. They had only imposed revised Ptolemaic techniques for certain solar, lunar, and planetary phenomena.[75]

Yang Guangxian's first accusation against the Jesuits came in a 1660 memorial, which included his treatise "On Choosing an Auspicious Date." In it, he exposed an alleged error in the Jesuit calendar because the latter contained three fortnightly periods in the twelfth month of 1661, which was impossible by traditional Chinese methods. Yang's charges were based on the different methods of calculation employed by each side to order the calendar.

The antiquarian astronomy that literati such as Yang Guangxian learned divided the tropical year into twenty-four equal periods of a little over fifteen days each. Since calendar months were either twenty-nine or thirty days long, none could wholly contain two of these so-called "solar fortnights." Usually parts of three overlapped, a point about which Yang was very confused. The Jesuits used instead the Yuan-Ming system of twenty-four equal divisions of the sun's apparent path on the ecliptic. Because the sun's speed varied through the seasons, these apparent fortnights differed noticeably in length from the average. It was possible for two complete fortnights, and parts of one or even two more, to fit within a month. This fine point did not matter, however. The receiving clerk rejected Yang's memorials outright because as a commoner he did have the privilege to submit them.[76]

Yang's second attack began in early 1661. He also accused the Jesuits of a serious error in adding an intercalary month to that year. Without access to the official computational system they had used, Yang relied on Wu Mingxuan's

Muslim technique to charge peremptorily that the Jesuits had selected the wrong month. The point was not purely technical. Like Yang's other charges, this one bore on many issues that figured in court debates such as the correct performance of ritual, siting of buildings and tombs, and astrology (especially hemerolgy—the choice of auspicious days). Not only the technical work of the Jesuit-controlled Astro-calendrical Bureau, but the cultural and political value of its products, was at stake.[77]

Overturning the Jesuit Calendar in 1664

In spite of Yang Guangxian's and Wu Mingxuan's efforts, Schall and the Jesuits continued to enjoy the support of the Qing court until 1664. Moreover, the Jesuits successfully accused Yang of intrigues against the dynasty related to the just-started *Ming History* project. The charge was that a preface to Yang's anti-Christian essays by a relative referred to the Ming dynasty by a phrase that betrayed loyalty to it. Yang made a confession of error before the Ministry of Punishments.

Yang Guangxian quickly struck back. Oboi (d. 1669) and the other Manchu regents in charge of the government while the Kangxi emperor was too young to rule received Yang Guangxian's third memorial. The court finally called for an investigation of Yang's charges that the Jesuits were plotting rebellion and spreading heterodox doctrines among the people. Yang was aware of the success the Jesuits in the field were having converting among the common folk. Because Yang tied some recent inauspicious deaths in the imperial family to Schall's divination to determine burial sites, the charges now caused a sensation in the imperial court.[78]

Yang began by challenging the date chosen for the burial of an infant prince in 1658. The Shunzhi emperor's concubine, who became the imperial honored consort in 1656, gave birth to a son on November 12, 1657, but he died unexpectedly after three months. Schall, in charge of the Astro-calendric Bureau, was ordered to select an auspicious date for burial and to report it to the Bureau of Sacrifices in the Ministry of Rites. The Chinese supervisor of the timekeeping section of the Astro-calendric Bureau determined a date based on the "Great Plan's Five Phases" (*Hongfan wuxing*) tradition, which emphasized the siting of the tomb rather than the date of birth of the prince. The Ministry of Rites then translated the time into Manchu and reported it to the emperor, but a Manchu clerk apparently mistranslated the correct time of the fifth double hour (7–9 A.M.) as the seventh double hour (11 A.M.–1 P.M.).[79]

The timekeeping section protested this error, but the date had already been forwarded to the emperor, and the prince was buried at the wrong time. Because the imperial house considered it inauspicious, the supervisor of the timekeeping section attacked the Ministry of Rites. In 1658, those officials responsible for the error were tried and dismissed from office, barely escaping the death penalty. Others were beaten. Schall and the Jesuits were not implicated. When the Shunzhi emperor and his empress died of smallpox in quick succession in 1661, however, his mother concocted a series of complaints alleging that the dead emperor had been critical of Han Chinese involvement in the government. Under the succeeding Kangxi emperor, his Manchu regents gradually diminished Schall's influence.[80]

Using his own treatise "On Choosing an Auspicious Date" as a guide for deliberations on hemerology, Yang Guangxian's renewed attack on Schall in 1664 charged that the year, month, day, and time had all been inauspicious for the burial of the young prince. Yang contended that the date should have been chosen on the basis of the prince's birth date by following the "Orthodox Five Phases" (*zheng wuxing*) method. He attacked as heterodox the "Great Plan's Five Phases" routine for siting the tomb that the bureau had used. In effect, Yang was accusing Schall of purposely selecting an inauspicious day in 1658 for the burial of the prince so as to cast a deadly spell on the emperor and empress as the parents of the deceased child.[81]

The political background for Yang Guangxian's use of hemerology and other esoteric techniques in his anti-Christian attacks makes it clear that the responsibility of the Astro-calendric Bureau for determining auspicious days was not lifted under the Jesuits, much to the dissatisfaction of those Jesuits in China who were opposed to astrology and hoped to Christianize imperial time. Longobardo, for example, had explained earthquakes in his 1626 "Interpretation of Earthquakes" as a warning from God when evil people outnumbered the good. Although Schall claimed ignorance of Chinese prognostication methods, as the bureau director he had no choice about supervising their use. When other missionaries attacked him for allowing their practice, Schall claimed that he was unifying the methods so that everyone would adhere to the same hemerological signs.[82]

In 1645, for instance, Schall had gone through the motions of submitting a report about the "waiting for the *qi*" (*houqi*) procedures used by the bureau to determine the "onset of spring" fortnightly period, whose precise timing was one of its charges. Waiting for the *qi* was a technique for measuring the earth's *qi* emanations. By using a standard set of musical pitch pipes of graded lengths, whose ratios were used for weights and measures, the onset of the fortnightly periods could be measured. The procedures involved burying the set of pipes in a sealed chamber after filling them with light ashes. The ancients had believed that when the sun entered the second fortnightly period of any month, the earth's *qi*, a seminal force, would rise and expel the ashes from the appropriate pipe.[83]

Ancient debates had discussed the correct location for the experiment, the quality of the soil, the selection of materials, and the placement of the pitch pipes. By the mid-Ming, the repeated experimental failures of this technique had led to skepticism. At the end of the Ming, Zhu Zaiyu, who had also made recommendations for calendar reform, published his *Essentials of Musical Harmonics* (*Lülü jingyi*, 1596), in which he rejected the method as an incorrect elaboration of the methodology for investigating things and exhaustively mastering principles to extend knowledge.[84]

Since the ritual of waiting for the *qi* was a bureau responsibility, Schall could not prevent its becoming one of the principal issues in his confrontation with Yang Guangxian over its efficacy. When Yang brought his 1664 suit against the Jesuits, for example, the Jesuit attitude toward waiting for the *qi* became a focus of Yang's attacks. Yang accused Schall of relying on his own calculations and ignoring the old system to determine the onset of spring. When Schall replied that his predecessors in the Ming Astro-calendric Bureau had abandoned it, however, the court noted that he had earlier used it to ascertain the onset of spring. The Oboi regents were unhappy with Schall's dissimulation, which made it uncertain that the timing of the onset of spring had really been verified with the old system.[85]

Similarly, Schall was in charge of the bureau when the Dalai Lama planned to visit Beijing in 1652. Prince Dorgon (1612–1650), one of the Manchu court ministers, initially had wanted the Shunzhi emperor to meet the Dalai Lama at the border of the capital region and escort him to Beijing. Seeking to supplant the political and religious alliance between the Mongols and the Tibetans, Han metropolitan officials, however, were opposed to this arrangement.

When the emperor announced he would go, some of the grand secretaries appealed to two astrological portents from the Astro-calendric Bureau to dissuade him: (1) the planet Venus rivaled the brightness of the sun on the day before; (2) a threatening meteor was sighted in the region of the polestar, which corresponded to the imperial court. Because of these signs, they argued, the emperor should send a high official to escort the Dalai Lama. This astrological intervention was successful, and the emperor cancelled his journey. Qing-Tibetan relations were not cemented until the 1690s when Kangxi emperor personally led military campaigns against the Zunghars, challenging their influence in Tibet.[86]

Because portents had political implications, Schall's bureau had obligingly supplied the grand secretaries with omens. Neither the daytime apparitions of Venus nor the meteor were rare phenomena. The rarer combination of the two within a day of each other was suspicious. Schall's goal seems to have been to lower public esteem for Lamaism in order to promote Christianity.

Before the Dalai Lama arrived, for example, Schall sent a memorial that sunspots (another common phenomenon) had appeared as an alert to the court that the Dalai Lama was obscuring the emperor's radiance. Schall had allied with the literati in the court who shared his reservations about the Dalai Lama's visit.[87]

Schall did not foresee that the intertwining of astronomy, waiting for the *qi* tradition, and yin-yang numerology in telling fortunes and determining auspicious days might ensnare the Jesuit mission in a battle where astronomical competence would not be the key issue. Nor did he anticipate that the previously supportive Manchu court would respond to charges that the Jesuits and their converts were teaching that the Chinese and other Asian peoples were lineal descendants of the Hebrews after the Great Deluge. These claims suggested that the Jesuits were plotting against the dynasty.

Noah and the Chinese

When Yang sent his 1664 memorial to the Ministry of Rites, he also attacked Schall's Chinese disciples in the Astro-calendric Bureau, such as Li Zubo, who had helped Schall translate a work on the telescope in 1626. Yang alleged that Li's short history of Christianity in China, titled *Overview of the Transmission of Heavenly Studies* (*Tianxue chuan gai*), was part of a missionary plot to foment revolt. The preface by the censor Xu Zhijian, a 1655 palace graduate, also implicated Xu in the charges against Li.[88]

Yang thus tied his accusations against Li Zubo and Xu Zhijian to Schall's fixing of an inauspicious day in 1658 for the burial of the infant prince to cast spells on the emperor and empress. Yang argued against a notion of an anthropomorphic God by appealing to heaven as the origin of all things and cited "heaven's hexagram" (*qian*) in the *Change Classic* as proof. There was no need for another ruler over heaven, Yang maintained, and he accused Christianity of being a by-product of Buddhism. Because Li had also linked Fu Xi, the reputed ancient founder of Chinese writing and the first sage-king of Chinese history early in the third millennium B.C., to the history of Judea in the Bible, Yang concluded that Li's 1664 pamphlet on Christianity was heretical. Yang also attacked the absurdity of Christian beliefs in the virgin birth.[89]

Jesuit-Chinese cooperation in mathematical astronomy was an exceptional accommodation project, which had obscured a larger clash between European universal time and history (told in light of Biblical chronology) and the likewise universal cyclical chronology of Chinese ancient texts and historical accounts that might have preceded the Great Deluge associated with Noah. The chronologies based on the Hebrew Masoretic texts, for instance, dated the flood to a time later than the dates of the early Chinese sage-kings. Archbishop James Ussher's (1581–1656) famous chronology published in 1650–1654 gave 4004 B.C. as date of creation. By this reckoning the deluge occurred in 2348 B.C., or 1,656 years later, and thus far later than the reigns of Fu Xi and other early Chinese sage-kings for whom the Chinese had continuous records (although only Europeans were interested in the spurious dates worked out for these legendary monarchs).

The Latin Vulgate version of the Bible prepared in the fourth century had been based on Masoretic texts that were no longer extant. The Septuagint version, which was prior to the Latin Vulgate, had been based on a Greek version of the Old Testament from Alexandria. It dated the creation to 5200 B.C., with the flood in 2957 B.C. This earlier date for creation allowed Christians in China and Europe to reconcile Noah's universal patriarchy after the deluge with Chinese records. Nevertheless, the Catholic church refused to entertain an infinite, eternal, and possibly uncreated universe, a heterodox position that had cost Giordano Bruno (1548–1600) his life. Yang Guangxian, like most Chinese literati, believed in a creationless universe.

Because the chronology based on the Latin Vulgate caused problems for the Jesuits in China, they conveniently received papal permission to use the Greek Septuagint chronology instead, which allowed Chinese historical accounts to date after the universal floods. The need to accommodate Chinese history with the Bible also entailed that the Christians identify the Chinese sage-kings with Old Testament patriarchs. Once the chronologies were rectified, the Jesuits could then correlate ancient Chinese history with the Old Testament. The Europeans thus had to make room for China in their universal histories, but they did so on their own universalist terms of time.[90]

Michael Ryan has noted that the Europeans tended to understand other peoples and cultures in terms of biblical genealogies and timelines, thereby making new worlds into old worlds by reducing their uniqueness. Philippe Couplet's (1623–1693) chronological tables, for example, dated the origins of Chinese history to the Yellow Emperor in 2697 B.C. Curiously, the Chinese, with their own universalist pretensions, developed a similar strategy—with no dates at all—by asserting that the origins of Western learning were originally in China.

Martinus Martini's 1654 history of China challenged European chronologies by accepting the Fu Xi reign. Hence, Martini also preferred using the Septuagint chronology to place the Great Deluge in China before 3000 B.C., although he was skeptical that the Chinese flood in Yao's reign was the same as Noah's flood. John Webb, on the other hand, drew from Martini's account to argue that Emperor Yao was in fact Noah. Others chose Noah's son Sem as the first to get to China, while Athanasius Kircher (1601–1680) chose Cham.

Kircher had already read into the Egyptian hieroglyphs esoteric truths about the world and God. In his 1667 magnum opus on China, Kircher accepted Fu Xi as the originator of all writing and contended that the origins of Chinese pictographic writing were the same as Egyptian hieroglyphs. Webb in his heralded *An Historical Essay Endeavoring a Probability that the Language of the Empire of China Is the Primitive Language* (London, 1669) maintained that the Chinese written language was fundamentally continuous with the Adamic language before the ancient languages diverged from an ancient purity.[91]

Based on his studies under Adam Schall, Li Zubo also linked early Chinese history with biblical time in his *Overview*. The work, a collaborative effort involving several Jesuits working with Li, claimed, "The first Chinese really descended from the men of Judea who had come to the East from the West, and the Teaching of Heaven is therefore what they recalled. When they reproduced and reared their children and grandchildren, they taught their households the traditions of the family, and this is the time when this teaching came to China."[92]

Li and others thus implied that the Chinese Classics embodied the same primitive revelations as the Christian Bible. The biblical creation was described with traditional Chinese terminology, and heavenly learning was equated with biblical learning, which had been carried to China by the first descendants of Adam and Eve. According to the *Overview*, the founding Chinese sage, Fu Xi, was one of these descendants. Because of the Qin dynasty book-burning policy, these teachings had been lost.

The impetus for publishing Li's *Overview* in 1664 came from Yang Guangxian's 1660 *An Exposure of Heresy* (*Pixie lun*), which had attacked Schall as head of the Astro-calendric Bureau. Consequently, Li Zubo's translation was a defense of Schall, as well as a restatement of early Qing Jesuit accommodation tactics on time and the calendar. Such claims infuriated Yang, who immediately wrote his anti-Christian diatribe, titled *I Cannot Do Otherwise* (*Budeyi*). Yang argued at the outset, "Formerly Li Madou [Ricci] used the sacred Chinese Classics and commentaries of the sages as texts to adorn his heterodox teachings. Today [Li] Zubo cites the sacred Chinese Classics and commentaries of the sages as if they were passages from the scriptures of the heterodox teaching itself. Zubo's crime deserves the greater punishment."

Louis Buglio (1606–1682) wrote a rebuttal published in 1665, which led to his implication in Yang's charges against the Jesuits and their converts. Although the Jesuits eventually emerged victorious in 1667, the costs of this confrontation proved so high to the mission that they refrained from presenting this sort of biblical argument in favor of Christian time publicly in China thereafter.[93]

Yang Guangxian in Charge

As a result of Yang Guangxian's third memorial and its litany of charges, Schall, Verbiest, and their Chinese colleagues were imprisoned and condemned to death. All Jesuits in the provinces were to be arrested and forwarded to Beijing. Despite an amnesty for the Jesuits following an earthquake on April 16, 1665, a day after their sentencing, five of the Chinese Christian astronomers, including Li Zubo, were executed. The Kangxi emperor's grand empress dowager had favored the amnesty because of Schall's past services and his poor health after a stroke while in prison. Jesuit errors were pardoned but not those of Li Zubo and their Chinese collaborators.

After 1664, the Sino-Western calendar was still used, but it was again complemented by the Ming Grand Concordance Calendar. Moreover, the Muslim section of the Astro-calendric Bureau was restored, with Wu Mingxuan back in charge. Although Western methods usually remained superior in predicting solar eclipses by from fifteen to thirty minutes, this accuracy was not enough to rescue Schall, Verbiest, and Buglio from Yang's cultural and political charges. In 1665, the Kangxi regents appointed Yang Guangxian director of the Astro-calendric Bureau despite his efforts to decline the position. Five times he pleaded lack of knowledge of mathematical astronomy, though he was, he claimed, knowledgeable about "calendrical principles." The court, however, held him to this task, and Yang was forced to put his ideas, such as they were, into practice.[94]

With Wu Mingxuan as associate director, Yang had the bureau set up a traditional calendar based on the onset of spring using calculations to be verified by "waiting for the *qi*." They built a special chamber in front of Yang Guangxian's residence, but the "waiting" rituals failed. When the Kangxi emperor took power from his regents in 1667, he remained concerned about astronomical irregularities and increasingly became involved in mathematical matters himself. In 1668, he asked Verbiest, who after Schall's death in 1666 became the Jesuits' chief spokesman, to review the calendar that Yang and Wu had prepared.

In late 1668 Verbiest made a case that "using reed pipes and flying ashes" was no way to measure the twenty-four fortnightly periods of the year. Verbiest argued that the spring equinox was determined not by *qi* but the sun crossing the intersection of its orbit (the ecliptic) with the equator. According to Chinese convention, the equinox was not the beginning of spring but its midpoint. Astronomers defined the fortnight "Onset of Spring," which began the season, by its distance in space and time from the same intersection. Verbiest further asserted that the earth's *qi* (*diqi*) varied according to the sun's location so that rituals based on it could not yield consistent results.[95]

From December 27 to 29, 1668, the emperor demanded a prediction of the sun's position on the next day, which tested Yang Guangxian and the Muslim astronomers in the bureau who were allied with Yang against the Jesuits led by Verbiest. Given the methods that Yang Guangxian and Wu Mingxuan used, Verbiest was able to itemize a series of errors in the Muslim approach, verified before the emperor by observations. Additional tests were called for, however, when Yang and Wu denied the accuracy of Verbiest's findings.

Moving carefully as a young monarch in treacherous political waters, the Kangxi emperor began to study mathematics and astronomy under Verbiest because he realized that the cultural legitimacy of the new dynasty was tied to a successful ephemiredes. He also authorized the Jesuits to review the 1669 calendar prepared by Wu Mingxuan. Yang Guangxian and Wu continued to appeal to the cultural conventions of the Chinese calendrical tradition, while Verbiest and the Jesuits asserted that what the emperor needed was an accurate calendar, best provided by the new "Western" methods. Verbiest again demonstrated many errors in Wu's calendar, due to misplacing the intercalary month and the fortnightly periods.[96]

The emperor then convened a Council of Deliberative Officials to measure two fortnightly periods, "Onset of Spring" and the "Onset of the Rains." During these deliberations, the Jesuits also had to adjust their calculations. Verbiest, for example, changed his calculation of an intercalary month from the first to the second month of 1670, prompting the Kangxi emperor to correct the position of the intercalary month on Wu's 1669 calendar. In effect, the emperor endorsed Verbiest's calculations, expecting Jesuits to follow traditions associated with the cosmological and ritual aspects of the calendar.

These court tests to measure the shadow of the sun relied on instruments calibrated to European methods. Although Jesuit predictions were slightly more accurate than Wu Mingxuan's, they were based on obsolete methods, and far from exact either. To conclude that the tests were settled by imperial endorsement rather than accuracy, however, overdetermines the battle between Yang Guangxian and Schall as simply a struggle over the rules of the game. It was certainly in part an ideological scuffle, but in the end, when compared to their opponents, the Jesuits, despite some errors, were able to predict eclipses within minutes rather than within a quarter or half an hour of the Muslim methods.[97]

When the deliberative council reported the Jesuits' findings were more accurate and proposed that the 1670 calendar should be prepared by them, the emperor accepted their recommendation. He was critical of the council, however, for not explaining why Yang Guangxian and Wu Mingxuan had been declared right and Schall wrong in 1664 but not now. The Kangxi emperor also demanded to know why Schall's methods were now acceptable when they were not before. To satisfy the emperor, Yang was thus found guilty by the council of making false accusations and was banished, but the emperor pardoned him because of his age. Yang apparently died on his way home.[98]

Ferdinand Verbiest and the Kangxi Emperor

Verbiest's appointment as administrator of the calendar of the bureau in 1670 restored the Jesuits to preeminence in the Astro-calendric Bureau. Wu Mingxuan initially saved his position by switching sides and endorsing Verbiest's calculations, but in August 1670 Wu was flogged forty strokes for his continued calendrical errors. Later, Verbiest presented the Kangxi emperor with a two-thousand-year calendar in 1678, which was called the Eternal Calendar (*Yongnian li*) to symbolize the longevity of the Qing dynasty. Earlier, Yang Guangxian had accused the Jesuits of preparing a two-hundred-year calendar to symbolize a short reign.

Verbiest also explained the difference between the Chinese lunar-solar calendar's intercalary month and the Gregorian solar calendar's leap years. He also defended the Shunzhi emperor's use of "Western methods" to subtitle the new Qing calendar in 1645 by arguing that the emperor—not Schall—was simply referring to the unique character of the "new" dynasty. Nevertheless, the Qing court retained the three main divisions of Chinese, Western, and Muslim sections in the bureau based on the Ming model. Only in 1725 were Jesuits again appointed as directors of the bureau, and only then with a Manchu counterpart.[99]

Although the Kangxi emperor allowed the missionaries to return from exile in Guangzhou in 1671, he refused their request to spread Christianity. He also ordered that Schall's phrase "in accordance with new Western methods" on the title page of the Qing calendar should be replaced by "Temporal Model Calendar" (*Shixian li*). The late Ming collection of Jesuit mathematics and astronomy, titled *Mathematical Astronomy of the Chongzhen Reign*, which had been renamed by Schall in 1645, was reissued in 1669 as *Mathematical Astronomy Following New Methods* (*Xinfa lishu*). The Qing government published the Temporal Model Calendar with imperial authorization. It was distributed annually empirewide and improved when French Jesuits helped the court in the 1720s update its translations of mathematical and astronomical works.[100]

Failure of the Studies to Fathom Principles *Project*

While Director of the Astro-calendric Bureau, Ferdinand Verbiest collected all preceding Chinese translations of the commentaries on Aristotle by the Jesuit University of Coimbra and called the result *Studies to Exhaustively Master Principles* (*Qiongli xue*). Verbiest started this project in 1678, perhaps to accompany his Eternal Calendar of 1678, which he presented on request to the Kangxi emperor. The emperor seemed to show some interest in such a collection during his personal studies with Verbiest, when Verbiest presented the *Studies to Fathom Principles* as a complete repository of Western, that is, Catholic, scholastic philosophy, which he contended was the beginning point for astronomy and mathematical studies.[101]

Tactically, Verbiest like Schall saw the calendar and astronomy as a means to imply the ultimate superiority of European philosophy. The sources of the Coimbra collection are unclear. About half seem to have come from *De Logica*, although Verbiest reworked this part without mentioning Aleni or Li Zhizao, the original compilers. Verbiest added translations related to medicine, philosophy, and natural theology, that is, logic, the soul, the mind and brain, creation, etc., before submitting the book to the emperor. For example, Verbiest appropriated Schreck's translation (completed by Bi Gongchen) of the Galenic account of the body, which elevated the importance of the brain for human memory and the perception of things. Because Chinese physicians believed that the heart was the center of cognition, this new medical claim embroiled the *Studies to Fathom Principles* in a major brouhaha that nipped Verbiest's clever plan in the bud.[102]

The compendium was intended to explain European philosophy to the court as an ordered collectanea in Chinese of Aristotelian moral and natural studies. Verbiest's ambitious goal was to receive imperial permission to require such learning for the prestigious civil examinations. This, he dreamed, would lead the throne and literati to the Church and its divine law. It was a daring line of attack for the Catholic enterprise in Qing China, where millions of male candidates aimed their educations at the official examinations, the gateway to officialdom.

Perhaps Verbiest had in mind complementing the Five Classics, Four Books, and orthodox Cheng-Zhu studies. Later, in 1767 the French Jesuit Alexander de la Charme (1695–1767) prepared his *True and Complete Explanation of Nature and Principle* (*Xingli zhenquan*) to present a Catholic gloss on the official canon. The title mimicked the official Cheng-Zhu compendium titled *Great Collection of Nature and Principle* (*Xingli daquan*), orthodox since the early Ming. Verbiest's collectanea was intended to enrich the Chinese classical tradition with supplementary interpretations that would recreate the religious character of ancient China as a blend of Christianity and the Chinese Classics.[103]

In his October 1683 memorial that accompanied the presentation of the *Studies to Fathom Principles* collection to the Qing throne, Verbiest contended that the study of European philosophy was the correct framework for the study of the calendar and astronomy. He stressed that the "essential method of inference" (*litui*) as taught by Aristotle to forecast principles was the foundation for all practical knowledge. In reply, the Kangxi emperor strategically followed the advice of the Hanlin Academy and the Ministry of Rites, declining to publish the collection under imperial auspices.[104]

The reason the Kangxi emperor gave for his rejection was that his Chinese advisors considered it an inappropriate view of learning. His disapproval, dated December 1683, addressed Verbiest's work as "Western studies for exhaustively mastering principles," thus pointing to its peripheral origins and its intellectual differences from native studies for mastering principles. Specifically, the Kangxi emperor complained, "The style of the book is very perverse, erroneous and illogical." The Manchu Grand Secretary Mingzhu (1635–1708) and others noted, "Saying that man's knowledge and memory belong to his brain completely contradicts the reality of principle." Because Chinese literati believed that "the heart-mind is the ruler of intelligence," they rejected Verbiest's notion that "all knowledge and memory is located not in the heart but in the brain."[105]

This imperial position accurately reflected the priority of the heart-mind over the brain in Chinese classical thought and medicine. Mencius, for example, had given priority to the heart-mind in thinking, "The organs of hearing and sight are unable to think and can be misled by external things. When one thing acts on another, all it does is to attract it. The organ of the heart can think. But it will find the answer only if it does think; otherwise, it will not find the answer." In the sixteenth century, however, the physician Li Shizhen had already argued that the brain was an important organ for conscious life, as had other late Ming physicians.[106]

According to I. Dunyn-Szpot, S.J., the Lithuanian compiler of the history of Jesuit mission circa 1700, ministers in the Hanlin Academy and Ministry of Rites—some likely recalling the Yang Guangxian era—had recognized Verbiest's strategy of disguising Christianity in the dress of dialectics and philosophy. The Kangxi emperor could not act against the advice of his Chinese officials despite his personal friendship with Verbiest. Nor could the emperor unilaterally change the core classical curriculum of civil examinations. The *Studies to Fathom Principles* collection that Verbiest had labored on was returned to him and never published. Much of it was later lost.[107]

Moreover, at the time the *Studies to Fathom Principles* was provoking such an interesting response about the brain from the Ministry of Rites, a Muslim scholar Liu Zhi (1660–1730) was working on a treatise that when published in 1704 was titled the *Arabic Discourse on Nature and Principle* (*Tianfang xingli*). In it Liu indicated that the brain integrated various perceptions. Liu, one of the Islamic scholars involved in the Sufi-inspired *Han Kitab* (Han Chinese books on Islam) project used to revive Muslim high culture in the Yangzi delta, held that the brain coordinated all the sense organs.

Liu—like Verbiest—was likely building on Galen's refutation of Aristotle's view that the heart was the ruling organ, which also contributed to the eclectic Muslim view of the brain's centrality. Liu added that Chinese literati had wrongly maintained the heart-mind as the venue for intelligence. This fascinating difference of medical views among Chinese, Jesuits, and Muslim-Chinese over the role of the brain may have paralleled their interactions concerning the calendar and astronomy.[108]

This medical difference between Qing officials and the Jesuit missionaries over the location of human agency in humans revealed deep political waters. In a February 1685 letter, Verbiest did not appeal to East-West incommensurability as the cause of his failure. Instead he bitterly pointed to a breakdown in the accommodation process: "Our philosophy cannot be published, because the Rabbis' [lit., *Rabini,* i.e., the members of the Hanlin Academy and Ministry of Rites] had convinced the emperor that the fundamentals of the Christian learning were embedded in it."

The emperor, however, still sent an aide to study Aristotelian philosophy with Verbiest in private, perhaps to assuage the latter. In addition, Verbiest was widely cited and read by Qing literati, some of whom were interested in his Christian teachings, such as Wang Honghan, and others who were interested in his *scientia* such as Li Guangdi and Mei Wending. The latter assimilated Verbiest's contributions in mathematics and astronomy into Chinese traditions for investigating things and fathoming principles. Verbiest's astronomical corpus in China was also influential in Europe.[109]

We will see that the Kangxi emperor also relied on Verbiest and the Jesuits to mediate between the Qing and Russian empires when their geographic borders expanded in the 1670s and 1680s. There were other signs, however, that Verbiest's power on the whole was limited to the Astro-calendric Bureau. In 1673 another attack on the Jesuit's methods had come from Yang Jingnan, a commoner from Suzhou, who had assisted Yang Guangxian earlier. Yang Jingnan also accused the Jesuits of errors in their official calculations. The Kangxi emperor realized that the charges were still predicated on the unpromising "waiting for the *qi*" procedures, which he considered unreliable. In 1681, when the divination for a tomb was disputed, the Kangxi emperor again intervened. He charged that in this field of learning there were no definitive findings, just fabrications by officials to suit their purposes.

Similarly, Zhang Yongjing's efforts to discredit European astronomy by petitioning for a restoration of native methods for computing calendars based on ancient cosmology fell on deaf ears in the late seventeenth century. Nevertheless, in 1685 the emperor authorized compilation of an imperial manual of hemerolgy as the single standard. In 1741, the Qianlong emperor ordered the bureau to produce a new guide for hemerolgy that approved the approach that both Yang Guangxian and Wu Mingxuan had used to assail Schall and his staff in 1664.[110]

After 1670, the Jesuits depended almost exclusively on Manchu imperial patronage and not on the support of Chinese literati. The opposite had been the case in the late Ming, when Xu Guangqi and Li Zhizao had been the order's stalwarts and sanctioned the quest for converts. Literati leaders' turn toward evidential studies in the eighteenth century recognized the importance of European studies, particularly mathematics and astronomy, but they remained suspicious of the Jesuits' motives in coming to China. While the Jesuits immortalized Yang Guangxian as a bigot and xenophobe in their accounts of the China mission, Yang's works were highly praised by first-rate critical scholars such as Qian Daxin (1728–1804) and Dai Zhen (1724–1777) for attacking Christian teachings in the name of China's imperial honor. Jesuit historians have claimed that Dai even accused Christians of paying high prices for Yang's works in order to burn them.[111]

Moreover, Yang's belief that the Jesuits were part of a European plan to invade and conquer China resonated in the nineteenth century when Protestants rather than Catholics flocked to China to spread a Christian agenda via science and medicine. These fears had been first voiced in a late Ming collection compiled by Xu Changzhi called the *Collection Exposing Heterodoxy* (*Poxie ji*), printed in 1639, which included literati and Buddhist monks' attacks on Christianity dating from the 1616–1617 persecutions initiated by Shen Que. Indeed, the Polish Jesuit Michael Piotyr Boym had called for a European invasion of China after the fall of the Ming. Later several anti-Christian sequels were prepared in the late nineteenth century to attack the Protestants in China.[112]

When, after the Rites Controversy, the Jesuits lost the support of their last major patron, the Kangxi emperor, their chief operational venue remained the Astro-calendric Bureau, where Indian and Persian astronomers in the Tang and Muslim calendricists in the Yuan and Ming had also been successfully sealed off from the rest of Chinese state and society. Because German and French Jesuits were increasingly tied to the emperor, his court, or the bureaucracy, rather than to provincial literati, they were now directly subject to dynastic power. The imperial politics of high risk and high gain, such as the 1667 calendar controversy, almost wrecked the Catholic mission in China. Besides controlling the Astro-calendric Bureau, Jesuits also served the throne as cartographers, clock makers, architects, glass makers, and optical lens makers. The last European to serve in the bureau, the Lazarist Monteiro de Sera, resigned in 1826, and the post reserved for a foreigner was then eliminated.[113]

Li Guangdi's part in supporting Chinese specialists in mathematical astronomy in the Kangxi court was particularly significant. Literati scholars who were part of this group served the Kangxi emperor in his private quarters, where they helped compile in the 1720s the mathematical and astronomy treatises that superseded the late Ming translations prepared under the auspices of Xu Guangqi. Mei Wending had helped to train many of them. By the time Li Guangdi introduced him to the Kangxi emperor in 1705, Mei already had the reputation of a polymath who had mastered both traditional Chinese mathematical astronomy and the new learning introduced by the Jesuits.[114]

Mei Wending and his successors contended that study of natural phenomena gave classical scholars access to the principles undergirding reality. In essence, Mei saw Jesuit learning as a way to boost the numerical aspects of Cheng-Zhu moral and metaphysical principles. At the same time, however, the imperial court and Mei Wending prepared preliminary accounts stressing the native Chinese origins of European natural studies.[115]

Mei's view of the history of mathematics combined aspects of both Song Learning and evidential studies. In many ways, his own research represented a transition from Cheng-Zhu learning to ancient learning. Such native origins made it imperative for Mei (and his highly placed followers in the Qing court, such as Li Guangdi) to rehabilitate and restore to their former glory the native traditions in the mathematical sciences.[116]

4

The Limits of Western Learning
in the Early Eighteenth Century

When Michel Benoist (1715–1774) belatedly introduced a brief but accurate account of Copernican cosmology in China after the Church's ban ended in 1757, the Jesuit society was still defending obsolete Counter-Reformation scientific orthodoxy. Anti-Jesuit polemics generated by the Jansenists led to suppression of the order, first in Portugal in 1759 and then by France, Spain, Naples, and Parma, before the pope dissolved the order worldwide in 1773. China's window on continental Europe and Europe's window on China were shattered by forces internal to both European and Chinese history. Benoist apparently died of a stroke or grief upon hearing in 1774 that the order had been suppressed.[1]

As Jesuit influence in the court faltered, Manchu emperors increasingly patronized their own imperial mathematicians, whom they relied on to lessen the dynasty's dependence on foreign experts for astronomical precision. The lowly social status of the native mathematicians who emerged during the Ming-Qing transition changed dramatically after 1700 when they were increasingly patronized by Han Chinese literati officials close to the Kangxi emperor. Mei Wending in particular emerged as a native mathematician who could challenge the credibility of the Jesuits in the Astro-calendric Bureau and at court. In turn, the emperor enhanced Mei's astronomical credibility empirewide by according him unprecedented honor as a model for court mathematicians. The social standing of literati mathematicians during the eighteenth century rose dramatically.[2]

The Kangxi Emperor and Mei Wending

After dealing with his regents in the 1660s, the Kangxi emperor recognized the need for astronomical and mathematical expertise in court and in the bureaucracy. Although the Ming had banned public knowledge of computational astronomy and astrology in order to monopolize it in the Astro-calendric Bureau, the Kangxi emperor initially opened them for official study by appropriate Han officials and trustworthy Manchu bannermen. His goal in part was to lessen the hold of the bureau's hereditary Han and Muslim families on the transmission of mathematical astronomy by encouraging learning among official students from the eight Manchu military banners. He also faced the monopoly of entrenched Catholic converts in the bureau.[3]

The court increasingly patronized mathematics (*suanxue*) to counter the Jesuit monopoly of European studies. Manchu openness to Western learning when compared to the late Ming should not be overstated, but the early Qing emperors prioritized such studies more than late Ming rulers. Court patronage meant that official mathematics (*guanxue suanxue*) was no longer under the sole purview of the Astro-calendric Bureau or its hereditary officials.[4]

The Qing bureau also had more students than the Ming bureau, although the same fields of study remained in place. In 1644 the bureau had sixty-six Chinese students. By 1666, the number increased to ninety-four studying astrology and mathematics. Banner students of mathematics had three choices for careers: they could be added to the bureau's students, they could participate in the provincial examinations and become generalist officials, and they could become instructors. Because the bureau employed more students (eighty bannerman) in astrology than in mathematics, the annual quotas were six astrology students from the Qing banners and four from the Han banners.

Nevertheless, sixty-six such places for Han students of astrology remained, which the emperor increased to 160 in 1666. In 1675 the quotas for banner students of astrology decreased to twenty-four, while the Han quota of eighty also decreased. Students of mathematics were given a better chance than astrologists for careers outside the bureau, especially those the emperor chose as his court mathematicians. Because the bannermen regarded civil examinations and mathematical training as secondary and complementary to their primary military skills, however, their mathematical achievements were limited. The Qianlong emperor stopped bannermen from training in mathematics in 1738.[5]

The Yang Guangxian affair had interrupted the Kangxi emperor's youthful studies under Ferdinand Verbiest. When several French Jesuits arrived in Beijing in 1689, the emperor began a second period of learning European mathematics and astronomy. The French used the Manchu language to instruct the emperor in the newer Western learning they brought. They also assisted the Qing dynasty in its 1689 negotiations with the Russians that would culminate with the Treaty of Kiakhta, delimiting the border between the Manchu and Russian empires. Enjoying the favor of the emperor, the French Jesuits prospered when the Kangxi emperor lifted the ban on proselytization in

1692. In 1693, after the French Jesuits cured the emperor's bout with malaria by administering "Jesuit bark" (*cinchona*), he asked Bouvet to invite other Jesuits to come from Europe to China to serve him.[6]

The interactions with the French Jesuits in court produced several outstanding native mathematicians, such as Minggatu (Ming Antu, d. 1765?), a Mongol bannerman, who had remained a lowly licentiate in the examination hierarchy. In his *Quick Trigonometric Methods for Obtaining the Precise Ratio for Divisions of the Circle* (*Geyuan milü jiefa*), Minggatu elaborated on Peter Jartoux's (1668–1720) work on the geometric power series expansion of trigonometric functions. Minggatu used the method of finding the chord to know the arc and added six procedures to Jartoux's original three. Such studies, which reformulated the traditional algebraic craft of dividing a circle, were carried on by his son and disciples. One of the latter, Chen Jixin, worked in the Astro-calendric Bureau and continued Minggatu's work on trigonometric methods, as did Zhang Gong.[7]

The manner in which the Jesuits introduced European mathematics determined how Chinese literati would assimilate it. In contrast to Euclidean geometry, which the Jesuits and their literati informants adapted more or less in its entirety, Minggatu and others reinterpreted the power series in light of the Chinese system of calculations. Jartoux, who arrived in China in 1701, failed to present the roots of the series giving π, the power series expansions of the sine, or the versed sine functions in the contemporary calculus of Newton and Leibniz. Although developed in the late seventeenth century and expanded in the eighteenth, these were not introduced to China until the nineteenth century.[8]

The Kangxi emperor's private sessions with the French Jesuits included training in surveying, mensuration, calculating, astronomy, geometry, and logical argument. The source for their lectures on Euclid's *Elements,* for example, was a French edition of I. G. Pardies's (1636–1673) *Elemens de Geometrie,* which the Jesuits and their Manchu aides translated into Manchu. Later, using Jean-François Gerbillon's (1657–1707) lectures at the Kangxi court, they translated Pardies's version of the *Elements* into Chinese. The emperor's court mathematicians included this version in the new official mathematical collection, which effectively replaced the incomplete version of the *Elements* by Ricci and Xu Guangqi at court and among literati. The French geometry gave different proofs from the Ricci-Xu translation.[9]

Besides Li Guangdi and Mei Wending, Wang Xichan (1628–1682) and Xue Fengzuo were also Kangxi-era students of mathematics and computational astronomy. Xue worked as an obscure Christian disciple of Smogulecki to better predict lunar and solar eclipses. Du Zhigeng (fl. ca. 1700), like Fang Zhongtong (d. 1698), a Ming loyalist and the polymath Fang Yizhi's second son, tried to integrate Chinese and European mathematics. Du tried to create a unified system based on the *Computational Methods in Nine Chapters* (*Jiuzhang suanshu*), one of the Ten Mathematical Classics. Du also wrote the *Abridgement of the Elements* (*Jihe lunyue*) in 1700, which left the basic structure intact. Aware of those literati who shared his passion for mathematics, Mei Wending admired Du's mathematics and Yuan Shilong's work on equations. He was also familiar with Wang Xichan's work, although they never met.[10]

Fang Zhongtong compiled a summary of mathematics early in the dynasty, but because of his father's Ming loyalist sentiments the work was not published until 1721. It presented a preliminary discussion of musical harmonics and numbers and a treatise on Euclid based on Ricci's and Xu Guangqi's partial translation of the *Elements*. Fang also discussed written calculations, using the abacus for computation, and Napier's rods for mechanically multiplying, dividing, and taking square and cubic roots of numbers. Like Du Zhigeng, Fang believed that European mathematics derived from ancient Chinese mathematical classics, particularly procedures for measurement based on the relational features for the sides of a right-angled triangle, that is, a kind of trigonometry (*gougu*).[11]

Mei Wending's Achievements in Mathematics

Mei Wending was so honored by the emperor that later accounts have tended to overlook the above-mentioned Qing students of Chinese and European mathematical astronomy with whom he interacted. Wang Xichan's work remained in manuscript form, for example, and never had the impact of Mei's publications. Like the others, however, Mei Wending's mathematical career might have gone unheralded. His father and grandfather had been book collectors interested in the *Change Classic* and numerology. Ming loyalists, they kept their distance from the Manchu regime. In addition, although Mei recognized the value of Jesuit learning, he was leery of Christian influence in the Astro-calendric Bureau.[12]

As a youth, Mei studied astrology and mathematical astronomy (*tianwen lixue*) along with the Classics and Histories. When he came to Beijing for the 1666 capital region examination at the age of thirty-three *sui* (roughly thirty-two years old), which he failed, he acquired several books on European mathematical astronomy. From 1669 to 1678, Mei met Fang Zhongtong four times in Nanjing when Mei and Fang took the provincial examinations. Mei used the time to collect books on astronomy and mathematics. While in Nanjing in 1675, for instance, Mei located works on logarithms of trigonometric functions and celestial motions by Smogulecki and Xue Fengzuo. By 1675, he had also seen works from the *Mathematical Astronomy of the Chongzhen*

Reign collectanea and thus had access to much of the mathematical tool kit introduced by the Jesuit mission. In 1678, Mei met the famous book collector Huang Yuji (1629–1691) and obtained a copy of the seminal *Computational Methods in Nine Chapters,* which although a mathematical classic was by then hard to come by in its complete form.[13]

As a result, a friend recommended Mei to work on the calendrical treatise for the official *Ming History* project initiated in 1679. Mei corrected the chapter on the calendar in the *Ming History* and prepared a draft of the calendrical treatise. Mei also met with Prospero Intorcetta in Hangzhou in 1688 to discuss mathematical astronomy before proceeding on to Beijing a second time in 1689, perhaps to meet with Verbiest, whose death in 1688 had dashed that hope. While there, Mei met Li Guangdi and received the latter's patronage for the four years Mei remained in the capital. When Li Guangdi printed nine of Mei's works on mathematics in 1701, Mei became an acknowledged expert on European and classical mathematical astronomy, whom the Jesuits regarded with suspicion.[14]

During the emperor's southern tour in 1702, Li Guangdi presented the Kangxi emperor with a copy of Mei Wending's *Queries on Mathematical Astronomy* (*Lixue yiwen*), which was written in the 1690s when Mei stayed in Li's Beijing residence. Li hoped that the emperor would be pleased with Mei's talents at a time when the Rites Controversy was beginning to force the court to recognize the limitations in relying on the Jesuits for astronomical and mathematical expertise. After reading the work, the emperor was impressed, although he noted, perhaps after consulting his French Jesuit minions, that it had too few problems to solve.[15]

Later in 1705, when the Kangxi emperor again traveled south, he asked to meet with Mei Wending. For three days, the emperor and Mei apparently discussed mathematical astronomy on the imperial barge, and the emperor was impressed that he had a native scholar who was as knowledgeable as the Jesuits, which might enable his court to rely less on the latter. Mei presented the emperor with a copy of his new work on trigonometry, which Li Guangdi had published. When the Jesuits proved to be unreliable intermediaries with the Catholic pope, the Kangxi emperor's discovery of Mei Wending and other Chinese mathematicians was timely.

Moreover, Mei's efforts to reconcile Chinese and European mathematical astronomy appealed to the emperor because it legitimated both traditions. In a preface for a work titled *Mastering Chinese and Western Mathematical Calculations Comprehensively* (*Zhongxi suanxue tong*), Mei wrote:

> During the Wanli period (1573–1619), Master Li [Matteo Ricci] entered China, and began promoting the study of determining quantities

[that is, geometry], which took points, lines, surfaces, and solids as the basis for measurement. He constructed instruments and prepared maps that were all the epitome of accuracy. However, his books were limited by the precision of the translations. Because they contained many sections, it was difficult to navigate through them. The style was rambling and filled with trifles, making it difficult for the reader to complete the work.

Moreover, they made Jesus' [life] into a religion, which was odd for literati. Those who studied their teachings tended to overly praise them. They never took time to inquire deeply into the origins and vicissitudes of Chinese calculation [methods]. They precipitously passed on the shallow techniques of their day and referred to them as the culmination of the ancient *Nine Chapters* [classic]. Consequently, they slighted ancient methods as not worth taking seriously. Moreover, others who clung vacuously to ancient gleanings attacked Westerners for their heterodoxy.

Both schools of thought are separated by a wide chasm because of the extremes of scholars. In my seeking the way to scholarly learning, I have sought to master them comprehensively, and no more. If others are more comprehensive than I, then they have successfully closed the distance between the ancients and moderns. What distance then remains between China and the West?[16]

Mei's unstated contrast between Xu Guangqi's overenthusiastic support of European learning in the 1630s and Yang Guangxian's foolhardy antiforeign xenophobia in the 1660s were presented as false choices for the emperor and literati to eschew in mastering mathematical astronomy.

After meeting with the emperor, Mei prepared his *Supplement to the Queries on Mathematical Astronomy* (*Lixue yiwen bu*), in which he appealed to the "Chinese origins" of Western computational astronomy to mediate between those who championed European methods and those who defended traditional approaches. For instance, in his discussion of the "vaulted heavens" (*gaitian*) cosmology mentioned in *The Gnomon of the Zhou Dynasty and Classic of Computations* (*Zhoubi suanjing*), Mei traced this ancient model back to its classical alternative, the "spherical heavens" (*huntian*) cosmology. In this way, he affirmed the celestial sphere in terms of classical mathematical astronomy. In Mei's view, although the vaulted heavens cosmology had not confirmed the sphericity of the earth, the celestial phenomena in the southern hemisphere could be derived from the same computational procedures used to demarcate the northern hemisphere.[17]

Despite such ecumenism, the Kangxi emperor began in 1689 to enunciate an official view that suggested European studies were derived from ancient China, a traditionalistic view that unexpectedly was reinforced by the emperor's studies of European mathematics under Bouvet and Gerbillon. The latter used ancient precedents in China to justify such learning. Later, Mei flattered the emperor by attributing this view to his majesty. Further refined, the new approach would yield in the late seventeenth century the second stage of a claim that "Western learning had its origins in China" (*Xixue Zhongyuan*).[18]

Thereafter, the prominence of the Mei family in Qing mathematics was widely recognized. Mei Wending's two brothers were also well versed. Although Mei's son died early, he was also trained in mathematics, which Mei's grandson, Mei Juecheng (d. 1763), continued. The latter became a prominent member of the Kangxi, Yongzheng, and Qianlong courts in the eighteenth century. Mei Juecheng's sons also mastered mathematics. The Kangxi emperor confirmed the Mei family's eminence by declaring Juecheng an Imperial School student in 1712, granting him a provincial degree in 1713, and the highest degree as "a palace graduate in mathematics" (*suanxue jinshi*) in 1715. Mei was allowed to take the palace examination without taking the preliminary metropolitan examination.[19]

The Mathematical and Astronomical Content of Mei Wending's Works

Mei Wending's first work, the *Superfluous Observations on Mathematical Astronomy* (*Lixue pianzhi*, 1662), was about the strengths and weaknesses of Yuan and Ming calendars. *On Simultaneous Linear Equations* (*Fangcheng lun*, lit., "Theory of rectangular arrays of numbers," 1672) rehabilitated ancient Chinese techniques for solving linear equations, which were later incorporated into the *Imperially Instituted, Treasury of Mathematics* (*Yuzhi shuli jingyun*, 1723). In Mei's view, the algebraic techniques the Jesuits introduced were foreshadowed in earlier Chinese works.[20]

Mei's impression was plausible because the Jesuits had devised a simplified algebraic system for the Chinese that permitted only polynomials using one unknown with numerical coefficients, which was called "borrowing of roots and powers" (*jiegen fang*; *gen* as the root denotes the unknown, and *fang* denotes any power of that root). In practical terms Jesuit algebra was not superior to Song-Yuan techniques, which could solve polynomials of higher degrees and with multiple variables. The latter had remained a craft stressing the results more than the formal process. The French Jesuit Antoine Thomas (1644–1709) later prepared Qing translations of more advanced arithmetic and algebra for the Kangxi emperor.[21]

Mei's goal was to synthesize calculating techniques (*suanshu*, i.e., arithmetic) and methods of measurement (*liangfa*, i.e., geometry) into a unified mathematics (*shuxue*). Unlike Fang Zhongtong, who reduced European mathematics to the ancient Chinese mathematical classics, Mei sought to re-habilitate traditional mathematics. In addition, he never denigrated what the Jesuits introduced because he recognized the limitations of traditional methods for solving problems in mathematical astronomy. He sought to use European achievements to advance traditional techniques, now reinvigorated with logarithms and trigonometry. In the late eighteenth century, this approach climaxed with the rehabilitation of the Ten Mathematical Classics.[22]

In mathematical astronomy, Mei sought a synthesis between traditional approaches and European methods that would make them commensurable. He built on Wang Xichan's studies, which had criticized the division since the Song dynasty between scholarly and specialized calendrics. For both Wang and Mei, this split had prevented interaction between literati scholars and bureau specialists. Literati did not understand calculations, and specialists lacked knowledge of principles. The result was poor understanding of mathematical astronomy and little innovation.[23]

Wang Xichan had earlier criticized some aspects of "Western methods" (*Xifa*) but took its practical applications as improvements on Chinese methods (*Zhongfa*). Wang's reaction to late Ming calendar reform was also critical, but in general he followed Jesuit writings on the movement of the five planets. In addition, Wang prepared a critical overhaul of the Tychonic system, which Mei Wending saw as an effort to ameliorate the European and Chinese systems. To work out the calendar, Wang and Mei had shifted from the numerical procedures of the Chinese computus to geometric models of successive locations of celestial bodies in space. Each realized that, unlike traditional numerical procedures, the mathematical models associated with the Tychonic world system enabled them to explain as well as predict celestial phenomena.[24]

Wang was also critical of the separation between classical studies and mathematical learning. He regarded Xu Guangqi's late Ming calendrical reform as too practical minded. Through the selective reception of European methods, Wang sought to rescue the traditional system. He challenged the newness of foreign methods by suggesting Chinese precedents. In a 1693 work, Mei Wending emphasized the cumulative development of astronomy and its improved precision, which encompassed both European and Chinese methods. Hence, he agreed with Xu Guangqi's efforts to amalgamate European methods within the traditional system. Mei's work influenced Li Guangdi and through Li the Kangxi emperor, who sought to ensure there would be an emerging pool of native talent in mathematical astronomy during his reign.[25]

Moreover, Wang's and Mei's accommodative approaches to the new computational astronomy encouraged the court to avoid the fatal choice of either returning to outdated Chinese or Muslim computational methods, an approach that was bankrupt after the Yang Guangxian fiasco, or replacing Chinese astronomy with that of Catholic Europe as the Jesuits wished. Mei Wending, for example, demarcated the cultural versus technical aspects of calendar making. For the latter, he commended European methods for dividing the celestial sphere into 360 degrees rather than the more cumbersome 365.2425 *du*. He also defended a spherical earth in the face of a classical tradition that assumed the earth was square. Mei stripped European techniques of their Jesuit cultural and religious accoutrements and harvested them for their skeletal efficacy.[26]

With regard to Chinese cultural features, however, Mei acknowledged that the Chinese method for dividing up the year into twenty-four fortnights and inserting an intercalary month best suited the imperial calendar. Adopting purely European methods would require doing without the cultural import of Chinese festivals, auspicious and inauspicious days, and the lunisolar calendar on which all rituals were based. If it was a matter of cultural rather than technical proficiency, Mei came down on the side of maintaining Chinese ways. The Jesuit computus could be assimilated to the imperial calendrical tradition, just as the Indian and Muslim computus were.[27]

Mei Wending's stress on the accumulated authority of astronomy had a long pedigree in China, although he based his views on the accrual of observational data in both China and Europe, and maintained that European input to Chinese developments was justified. For instance, Mei outlined the development of astronomy from Ptolemy to Brahe and stressed the role of the telescope in improving on Brahe's observations. New information, such as Kepler's laws and Newton's improved methods for predicting eclipses, were later adopted in the *Supplement to the Compendium of Observational and Computational Astronomy* (*Lixiang kaocheng houbian*, 1742) by Ignatius Kögler (1680–1746) and Andre Pereira (1690–1743), with the help of Mei Juecheng and He Guozong (d. 1766). The new computus stabilized eclipse prediction until the end of the dynasty.[28]

Mei favored a quantitative approach to dealing with astronomical principles. Computations for grasping heavenly phenomena permitted numbers (*shu* = "heavenly correspondences") and principles (*li*) to be linked:

> Someone might ask Master Mei: "Is mathematical astronomy a matter of concern for the classical scholar?" I reply: "Of course. I have heard that it is the classical scholar who masters the comprehensiveness of heaven, earth, and humans. Is it permissible to be enveloped by but not know the height of the heavens?"
>
> He might ask: "The scholar in knowing heaven knows only its principles. What use does he have of astronomy?" I reply: "Mathematical astronomy requires numbers. Outside of the numbers there are no principles,

and outside of the principles there are no numbers. Numbers are the orderly demarcation of principles. Numbers cannot be spoken of arbitrarily, but principles at times can be talked about via vague images. Hence, arbitrary views have been associated [with principles], which have deluded the people and brought chaos to heaven's regularity. All this results from not obtaining the true principles and numbers and instead maliciously overturning reality.[29]

For Mei it was permissible to ban private astrology as long as scholars mastered mathematical astronomy, which would allow increased precision in the calendar. Mei Wending appreciated the link between computational astronomy and the new instruments Verbiest had built. Problems in measurement that had remained unresolved despite earlier progress, if properly conceptualized, could be solved.[30]

In Mei's view, the cumulative development of European methods was analogous to those in China. Because the two were parallel, they could be unified. European learning demonstrated that the prediction of eclipses required knowledge of the time and position of objects in space, which it had developed beyond the limits of traditional calculation. Equivalent developments in Europe and China demonstrated that the two traditions were commensurable. For example, Mei noted that Guo Shoujing during the Yuan dynasty had preceded Tycho Brahe in recording regularly the positions of planets.[31]

Mei regarded geometry as the most systematic aspect of European studies and the root of its mathematics. He was upset, however, that Ricci and Xu had only translated the first six of fifteen chapters of Euclid. Mei prepared his *Essentials of Geometry* (*Jihe zheyao*) as a commentary on Euclid's chapters. In a second work, Mei prepared the *Appended Sections on Determining Quantities* (*Jihe bupian*, 1692), which he regarded as a supplement to the Xu-Ricci translation of the *Elements*. There Mei attempted to reconstruct the parts of Euclid that Ricci and Xu had not translated by looking for possible hints in translations from the late Ming.

Although he perceived geometry as an outgrowth of traditional mathematics, Mei Wending acknowledged that European spherical geometry provided new universals that even the ancient sages did not anticipate. Mei explained geometry by using traditional solutions for the sides of a right triangle. His interest in geometry was selective, but his approach to computational astronomy was thoroughly geometrical. He seems to have thought that Euclid could be reduced to traditional mathematics.

Mei Wending was the first to explicitly use the traditional term for determining quantities (*jihe*) to mean geometry. In addition, Mei used trigonometry to apply the equivalent of spherical geometry, which he regarded as a simplification of traditional methods for transforming equatorial to ecliptic coordinates, etc. His writings on Napier's bones and the proportional compass argued that the new calculating devices developed by Napier and Galileo were essential aids for computational astronomy. Mei tied them to traditional counting rods and the late Ming use of logarithms.[32]

When his grandson compiled Mei's collected works in 1761, forty years after his death, Mei Wending was recognized as the leading literatus who had mastered all the fields of European mathematics introduced to China. Moreover, his *Complete Works on Mathematical Astronomy* (*Lisuan quanshu*)—actually not a complete collection—became the starting point for subsequent efforts until 1850 to reinvigorate traditional methods with the sophistication of new European approaches. Robert Morrison (1782–1834) of the London Missionary Society, for example, had Mei Wending's complete collection on mathematical astronomy in his library.[33]

Although he had rehabilitated ancient Chinese techniques, Mei was unable to locate all of the works included in the Ten Mathematical Classics because many were not widely available. While mid-Qing evidential research scholars included mathematics in their classical fields of interest as a result of Mei Wending's influence, none of them followed up on later European mathematics. That was inevitable, for new developments such as the kinematic solutions and moving models of Newton's fluxional calculus and Leibniz's infinitesimal calculus, which directly linked time and motion, were not transmitted to China in the eighteenth century. Mei and his successors had only the static, essentialist geometry of Euclid and the qualitative motions of Aristotle to work from.[34]

Mei's and the Jesuits' mathematical astronomy survived in eighteenth-century classical studies through Jiang Yong (1681–1762), Dai Zhen, and Qian Daxin (1728–1804), who wrote on computational astronomy. When evidential research scholars after Mei Juecheng rediscovered the Song "single unknown" (*tianyuan*, lit., "method of celestial origin," to solve polynomial equations with a single variable) craft of algebra, they had no reasons for preferring the more formulaic Renaissance science of algebra (= *jiegen fang*) and ignored it on nativist grounds in favor of Chinese techniques. Nevertheless, Mei's canonization by the Kangxi emperor was responsible for the change in status of mathematical studies during the Qianlong era.[35]

The Rites Controversy and Its Legacy

The celebrity Mei Wending achieved in 1704 via promotion of his mathematical studies by the Kangxi emperor occurred when the political situation of the Jesuits in the Qing court decisively changed. The French Jesuit John

Baptist de Fontenay (1643–1710) noted at the time, for instance, that the emperor was no longer as interested in European mathematics as before. Several months after the emperor and Mei met during the emperor's southern tour, the Chinese Rites and Term Controversy challenged Jesuit claims that Qing subjects who adopted Christianity did not have to reject cultural practices such as ancestor worship.[36]

The Jesuits believed that Chinese names for God and the rituals converts used to honor their ancestors were acceptable. Ricci and his followers, with misgivings from some such as Longobardo in 1615, thought Chinese rites for ancestors were of social and moral signficance, but that they did not compromise an individual's belief in the Christian God. Because of the "Rites Controversy," the theological battles among the Jesuit, Franciscan, Dominican, and Augustinian orders in Rome exceeded its papal guidelines. A chill ensued in imperial and literati attitudes toward the Jesuits and their missionary work in China.[37]

A papal legate sent to Beijing by 1705 announced that the Church had ended the Jesuit policy of accommodating Chinese rituals and their names for God. This action began the decline of the Catholic mission in Qing China and echoed ominously Yang Guangxian's view of the Jesuits in the 1660s as treacherous foreign agents. As a result, the Kangxi emperor overturned his Edict of Toleration for Christianity issued in 1692. In the late eighteenth century, classical scholars rehabilitated and praised Yang Guangxian for his cultural astuteness while they, like Mei Wending, dismissed his ignorant views of astronomy and the calendar. Qian Daxin, for example, praised Yang for his attack on Christianity in his colophon for the 1799 reissuing of Yang's *I Cannot Do Otherwise:*

> Mr. Yang was not an expert of calculation, nor did he have adequate help. Therefore, in the end he had to give up. However, he denounced Jesus's heterodox religion and prevented its diffusion, and this cannot be anything but meritorious for [our] renowned doctrines. In the judgment of posterity, then, Yang's merit lay more in his opposition to Christianity than in his opposition to Western studies.[38]

The Breakdown of Jesuit Consensus in China

Conflicts among Catholic missions worldwide were exacerbated by long-standing competition between Spain—which sponsored the Dominicans, Franciscans, and Augustinians—and Portugal, which trained and sent the Jesuits abroad. Later, French Jesuits would be justly accused by the German Jesuit Kilian Stumpf (1655–1720) and the Portuguese in the Astro-calendric

Bureau of trying to glorify French science. Each denomination had followed different strategies to convert the peoples of the world and please their patrons. Initially, the better-educated and more urbane Jesuits were able to put in place an accommodation policy to facilitate their dealings with knowledgeable Chinese literati and the sophisticated courts of the Ming and Qing empires. This policy was not accommodation at any cost, however. Ricci refused to baptize converts who clung to their concubines or approved homosexuality. He also attacked Song metaphysics as atheism.[39]

From the beginning, Franciscan, Dominican, and Augustinian orders, as well as the secular priests of the Missions Étrangères de Paris, voiced opposition to the Jesuit mission. They all contended that the accommodation policy of Ricci, Schall, and Verbiest violated Christian teachings. The Dominican and Franciscan missionaries believed their orders in China had greater legitimacy because the Jesuits through accommodation had slipped into heterodoxy.

Although later demonized in pro-Jesuit accounts, Domingo Navarrete published an account of politics, ethics, and religion in China in 1676 in which he exposed the dangers of the Jesuit conversion of the Chinese and excoriated the order's monopoly of influence in the Manchu court. Navarrete had been the Superior of the Dominican mission in Fujian when the 1664–1667 anti-Christian policies were promulgated. He resided in Guangzhou with several Jesuits while the Yang Guangxian affair took its course. Later Navarrete's report of his Jesuit colleagues and their tolerance of superstitious practices in China influenced the pope to limit Jesuit influence on the court, but to no avail.[40]

Navarrete was taking a position in the dispute between the Jesuits based in China and the friars (Dominicans and Franciscans) based in China and the Philippines. Although Jesuits' opinions differed, such as Longobardo's late Ming criticism of Ricci, the Rites Controversy at first was a conflict between the religious orders, not a conflict of Chinese versus Christian culture. Like the Jesuits, the first Dominicans in China were also well versed in at least one Chinese language after working among the Fujianese living in Manila.[41]

In March 1693, however, the conflicts among the Catholic orders became public when the Apostolic Vicar of Fujian, Carolus Maigrot (1652–1730), issued the *Mandatum seu Edictum* (Mandated edict). A Parisian prelate who served in the Missions Étrangères de Paris, Maigrot represented a congregation of secular priests authorized by the Sacred Congregation (Propaganda Fide) established by Pope Gregory XV in 1622 to promote the foreign missions. Its clerics, like Navarrete and the Dominicans, challenged the Jesuits by establishing a clerical hierarchy in China independent of Portugal.[42]

Maigrot declared seven prohibitions effective immediately for all missionaries in Fujian. They pointedly disallowed use of "Heaven" (*tian*) or "Lord on High" (*shangdi*) as appellations of God, denied that Christians could engage in sacrifices for Confucius and their ancestors, rejected the "Supreme Ultimate" (*taiji*) as an alternate term for God, and discarded the *Change Classic* as a repository of acceptable moral teachings. Maigrot also sent an envoy to Rome to obtain the pope's endorsement.[43]

Although not effective outside Fujian, Maigrot's unilateral action—communicated only to Rome—frightened the Jesuits when they learned of it in 1698. They called on the Kangxi emperor to resolve the conflict in their favor. This act brought the internal battle among the Catholic orders into the court. In 1700, the emperor declared that ancestor worship and the veneration of Confucius were secular rites, as his Jesuit favorites maintained. The latter triumphantly forwarded the emperor's decision to Rome, seeking a papal intervention to counter Maigrot's edict. Philippus-Maria Grimaldi (1632–1712), the leader of the Jesuits in Beijing, called upon officials in the Astro-calendric Bureau and Chinese converts to weigh in by communicating their views to Rome.[44]

Pope Clement XI (1649–1721), however, faced a powerful anti-Jesuit clergy in Europe, particularly among the Jansenists in the Missions Étrangères at the Sorbonne. Between 1696 and 1700, the latter initiated a series of public debates to humiliate the Jesuits for their advocacy of probabilism and for the misguided policy of accommodation toward the Kangxi court. They were incensed by the claim that Chinese classical texts and basic Christian doctrine agreed. Le Comte's memoirs, published in 1696, later provoked a letter to the pope by opponents and denunciations of the Jesuits for two months by the Sorbonne faculty of theology in 1700. The Jansenists also opposed excessive papal power and Jesuit complicity in that authority.[45]

Pope Clement, unlike his predecessor, sympathized with the purists in the Catholic Church, although before 1702 he did not side with the Jansenists. To do so Clement had to overturn a papal decree issued by his predecessor that condoned Jesuit policy in China. In 1701, however, he appointed an opponent of the Jesuits, Charles-Thomas Maillard de Tournon (1668–1710), as his legate to China to ensure that the prohibition of Chinese rituals among Chinese converts was enforced. De Tournon set out in 1702 and arrived in China at the end of 1705. The pope also issued an unpublished decree, which approved Maigrot's mandate as Church policy.

When de Tournon arrived in China, the Kangxi emperor treated him cordially. The emperor summoned the legate for several audiences in 1706, and de Tournon brought Maigrot with him each time. The naive legate's meetings with the emperor touched off a new phase in the Rites Controversy, resulting in the decline of all Catholic missions in China. In accord with the French Jesuits at court, the emperor instructed de Tournon that Christianity had to be compatible with Confucius's teachings, as Ricci had shown it was. If not, he would not allow the Europeans to remain in China.[46]

Maigrot was asked at the audiences to demonstrate what passages in the Chinese Classics were incompatible with Christianity. When Margot's reply indicated that his abilities in classical Chinese were dubious, the emperor accused him of illiteracy and further charged him with inability to distinguish between the original writings of Confucius and Song literati writings, as Ricci and his Jesuit followers had done. Disenchanted with Maigrot's comportment, the emperor decided to deport all missionary troublemakers.

To the dismay of the Jesuits, the new policy rolled back the emperor's 1692 Edict of Toleration. After 1706, he required all missionaries to have an imperial certificate to maintain residence in Qing China. Even though the court supported their policy of accommodation, the Jesuits feared that the required certificates would give provincial governors discretion to deport all missionaries. Indeed, the Jesuits in Beijing were forced to pledge loyalty to the Qing throne to acquire such certificates. Because members of the Missions Étrangères de Paris refused to follow suit, they were refused certificates and deported. Maigrot was ordered out of the country when de Tournon left the capital in 1706. A Jesuit who had served on de Tournon's staff was imprisoned for his role in the fiasco.

Suspicious of Jesuit complicity with the Qing emperor, Maigrot left for Europe in 1707 and arrived in Rome in 1709. The pope appointed him as Assistant to the Papal Throne. This selection was followed by a 1710 papal decree that reconfirmed Pope Clement's antiaccommodation decree of 1704. Starting as an internecine squabble among the Catholic orders, the Rites Controversy reopened the Christianity controversy of 1660s in China and Rome, which had ended with the demise of Yang Guangxian in China and a 1669 papal decree granting relative freedom to the different orders for missionary work.[47]

Despite the furor, Maigrot's 1693 Mandate and Rome's condemnation of 1704 remained unpublished until 1707 when, as a result of his imperial audiences, de Tournon decided to include them in his Nanjing Decree. As the pope's legate, de Tournon supported Maigrot in China. When the Kangxi emperor censored Maigrot in 1706, he also demanded that de Tournon present his credentials of appointment. Upon de Tournon's refusal, the emperor ordered him deported and retrieved all imperial gifts given to the Legate to transmit to Pope Clement. De Tournon in a letter from Macao dated July 27, 1707, instructed the superiors of the Dominican, Franciscan, and Augustinian orders to honor his excommunication of the Bishop of Macao for refusing to recognize his authority.[48]

The Jesuits responded by sending a second mission to Rome in 1707, after one that left in 1706. The first trip had met with shipwreck and drowning off Portugal. Not knowing this, it thus seemed to the Jesuits that de Tournon's 1707 decree ignored the emperor's legitimate request of clarification from the pope in the 1706 mission. For his part, de Tournon felt that the Jesuits were manipulating the emperor, as Dominicans and Franciscans had long claimed.[49]

The legation of 1705–1710 and the audiences with Maigrot and de Tournon turned the emperor against the missionaries, who he now understood were agents of the papal authorities in Rome. The eldest prince thought the missionaries were simply spies for the Spanish or Portuguese, who were preparing an invasion of China. In 1717, the emperor prohibited all missionary work in China. On the papal side, de Tournon's *Regula* became the first condemnation of the Chinese rites that publicly affirmed Rome's authority on the issue. The Decree of the Inquisition signed by Clement in 1710 imposed an absolute ban on writing on the rites issue. Those who failed to heed the injunction, particularly the Jesuits, would face excommunication. Later papal bulls of 1715 and 1742 silenced all discussion of the controversy in Europe and cemented the triumph of the anti-Jesuit clergy.[50]

The emperor's anger on receiving the 1715 papal bull led directly to the "Red Manifesto" (*Hong piao*) of 1716, which was printed in vermilion in Latin, Chinese, and Manchu and sent to Europe. The emperor declared he would recognize only documents from Rome that were delivered by his own envoy—on his own terms—to Europe, Father Joseph Anthony Provana (1662–1720). Fan Shouyi, a Jesuit priest, was influential in the Kangxi emperor's decision to limit Jesuit activities in the empire. Fan returned to China from Rome in mid-1720 and explained papal politics to the emperor, namely that Europe was militarily very powerful but that the authority of the pope was relatively weak. The papacy in the sixteenth century had been a "precursor of the modern state" with a standing army, direct taxation, sale of offices, and the first permanent diplomatic corps. By the seventeenth century, however, the nation-states of France and Spain eclipsed the Papal States in political power and were competing to use the papacy.

In the late 1720s, the Kangxi emperor understood that the pope could not command the English or Dutch and that the Jesuit claim that their order represented the strongest ruler in Europe was false. Louis XIV (1643–1715) had occupied papal territories in Avignon and Venaissin. To reconcile him, Clement IX granted the French monarch control of all Church appointments in his realm. A century of Jesuit portraits of the pope as emperor of the Church (*jiaohuang*) had unraveled. The legation to China led by Carlo Ambrogio Mezzabarba (ca. 1685–1741) in December 1720 sought to mollify the emperor by proposing eight permissions for certain Chinese ceremonies.

The emperor's distrust was confirmed, however, when Mezzabarba also tried to gain the unconditional obedience of the Jesuits to the pope.[51]

When the Kangxi emperor died on December 20, 1722, he was succeeded by the Yongzheng emperor, whom the Jesuits in the Manchu court had opposed in the late Kangxi succession struggle. The emperor's eldest son had earlier interrogated the missionaries during the de Tournon mission. The younger son, who succeeded to the throne, hated the missionaries and compared them to the White Lotus sect, whom he suspected of plotting to overturn the dynasty. In June 1723, just after the new emperor took the throne, a Christian literatus in Fujian apostatized and wrote to the local magistrate denouncing the missionaries. Most of them were exiled to Macao. Several Jesuit astronomers, artisans, and artists who professed their loyalty to the Qing stayed in Beijing.[52]

In the Qianlong period, Jesuits could not preach Christian doctrines, but some were involved in clandestine Christian activities outside the capital. Compromise between the Church and the Qing court was impossible. In a decree of July 1742, for instance, Pope Benedict XIV demanded oaths of loyalty from all missionary orders in Asia to enforce his prohibition of Chinese and Indian rites. The legations of 1705–1710, 1720–1721, and 1724–1725 to the Manchu court were sent to establish permanent diplomatic relations between the Qing and Rome, but they only widened the gap.[53]

The Rites Controversy also backfired on the Jesuits in Europe. As the Jesuits lost influence in China after 1705, their society also lost its legitimacy in Catholic Europe. In 1749 there were 22,600 members of the society and some 669 Jesuit colleges and 176 seminaries worldwide. After 1773, the pope sequestered the society's property, and its rich library collection in the Jesuit College of Clermont was sold *en bloc* in 1764. According to the 1773 decree, in which Pope Clement XIV dissolved the Society of Jesuits, the key issue was the Jesuit accommodation policy because it placed China and its rituals on equal footing with Europe and Christianity.[54]

We will discuss below Jesuit opposition to Newtonianism in their colleges during the eighteenth century, which lost them the esteem of French philosophes such as Condorcet (1743–1794), who championed the new sciences. Although the Jesuits were the schoolmasters of France and Europe, Condorcet, from personal experience, saw Jesuit education as a disaster. In the 1740s, the curriculum of nearly 400 French colleges slowly shifted from Cartesianism to theoretical and applied Newtonianism. Despite the strong emphasis on mathematics in French colleges, "if Newton finally triumphed in France it was probably over the corpses of the Jesuit order." French Jesuits were expelled in 1762.[55]

The Elimination of Natural Studies in Late Kangxi Civil Examinations

The collapse of the Ming dynasty initially created opportunities for Jesuit experts in mathematical astronomy to break out of their subordinate positions in the Astro-calendric Bureau and to challenge a discredited officialdom for power under the new Qing dynasty. The increased cultural importance of astronomical expertise, when the new dynasty had to reformulate in expert terms its raison d'être, challenged for a time the preeminence of literati who had mastered Song-Ming classical studies.

The Kangxi emperor, in particular, relished learning mathematics and surveying techniques and encouraged his sons and bannermen to master them. After the 1680s, when the Manchu dynasty expunged its enemies by putting down the Revolt of the Three Feudatories, the political fluidity of the early Qing waned, leaving Han Chinese literati and Manchu elites in a precarious balance at the top of the political and social hierarchies. This delicate balance was strengthened by the Rites Controversy and the Jesuit debacle.

One of the ironic consequences of the Rites Controversy was the slow but steady encapsulation of European studies within the palace. French Jesuits continued to wield influence in the Astro-calendric Bureau and court generally in the early eighteenth century, but Chinese literati domesticated the new learning by transforming its study into the recovery of ancient learning. Literati who followed Mei Wending's pioneering mathematical studies now aimed to retrieve the classical mathematical and astronomical traditions in which the Jesuits had piqued their interest.

For example, around 1711 the Kangxi emperor already knew about the Paris Academy of Sciences, which the French Jesuit Foucquet translated into Chinese as the "Academy for the Investigation of Things and Exhaustively Mastering Principles" ("Gewu qiongli yuan"). The emperor also modeled the court's "Academy of Mathematics" (*Suanxue guan*) after it, but this was a temporary organization to meet the emperor's immediate needs. Its influence was confined to the court. French Jesuits worked mainly on calendar reform and the empirewide land survey commissioned between 1708 and 1717.

The emperor preferred to send Manchu bannermen rather than Chinese literati to study astronomy with the Jesuits, with a quota of ten students per banner. He did not encourage a broader focus on natural studies, as was increasingly the case in early modern France. Although literati mastered the new European learning in the eighteenth century, the end of the Jesuit mission in China precluded knowledge of the Newtonian century in Europe.[56]

Simultaneously, policy questions on the triennial Qing provincial and metropolitan examinations, which drew civil candidates from every corner of

China's 1,300 counties, virtually ceased to include topics in natural studies, as did the 1525 policy question on mathematical astronomy discussed earlier. In the Ming, questions on the calendar crisis, for instance, were not uncommon. In the eyes of the astute Kangxi emperor, the Rites Controversy confirmed the dangers of making imperial accommodations with European mathematical astronomy part of the civil examination curriculum.[57]

The French Jesuits sent by Louis XIV to the Kangxi court as royal mathematicians and as members of the Paris Academy of Sciences soon stagnated. Their combination of scientific and religious objectives proved as problematical for the Jesuits and their Church critics as for the Kangxi court. Although a Jesuit—or ex-Jesuit—remained director of the Astro-calendric Bureau until 1805, once the calendar and mathematics translation projects of the late Kangxi era were completed, the emperor felt that enough Manchus and Chinese were trained in the new astronomy, mathematics, and surveying techniques.[58]

In 1713, the Kangxi emperor proscribed questions in the civil examinations on astronomical portents and the calendar because they pertained to Qing dynastic legitimacy. The emperor decreed that examiners assigned to serve in provincial and metropolitan examinations were forbidden to prepare policy questions on astronomical portents, musical harmonics, or calculation methods. The latest works in Qing natural studies and court translation projects on mathematical harmonics and astronomy, which had employed Jesuit experts, were off-limits to examiners and examination candidates. The court tolerated some access for literati outside the civil examination bureaucracy, however.

The ban on examination questions dealing with natural studies occurred within an effort by all dynasties to keep divination and portents out of public discussion. The Ming dynasty shared this concern and the Qing reinstated imperial control over the public uses of European learning. The court's translation projects were designed to shift control of this technical learning from a foreign religious order to native literati and scholars in court without making them public.[59]

In place of the banned natural studies and mapmaking, historical geography in particular prospered as an acceptable examination field of Qing scholarship. The court and literati agreed that the dynasty must not depend on a religious group whose technical prowess was inseparable from their ties to the papacy and European expansion.[60]

The Kangxi emperor's meetings with Mei Wending had been fortuitous because they signaled that the emperor could rely on literati mathematicians to assimilate any new mathematical astronomy the dynasty would need to maintain the calendar. The Yongzheng era began a closed-door policy that lasted until the Opium War (1839–1842). Nevertheless, scholarly work on

European mathematics continued, particularly geometry and algebra, which made it to China before 1723. Relative openness toward needed astronomical reforms continued even under the Yongzheng emperor. For example, Minggatu was allowed to revise the *Compendium of Observational and Computational Astronomy,* which when published included Newton's improved methods for predicting eclipses.[61]

The Yongzheng emperor and his court could not roll back the impact of European studies in the eighteenth century. Qing scholars built on the European learning translated into Chinese in the late Kangxi era. Those brought by the French Jesuit mission provided an important transition in Chinese classical learning from the Kangxi to the Qianlong reigns. We will address the full magnitude of the direct and indirect Jesuit impacts on literati in the eighteenth century below.[62]

French Jesuits in the Kangxi Court

The members of the French mission left Europe from La Rochelle for China in 1685. The declining political fortunes of the Jesuits in the early eighteenth century had led to decisions in Rome and at the Sorbonne, as we have seen, to reject the Jesuit accommodation strategy. Such policies in Europe produced through the Rites Controversy an increasingly hostile environment within China itself. By 1706, the Kangxi emperor was ambiguous about the new learning.[63]

Fontenay's return to France in 1703 ended French Jesuit scientific work for the mission in China. By 1716 members of religious orders were no longer nominated for honorary status in the Paris Academy or otherwise affiliated. Nevertheless, in 1717, 28 of the 120 Jesuits in China belonged to the French mission. Moreover, Chinese Jesuits constituted the largest non-European group in the entire Society of Jesus in the middle of the eighteenth century, comprising about one-third of the total.[64]

Fontenay, who was the only well-trained mathematician in the 1685 group, had been responsible for astronomical observations in China. Bouvet initially studied natural history and medicine, and Le Comte focused on history and the mechanical arts. Unlike their Italian and German predecessors, the French Jesuits shifted from collaborative work to individual study and authorship. Because of the Kangxi emperor's focus on mathematics for translations, natural history took a back seat to astronomical work. When Fontenay returned to Paris in 1703, Bouvet and Le Comte in China rarely interacted, while the Rites Controversy took up much of the French Jesuits' time in Paris.

Fontenay had no immediate successor for astronomical work, and Bouvet increasingly worked on recovering ancient Chinese wisdom. Although Antoine Gaubil (1689–1759) tried to revive the mission's scientific work after arriving in China in 1722, his proposals to the Society of Jesus in Paris to fund an observatory in Beijing fell on deaf ears. Discouraged by the lack of response, Gaubil watched as the Jesuit astronomers associated with the Portuguese mission in Beijing took over the leadership role in European astronomy. He turned his attention to the history of Chinese astronomy, producing the best reports before the twentieth century.[65]

Bouvet and Figurism in Europe

Bouvet's accommodation policy and his studies of the *Change Classic*, for example, generally followed Ricci's precedent of mastering the Classics as a tactic to convince literati of Christian doctrines. Unlike Ricci, however, Bouvet regarded the *Change Classic* as the oldest written text in China and the source of all her natural studies and customs. He also viewed it as a prophetic book that concealed the mysteries of Christianity, a position that contributed to the early eighteenth-century Figurist controversy in Europe and intensified European Catholic opposition to the Jesuit accommodation policy in China.[66]

For Bouvet, written Chinese characters were hieroglyphs or pictorial writing that should be interpreted figuratively and allegorically rather than as literal historical records. In Bouvet's study of the *Change Classic*, sponsored by the Kangxi emperor, he explained the line changes in the hexagrams in terms of mathematics and Leibniz's binary numbers. This view diametrically opposed Ricci's earlier critique of the *Change Classic* as a fount of Song atheism and materialism, a charge that Maigrot later also made. Leibniz, following Bouvet, was willing to concede that, once recovered, ancient Chinese learning based on the *Change Classic* would prove to be the precursor of his binary mathematics.[67]

For example, Bouvet cited the *Systematic Treatise on Computational Methods* by Cheng Dawei, which had explained the mathematics of the magic squares in 1593. The Kangxi emperor and his court then became interested in Cheng's work and had it reprinted. Bouvet argued that one of diagrams in Cheng's work was equivalent to European algebra and explained the 3–4–5 right triangle. The emperor mediated between Bouvet and Li Guangdi, who read Bouvet's studies of the *Change Classic*, although Li confessed to the emperor that he couldn't understand what Bouvet was doing. Neither the emperor nor Li Guangdi ever saw the Figurists' claims in French and Latin in which Bouvet and others argued that if one went back far enough in Chinese history one could find the point at which China diverged from God's teachings.[68]

The later French Jesuits privately identified Chinese history with biblical history and thus came into conflict with Biblical claims of the high antiquity of the Jews. Thus, the French Jesuits reintroduced Schall's and Li Zubo's earlier volatile efforts to link the Chinese to Noah and his sons in new ways. Bouvet, for instance, questioned Chinese claims that Fu Xi was the ancestral founder of China. For Bouvet, Fu Xi represented the universal lawgiver, while the sage-kings Yao, Shun, and Yu represented the first rulers of the Chinese people some six centuries later.

In contrast to his Italian, German, and Flemish Jesuit predecessors, who had stressed Confucius and the Four Books, Bouvet emphasized the Five Classics of antiquity and the *Change Classic* as the oldest surviving relic of a primitive universal language before the Great Deluge. Consequently, for Bouvet, the hexagrams attributed by the Chinese to Fu Xi had been obscured by Confucius and his followers. When stripped of their later perversions, the hexagrams in the *Change Classic* revealed a binary mathematical order that allowed access to God's mathematical vision of all things in the world in light of numbers, weights, and measures.[69]

Bouvet also agreed with the Chinese view of the *Change Classic* as a record of the regularities of heaven and earth, but he regarded the trigrams that made up the hexagrams as abstract principles in a philosophy of numbers identical to those of Pythagoras and Plato. Bouvet's efforts to restore the original and lost meanings of the *Change Classic* also stressed the Pythagorean link between numbers and music. Because of their fascination with symbols such as the trigrams, Bouvet and other French Jesuits in the early eighteenth century were "Figurists." They read into the *Change* a protolanguage and protomathematics of all mankind.[70]

Similarly Bouvet believed in Hermetism, which was based on the *Corpus Hermeticum* that Neo-Platonic scholars claimed was written by Hermes Trismegistus, an Egyptian priest who lived after Moses. This curious blending of pagan philosophy with Christianity had turned Hermes Trismegistus into a quasi-Christian sage. Isaac Casaubon disproved the authenticity of the *Corpus Hermeticum* in 1614, claiming it was a forgery that reflected Greek teachings of the early Christian era drawn from a blend of Platonism and Christianity.

Bouvet added to the Hermetic tradition by focusing on the *Change Classic* and developing correspondences between Hermes Trismegistus and the sage-king Fu Xi, situating the father of writing within ancient divine wisdom from which all cultures had drawn. The Figurists thought the hexagrams and cosmic diagrams in the *Change Classic* contained the key to reducing all phenomena to quantitative elements. According to Bouvet, just as Hermes Trismegistus had preceded Jesus' revelation, the *Change Classic* contained later Christian mysteries in a prophetic manner.[71]

Others such as Gaubil remained more sober minded, but Joseph de Prémare (1666–1736), like Bouvet, believed that the sacred content of Chinese characters would reveal Christian doctrines that the Chinese had lost. The Figurist approach to the Classics reached a climax when Bouvet linked ancient Chinese writings to the Paschal Eclipse associated with the Resurrection of Christ. According to Bouvet, the Paschal Eclipse was a supernatural event allegedly recorded in Chinese records and tied to Jesus' death. Gaubil refuted this position by citing Chinese textual evidence that the Chinese record of the alleged Paschal Eclipse was of A.D. 31, too late for the crucifixion.[72]

Jean-Alexis de Gollet (1664–1741) believed that pagans such as the Chinese had been recipients of ancient revelation. He equated the sage-king Yao with Jectan, a descendant of Shem, and systematically identified biblical and Chinese personages. In this view, Chinese observational records came from the Jewish patriarchs before the deluge. Gollet, although a Figurist, made biblical correlations in line with those of the 1660s in China, which we have seen cost Chinese Christian collaborators who published such views their lives. Bouvet and the other French Jesuits in China had the savoir-faire not to translate such aspects of Figurism into Chinese.[73]

They were not sufficiently careful about publishing their views in Europe, however, where anti-Jesuit feelings reached a peak after 1700. The Jesuits lost their credibility and their influence for expressing their Figurist views in eighteenth-century Europe. Foucquet presented Figurism as an alternative to the two opposing European camps in the controversy. Instead of profane literature filled with atheism and idolatrous rites (as the Jesuit opposition interpreted Cheng-Zhu learning) or Classics containing the principles of natural theology (as most Jesuits did), Foucquet interpreted the Classics as sacred writings containing the mysteries of Christianity in prophetic form.

Because he reexamined the rites and terms question from a Figurist point of view, Foucquet was recalled from China for disobedience. The papal ban on further discussion of issues relating to the Rites Controversy went into effect in Europe in 1710. Hence, the full appreciation of Bouvet's and the Figurists' comprehension of Chinese classical literature, suppressed by the Rites Controversy, has only come about quite recently.[74]

The Second Stage of the Claim for the "Chinese Origins of Western Learning"

Most eighteenth-century Catholic Europeans found the Jesuit claim that the ancient Chinese had received the Christian Revelation blasphemous. Others such as Gaubil dismissed it as historiographic bluster. For the most part, however, Europeans discussed the transmission of culture, science, and religion to

China, pros and cons, in terms of the priority of biblical events. On the Chinese side, however, the accommodation project provided an unforeseen ally for the Chinese observation that if the Classics were indeed repositories of ancient wisdom—Chinese and European—then all European learning, including the mathematical and natural history fields of *scientia*, originated in China and was later transmitted to the West.

The Chinese could agree with the Jesuits about cultural transmission in theory, but they were free to change the direction of that transmission in practice and make themselves central. Although he was more favorably disposed to Western works, Mei Wending enunciated an early version of this narrative. According to Sima Qian some ancient Chinese practical mathematicians (*chouren*) had dispersed toward the Western regions during the chaotic transition in the eighth century B.C. from the Western to Eastern Zhou dynasty. Mei used this passage to explain how astronomy in Europe had its origins in China. In the official account of the *Ming History*, prepared in the mid-eighteenth century, for example, the Catholic centrality of the Great Deluge and the Levant was turned inside out:

> Those who have come to the Chinese land from the West, all call themselves Europeans. Their mathematical astronomy is the same as the Muslims, but they are more precise. In reviewing past eras, we find that many of those who came from far off countries and are conversant with mathematical astronomy reside in the Western regions. We have heard nothing [of such knowledge] from the southeast or north. It is likely that when Yao ordered the brothers Xi and the brothers He to reside separately in the four cardinal directions, Xi Zhong, Xi Shu, and Ho Shu were restricted to Yuyi [the area to the east], Nanjiao [in the south], and the northern area. Only He Zhong was said to "reside in the West" and was not restricted to a particular area. At that time the fame and teachings of the West were far off indeed!
>
> At the end of the Zhou dynasty [in the third century B.C.], the sons and disciples of the practical mathematicians [*chouren*] had dispersed. In the Western region, the Islamic countries were all linked continuously to the West, unlike the southeast where the great sea was a barrier. In addition, there was no [barrier in the West] like the fearsomeness of the north's extreme cold. Hence, it was convenient to transmit books and implements in expeditions to the West.
>
> The Europeans are to the West of the Muslims. Their customs are similar, but the Europeans surpass them in their quest for novelty, and in their desire to excel over others. Their mathematical astronomy has the same origins as the Muslims, but they have improved on them genera-

This view later influenced Mei Juecheng, who accepted the emperor's interpretation, and the claim became a cliché among a generation of evidential scholars.

Mei Juecheng affirmed the origins of algebra in the *Zhou Dynasty Canon*, and he regarded the heavenly unknown notations (*tianyuan shu*) craft for expressing and solving quadratic and higher algebraic equations of several unknowns as equivalent to Jesuit algebra (*jiegen fang*). This view was later expanded with a new term for algebra (*daishu xue*) by Protestants in the nineteenth century. Succeeding evidential research scholars explored the mathematical sciences, but in the absence of the calculus and mechanics (see below) they strictly adhered to traditional mathematical astronomy.[78]

In the 1690s the emperor used the Chinese origins claim to mediate disputes among literati in his court over the priority of European versus Chinese learning. This approach allowed the emperor and the Jesuits to justify study of European *scientia* as a native product returning to China after development abroad. Mei Wending advocated the Chinese origins theory at about the same time, but since he was employed by the court, it is likely that he flattered the emperor with credit for the claim.[79]

A good example of how court scholars deployed the claim of Chinese origins occurred with the 1723 publication of the *Imperially Instituted, Treasury of Mathematics (Yuzhi shuli jingyun)*. The claim was included in the collectanea of mathematics and astronomy compiled by Wang Lansheng and Mei Juecheng in the Studio for the Cultivation of Youth (*Mengyang zhai*) since 1712. They had worked with or been influenced by French Jesuits such as Antoine Thomas. The *Treasury,* for instance, opened with discussion of the "River Chart" (*Hetu*), "Luo Writing" (*Luoshu*), and *The Gnomon of the Zhou Dynasty and Classic of Computations.* This was followed by the first six chapters of a Chinese translation drawn from the French edition of Euclid's *Elements of Geometry* by Pardies, which was thought to be simpler and easier to understand than Ricci's and Xu Guangqi's version. The first part of the *Treasury* opened with a brief discussion of the "Origins of the Principles of Mathematics" ("*Shuli benyuan*").[80]

The "Lo Writing" was a three-order magic square in which three numbers along any diagonal line or column added up to fifteen, while the "River Chart" was arranged in such a way that the outside odd and even numbers all added up to twenty. Each received prominence of place in Zhu Xi's commentary on the *Change Classic*, which was still the orthodox interpretation when the *Treasury* was compiled. In particular, the magic squares were regarded since Song times as the primal world-ordering instruments used by the sage-kings and were also thought to be linked with the eight trigrams. In addition, as magic squares, they were considered the veritable origins of writing and mathematics. By tradition, the sage-king Yu had been presented these two magic charts by fabulous animals after he had tamed raging floods, which the Jesuits dated to the Biblical deluge (see Figure 4.2).[81]

In the "Origins of the Principles of Mathematics," the compilers affirmed the mystical origins of the cosmograms (i.e., charts of symbolic correspondences between the natural world and human affairs), by which "the sages illuminated mathematics and exhaustively fathomed the principles of all things." The sages composed the early mathematical Classics, and the Duke of Zhou had made mathematics one of the "six arts" (*liuyi*): rites, music, archery, charioteering, calligraphy, and mathematics. They added, "We know that the study of numbers and calculations had in reality been an important task for investigating things and extending knowledge."

Wang Lansheng and Mei Juecheng, among others, contended the shapes, sizes, distances, and heights in Euclid's *Elements* were rooted in the ancient magic squares, which Verbiest had earlier in 1668 and 1671 tried to work out as a series of geometrical diagrams. Verbiest tied the *Change Classic* to geometric forms of circles inscribed in triangles, circles inscribed in squares, and multisided polygons inscribed in circles. In the *Treasury of Mathematics,* the compilers bemoaned the fact that "mathematics [in China] had not been transmitted for a long time." Only the *Zhou Dynasty Classic* had survived from antiquity. They added that fortunately Ricci and other Europeans had helped restore mathematics and calendrics to a higher level of expertise, while insisting on Chinese origins: "Still, in inquiring into their achievements we can say that originally they were transmitted from China."

The Kangxi emperor and his court mathematicians believed that their Jesuit colleagues were helping them restore the ancient learning of the sages, which had been transmitted to the rest of the world. European learning was based on Chinese. Presented in this strategically orchestrated order after the cosmograms and the *Zhou Dynasty Canon* mathematical classic, the second Chinese version of Euclid's *Elements* was successfully domesticated in the eighteenth century as an outgrowth of native learning, unlike Ricci's and Xu Guangqi's version in the seventeenth century, which had in the preface championed a new Western learning and an unprecedented way of reasoning.[82]

The Academy of Mathematics in Beijing

In the 1670s, Verbiest emphasized his role in China as an astronomical official serving in the Qing imperial bureaucracy. Bouvet, on the other hand, found a place for himself in the direct service of the ruler. The French mission's first superior, Jean de Fontaney, equated the Kangxi emperor with Louis XIV and hoped to introduce the institutional framework of contempo-

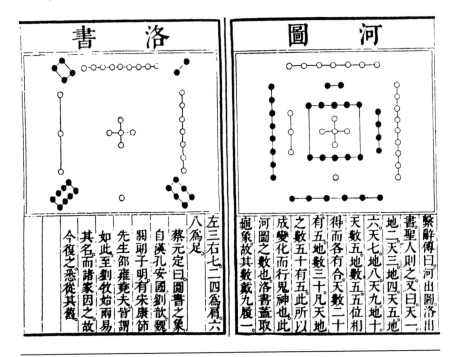

Figure 4.2. The River Chart (*Hetu*) on the right side and the Luo Writing (*Luoshu*) on the left. The *Loshu* is a magic square in which numbers along any diagonal, line, or column add up to fifteen. The *Hetu* is arranged so that when the central five and ten are disregarded, both odd and even number sets add up to twenty.

Source: Zhu Xi, *Zhouyi benyi* (Taibei: Huanlian chuban she, 1971).

rary French science in China. He had held the mathematics chair at the Paris Jesuit Collège Louis le Grand between 1676 and 1685 before leaving for China. Bouvet similarly expected, for instance, that the Kangxi emperor would establish his own Academy of Science, which would emulate the Academy in Paris.[83]

The Studio for the Cultivation of Youth was established in the suburban Lofty Pavilion Garden-Palace (*Yuanming yuan*) in 1712–1713 for astronomical and mathematical work by the emperor's mathematicians. Earlier in 1711 Kangxi had noted mistakes in the summer solstice based on Western computations, and he wanted to employ French Jesuits to improve the calendar despite his dissatisfaction with Rome's papal policies toward China after the Rites Controversy. He invited the French Jesuits to work for him in the same way as they worked for the French Academy.[84]

The Kangxi emperor also formed his own Academy of Mathematics (*Suanxue guan*) loosely on the model of the Parisian Academy of Sciences, but it was named after the Tang school for mathematics and was not autonomous. The Academy was established in 1713 as part of the Studio for the Cultivation of Youth for mathematical astronomy, but only Qing literati and bannermen were appointed to it. No Jesuits were allowed in this inner coterie of imperial scholars, which included the third prince, Yinzhi. This post–Rites Controversy policy ensured that the Jesuits would not unduly influence court mathematics. The emperor wanted his own men in charge of such technical knowledge.[85]

Li Guangdi, supporting a coterie of native specialists on mathematical astronomy in the Kangxi court, also sought to avoid relying on the Jesuits. Li's group included Wang Lansheng, who was granted the highest civil service degree in 1721 by the emperor because of his mathematical abilities and called a "palace graduate in mathematical astronomy" (*chouren jinshi*). Wang then entered the Studio for the Cultivation of Youths and helped the French Jesuits compile newly translated works on mathematical astronomy, which Mei Juecheng and Chen Houyao (1648–1722) also worked on starting in 1712.

In 1712, Chen proposed a new compendium of European mathematics to replace the Ming *Mathematical Astronomy of the Chongzhen Reign*. The result was the *Sources of Musical Harmonics and Mathematical Astronomy* (*Lüli yuanyuan*) collectanea, which included the *Treasury of Mathematics*. In 1713, the Kangxi emperor charged Mei Juecheng and Chen Houyao with supervising He Guozong, Minggatu, and others to complete the project. The *Sources* was printed in 1723. This special group of mathematical and astronomical specialists included Wei Tingzhen and those Mei Wending had helped train before he died in 1721. All had benefited from Li Guangdi's court patronage.[86]

Altogether the emperor recruited more that one hundred scholars—regardless of their civil examination status—to join the Academy of Mathematics. Mei Juecheng was chief and Minggatu assistant editor for preparation of the *Treasury*. In addition to those in the Academy of Mathematics who studied mathematics, astronomy, and music, a large number of instrument makers were hired for the technical needs of the new academy. A team of fifteen calculators verified the computations based on the theoretical notions, mathematical techniques and applications, and numerical tables in the first part of the *Treasury*.

Patterned after mathematical textbooks used in Jesuit colleges, the *Treasury* introduced European algebra. The last part had a section on logarithms to the base 10 and drew on the methods used in Briggs's 1624 *Arithmetica Logarithmica* to compute decimal logarithms. Although Briggs's work had been introduced in 1653, the *Treasury* explained the use of logarithms in greater detail, and it also included tables for sines, cosines, tangents, cotangents, secants, and cosecants for every ten seconds up to ninety degrees, as well as a list of prime numbers and a log table of integers from 1 to 100,000 calculated to ten decimal places. The emperor had the *Treasury* distributed empirewide. It was reprinted in 1875, 1882, 1888, 1896, and 1911, which suggests a more limited audience until the nineteenth century.[87]

The Chinese mathematics that informed the *Treasury* included traditional equations for mensuration (*fangcheng*) and techniques for computing the sides of a right triangle (*gougu*), which were based on Mei Wending's reinterpretation of traditional techniques to solve simultaneous linear equations. The French Jesuits in the court significantly influenced the compilation of the *Sources of Musical Harmonics and Mathematical Astronomy* and its staff of over one hundred working under Wang Lansheng and Mei Juecheng.

The new and influential *Sources* collectanea included (1) the *Compendium of Observational and Computational Astronomy* (*Lixiang kaocheng*); (2) the *Treasury of Mathematics;* and (3) the *Orthodox Meaning of the Pitch-pipes* (*Lülü zhengyi*), which were all compiled in the Studio for the Cultivation of Youth starting in 1712. The collection was intended to be a series of textbooks for the studio and for students in the Imperial College's own Academy of Mathematics.

The *Compendium of Observational and Computational Astronomy* in particular was compiled for fear that the Jesuits had not divulged all their astronomical techniques in the earlier Ming version of the *Mathematical Astronomy*. Mei Wending had been critical of the Jesuits, particularly Ricci, for not translating all fifteen books of Euclid's *Elements* and charged them with deliberately keeping the last nine books secret from the Chinese. In the aftermath of the Rites Controversy, such charges seemed credible.

After the *Treasury* was printed in 1723, no further European mathematics was introduced into China until after the Opium War (1839–1842). Notably missing in China was the European discovery of the dynamic calculus. Moreover, the version of Euclid's *Elements* in the *Treasury* remained the official version until 1865, when Alexander Wylie and Li Shanlan translated the remaining nine books of Euclid and added them to the Ricci-Xu translation of the first six. This lack of innovation was due in part to the way the Jesuits introduced the new mathematical currents of the early eighteenth century. Foucquet, for example, unsuccessfully introduced algebraic symbols using Descartes's new notational forms in a treatise prepared for the emperor. The Kangxi emperor could not grasp the benefit of general symbols that described the algebraic process. He preferred the calculating craft of algebra, which lead to the precise solution. By 1730, moreover, astronomical books of the Kangxi era were out of date by European standards.[88]

The Kangxi Era Compendium of Observational and Computational Astronomy

The French Jesuit role in calendrical reform during the Kangxi reign paralleled their predecessor's role in late Ming improvements of the Astro-calendric Bureau detailed in Chapter 2. Kangxi-era efforts at reform culminated in the 1724 promulgation of the *Compendium of Observational and Computational Astronomy* and a sequel. The European astronomy in the *Compendium* was mostly a century old, but the 1742 *Supplement* adapted more recent European discoveries, such as Kepler's elliptic orbits, for traditional calendar reform.[89]

Hashimoto Keizō has noted the use of astronomical modeling by the group that compiled the *Compendium of Observational and Computational Astronomy*. Such new techniques were wedded to calendar reform and not developed further, however. The *Compendium* did lead to Chinese observations and computations during the Yongzheng reign that no longer depended on Jesuit help. There was no turning of the clock backward, as Yang Guangxian had attempted in 1664, since mastery of "Western methods" (*Xifa*) by Chinese and Manchu specialists had demonstrated to them the observational and theoretical shortcomings of Chinese methods (*Zhongfa*).[90]

Overall, the Kangxi emperor's experts followed Mei Wending's lead in rejecting Jesuit efforts to insinuate Christianity into their astronomy. Mei's mathematical work fit in with the court's efforts to have a calendar that would fuse European and Chinese techniques into a comprehensive system and end wrangling on both sides. When the *Compendium* was drafted in 1722 and promulgated in 1724, for instance, it followed European models but was prepared by Chinese in the court with only indirect Jesuit input. The key figure was Mei Juecheng, who had served on the staff since 1713 and had links to his grandfather's patron, Li Guangdi. Along with Mei Juecheng, He Guozong and Chen Houyao were also instrumental.[91]

The first part of the *Compendium* dealt with general mathematical astronomy. Through its use of physical and mathematical models, the work was not simply a continuation of Schall's *Mathematical Astronomy According to New Western Methods,* which had been based on the late Ming calendar reforms. Mei Wending's study of the moon's motions allowed Mei Juecheng and his group to make improvements in lunar eclipse prediction. The results enabled the Qing court, following Jean-Dominique Cassini's (1625–1712) correctives, to adjust Verbiest's earlier eclipse prediction in 1676 that had been slightly in error due to atmospheric refraction. As the head of the Paris Observatory, Cassini had enlisted Fontenay's help in China to provide him data for the French project to remap the world, and had provided in return his corrected tables for Jupiter's moons.[92]

The early Qing calendars produced by Schall and Verbiest were based exclusively on European models, but the new system cobbled together in the *Compendium* fused European with Chinese methods. Mei Juecheng and his group of native specialists affirmed Mei Wending's efforts to reach a synthesis that would supersede past European and Chinese systems. During the 1720s, the court specialists mastered Jesuit astronomical methods and made them part of the imperial repertoire for computational astronomy.

However, Qing specialists had no domestic incentive to go beyond the immediate needs of the Qing calendar, now successfully reformed. Nor did the Jesuits press them to do so. By 1725, the latter were themselves no longer on the cutting edge of the early modern sciences, and their mathematics, no more than simple algebra, trigonometry, and logarithms, had been domesticated by late Ming and early Qing specialists. In the eighteenth century, a larger community of Qing classical scholars associated with evidential studies would restore traditional Chinese mathematics to a level of intellectual prestige commensurate with the Jesuit version of European studies.[93]

Despite the Yongzheng rejection of Jesuit learning in court, the Kangxi-era advisors under Mei Juecheng were influential in the succeeding Qianlong court. Although the Studio for the Cultivation of Youth was discontinued, the Yongzheng emperor expanded official studies of mathematics by selected bannermen in 1734. This act allowed mathematical studies to grow outside the Astro-calendric Bureau. In 1738, however, the Qianlong emperor ended this program of official studies in mathematics because the students were poorly trained. Instead, he established mathematics as a course of study for the bureau. In 1739, the emperor placed mathematics under the purview of the official school system. Han Chinese students outside the Astro-calendric Bureau could now officially study mathematics.[94]

When compared to eighteenth-century developments in Europe, however, the fate of the Qing dynasty Academy of Mathematics is instructive. In France, the Paris Academy of Sciences became a building block for an increase in men of science and the institutions that supported them. Such institutional changes encouraged the eclipse of the more general learned societies and favored the rise of more technical institutions. The establishment of strictly scientific disciplines by the late eighteenth century was accompanied by the expansion of universities and research institutes where professionalized science slowly incubated in institutions of higher learning, and specialized laboratories eventually replaced gentlemanly academies. Not until the nineteenth century would such developments commence in China.[95]

The Newtonian Century and the Limits of Scientific Transmission to China

Initially, the French Jesuits played a significant role in the scientific developments associated with the French Academy of Science. Similarly, the Jesuits introduced Isaac Newton (1642–1727) in China in the 1742 edition of the *Supplement to the Compendium of Observational and Computational Astronomy*, which mentioned Newton by name but gave no systematic presentation of his theories. By the late eighteenth century, however, the Jesuit order had been abolished and ridiculed by the likes of Lagrange (1736–1813), D'Alembert (1717–1783), Condorcet (1743–1794), and Delambre (1749–1822). The latter were the leaders in the integration of mathematics and science and the practical application of the differential and integral calculus to mechanics and probability theory. The calculus of probability, for instance, provided a mathematical model for evaluating the validity of individual opinions and determining the probable outcomes of individual actions.[96]

An Analytic Style of Mathematical Reasoning in Europe

The rise of the analytic style in European mathematics and probability, which stressed a formal examination of the steps used in algebraic reasoning, facilitated the application of mathematics to late eighteenth-century public policy. The triumph of analysis also provided a firm foundation for the growth of statistics during the Enlightenment, when the mathematization of the social, economic, and political world in the name of political economy became de rigueur.

Similar to the calculus, these social science trends did not penetrate Qing China until a century later. The widespread attempt to apply scientific thinking to all aspects of public affairs defined the Enlightenment. Educated at a time when French science confirmed and elaborated Newton's mathematical principles, Condorcet, for instance, sought to extend that mathematical vision into a universal social mathematics. A nobleman who refused a military career, Condorcet turned instead to pure science and mathematics.[97]

Newton did not use fluxions in his 1687 *Principia,* which relied on formal geometric proofs that his readers would accept. Fluxions represented a notational form for a calculus of flowing numbers (i.e., in "flux"). Newton's geometric proofs were passable for presentation, but useless for research. Descartes's (1596–1650) analytic geometry had translated and extended the classical art into a more flexible algebraic form that became part of the new approach. His equations did away with the special status of celestial circular motion, sublunar natural motion, and many other fixtures of the Greek tradition. Hence continental European scholars checked, verified, and extended Newtonian natural philosophy using the more effective calculus notations of Leibniz based on an infinite series of small differences (i.e., infinitesimals). Continental specialists explored the mathematical physics of the *Principia*, while British scholars (pro Newton and contra Leibniz) focused on the experimental physics in Newton's *Optics*.

Laplace (1749–1827) insisted in his celestial mechanics that the perfection of Newtonian mechanics was mainly due to French scholars and the essays supported by the French Academy's prizes. In French hands, the sciences of algebra and calculus provided analytical tools that no longer needed classical geometric presentation. The task was to translate mechanics into a quantifiable algebraic language that explained movements in space over time.[98]

In the middle of the eighteenth century, for instance, Leonhard Euler (1707–1783) first expressed Newton's three laws of motion through the calculus to defend universal gravitation as a physical theory. Mathematically inclined astronomers filled in the details of the Newtonian solar system. Adapting the calculus required differentiation and integration, and the summing of divergent series. Euler's success created a sense of progress in explaining physical operations of the solar system, and the equations that Laplace, Lagrange, and others devised augmented the engineering tool kit of Newtonian mechanics.[99]

The application of Newtonian science created laws of mechanics that superseded Renaissance mechanics, which had existed as a body of principles explaining the use of levers, wedges, and pulleys through rules of practice. Artisans had understood the workings of machines but never considered an ordering theory or a set of principles to give their mechanics coherence. Classical mechanics had included the parallelogram of forces, the laws of the lever, principles of virtual work, and other diverse principles, but the Newtonian theory of force unified them into a new system in which mechanical devices were models for processes in nature. After the *Principia* was published, continental mathematicians regularized and systematized mechanics, pneumatics, hydrostatics, and hydrodynamics.[100]

Artisans played an important role in mechanical expertise by recognizing the transformations in the mechanical philosophy, particularly in England. Similarly, engineers and mechanical artisans popularized and actually implemented the reform of natural philosophy in eighteenth-century England. By 1750, textbooks made the application of mechanical principles accessible to anyone literate in English and French, and artisans and engineers applied them.[101]

The Jesuits never made the jump to the mathematical analysis of changes in practical mechanics because they upheld the Aristotelian notion of movement based on an object's elemental makeup. Typical accounts of the introduction of geometry to China by the Jesuits overlook the fact that by the late eighteenth century geometry had been superseded in Europe by experimental physics and mechanics. Moreover, traditional Chinese mathematical crafts, once the mathematical sophistication in the Ten Mathematical Classics was reconstructed using Jesuit models for algebra, provided a convenient tool kit of numerical methods that facilitated mastery of the calculus in the nineteenth century by Chinese mathematicians such as Li Shanlan, Hua Hengfang, and others.[102]

Newtonian Science in England

As Newton's *Principia* was extended, entrepreneurs and engineers employed mechanics to link natural philosophy with the practical arts, science and technology. The engineer's tool kit that emerged mathematized the laws of nature via the elaboration of the calculus by Newton's British and French champions in the eighteenth century. The new science was integrated differently in the British social and cultural landscape than on the continent. From the 1720s, British schools taught basic mathematics, including algebra, geometry, surveying, mechanics, and astronomy. Arithmetical and mathematical texts doubled in number during the first half of the eighteenth century. Young artisans in the 1720s already knew rudimentary mathematics and mechanics. At the same time, Oxford and Cambridge in their somnolence included only classical geometry in the curriculum.[103]

Rather than contrasting the pure French sciences of Laplace and d'Alembert with practical engineers such as James Watt (1736–1819) in England, Margaret Jacob has focused on the multiple cultural values that fostered disciplined curiosity among natural philosophers, capitalists, entrepreneurs, educators, engineers, and industrialists in both countries. British and French missionaries subsequently introduced their own regime for science to China in the nineteenth century. The Jiangnan Arsenal in Shanghai, for instance, was organized largely on the English model, while the Fuzhou Shipyard followed a French approach to technological development.

The popular assimilation of science through schools, lecture halls, textbooks, and newspapers represented the educational beginnings of the industrial revolution in Britain. By 1750 England had many male entrepreneurs who approached the production process mechanically. Production could be mastered by machines and conceptualized in terms of weight, motion, and the principles of force and inertia, which was then translated and practiced in light of work and workers in factories. Watt, for example, knew that the practical applications of mechanics allowed engineers using the calculus to quantify the motion of fluids and solids, measure the weight and pressure of different substances, and create mechanical devices such as pumps, pulleys, levers, weights, as well as utilize electricity and light.[104]

In contrast with the doctrinal inflexibility of the French Catholic church, Jesuit colleges, and fundamentalist Calvinism, Newtonian physics quickly became the scientific core of natural religion. Mechanical knowledge was at center of the curriculum of Dissenting academies in Britain. By the 1720s, Newtonianism became the body of natural learning for laymen in Britain to master, practice, and apply. One of them, James Watt, modified and improved the simpler steam engines of the eighteenth century and made them into most advanced technology of the age. The steam engine linked Newtonian mechanics and engineers in a cycle of manufacturing and industrial changes that made efficient use of natural deposits of fossil fuels to provide an unprecedented amount of new energy for practical work in factories, arsenals, and on ships.[105]

Watt's engine became a model for industrial change, and his own intellectual life reflected the cultural origins of the first industrial revolution. Watt came from family of mathematical practitioners who were well informed about instruments and machines. He understood the latest mechanical science and turned to mechanized industry as an engineer and entrepreneur. Jacob contends that a new cultural paradigm emerged in eighteenth-century Britain, which was marked by a set of recognizable values, experiences, and learning patterns among engineers and industrialists that formed an "industrial mentality." The autonomy of the press in England and its independent public favored the interests of practical-minded scientists and merchants with industrial interests. After 1750, laymen and civil engineers communicated through a common scientific heritage that fashioned the mental world of the Industrial Revolution.[106]

The French Century in Science and Engineering

By the 1790s, the French emulated the British educational system, but earlier in midcentury they had already become leaders in applied mechanics requiring mathematical training and basic geometry. The Ministry of Turgot drew upon science and systematic knowledge to revitalize the French monarchy in 1774. According to Charles Gillispie, the Paris Academy of Science in the eighteenth century attained a higher standing in science when compared to the London Royal Society. Science in France had evolved further toward the formation of a profession whose expertise presupposed mastery of a technical body of knowledge that granted social prestige. As the profession of science gained its own jurisdiction over the education, qualifications, and conduct of its members, its disciplines approached the levels already reached by the divinity, law, and medical faculties in French universities.[107]

Loyal Aristotelians, Jesuits fought Newtonianism in their colleges into the 1740s and beyond. After that the curriculum of nearly four hundred French colleges shifted from Cartesianism to theoretical and applied Newtonianism. If French Cartesianism was used by its clerical supporters to glorify Louis XIV as the Sun King, the negative associations of Cartesianism for its links to absolutism and atheism were overcome in France when Voltaire (1694–1778) and the philosophes offered Newtonian science and English society as models of enlightenment. In addition, they attacked Cartesian science as insufficiently experimental.

Willem Jacob s'Gravesande (1688–1742) in Holland was noteworthy in popularizing Newtonian science among the scientifically literate by presenting a mathematical exposition of Newtonian science in textbook form. After s'Gravesande secured a professorship in astronomy at Leiden University, with Newton's intervention, he educated a generation of students in an applied version of Newtonian mechanics through his collection of mechanical instruments and illustrative devices. His *Mathematical Elements of Natural Philosophy* (Latin edition, 1720–1721; English, 1720–1721; French, 1746–1747) replaced the *Principia* in the transmission of Newtonian mechanics for basic scientific education.[108]

Unlike Britain, major industrialization did not occur in France until the nineteenth century, despite Newton's impact on the emerging French scientific community. Abbé Jean Antoine Nollet (1700–1770) was one of chief French promoters of the new science and its mechanical applications. He learned the techniques of demonstration from s'Gravesande and the Dutch Newtonians and then established a course of physics as a series of lectures while traveling in France. Such French popularizers of mechanics provided an alternative to the Cartesianism dominant in colleges and the University of Paris since the 1690s.

While England evolved toward a more balanced polity, the French philosophes justified the linkage between the absolutist state and support for scientific inquiry. Condorcet saw science academies as beneficial for the monarchy. French applied mechanics popularized by Nollet and others filtered into Diderot's (1713–1784) *Encyclopédie* in its 1750s volumes. Although the inspiration was Baconian, the volumes were filled with drawings and descriptions of mechanical devices. Some twenty-five thousand copies circulated before 1789. Diderot and d'Alembert, for example, urged that mechanics be the first science studied for its utility, but this was realized only after the revolution overturned the aristocratic domination of science academies and military engineering schools.[109]

In the 1750s the French monarchy focused on steam-powered boats for military use. In the 1770s and 1780s, the state encouraged mechanical devices for controlled farming. In the French pattern, engineers and men of science served the state. In Britain science had a greater affinity with public entrepreneurs. An emphasis on mechanics in French education began in French colleges in the 1770s and 1780s, but French military engineers were the only cluster with sufficient mechanical knowledge. Their military values prioritized military needs over commerce. Laplace, for example, arrived in Paris in 1769 to take up first teaching post at the École royale militaire.[110]

Civil and military engineering were introduced as fields of study in schools for the state's corps of civil engineers, which was given royal patronage in 1775. The corps of engineers consisted of 230 commissioned engineers in the 1780s, whose work focused on designing, drafting, and verifying structures. The École des pontes et chausées became a professional school with a technical curriculum, which was introduced via French engineers to China at the Fuzhou Shipyard in the 1860s and 1870s. Subjects, arranged in descending order of importance and difficulty, included

1. *Mathematics:* (1) mechanics, hydraulics, differential and integral calculus; (2) algebra and conic sections; (3) elements of geometry.
2. *Architecture:* (1) bridges; (2) docks, jetties, locks, dykes, canals; (3) civil building; (4) stonecutting (stereotomy).
3. *Style and method:* (1) drafting; (2) leveling and volumetric calculations applied to earthworks; (3) estimating work-tasks.
4. *Drawing:* (1) maps (geographical and topographical); (2) figures; (3) landscapes.
5. *Writing:* (1) block lettering; (2) penmanship.[111]

In 1793 at the height of the French revolution, all scientific academies were abolished. The Paris Academy, founded by Jean-Baptiste Colbert (1619–1683), was revitalized in 1795 with largely new personnel. It had lost half of its members. New populist versions of science filled the void left by the disestablishment of Enlightenment science. Laplace was appointed in 1795 to give mathematical lectures at the École normale, where he ridiculed Leibniz's proof for the existence of God.

The triumph of engineering as the quintessential practical science was evident in the fanfare surrounding the founding of the L'École polytechnique in 1794. It became an incarnation of revolutionary ideals and a new vision of science to change the world. The leaders of the revolution and Napoleon all embraced an industrial vision for France that drew on the power of science to change society and nature. French civil engineers were prized a generation after their English counterparts in the 1760s and 1770s. In the nineteenth century the Napoleonic success in science and technology was studied and transferred to Germany.[112]

The typical historian's view of Chinese intransigence and imperial arrogance has relied on a narrative dating from the 1793 Macartney mission, which we will reassess in the next chapter. When Lord George Macartney led his diplomatic mission to China, he commented in his journal on the failure of the Jesuits to grasp the Newtonian sciences. As Joanne Tong has noted, he pointed out the deterioration in scientific understanding from the leading figures of the first generation of Jesuits in China, such as Ricci, Schall, and Verbiest, to those of his own time: "The truth indeed is, that the present missionaries are very little conversant in algebra, or fluxions, and but poor proficients in any other branches of science."[113]

When Xu Shou had his first intellectual contact with modern science as it appeared in Benjamin Hobson's *Treatise of Natural Philosophy* (*Bowu xinbian*, lit., "Broad learning of things newly compiled," 1851), it awakened him to how far beyond the Jesuits natural studies in Europe had gone. His eighteenth-century predecessors in China, however, never knew what had transpired in Europe after the demise of the Jesuit order. They were content to domesticate the Western learning that the Jesuits had introduced and note its similarity to ancient astronomy and Song-Yuan mathematical innovations.[114]

5

The Jesuit Role as Experts in High Qing Cartography and Technology

This chapter explains why the Manchu court still valued its Jesuit minions during the eighteenth century despite the anti-Western policies of the Yongzheng reign. We will see that the Jesuit-supervised imperial factories in China before 1800 survived principally as venues for luxury arts and crafts. Meanwhile, China's immense porcelain industry produced millions of pieces for sale in the eighteenth century, which suggests an interesting parallel to the rise of Wedgwood porcelain in England in the early nineteenth century during the Industrial Revolution.

When Manchu forces captured a Ming artillery division using cannon based on Jesuit prototypes, their firepower proved decisive in the siege warfare that enabled the Manchus and their Han Chinese allies to subdue fortified Ming cities. During later Qing campaigns to subdue the southwestern revolt of the Three Feudatories, the emperor ordered Verbiest in 1674 to cast 132 heavy cannon. The court also ordered 320 light cannon for the military. Verbiest alone supervised the casting of more than half of the 905 cannons produced during the Kangxi reign, for which he also provided detailed translations on the theory and method for their firing. Manchu military banners were reformed to include artillerymen.[1]

Despite setbacks during the Rites Controversy, the Jesuits remained not only as important participants in the Astro-calendric Bureau and the Academy of Mathematics in Beijing but as supervisors in Qing imperial workshops. They also introduced precision instruments that made accuracy a realistic goal for mathematical astronomy in China. This linkage of precision and accuracy was then extended to cannon making, surveying, and mapmaking along the Manchu-Russian frontier. After Ricci, Adam Schall and Ferdinand Verbiest championed the role of mathematics to Christianize literati elites. They also produce astronomical instruments and military weapons whenever the court needed them.[2]

In addition, Jesuits participated in building new imperial palaces and gardens at a time when the Manchu court introduced the innovative garden-palace complex in picturesque suburban surroundings. Before 1700, the Ministry of Public Works and the Imperial Household created temporary palace-building agencies of craftsmen, supervisors, and planners whenever needed. In the late Kangxi era, the court introduced what Cary Liu has described as "the architect as holistic designer, planner, and builder" to develop what eventually became the Lofty Pavilion Garden-Palace in the northwestern suburbs of Beijing. There Chinese imperial designers, particularly the Lei family, and the Jesuits helped develop and build bucolic complexes of whimsical mountains and lakeside villas that reiterated regal authority when the rulers resided outside the Forbidden City.[3]

While in Beijing, for example, the French Jesuits Michel Benoist and Jean-Joseph-Marie Amiot (1718–1793) interested themselves in electricity. In 1755 they sent a report of their tests to St. Petersburg. Such experiments were kept secret allegedly so as not to upset the Chinese. In 1773, Benoist demonstrated an air pump to the Qianlong emperor, who decided that the proper name for the device would be a "pipe to wait for the *qi*" (*houqi tong*), a reference to the fitful efforts by the Astro-calendric Bureau to determine the onset of spring by timing the *qi* emanating from the earth. The Kangxi emperor had ridiculed this ritual early in his reign, but the Qianlong emperor was less critical about such matters. Other Europeans and Americans in China compared native technologies in China with their own, which continued the traditions of Jesuits such as Terrenz Schreck's 1627 *Diagrams and Explanations of the Marvelous Devices of the Far West* (*Yuanxi qiqi tushuo luzui*).[4]

Mensuration and Cartography in the Eighteenth Century

The late Kangxi and Yongzheng bans on propagating European natural studies did not sink literati learning, where a decisive sea change in classical learning was occurring. Imperial policies lost much of their intellectual influence. Eighteenth-century evidential research (*kaozheng*) scholars stressed painstaking research, rigorous analysis, and impartial evidence drawn from ancient artifacts and historical documents and texts. They made verification a central concern for the emerging empirical theory of knowledge they advocated. Their slogan revived the Han dynasty expression "to seek truth from actualities" (*shishi qiushi*).[5]

In geography, moreover, evidential scholars during the late seventeenth and eighteenth centuries reacted to what they knew of world geography by domesticating new knowledge in the midst of Qing empire building, as the court used Jesuit surveying techniques to measure its domains. At the same time that the Manchu dynasty took advantage of international changes in central Asia, Chinese literati in their new geographic works took an inward turn intellectually by focusing on native topics.[6]

The Seventeenth-Century Turn Inward

Chinese naval power had revived when the Ming defended the south China coast from Japanese pirates (and their Chinese collaborators) in the mid-sixteenth century. Led by their naval commander Qi Jiguang (1528–1588) in the 1560s, hundreds of Ming coastal ships equipped with large European-style cannon destroyed the pirate ships used to maraud on land. Later, the Ming fleet Qi had trained joined forces with the Korean navy to resist Japanese invasions of the Korean peninsula in 1592 and 1597.[7]

Modeled on their Ming ally's large warships, Korean turtle-boats (i.e., sailing vessels also propelled by oars that were reinforced with iron plates, ringed with spikes, and equipped with cannon) under Admiral Yi Sun-sin (1545–1598) helped gain control of the sea war. Their combined forces of some five hundred ships and fifteen thousand men were decisively superior in tactics and technology and continually threatened Japan's land-based supply lines. Despite mobilizing a Japanese fleet of twelve thousand men on five hundred ships for the climatic battle at the Noryang Straits in December 1598, Hideyoshi (1536–1598) failed in his grandiose plans to use Korea as a stepping-stone to conquer China. Some 300 Japanese ships with ten thousand sailors were lost.[8]

Despite the increased awareness of foreign nations as a result of the Korean wars, contact with Jesuits, or border treaties between the Qing and Russian imperial governments in 1689 and 1727, Han Chinese scholarly interest in the maritime world waned in part due to Japan entering a period of relative seclusion. During this time, there was wide travel in North and South China by many eminent scholars in the seventeenth century, particularly as a response to the Manchu military conquest in the 1640s to 1660s. Research on geography in the early Qing returned to traditional questions of regional military strategy and local, coastal defense. Meanwhile, the Dutch colonized Taiwan from 1623 when the Dutch East India Company increasingly contracted Chinese traders and farmers from southeast China to settle the island.[9]

Subsequently Southern Ming loyalists led by Zheng Chenggong (Koxinga, 1624–1662) resisted the Manchus in major naval and land battles along the Fujian coast in the 1640s and 1650s. Zheng's land and sea forces took heavy losses, however, when they moved up the Yangzi River to Nanjing in 1659, and were forced to retreat to Xiamen (Amoy). Southern Ming naval forces then challenged the Dutch garrison in northern Taiwan (called "Formosa" by Europeans) at Castle Zeelandia in April 1661 with a force of six hundred ships and twenty-five thousand sailors. The Dutch capitulated after a bitter nine-month siege. Zheng unsuccessfully demanded via a Dominican missionary that the Spanish in Manila recognize his suzerainty.[10]

For its part, the Qing government in 1662 ordered coastal inhabitants from Shandong in the north to Guangdong in the south to move inland to cut Zheng's supply lines and to negate the value of the coast as a battleground. Earlier, Manchu military forces had captured Ming cannon, which they used to pound walled Chinese cities into submission. They also turned to a naval fleet to defend the coastline, after Shi Lang (1621–1696), one of the Southern Ming's most capable admirals, joined the Manchus in 1646 because of a dispute with Zheng. Shi commanded Qing naval forces in the 1650s and 1660s along the Fujian coast. In July 1683, he led the Qing fleet of some three hundred warships and twenty thousand sailors, which subdued the Pescadores. Taiwan fell to the Qing navy in October, and for the first time the island became part of China.[11]

The Qing navy no longer remained on a war footing after Taiwan was annexed, and the Manchu emperors became increasingly preoccupied with the land-based expansion of the Russians from Siberia into the Manchu homelands and the renewed dangers posed by the Zunghars in Central Asia. In addition, the Qing expanded its empire in Tibet and Turkestan. By the end of the eighteenth century, it had more than doubled the size of China. When the Opium War with England broke out in 1839, therefore, the Qing fleet was again mainly a coastal navy used principally for defense against pirates and local marauders.[12]

Furthermore, even before the conquest of Taiwan, maritime considerations evoked less interest among Chinese literati. Because the Manchu conquest emanated from the northern steppe, Gu Yanwu (1613–1682), a leading voice of the early Qing turn toward precise studies, emphasized China's strategic positions in relation to her traditional foreign neighbors to the north and northwest in his influential geographical treatise titled *The Strategic Advantages and Weaknesses of Each Province in the Empire* (*Tianxia junguo libing shu*), compiled between 1639–1662. Ironically, such foreigners, particularly the Manchus, Mongols, Tibetans, and Uighurs, would become part of the political-cultural unit called China by Europeans in the nineteenth century.

Concerned with texts as much as he was with maps, Gu Yanwu did not even mention Ricci's *mappa mundi*. He was concerned with the effects of topography on political and economic development within China. His findings, however, were based on his wide travels, careful firsthand observations, and the study of written materials. Likewise, Gu Zuyu's (1631–1692) *Essentials of Geography for Reading History* (*Dushi fangyu jiyao*), written 1630–1660, was a study of native historical, administrative, and natural geography, with an emphasis on the importance of topography for military strategy. Literati disregarded Ricci's world map, and late Ming interest in the maritime nations waned.[13]

Zuo Zongtang (1812–1885) read Gu Yanwu's and Gu Zuyu's geographical treatises in 1829. They inspired in him a lifelong interest in Chinese topography and military strategy. His interest in Chinese Turkestan underscored his insistence in the 1870s that troops be sent to Northwest China to prevent that area from falling permanently into Russian hands. Zuo's campaigns provoked the opposition of Li Hongzhang (1823–1901) and other Beijing officials who regarded naval power for coastal defense and protection of Korea from Japan as more pressing needs than the recovery of territory in the distant interior.

The compilers of both the *Ming History* and the *Comprehensive Geography of the Great Qing Realm* (*Da Qing yitong zhi*) had access to Ricci's and other Jesuit geographical works, but they dismissed as fictitious many of the Jesuits' claims and much information on their *mappa mundi*. Nevertheless, Xu Qianxue (1631–1694), who chose scholars to work on the *Comprehensive Geography* project when he was appointed director in 1687, had high regard for the emerging evidential scholarship of his times, which we can see in his choice of leading textual scholars as editors of topographical material for the geography project. Xu's appointment of Gu Zuyu, perhaps the most qualified student of historical geography in his time, indicates the caliber of experts who carried out the *Comprehensive Geography* project.[14]

Similarly, the late Qianlong compilers of the 1787 edition of the *Comprehensive Analysis of Archival Sources During the August [Qing] Dynasty* (*Huangchao wenxian tongkao*), which included documents and materials covering the period 1644–1785, demoted the mention of Europe to a minor section on Italy within the category of the "Four Frontiers." Twenty-four chapters in the traditional category of "Geography and Lands" dealt with imperial domains, while only eight covered the borderlands. The compilers of the "Four Frontiers" focused on Korea, the Ryukyu Islands, Vietnam, and other neighboring tributary countries.

The section on Italy mentioned that "what the Italians [i.e., Matteo Ricci] had said about the division of the world into five continents followed from Zou Yan's Warring States theory of the Sacred Ocean [*Shenhai*], although the Italians dared to add that the land of China was but one of the five continents." Such claims infuriated the compilers, who dismissed the Italians as too grandiose, so much so that they failed to mention Zou's nine continents, in which China was central. The Italians could not be taken seriously because they were simply trying to impress the Chinese with European customs, goods, governance, and education. Still, the *Comprehensive Analysis* included detailed geographic discussions of Europe and Russia.[15]

Moreover, the dramatic political impact that European surveying methods had in China early in the eighteenth century piqued the interest of the Qianlong emperor and his court when the Kangxi map of the empire was updated using European surveying methods. European geographical content may have been overlooked, but European methods were still admired and copied in official geography as well as astronomy.[16]

We saw in Chapter 3 that Xue Fengzuo studied astronomy and mathematics under the Jesuit Jean-Nicholas Smogulecki. Xue then applied the techniques of spherical trigonometry and logarithms to surveying, which was appreciated by the Qianlong Imperial Library editors in their review of Xue's compendium on the Yellow River and Grand Canal. The editors noted that Xue's mathematical expertise was an invaluable aid in analyzing problems related to flood control and canal upkeep. His use of European trigonometry was recognized as a clear improvement over the native forms of trigonometry known as "double application of proportions" (*chongcha*, i.e., solving right triangles by using the proportions of their sides), which dominated traditional Chinese surveying techniques.[17]

By comparison, the Imperial Library editors writing in the late eighteenth century singled out a 1557 study of West Lake in the Hangzhou area by Tian Rucheng (1500?–1563?) to make a methodological point: "Because of the lack of corroborating evidence, no one can evaluate the accuracy of its claims. This [imprecision] is a blunder that Ming scholars all shared." Since Song times, increases in precision informed greater accuracy in mathematical astronomy, an expectation that the Jesuits used to enter and control the Astrocalendric Bureau. Qing literati expected similar levels of accuracy in evidential learning due to overlaps between classicism, geography, and mathematical astronomy.[18]

Evidential Research and Geography

Literati scholars who compiled imperial gazetteers such as the *Comprehensive Geography* and its Yuan and Ming precursors increasingly presented the history of geography as the study of topographical change. As scholars, they avoided the moral correspondences between heaven and earth that still informed the traditional cosmography popular in unofficial life. Within the academic community that promoted evidential research, historical and physical geography took precedence over the application of the idealized geopolitical paradigms popular in Han, Tang, and Song classical cosmography.

Geographical historians recognized the physical difficulties in using the ancient mensural order to demarcate existing mountains and rivers. Evidential scholars, for instance, no longer accepted uncritically the portent astrology associated with the field allocation system (*fenye*, lit., "allotted countryside"). Specific celestial fields associated with the twenty-eight lunar lodges (*xiu*) indicated the positions of celestial objects on the celestial sphere, which corresponded to terrestrial regions presented in the *Documents Classic*.[19]

Since the Former Han dynasty, the lodges were associated with twelve ancient feudal states. Analogies were drawn between the celestial realm and the imperial court. Constellations and stars were celestial counterparts to governmental bureaus and their officials. Astrological prediction based on this theory of the continuity between the astronomical and geographical realms remained a common feature in official astronomy through the late Ming and early Qing, even after the Kangxi emperor charged that officials manipulated the major traditions in portent astrology to suit their political purposes.[20]

Although many, Zhu Xi included, had questioned the validity of the field allocation system, and although astrology had not interfered with the accumulation of geographical knowledge, the field allocation was not attacked as an outdated notion until the seventeenth century. During the late Ming, Fang Yizhi noted that the southern region, mentioned in the *Documents Classic* as Yangzhou, was a large portion of current China. In antiquity, when the capital was in the northwest, however, Yangzhou only had three of the twenty-eight stations. Huang Zongxi pointed out that field allocation was bound to a time when the northwest was the core of China. Yan Ruoju discussed the system and noted that it took no notice of foreign regions. He asked rhetorically, "How can it be that the sun and stars did not look out over the Man, Yi, Rong, and Di [peoples]?"[21]

Scholars recognized the gradual displacement of the classical center of ancient culture in the northwest since medieval times. They also perceived the concomitant enrichment of the southeast after the Yangzi delta emerged as the cultural nexus of China since Song times. Gu Yanwu and Huang Zongxi judged this development the result of historical growth in which circumstance and human effort had transformed the Chinese empire. Hu Wei (1633–1714), writing in the second half of the seventeenth century, confirmed the recent rise of the Yangzi delta and concluded that the ancient "nine regions" correspondence order could no longer account for historical changes. Historicizing cultural geography meant that the old "cosmograms" were no longer applicable. The enlargement of geographical horizons led to discrediting the idea of local applicability of portents.

Such pioneering views were more critical than those of the Jesuit Figurists, who naively tried to add the Bible to the cosmological correspondences Chinese literati had woven together for centuries. Such correspondences, re-

worked by Mei Juecheng and others to favor the priority of Chinese traditions, were included in the *Treasury of Mathematics*. Evidential learning
(based on a community of scholars) and imperial ideology (propagated in the
court) had parted ways. Thereafter, imperial rhetoric based on the Classics
was ever more undercut by scholarly criticism of the Classics.

The reevaluation of traditional cosmograms (*tu*) as mensural models for
numbers (*Hetu*) and mathematics (*Luoshu*) was part of the return to antiquity (*fugu*) favored by Qing evidential scholars. They made the cosmogram
an object of analysis rather than the accepted grounds for discussion. They
understood the well-field system and nine regions as specific responses to particular historical and geographical conditions rather than universal, exem-
plary models.[22]

Scholars who used empirical methods in their geographical research during
the Ming-Qing period rejected the symbolic geography in cosmograms and
magic squares (see Figure 4.2, page 178). They turned instead to precise
fields of descriptive and historical geography. Criticism of the River Chart and
Lo Writing was so intense that even champions of Cheng-Zhu learning felt
compelled to dissociate the mysterious charts from the teachings of their Song
dynasty masters.

Embarrassed by the inclusion of the charts in Zhu Xi's commentary on the
Change Classic, orthodox scholars played down the importance of the charts.
Huang Zongxi denied the cosmological significance of the charts and maintained instead that they were originally primeval geographical maps, charts, and
registers, not cosmograms of transcendental significance. Continuing Huang's
efforts to historicize the universal charts, Hu Wei noted the Daoist origins and
associations for the charts in the *Change* in his critique of their purported
mystical correspondences. Their heterodox origins also placed into doubt the
legitimacy of Song literati as rightful transmitters of the classical Canon.[23]

We have see above that Han and Song literati linked these cosmograms to
the origins of Chinese writing and mathematics. The *Treasury of Mathematics* compiled in the Kangxi court reaffirmed such claims. The Qing literati
attack on the authenticity of these charts thus rejected the traditional cosmological ordering of the heavens and the earth and, unknowingly, Jesuits like
Bouvet who associated these cosmograms with Biblical narrative. Jesuit surveys
of the empire completed in 1718 also encouraged this practical emphasis.[24]

Separated since the late Song period, the reunion of classical and technical
studies during the Qing produced specialists who adopted an active, inter-
ventionist approach to problems in river and flood control. The empirical
emphasis of evidential studies had implications for imperial statecraft that
was felt a century before the forced introduction of European technology.
Dai Zhen, one of the leading evidential scholars in the late eighteenth cen-

tury, helped compile a prefectural gazetteer in 1769, an opportunity he used to spell out precise standards for accurate maps and reliable descriptions: "Those who produce maps should emphasize exact scholarship. Upon examination, maps that do not [accurately] represent direction and distance or differentiate names are like not having maps at all. Even worse, they perpetuate errors."[25]

The leading Qing literati-scholars followed Dai Zhen's lead and turned historical geography into a field of evidential inquiry. Place-names were historicized as a sign of their time and specific location. They were employed by scholars as the empirical basis for discerning geographical anachronisms in forgeries. Geography was an important technical element of concrete studies (*shixue*) and was useful in land reclamation and hydraulic works, projects used to order physical space in the eighteenth century.[26]

More and more, diagrams (*tu*) and tables (*biao*) were used as aids in discussion, explanation, and classification. One of the most prominent features of eighteenth-century historical scholarship, for instance, was the increased use of tables of persons (*renwu biao*), supplements, and supplemental tables to help make the Dynastic Histories more accessible as research tools. Many evidential scholars produced important works in this genre. Interested in the organization of knowledge more than in evidential studies, Zhang Xuecheng (1738–1801), for instance, insisted that historical writing should include documentation tables and tables of persons, which would summarize the institutional forms and workings of central and local government.[27]

Gu Donggao's (1679–1759) *Table of Major Events in the Spring and Autumn Annals* (*Chunqiu dashi biao*, 1748) was a model for the collection of chronological, geographical, genealogical, and economic information concerning the late Zhou period from 722 to 481 B.C. Gu arranged his work in tabular form under fifty topics, and he placed supplementary notes by other scholars after each topic whenever there was an element of dispute or doubt. Gu also attached maps that gave ancient and present forms of place-names.

The use of diagrams as the first step in the graphic reconstruction of ancient relics by Dai Zhen and others indicates that leading Qing scholars changed the meaning and use of such diagrams from cosmogram to explanatory model. Geometrical diagrams were so abundant in Mei Wending's writings because Mei was depicting the mathematical nature of astronomy through such diagrams. In their attempts to comprehend celestial motions, Chinese astronomers such as Wang Xichan shifted from strictly numerical procedures to geometric models of successive locations in space.

Rather than *imago mundi,* however, diagrams became for evidential scholars such as Qian Daxin (1728–1804) simply ingenious representations. Unlike Mei Wending, Qian Daxin would not accept abstract diagrams as possible keys to understanding natural phenomena. The mathematization of the world, which in Europe was dependent on Newtonian mechanics and the calculus in the latter half of the eighteenth century, was unavailable to evidential scholars in China until the aftermath of the Opium War.[28]

Although the inward turn in Qing evidential research did not produce a cartographic reconceptualization of foreign lands, we should not underestimate Jesuit and European cartographic influence in China. Through systematic gathering of materials that they would then critically scrutinize and in some cases quantify, Qing scholars combined evidential research methods with data collection and organization. As research pushed forward in the eighteenth century, Asian and Chinese continental geography became a key discipline.[29]

Despite the inward turn of this research away from concern with maritime lands far from China, achievements in geographical knowledge during this period were still evident in military defense and historical and descriptive geography, particularly along the Qing borders with Russia, Zungharia, and Kashgaria in Siberia and central Asia. In addition, the cultural construction of Mukden (Shenyang) and its environs as the exclusive homeland of the Manchus was achieved in part through the mapping of the area under Jesuit direction. The mapping of Manchuria began circa 1690 in the aftermath of negotiations with the Russians to determine the boundaries of the Amur River in northeast Asia.

Such achievements lent themselves to the cumulation of geographical knowledge. Cumulative progress was possible because evidential scholars, building on the efforts of their predecessors, stressed an empirical epistemology and focused on research topics that allowed for continuity in geographical research. As a result, geography emerged as an exacting discipline during the Ming-Qing transition period. We will see that Song-Yuan mathematics also became an important aspect of the revitalization of classical learning.[30]

To be sure, the Americas were still depicted as parts of the Asian land mass north of Great Wall in a Chinese world map circa 1743, which was based on Liang Zhou's 1593 world map. Moreover, plans for the Manchu summer capital in Chengde and for other imperial sites instantiate how European cartographic technologies coexisted with earlier Chinese geographic practices such as cosmographic siting, landscape design, and urban arrangements. The geography of the Manchu empire was intertwined with Tibetan Buddhism, an essential component of Manchu expansion into central Asia. That expansion, as we will see below, also required a substantial investment in state-of-the-art mapping techniques from Europe to delineate accurately the Russo-Chinese border in the eighteenth century.[31]

Cartography, Sino-Russian Relations, and Qing Imperial Interests

Early Manchu rulers recognized they needed better records for land use and taxation. In 1646, a cadastral survey was undertaken, but officials recognized its geographic inadequacies. When the Russians appeared in force along the northern frontier in the seventeenth century, the Manchu court required more accurate geopolitical information to deal with this threat to the empire. After the Tungusic chief Gantimur defected from the Qing in 1670 to allegedly Russian territory, the Russians quickly took advantage.

A crisis in Sino-Russian relations ensued from the 1670s to 1690s, when the Manchus learned that the Russians had already built a fortress in 1654 along the Amur River at Nerchinsk in Gantimur's native region. The changing borders threatened the Manchu homeland. The Kangxi emperor refused any further trade or diplomatic relations with Russia until the deadlock was resolved. Meanwhile, the Russians and Zunghar Mongols both expanded their interests in the northwest while the Qing was preoccupied in the south and southwest during the Revolt of Three Feudatories from 1673 to 1681. Much like late Ming calendar reform, Qing recognition of its geographic needs preceded the European contributions to Chinese cartography.[32]

During the Rites Controversy, the Manchu court was embroiled simultaneously in military threats from Zunghars and Russians along the borders of the empire in central Asia, which introduced new elements into the storm over the Jesuits and their loyalty to the Qing dynasty. For example, when the Russian mission was allowed into Beijing in 1676 to negotiate trading agreements and population movements, Ferdinand Verbiest was involved. The lack of a clear boundary in the Amur River area and the resulting ambiguous claims to sovereignty later led to the treaty of Nerchinsk in 1689, negotiated by the French Jesuit Gerbillon, which demarcated the frontiers between the Qing and Russia.[33]

Jesuits and the Mapping of the Qing Empire

The Jesuits and others were commissioned to provide data that would enable Qing leaders to stem the tide of Russian infiltration into Manchu and Mongolian homelands. The geographical knowledge that accrued during this time was an important addition to earlier information about foreign lands. In the process, the Kangxi court's awareness of the actual geographical divisions of the Sino-Russian frontier slowly caught up with their knowledge of Southeast Asia.

In addition, the Kangxi and Qianlong emperors used new techniques for trigonometric surveying to map Qing dominions in the eighteenth century. Surveyors first carefully measured a base line for one known location. They then used angular measurements of the stars to determine the latitude. By actually measuring distances using a calibrated chain, telescopes, and angle-measuring instruments, they also resolved the precise longitude. When they knew the location of the second end of the base line, they could reckon the third point by triangulation. A series of triangles could be extended trigonometrically for any distance in an entire area or region.[34]

The Jesuits produced their first survey of Beijing in 1700, which the Kangxi emperor checked. Later, he asked for a survey of portions of the Great Wall in 1707. In 1710 further surveys along the Amur River helped mark the strategic bases on the border with Russia. The Manchu homelands were surveyed between 1709 and 1712 and a complete map of greater Mukden, i.e., Manchuria, was produced. The text and maps included in the 1733 edition of the *Collected Statutes and Precedents of the Great Qing* (*Da Qing huidian*) were concerned with military deployments and garrison towns. They were compiled under the auspices of the Ministry of Military Personnel. The maps that the Jesuits prepared for the Manchu homelands became the starting point for later Japanese and European maps of the region.[35]

The geographical ambitions of the empires of the Qing dynasty, Russia, and the Mongol Zunghars in central Asia led to the redrawing of the frontier boundaries between Russia and the Qing in the region and the crushing of the autonomous state of Zungharia in 1760 by Qing armies. The Qing, Zungharia, and Russia each produced important new maps of unprecedented scale and accuracy as political and ideological weapons in their struggle for control of central Asia. Peter Perdue has noted how eighteenth-century central Asian borders were constructed through three stages: military confrontation, negotiated treaties, and symbolic representation on maps and instantiation in imperial documents. All sides added stelae inscriptions, military pageants, and commemorative paintings to their repertoire of empire building. The new maps were products of preliminary surveys that had preceded and often made the warfare possible.[36]

As Russian expansion in Siberia challenged Qing power, the Jesuits, when carrying out the survey for the Kangxi Atlas, had limited access to the most sensitive frontier areas. Dynastic security and territorial claims compelled the Qing court to hire only those Jesuits who did not intend to return to their native lands. The dynasty avoided circulating such information too widely inside and outside of China. Seeking to open an overland route from Rome to China via Russia, Verbiest may have secretly provided Russian missions with maps and descriptions of the border region with Siberia, which included the locations of Manchu forces he obtained from Russian deserters. By 1727, Qing surveyed the region of Amuria in light of the realities of Russian penetration into Siberia.[37]

When Russia and China defined their common borders in the treaties of

Nerchinsk in 1689 and Kiakhta in 1727, both applied new surveying techniques for cartography to the newly defined borders. In addition, they compiled new classification systems and ethnographic atlases to control the movements of refugees, nomads, tribes, traders, soldiers, and other mobile groups across the borders. Both sides used tax and land registers, censuses, border patrols, passports, and visas to keep people from moving freely across the borders. Each also applied seventeenth-century European technical knowledge transmitted through the Jesuits to survey their new territories.[38]

The atlas and its subsequent Qianlong-era revisions shared features that were consistent with contemporary European maps. They left out pictures and drew on astronomical observations to calculate longitude and latitude based on a precise scale. Hostetler interprets such developments in light of Qing evidential studies and the change in research epistemologies that affected scholarly views of geography in the late seventeenth century, a time when government interests increasingly focused on internal military defense and historical and descriptive geography. In the original maps, however, China of the Han Chinese and the Qing empire were not coterminous. China was presented as one distinct part of the Qing empire, and the Manchus homelands were another. Two other versions of the map from the same surveys, however, were entirely in Chinese with no Manchu script, perhaps to avoid offending Han Chinese cultural sensibilities. These Chinese language maps elided the Manchu view that the maps included distinct administrative and cultural spheres. Zungharia and Tibet were later added as distinct spheres of the Manchu empire.[39]

French and Russian Imperial Cartography

France became a leader in cartographic activity under Louis XIV after he appointed Colbert as minister for home affairs. Colbert made France a center for science and solidified that role in Europe by founding the Academy of Sciences in 1666. Louis XIV also promulgated topographical surveys for territories based on astronomical observations for triangulation that were initiated under his chief of astronomy, the Italian Jean Dominique Cassini. Colbert invited Cassini to France in 1669 to make astronomical observations crucial to improved navigation and mapping. Cassini communicated with the Jesuits in China because he sought such observations globally.[40]

Enlarged in 1676, Cassini's *Ephemerides* permitted astronomers to determine the latitude and longitude of the point from which they made their observations. In 1679, France began a national survey relying on Cassini's tables for accurate measurements. The French Academy required observations from around the globe, which, among other things, led to Louis XIV sponsoring the French Jesuits in China under the Missions Étrangères in 1663. Once mapmaking became a vital component of imperial expansion, the cartographic technology to carry out accurate geodetic surveys spread quickly. France, Russia, and the Qing employed experts regardless of their origins. Colbert had recruited the Italian Cassini, Kangxi employed the French Jesuits, and Russia engaged the Swedish officer Strahlenberg, who was taken prisoner by Russians in 1711, to collect information about Siberia, Mongolia, and neighboring regions.

Peter the Great (r. 1682–1725), likewise used maps to measure the growth of the Russian empire and to legitimate its claims. In 1698, Peter had already commissioned a survey of his new territories. New maps for an atlas were completed in 1701, although full surveys of the empire were not formally initiated until 1727 by Catherine I, Peter the Great's widow. For France, the Qing, and Russia, the requirement of better maps was tied to imperial expansion. New surveying and mapmaking techniques were essential. In the midst of Russian expansion into Siberia the Qing empire more than doubled in size from 1660 to 1760 in a global context of population growth, exploration, and colonial expansion.[41]

The Kangxi surveys were quickly completed by 1717, while French surveys of roughly the same precision took until 1744 to accomplish. The Qianlong revisions of the 1717 survey were finished in 1755, while the second edition of the French survey appeared in 1788. Similarly, the Russian imperial atlas, which emulated the French national survey, appeared in 1745. Peter the Great used cartography and his European experts, who were also hired to explore the North Pacific, to put Russia on the map of eighteenth-century Europe. The Kangxi Atlas decisively changed the content of European mapmaking when the Jesuit maps first arrived in France in 1725 and the new information was digested in Paris, London, and elsewhere.

Similarly, the latest mapping technology was effective for the Qing in legitimating and consolidating the empire, and became the basis all of China's territorial claims in the twentieth century. The Kangxi emperor's gift of his survey to Peter in 1721 aimed to apprise Russia of Qing sovereignty and cartographic sophistication. It pointedly did not record the strategic information the Qing had about the northern border areas.[42]

More importantly, the Kangxi emperor sought peace with the Russians to free his hand in wars with the Zunghar Mongols in Central Asia. By neutralizing Russia, the Qing court prevented a possible Russo-Zunghar alliance against them (see Map 5.1). Hence, when the Russians demanded equal political status with the Qing at the Nerchinsk peace negotiations, the Manchus did not allow what they considered ceremonial procedures to interfere with their primary diplomatic task. The emperor relented on the usual Qing

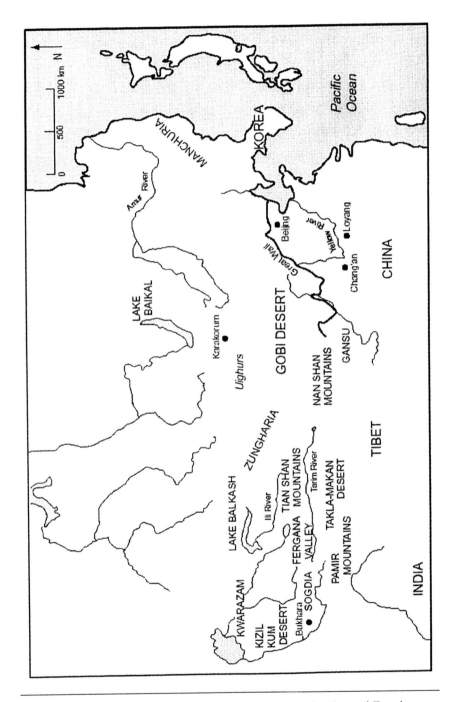

Map 5.1. Map of Qing dynasty territories in proximity to Russian and Zunghar lands in Central Asia

ceremonial claim of imperial superiority when dealing with bordering states. The Treaty of Nerchinsk represented a compromise in which the marking out of the frontier was more favorable to the Manchus, while the Russians kept Nerchinsk. In addition, the Manchus conceded that trade could be initiated by either side, and each could cross the border with passports. Furthermore, they settled the problem of the repatriation of fugitives.

As an instrument of diplomacy, the economic concessions made by the Qing government in the 1689 treaty proved their political worth when the leader of Zungharia, Galdan, proposed an alliance with the Russians in 1690. Joint military action against the Manchus was now impossible, however, because the Russians were bound by treaty with the Qing. The Kangxi emperor was left free to eliminate the Zunghar threat, arguably the "last nomadic empire," which he did in 1696. Galdan's death in 1697 reduced the Mongols as a potentially divisive third force in central Asia.[43]

In 1718, the Russians contemplated full normalization of Sino-Russian relations during the Rites Controversy, which was damaging Jesuit and Catholic interests in China. Peter the Great, for example, expelled the Jesuits from Russia in 1719 and tried to install a Russian Orthodox bishop in Beijing in 1722. Russian authorities unsuccessfully kept this effort secret from the Manchus and the Jesuits in Beijing, but Qing suspicions prevented the appointment. In 1728, the Zunghar threat against the Manchus in Turkestan and Tibet revived. Again, the Manchus eliminated the threat through the Zunghar wars in the 1750s, which were facilitated by the Treaty of Kiakhta that ended Russian interference. The Kiakhta treaty established officially supervised trade in Amuria that stabilized the Russian-Qing frontier from 1727 until the nineteenth century.

The Qianlong reign brought a complete victory over the Zunghars and the incorporation of Ili in the far northwest by the Qing in 1755. Manchu military victories led to Qing overconfidence with the Russians, which generated a ban on trade caravans to Beijing from Russia after 1755. Qing success in central Asia in the eighteenth century thus occurred in the context of Russian expansion into Amuria. Through compromise and accommodation, Russian interests in trade and Manchu interests in central Asia were protected. Diplomacy, warfare, and timely mapping of strategic frontiers enabled the Qing dynasty to incorporate major portions of Amuria, Zungharia, and Kashgaria at the expense of the Mongols, Uighurs, Kazaks, Tajiks, and Russians.[44]

The Jesuit Role in Qing Arts, Instruments, and Technology

Before the Napoleonic Wars were resolved in Britain's favor, the Qing court and its literati continued to co-opt certain aspects of European learning and artistic expertise. For example, Jesuits remained as court clock makers and geometers, and a Jesuit, usually Swiss, was always in charge of the imperial clock collection in the eighteenth century. When compared to the industrializing tendencies we identified in Britain and France, the imperial workshops in Qing China, like the Academy of Mathematics under the Kangxi emperor, were never breakthroughs to European-style science and technology.

Qing artisans were, however, important in manufacturing clocks, porcelain, and glass, and in building pavilions and gardens during the late empire. Interestingly, when the Macartney mission presented clocks and watches and a telescope and planetarium to the court to demonstrate European inventiveness, the Qianlong emperor was not particularly impressed and rejected any official trade with Britain. His haughty imperial tone belied his extensively collecting and reproducing European manufactures and clockwork at home.[45]

Clock Making in the Kangxi Era

The French and Swiss Jesuits' expertise in mathematics, astronomy, cartography, and clock making was connected to the unprecedented degree of accuracy in European time and space measurements. When the Kangxi emperor commissioned the Jesuits to survey the entire empire between 1708 and 1717, the required mapping was accomplished in collaboration with official Chinese astronomers and Manchu officials. Chinese clock makers were trained because the Kangxi emperor also thought clock-making skills were worth appropriating. Chinese who worked under Jesuits were expected eventually to make their own mechanical clocks.[46]

According to Catherine Pagani, mechanical clocks in early modern Europe were a metaphor for God's maintenance of the universe. In an age when a mechanistic view of nature was emerging, Europeans believed that the mastery of clock mechanics would lead to better understanding of God's design of the world. The Jesuits accepted this view, but they also recognized that when mechanical clocks were presented to potentates at home and abroad, this act gained them access to high places in the papal court as well as in the Chinese empire. Their skills as architects and glassmakers paralleled their careers as computists and clock makers in late imperial China.

Matteo Ricci believed, for example, that he had gained admission to the Wanli court by presenting a clock and repeating watch to the Ming emperor, a tactic which followed established European gift-giving practices. When Johann Schreck and Wang Zheng prepared their 1627 translation titled *Diagrams and Explanations of the Marvelous Devices of the Far West* (*Yuanxi qiqi tushuo luzui*), it represented the first work in Chinese that provided information about European escapement techniques for delivering power at regular intervals to move the gear train inside a timepiece. Subsequently, the Wanli emperor had his eunuchs work with the Jesuits to master the art of repairing the clocks.

In the late seventeenth century, the Kangxi emperor established a number of workshops that manufactured luxury items under the auspices of the Office of Manufacture in the imperial household. These were modeled after the studios in the French Academy. Mentioned in palace documents of 1689 and 1692, the Kangxi workshops were at the outset staffed by Jesuits and Guangdong workmen. After the imperial workshops were formally established in 1693, there may have been as many as thirty-one shops. Most clocks in the eighteenth century were completed in shops inside the Forbidden City. An Office of Clock Manufacture was subsequently created in 1723, which lasted until 1879, when the last list of clocks in the court was compiled.[47]

Three major sites were operating in 1756 under the Office of Manufacture within the Imperial Household Department. There, clock makers, painters, and engravers worked in the decorative arts, using enamels and glass for the clocks. The Lofty Pavilion Garden-Palace (*Yuanming yuan*) also served as a venue for clock making. In 1752, for instance, Amiot mentioned that the Jesuits prepared a hemicyclical mechanical theater clock with three scenes, which was built for the empress dowager's sixtieth birthday and remained in the Lofty Pavilion. Manchu and Chinese interest in elaborate mechanical clockwork, as in Europe, clearly reflected the linkage of imperial power and prestige.[48]

Five European horologists were working in Beijing in 1701 during the Kangxi reign. Under the Qianlong emperor, eleven Jesuits built clocks in the imperial workshops. By 1800, with only a few ex-Jesuits remaining, Chinese artisans were proficient enough to make mechanical clocks themselves, so the missionaries became less involved in production. Outside of Beijing, for instance, imperial clock-making workshops were also established in Suzhou and Hangzhou.

The native industry for clocks provided a commodity that served as an effective form of tribute gift giving among Qing officials. Modest levels of industrial production by native artisans, when compared to the massive porcelain industry, were monopolized by the court and its literati elites. Production in the provinces was controlled by Chinese merchant guilds. The popularity of mechanical watches among Chinese elites mirrored imperial taste in the eighteenth century. Mechanical clocks were mentioned in novels such as *Dream of the Red Chamber* (*Honglou meng*), where owning such a clock reinforced the high status of the Jia family.[49]

The Qing court and its Jesuit clock makers did not realize, however, that an urgent demand for more accurate determinations of longitude for ships at sea was now driving precision clock making in Europe. Small errors for seagoing vessels using dead reckoning could prove fatal when landfall was unexpected. Captains who followed only well-traveled paths on the ocean also knew their ships were vulnerable to pirates. By 1714, the Westminster Parliament established a sizeable prize "to determine longitude to an accuracy of half a degree." Subsequently, inventors presented increasingly accurate and synchronized timepieces to the prize committee. When placed on board these timepieces could determine a ship's distance from a standard zero-degree meridian.

Greenwich, England became the center of eighteenth-century efforts to measure a ship's distances by symphonizing its movements over time. A by-product of Newtonian mechanics and the calculus, which had linked time and space to map movement, the cutting-edge timepiece on Captain James Cook's triumphal second voyage of 1772, for instance, was not an idle curiosity, as mechanical clocks were in China. It was the key to plotting accurately the location of his ship throughout the voyage. The British soon discovered that another important by-product of accurate timepieces was that their agents abroad could adjudicate contested boundaries in their favor using impartial longitude measurements.[50]

Imperial Factories for Glassware

The French Jesuits remained heavily involved in the glass workshop that the Kangxi emperor had established in 1696 in the palace. The German Johann Kilian Stumpf worked with the French Jesuits to produce decorative glass under imperial auspices. To please the emperor, the Jesuits arranged to have two European glassworkers sent to Beijing in 1699 to work in the imperial workshop.[51]

The Beijing workshop was producing high-quality glassware from the beginning of the eighteenth century. Decorative snuff bottles were fabricated in a wide variety of colors and shapes. The emperor presented such glassware as gifts to high officials on his southern tour to the Yangzi delta in 1705, for instance. He also presented the papal legate de Tournon with an enameled glass snuff bottle at the outset of what proved to be a disastrous series of meetings in 1706 to discuss the rites controversy.

Matteo Ripa (1682–1745), an Italian Jesuit who arrived in China in 1710, noted that by 1715 the glass workshop consisted of several furnaces for glass making, which required a large number of skilled craftsmen under Stumpf's supervision. In 1719, Jean-Bapitiste Gravereau (1690–ca. 1758–1769), a French Jesuit expert in enamel, arrived in Beijing. The enameled glassware produced in the workshop was of high enough quality that in 1721 the Kangxi emperor sent the pope two large cases of enamelware, in addition to 136 pieces of Beijing glass.[52]

After the Yongzheng emperor's anti-Jesuit policies took effect, the imperial workshop, like the Academy of Mathematics, increasingly relied on native glass-making talent, particularly after Stumpf died in 1720. The court also encouraged the workshop to manufacture enamel colors independently of the Jesuits. In addition, enamel colors were sent to the imperial pottery kilns in Jingdezhen, where new, enamel-decorated porcelains appeared for the first time. The overglaze blue of the Kangxi era had been introduced to Jingdezhen in 1700, and under the Yongzheng reign a translucent pink enamel made from colloidal gold was used in porcelain. The opaque pink glaze with specks of metallic gold developed by the imperial workshop drew on a recipe for ruby glass in a clear blue matrix. The ruby glass derived from an ancient Venetian formula, rediscovered and developed in Germany.[53]

Despite his animosity towards Christianity and the Jesuits, the Yongzheng emperor established a branch of the glass workshop at the workshops in the Lofty Pavilion Garden-Palace in the northern suburbs of Beijing. Production also continued at the Jesuit glass workshop into the Qianlong era, when two additional Jesuits joined the imperial glass workshops in 1740. Glass production reached its high point when the Jesuits became involved in constructing European-style palaces and gardens for the Lofty Pavilion in the 1750s.

European-style glassware was produced for display in the elaborate buildings that the Qianlong emperor asked Giuseppe Castiglione (1688–1776) and others to design for him. Castiglione's designs for Gabriel-Leonard de Brossard's (1703–1758) carved glass and enamels were also prominently displayed in the palace complex of the Forbidden City (*Gugong*). Emily Curtis has noted how Chinese decorative themes were connected with European illustrative skills for shading and perspective to adorn glassware. By 1766, the Qianlong emperor had turned the Lofty Pavilion gardens into a treasure house of gardens, pavilions, paintings, glassware, porcelains, and furniture, whose artistic merits and technical prowess were of the highest standards.[54]

Jesuits and Garden Architecture

Due to the Qianlong emperor's patronage, the Jesuits found new life in the Qing cultural world of the 1750s through their expertise in painting, designing, and building the lavish European-style palaces and interiors that the emperor sought. Initially impressed with the grandeur of European-style fountains, the emperor asked Castiglione to join his imperial design offices to draw up plans for such fountains in the Lofty Pavilion gardens. Castiglione in turn sought the help of Michel Benoist, who had arrived in China in 1744 and later presented the court with an accurate account of Copernican cosmology. Because Benoist was knowledgeable in mathematics and hydraulics, as many Jesuits were, he was able to present the emperor with a model fountain, which the court quickly authorized Castiglione to build alongside the Baroque-style palatial buildings.[55]

The European-style garden designers were Jesuits, with Castiglione playing the most prominent role from 1745 to 1759. They were well prepared for their task in Beijing because in contemporary Rome the popes had glorified themselves and the Church via an ambitious scale of palaces, piazzas, and fountains throughout the old city, supervised and built by the clergy. The mid-eighteenth century was also a time when the impact on European rulers and aristocrats of chinoiserie-style architecture, gardens, and lakes led to the construction in Europe of Chinese-style rooms, gardens, and pagodas, such as the 1762 garden and pavilion in Wales designed by the architect Sir William Chambers (1723–1796) for the Princess Dowager Augusta.[56]

When the first European-style pavilion was completed in 1747, the Qianlong emperor indicated his pleasure, according to Benoist. Located on sixty-five acres, the European section (*Xiyang lou*) of the Lofty Pavilion gardens was 750 meters long and 70 meters wide when completed and stood at the northern end of the Long Spring Garden (*Changchun yuan*) inside the Lofty Pavilion's massive grounds (see Figure 5.1). The second phase of the European section inside the Lofty Pavilion took another eight years to complete.[57]

A three-story palatial building designed by the Italian Jesuit architect Fernando Bonaventure Moggi (1684–1761) was known as the Belvedere (*Fangwai guan*) when completed in 1759. Built in a crescent shape evoking Islam, it had marble balustrades enclosed by a moat. After entering from the bronze outdoor stairway that rose up to the second floor, one encountered two tall stone tablets with Arabic inscriptions. The front door resembled that at St. Andrea Al Quirinale in Rome and faced a marble bridge with flamboyant balustrades across from a moat that led to a smaller garden.

To the east a thirty-six room pavilion was named the Sea Calming Hall (*Haiyan tang*) when completed in 1781. The largest structure built in the European section, it was likely modeled on the Court of Honor at Versailles. It housed a reservoir with goldfish under a glass ceiling. A room on each side contained hydraulic machines to pump the water in the reservoir and feed the outside fountains and cascades. Nearby stood a triple gateway each with triumphal arches that resembled the Triumphal Arch in Paris.[58]

Young-tsu Wong has explained that the pavilions, fountains, and arches designed by the Jesuits made up about one-fifth of the Lofty Pavilion. For such a large-scale project, they were required to supervise many Chinese artisans, builders, and masons, whose architectural talents using Chinese materials and tools were indispensable for constructing the European section. Hence, the Baroque models that the Jesuits designed melded with Renaissance, French, and Rococo styles that the imperial laborers also reproduced. Workers added

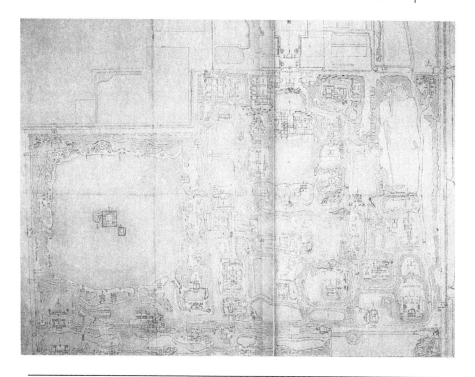

Figure 5.1. Layout of the Lofty Pavilion (*Yuanming yuan*) gardens.
Source: Le Yuanmingyuan: jeux d'eau et palais européens du XVIIIe siècle à la cour Chine, edited by Michèle Pirazzoli-t'Serstevens (Paris: Éditions Recherche sur les civilizations, 1987).

elements of Chinese garden architecture, from the inclusion of "Lake Tai stones" (*Taihu shi*) and bamboo pavilions, to the mix of buildings in the European section, which included roof tiles in native yellow, green, and blue colors.[59]

Overall, the Lofty Pavilion represented an eclectic architectural style that the Manchu court favored as part of their efforts to create a universal vision of their power first in Asia and then in the world. The most representative building of this syncretic style was the small Belvedere, which combined the design of a Florentine building with a classic double Chinese roof. The interior designs for the pavilions were also inspired by European models with glass windows, wood-plank floors, handrails, flower terraces overlooking lawns, mechanical clocks, hanging lamps, and oil paintings. Even Gobelins tapestries of French beauties presented by Louis XV were placed throughout the European section. Large nude figures were not included, however. Chinese garden elements outside and traditional scroll paintings inside added to an ambience of Manchu pretension as global style setters.[60]

Porcelain Production in Jingdezhen

Relatively unknown in Europe until 1675, porcelain was a significant feature in aristocratic households by 1715. Europeans avidly sought the secret for making porcelain but were unsuccessful until the Meissen factory in Germany produced the first European hard-paste porcelain after its discovery in Dresden in 1710. Englishmen such as Francis Bacon had at one time thought that porcelain developed from an "artificial cement" buried in the earth for a long time, not unlike Chinese views of the formation of amber (see Chapter 7). John Donne considered buried clay the source for porcelain. Song Yingxing's 1637 *Heavenly Crafts Revealing the Uses of Things* (*Tiangong kaiwu*) had included a section on porcelain making, portions of which were reproduced in the eighteenth-century *Synthesis of Books and Illustrations Past and Present* encyclopedia.[61]

In early modern Europe, Chinese porcelain was called "china." Queen Mary II of Britain, for instance, had a porcelain room designed by the Dutch decorator Daniel Marot. Another such room was restored in 1688 in the Oranienburg Palace by Frederick III, elector of Brandenburg and Prussia, which consisted of a salon decorated with cabinets and pyramids of Chinese porcelain. In addition to founding the Meissen porcelain factory, the king of Saxony, Augustus the Strong (1670–1733), had over twelve hundred blanc de Chine wares made in Dehua, Fujian. White-glazed porcelain was more fragile due to the brittle clay used in Fujian. Until their popularity waned in the eighteenth century, blanc de Chine wares were also collected at the English royal palace at Hampton Court and showed up in the 1688 catalog of holdings in the Cecil family's Burghley House. By 1700 over 150,000 pieces of porcelain might be unloaded from an arriving ship in England, and the British East India Company took some 400,000 orders by 1722 to satisfy the obsession for "chinaware" among upper-class women.[62]

Dating from 1004 to 1007, the Jingdezhen kilns were burned and plundered during the Ming-Qing transition. In the late Ming, moreover, major changes in the production of porcelain occurred, which continued during the early reign of the Qing dynasty. For example, the Jingdezhen kilns lost their imperial patronage when the Ming court became financially strapped for military funds needed to cope successively with the Japanese invasion of Korea, internal rebellions, and the Manchu threat in the northeast.

Managers of the kilns sought new patrons in and outside China and experimented with ceramic techniques and designs that satisfied new customers in export markets in Japan, Europe, and Southeast Asia. Without the imperial funds earlier earmarked for pottery production, potters were no longer dependent on imperial taste. After 1620, the Jingdezhen kilns increasingly produced underglaze landscape designs and depictions of popular scenes from native poems, novels, and dramas on blue-and-white ware. They also responded to Japanese tastes for the understated, rough pottery used in the tea ceremony, as well as to Dutch avidity for blue-and-white porcelain designs that the East India Company bought as foreign wares in enormous numbers (see Figure 5.2).

After the Revolt of the Three Feudatories was put down in Jiangxi, during which Jingdezhen was pillaged in 1675, the Kangxi emperor reestablished imperial control of the kilns. Although the new elements that had emerged in the transitional styles of pottery during the late Ming were continued in the early Qing, renewed imperial funding again became the potters' priority. Manchu emperors, moreover, took a personal interest in the industry, thereby imposing imperial tastes in porcelain during their reigns. The Kangxi emperor, for example, favored pure, bright sapphire-blue pieces made from local cobalt oxides, which replaced the "Mohammedan blue" with purple reflections popular during the Ming.[63]

The Jingdezhen imperial factory flourished under the Yongzheng and Qianlong emperors due to favorable economic conditions and effective directors who developed famille rose ware, which could produce any shade of pink by creating a solution of gold chloride and stannous chloride for new enamel. Qing potters also created expert imitations of Song crackled glazes and ware with purplish blue splashes, to which the Qianlong emperor often added his poetry. Qing artisans also exhibited a technical virtuosity for producing monochrome wares by developing a wide range of muted shades such as lavender blue.[64]

In addition to Jingdezhen wares, porcelain was also produced at Dehua in Fujian province. Dehua was noted for blanc de Chine ware made with closely packed snow-colored clay, which was more popular in the eighteenth century than polychrome wares. Guangdong stoneware was popular in the nineteenth century, while Yixing kilns northwest of Suzhou produced brown stoneware teapots, whose eccentric shapes became de rigueur among literati savants. In addition, polychrome ware in the "five colors" (wucai) of red, yellow, green, aubergine, and blue was produced for the domestic market at a time when trade in porcelain to Europe was dominated by blue and white ware. In the nineteenth century, "five colors" ware also became popular in Europe.[65]

After the Kangxi emperor ordered its rebuilding in 1682 or 1683, the manufacture of porcelain at the Jingdezhen factory drew the most attention from Europeans who wished to divine the Chinese secret of fabrication. The kilns were visited by French Jesuit Father Francis Xavier d'Entrecolles in 1712, for example, and he gave a detailed account of it, which touched off a European search for comparable ingredients to manufacture porcelain at home. Lang Tingji (1663–1715) assumed control at Jingdezhen while he was governor of Jiangxi from 1705 to 1712. The porcelain styles he promoted included *sang de boeuf* and apple green, which were imitations of mid-Ming wares. From 1693, metal cloisonné pieces, which required fusing pulverized multicolored glass to metal, were also produced during the Kangxi era under imperial patronage in imperial workshops.[66]

During the Yongzheng reign, the bondservant Nian Xiyao (1671–1738) became superintendent of customs and director of Jingdezhen factory in 1726. He began to include European designs as result of Castiglione's influence at court, producing a synthesis of Chinese and European art. Nian had written several works on mathematics, especially one on trigonometry that was printed in 1718 when Jesuit influence at the Academy of Mathematics was still important. In addition, he was impressed with techniques of perspective in European painting, which he had learned from Castiglione, and authored a treatise titled *Studies of Perspective* (*Shixue*) that was published in 1729 and reprinted in an enlarged edition in 1735. Due to Nian's tutelage the purple of Cassius and famille rose wares became prominent at Jingdezhen, as did blue-and-white ware and eggshell porcelain.[67]

Figure 5.2. Blue-and-white porcelain bowl from the Qing dynasty.
Source: Permission from China Institute Gallery.

Under the Qianlong emperor, Tang Ying (d. 1756) was initially appointed administrator of Jingdezhen in 1724 and again after 1737 where he remained in charge until 1749. Tang also promoted polychrome and blue-and-white ware. Under him, eggshell wares with European decoration were also produced. The Qianlong period in particular lead to a significant increase in the mass production of porcelain at Jingdezhen. As in other arts, this era was marked by the impact of Jesuit enamel-painting techniques on native designs for pottery and cloisonné.[68]

As they progressed since antiquity from earthenware to stoneware, the Chinese used hard-paste clays. Genuine hard-paste porcelain (*ci*) resulted by combining "kaolin," a fine, white refractory clay, with petuntse (*baidunci*, i.e., pulverized feldspathic rock also called "china stone," lit., "small white briquettes") from a rock known as pegmatite in a village near Jingdezhen, which fused when heated at a temperature of 1,280 to 1,350 degrees centigrade. Kaolin (from the village of *Gaoling*, lit., "high ridges") was derived from quarries outside of Jingdezhen. Their clay chips were crushed by hydraulic hammering devices, and the white powder was placed into water, stirred and decanted. When a solution of fine particles rose to the top, it was repeatedly collected until only paste remained at the bottom, which was then transferred to large molds to harden.

The molds were fashioned into lumps and transported to Jingdezhen where the petuntse was trampled and crushed into a powder. After the powder was transformed into soft paste, it was purified. The lumps of clay were then delivered to the potter's workshop for the fashioning and shaping process on the wheel. After the piece dried in the open, the potter finished it with a turning tool by carving or incising designs. Another craftsman prepared a mold using yellow clay for the required shape. For complicated pieces several sections were required for the mold.[69]

An underglaze of blue decoration was initiated in the Kangxi era. After drying, the pieces were coated to prevent coloring oxides from being absorbed. Cobalt oxide was then diluted in water and painted on for decoration. A final coating of glaze was applied until the piece was white. During the enameling process, potters applied another coating of pentuntse mixed with gypsum, fern ash and quicklime to the unfired body once it had dried, which made it white, but sometimes it was also stained with a metallic oxide. Firing was done through a box made of refractory earth or porous clay, and each piece was covered with a layer of sand including powdered kaolin to keep it from sticking to the box. More-delicate porcelain was placed in the middle of kiln. Smaller pieces could be fired for five days. Larger pieces might take up to nineteen days, including the cooling periods, to bring out the decoration.

The firing floor was covered with gravel, and had a spy hole through the door to view the firing. Once the coating of glaze had melted into the body of the piece and cooled, the potter applied the genuine decoration of overglaze to the surface using colorants of iron oxides combined with cobalt or cobaltiferous manganese. This process produced willow-green celadon, the shiny blue and mirror-black wares of Kangxi period, as well as the yellow, rust, and "tea-dust" wares of the Qianlong period. Copper oxide produced red and green colors, as well as the peach-bloom and turquoise-blue glazes.

Once the piece was decorated, it was fired a second time to vitrify and melt the coloring agents onto the coating. The temperature for the second firing never exceeded 800 degrees centigrade. Upon completion, the kilns were opened up in the presence of a official who took 20 percent of the contents as tribute for the Qing dynasty. Touching up was completed at the workshop before the pieces were sorted according to quality. Up to 50 percent could be rejected. A certain number of pieces were sold immediately, while others were transported in wooden casks from the remote area of Jingdezhen through the river and lake systems of the middle Yangzi region to major urban centers and then on to Beijing.

International trade was directed through Guangzhou to Southeast Asia and Europe. Aeneas Anderson, who accompanied Lord Macartney on his 1793 mission, wrote, "There are no porcelain shops in the entire world which can compare in size, richness or delivery with those in Canton." As imperial patronage lessened in the nineteenth century, however, the porcelain industry declined significantly, and Wedgwood in England and Japanese potters replaced Jingdezhen as the best-quality kilns in the world.[70]

European Technologies Depicted in Qing Art

A 1741 painting by Castiglione titled "Portrait of the Emperor Troating for Deer," which was commissioned by the Qianlong emperor, presented the Manchu ruler on an autumn deer hunt. Dorothy Bernstein has described the role of European perspective and technology informing the painting. First, it used a European pictorial technique. Second, both a gun and a telescope were included. The gun was carried by the emperor's companion, while the emperor carried a bow and arrow, indicating that the gun, while an essential hunting tool, was subordinated to the bow as a symbol of Manchu prowess in archery.

Like the Qianlong emperor, the Kangxi emperor while on hunting expeditions had used telescopes to impress his entourage with his knowledge of trigonometry and surveying. In the painting of the 1741 hunt, a large telescope was strapped in its case on the back of a horse in the procession. It represented the Qianlong emperor's surveillance of his empire. The painting thus affirmed him as master of his empire, Jesuits included. The European pictorial techniques that Castiglione integrated into the picture depended on perspective and confirmed the emperor's privilege to control space and time.[71]

Similarly, Jin Nong (1687–1764) and Luo Ping (1733–1799) represented eighteenth-century Chinese painters living in an era when a multiethnic society under the Qing imperiled Han Chinese cultural dominance. Their paintings showed a nominal awareness of European learning. For example, Jin Nong's visual imagery in his 1761 painting *The Radiant Moon* alluded to Sino-European diagrams included in the *Synthesis of Books and Illustrations Past and Present (Gujin tushu jicheng)* of 1728 (see Figure 5.3). Jin drew on diagrams there for the moon's rays (used for the sun in the *Synthesis of Books and Illustrations* diagram) and for the surface topography of the moon. Jin also read the moon hare of Chinese mythology into his painting.[72]

Luo Ping's intriguing handscroll known as the *Fascination of Ghosts* ("Guiqu tu"), painted circa 1766–1772, portrayed a series of eight ghosts, which ended with two anatomically accurate representations of human skeletons (see Figure 5.4). The source for the skeletons was Andreas Vesalius's (1514–1564) work on anatomy titled *De Humani Corporis Fabrica* of 1543 (see Figure 5.5). Nicolas Standaert believes that the proximate source was a selective late Ming translation into Chinese by Giacomo Rho of Ambroise Paré's *Anatomie,* which was based on Vesalius. Rho's translation was titled *Illustrated Explanation of the Body (Rensheng tushuo).*[73]

Although the anatomy of the human skeleton had little impact on traditional Chinese medicine until the late nineteenth century, Luo Ping copied the illustrations of skeletons in European anatomy to depict more accurately in the eighteenth century the world of ghosts after the body had decayed. Later, when Luo reworked the images in the 1790s for the final version of his handscroll, he depicted a menacing ghost as a skeleton holding a threatening arrow in the left hand and an hourglass in the right, suggesting images derived from the European depiction of death.[74]

Such graphic appropriations of European *scientia* and medicine by Castiglione, Jin Nong, and Luo Ping reflected the vast conceptual distance between the uses of science, medicine, and technology in contemporary Europe and eighteenth-century China. Jesuits such as Castiglione still looked at science and mathematics as a means of communicating Christianity. Manchu rulers and Chinese literati co-opted the new learning to suit their political and cultural agendas as well as their curiosities.

The conversion of the human skeleton from a medical artifact of sixteenth-

Figure 5.3. Jin Nong's 1761 painting *The Radiant Moon.*
Source: Palace Museum, Beijing.

century anatomy in Europe, which had little to offer in terms of actual therapy, to a physical rendering of a ghost tells us that the Chinese still saw in
Western learning a means to enhance their own fundamental perceptions of
the world. We will later see that such cultural refractions of human anatomy

Figure 5.4. Luo Ping's handscroll, *Fascination of Ghosts.*
Source: Luo Ping guiqu tu juan (Hong Kong: CAFA Co. LTD., 1970).

by the Chinese continued when Protestant anatomies were introduced in the mid-nineteenth century.

When members of the Macartney mission visited the imperial glass-making workshop in 1793–1794, one of them noted that the site was neglected. After de Brossard's death in 1758, production at the Jesuit workshop had declined. Subsequently, in 1827 the court confiscated all missionary property, and in 1860 Lord Elgin's troops, who had taken Beijing and forced the imperial court to flee to their summer retreat in Chengde, destroyed the Lofty Pavilion gardens and palaces. The German-made planetarium Macartney delivered had attracted some interest, but his notion of Chinese ignorance convinced him that diplomatic success would naturally follow once British cultural superiority and scientific expertise were demonstrated to the emperor. That presumption proved as premature for London as it had been for Rome.[75]

Before the arrival of the Macartney mission in 1793, however, the inward turn among Qing scholars to native traditions of classical learning remained

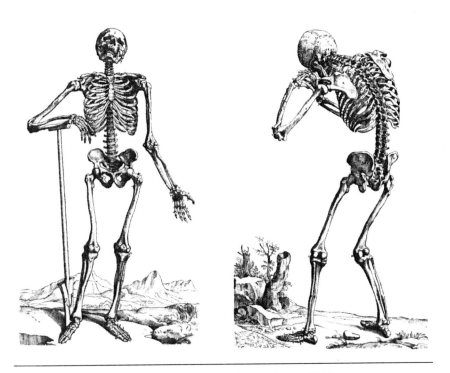

Figure 5.5. Vesalius's first and third plates of the skeleton, 1543.

Source: The Illustrations from the Works of Andreas Vesalius of Brussels, ann J. B. de C. M. Saunders and Charles D. O'Malley (New York: Dover Publications, 1973).

in place despite Jesuit influence in the arts and crafts. In Chapter 6, we will illustrate how during the Newtonian century in Europe Chinese scholars simultaneously focused on restoring native medicine, mathematics, and astronomy to admired fields of classical learning worthy of literati attention. We will describe how the Chinese medical and mathematical classics, in particular, became important fields for nonspecialists. These developments were not challenged until the middle of the nineteenth century, when modern Western medicine and technology became insuperable and irresistible.

III

Evidential Research and Natural Studies

6

Evidential Research and the Restoration of Ancient Learning

Chapter 1 described how literati in imperial China collected, studied, and classified antiquities using the epistemology for investigating things and extending knowledge. Ming scholars and printers amply recorded their passion for things, artifacts, utensils, and odd phenomena in encyclopedic collections that stressed the investigation of things, such as the fifteenth-century *Key Issues in the Investigation of Antiquities,* which discussed items retrieved from early Ming naval expeditions.

Here we emphasize how and why eighteenth-century Chinese scholars restored ancient medical and mathematical classics. They remind us of earlier European scholars who in the twelfth century translated ancient Greek and medieval Arabic mathematical and medical texts into Latin and helped forge the Scholastic synthesis that the Jesuits introduced in China as their natural philosophy. Even as mid-Qing literati recovered their classical mathematics and medicine, Europeans went beyond their ancient masters in the eighteenth century. In the nineteenth century, Europeans in China showed little patience with Chinese efforts to restore the past, forgetting their own debt to Renaissance schoolmen.[1]

In Chapter 8, we will see that beginning in the nineteenth century Chinese literati-physicians started to take seriously the challenges presented by Western medicine. In this chapter we therefore prepare the ground for understanding that confrontation. When traditional Chinese and modern Western medicine confronted each other after 1850, Chinese medicine had not remained a monolithic tradition. Its intellectual and therapeutic developments increasingly focused on "heat factor" illnesses as a new category of disease, which could no longer be subsumed as a category under perennial "cold factor" illnesses.

Early Qing Critiques of Zhu Xi and Wang Yangming

Huang Zongxi, for instance, had insisted on a more careful rendering of textual issues before the classical canon could be properly interpreted. In his influential *Scholarly Cases of Ming Classical Scholars* (*Mingru xue'an*), completed in 1676, Huang asserted that the basic difference between Zhu Xi and Wang Yangming was their interpretation of the investigation of things. Huang articulated a position, that seemed to favor Wang over Zhu on the matter. But what was needed, Huang thought, was a return to book learning and precise scholarship, not the mind-centered approach emphasized by Wang Yangming's immediate followers. This approach would recover the exact meanings of the texts themselves, rather than wasting time on speculation. The ancient content of the classical tradition could be revived, Huang thought, through exacting research and analysis.[2]

Similarly, in the early Qing Gu Yanwu lambasted those who since medieval times changed the classics according to their own whims. In particular, Gu rejected Zhu Xi's claims that the original version of the Great Learning had missing parts. Gu Yanwu blasted earlier scholars for their incompetence at classical phonology, which had led them to emend characters arbitrarily. Gu Yanwu linked the Song-Ming penchant for careless scholarship to the Song-Ming pure discussion approach, which he thought betrayed the influence of Chan Buddhism on classical discourse. He also played a role in redirecting literati interest away from the mathematics and astronomy introduced by the Jesuits, because he felt that astronomical findings would always remain indeterminate.[3]

The inward turn among Gu Yanwu and other evidential scholars of geography, medicine, and mathematics, (described in Chapter 5), was also prominent in early eighteenth-century encyclopedias and their invocation of investigating things and extending knowledge in the post-Jesuit era. Few classical scholars had any contact with the Jesuits who remained in China and were confined to the Qianlong court as lower-level architects and glassmakers. Most intellectual issues addressed by Han Learning scholars, for instance, revolved around the complexities of the native classical heritage and its misrepresentations in accounts by followers of Song and Ming "Learning of the Way." In the eighteenth century, the Jesuits were reduced from the intellectual equals of classically trained literati to foreign experts whose special skills were useful to the court.

Scholars such as Zang Lin (1650–1713), Li Gong (1659–1733), and Yan Yuan (1635–1704) also criticized the Cheng-Zhu interpretation of the investigation of things in the early Qing. Zang contended that originally the Great Learning had not been divided into a classic and commentary, as Zhu

Xi had maintained. Li and Yan attacked the airy speculations that had emerged from Cheng-Zhu learning, which for them betrayed the concrete teachings of the sages. Similarly, Li Fu (1675–1750) published a work titled *Complete Discussion of Zhu Xi's Views Late in Life* (*Zhuzi wannian quanlun*) in the 1730s, which declared that Zhu Xi had in the end changed his mind and agreed with his rival Lu Jiuyuan (1139–1193) that morality came prior to the investigation of things.

From Li's point of view, Wang Yangming was right when he returned to the original, older version of the Great Learning and reaffirmed Lu's insights over Zhu's. Although Li affirmed the Old Text version before it was altered by Zhu Xi, his views did not become mainstream until the eighteenth century, when Han Learning and evidential studies flourished. Wang Shu's (1668–1739) *Collection of Cheng-Zhu Methods for Investigating Things* (*Ji Cheng-Zhu gewu fa*), which favored Zhu Xi's views, was more representative of the Kangxi era. Wang blasted Wang Yangming's views as quietist and misguided. His studies of the Great Learning and Doctrine of the Mean were part of a revival of Cheng-Zhu learning during the early Qing.[4]

In the mid-Qing, this passion for antiquity and classical learning continued. Scholars reappropriated ancient medicine, the mathematical classics, and early astronomy for the millennial quest for ancient wisdom. The Qing court was buffered by European wars then preoccupying Britain and France. At a time when Europeans were preoccupied with Europe, Qing literati sought to equate what they knew of European learning with native learning. The restoration of ancient learning allowed Manchus and Chinese to domesticate European contributions in the arts, premodern sciences, and technology.

Medical Works and the Recovery of Antiquity

Qing physicians could not learn enough about contemporary European medicine to take it seriously until the middle of the nineteenth century. Instead, most were determined to pierce the veil of Song metaphysical and cosmological systems that since the Song and Jin periods had informed the theory and practice of classical Chinese medicine. They sought to recapture the pristine meanings formulated in the medical classics of antiquity.[5]

The medical classics provided scholars and physicians with a set of general assumptions about the application of yin-yang, the five phases, and the system of circulation tracts (*jingluo*) to understand the human body and its susceptibility to illness, which was defined as a loss of harmony in the body's operations. The system of circulation tracts, for example, allowed doctors to map acupuncture and moxibustion points on the skin according to the internal flow of *qi* through a series of main and branch conduits in the body. Since

Han times, physicians thought of such circulation as the "body's vital currents," in which there was no center (such as the heart) or starting point. They did not distinguish between the pulse and palpitations associated with the nervous system as Greek physicians had done since Herophilus.[6]

From the Song, the methodologies for medical learning and classical studies depended on the investigation of things. Similarly, for Qing literati-physicians, textual mastery of the medical classics was more than an auxiliary tool. Classical learning was required to recover medical principles and practice from ancient writings and their commentaries. The formation of evidential scholarship and the return to antiquity (*fugu*) in medicine reinforced each other. Reemphasis on ancient texts such as the *Treatise on Cold Damage Disorders* (*Shanghan lun*) in the eighteenth century stimulated the reexamination of pre-Song therapies for cold and heat factor illnesses. Instead of using warming medicines, physicians increasingly prescribed cooling drugs and methods for infectious illnesses.[7]

Reconfiguring the Medical Classics

The oldest and most important medical classic was the *Inner Canon of the Yellow Emperor* (*Huangdi neijing*), which was completed around the first century B.C. Set in its orthodox form during the Northern Song, it focused on anatomy, physiology, and hygiene in the part called the *Basic Questions* (*Suwen*), while presenting a basic understanding of acupuncture and moxibustion in the *Divine Pivot* (*Lingshu*). Treatments using drugs were rare, and the focus was on preventative medicine. The two parts of the *Inner Canon* were regarded by physicians and classical scholars since the Song as the basis for theoretical medicine and medical practice.[8]

Later the *Treatise on Cold Damage Disorders* by Zhang Ji (150–219) applied the *Inner Canon* to drug therapy for a starting point in clinical practice. Zhang wrote his book between A.D. 196–219 in response to contemporary epidemics. He called it the *Treatise on Cold Damage and Miscellaneous Disorders* (*Shanghan zabing lun*). From the Song, literati considered the *Inner Canon* the fount of medical doctrine, for which the *Treatise on Cold Damage Disorders* provided a guide to clinical practice. During the Northern Song, it became the guiding work to deal with infectious diseases brought by the winds that were linked to cold damage disorders, which were thought responsible for the increase in southern epidemics from 1045 to 1060.[9]

Zhang Ji's original version of the *Treatise on Cold Damage and Miscellaneous Disorders* was badly damaged, however, when the palace physician Wang Xi Shuhe (180–270) revised it. Beginning in the late Ming, scholar-physicians believed that Wang failed to even approximate Zhang's original.

The general Gao Jichong had submitted a reedited version of Wang Shuhe's version circa 975–986, and the Song Bureau for Revising Medical Texts did so in 1064–1065, when it was finally divided it into three separate books:

- *Treatise on Cold Damage Disorders* (*Shanghan lun*)
- *Canon of the Gold Casket and Jade Cases* (*Jinkui yuhan jing*)
- *Essentials and Discussions of Formulas in the Gold Casket* (*Jinkui yaolue fanglun*)

As treatises attributed to Zhang Ji on epidemic diseases, these works classified illnesses according to their chief symptoms and divided the symptoms into six stages, with three categories belonging to yang and three to yin.[10]

The *Treatise* concentrates on the cold damage group of disorders, most of which today's physicians would diagnose as acute, infectious, febrile diseases. The members of this group differed according to their etiology, classified abstractly as abnormal heat (*rebing*), cold, and analogous factors. To complicate the picture, one of the six cold damage (*shanghan*) disorders was itself called "cold damage." Concrete issues in both the *Inner Canon* and the *Treatise* varied from abnormal weather to undisciplined emotion. The classification was further elaborated in three books written between A.D. 100 and 280 that attempted to digest the inconsistent and sometimes dissonant ideas in the *Inner Canon*.[11]

Zhang Ji's medical works exerted only subsidiary impact between the Han and Song dynasties. Revival of interest began in the Northern Song during a wave of epidemics in south China from 1045 to 1060. In 1172, the physician Cheng Wuji continued to focus on Zhang's *Treatise*. Cheng contended that the *Treatise on Cold Damage Disorders* should be reorganized to reveal its relation to the *Inner Canon*. For example, he linked drug therapy to specific circulation tracts. Moreover, he employed the five phases theory of the *Inner Canon* to elucidate the pharmacology in the *Treatise on Cold Damage Disorders*. The *Treatise* thereby became a focal text, and the lineage of practice based on these texts became more closely aligned. A number of other scholars after the Song tightened the intellectual linkages. When Ming-Qing scholar-physicians reviewed the texts, however, they contended that scholars had not based their works on the authentic version of the *Treatise on Cold Damage Disorders,* which was lost.[12]

The literatus-physician Liu Wansu prepared his *Direct Investigations of Cold Damage Illness* (*Shanghan zhige*) during the twelfth century. When finally published in 1373, the work questioned the relation between the *Treatise* and the *Inner Canon*. It contained a thread of skepticism toward the *Treatise* that later authors enlarged. In contrast to the cold damage treatises popular in the Song, Liu stressed the phase of fire and accompanying heat as a facet of the pathogenic potential of climatic influences and the wind. Liu's most striking novelty was his emphasis on building up the *qi* of the splenetic system through cooling instead of the heating medicines recommended by the Song Imperial Bureau of Medicine.[13]

The Song government's domination of published medical texts and handbooks steadily declined between the fall of the Northern and Southern Song dynasties in 1125 and 1279. Moreover, the period of instability during the Song-Jin-Yuan transition, which lasted until 1368, ruled out a civil service career for most literati. They increasingly took up medicine, among other non-official careers. Out of direct control of the government, medical traditions, like art and literature, achieved extraordinary autonomy under Mongol rule. The imperial government could no longer dictate preferences in art or medication to painters and physicians. Thereafter, the influence of medical books written by literati-physicians and published privately became an important means of transmission for medical traditions. This was most obvious in the Yangzi delta and south China where the financial resources for a publishing resurgence were abundant.[14]

From Song-Jin Masters to Ming-Qing Ancient Learning

Physicians such as Liu Wansu and Zhu Zhenheng found the Northern Song medical handbooks useless when they struggled with the epidemics that surfaced during the Jurchen and Mongol conquests of north China in 1126 and 1234, respectively, and the Mongol subjugation of south China in 1279. Due to the massive numbers of Northern Song Chinese who migrated to the south after the fall of the north to Jurchen invaders in 1126, the new population from the north continued to face dramatic medical adjustments to survive in the warmer and wetter ecology of south China.[15]

Dense populations had already emerged in south China after the eighth century, particularly during the An Lushan Rebellion in the northwest from 756–763, when some 40 percent of the populace lived in the south. About 60 percent of the population resided in south China, many as émigrés to its urban centers, when the Northern Song capital of Kaifeng fell in 1126. These Southern Song numbers further increased when the Mongols invaded the south in 1234. Thereafter, the population in China declined under the Mongols. The fourteenth century was also the era when the bubonic plague likely began its dreadful march by land and sea during the *Pax Mongolica* from southeastern Asia through central Asia and the Middle East to Europe.[16]

The diseases of the south, with their richer variety of climates, infection, and infestation, led to new medical frameworks in Jin, Yuan, and Ming China, challenging confidence in standard approaches to acute fevers derived from the canonical cold damage therapies based on the *Treatise on Cold Damage Disorders*. Increasingly, Jin and Yuan physicians questioned Zhang's etiology and reevaluated Song medical orthodoxy based on the government's formularies. They also historicized the concept of illness. Liu Wansu and his immediate followers interpreted the changing clinical landscape in terms of epidemics such as smallpox, which they saw as evidence of long-term changes in diseases brought by the winds.[17]

This new theorization of the materia medica in light of climate and wind appropriated a correlative analogy of the body with the cosmos, which relied heavily on Zhu Xi's Learning of the Way and its explanations for why things were the way they were (*suoyi ran*). Zhu Zhenheng, for instance, presented the tendency of the body toward yang excess and yin deficiency. The root of illness could be traced to fire as the fundamental principle animating life through desires. In this metaphysical assessment of the human condition, the moral management of human emotions through "cooling" cultivation was required to maintain a rectified heart and thereby prevent sexual desire from turning the body's generative powers into self-destructive forces.[18]

Jin-Yuan scholar-physicians oriented toward Zhu Xi's focus on the investigation of things, such as Zhu Zhenheng, an initiated member of the main Zhu Xi lineage of teaching, were concerned with building symbolic structures of meaning to which they could relate all aspects of bodily experience. Song symbols of correspondence and cosmological parallels, like their Han dynasty predecessors, were based on charts or cosmograms of such symbolic correspondences. In medical doctrine, there were no purely theoretical issues. Doctors used abstract entities such as yin and yang or the five phases to diagnose and deduce programs of therapy. In the growing case-record genre, late Ming and Qing scholar-physicians began with the concrete signs and symptoms of individual cases, revealing the abstract categories they used only in their specification of prescriptions.

Jin-Yuan medical masters created a bookish tradition that emphasized Cheng-Zhu theory and favored pulse reading and drug prescriptions based on the correlation between cooling therapies and moral cultivation. They recorded their views of the materia medica in their interpretations of ancient texts such as the *Inner Canon* and the *Treatise on Cold Damage Disorders*. They also applied the moral concepts of Cheng-Zhu learning to the scholarly medical tradition. These new ideals of moral cultivation appeared in Ming introductory medical texts and manuals. Although Jin and Yuan medical nosology paid more attention to the local and regional environment, the medical discourses that Zhu Zhenheng and others had elaborated also stressed internal etiologies focusing on undisciplined emotion and other failures of self-control.[19]

To understand the seasonal variability of illnesses in south China, and to grasp the role of warming or cooling as effective treatment regimes, Song-Yuan-Ming physicians from Zhu Zhenheng on moved away from one-sided cooling and warming therapeutic strategies toward a more flexible approach that drew on both. The epidemics of the late Ming led to reexaminations of both the Song materia medica tradition and the ancient medical classics.[20]

Li Shizhen's late Ming *Systematic Materia Medica* was part of the general reevaluation of Song materia medica compilations and of classical works such as Zhang Ji's *Treatise on Cold Damage Disorders*. The outcome was new ideas on the diagnosis and treatment of illnesses. Li Shizhen should be counted among the many who in the late Ming went beyond the medical revisions of Jin and Yuan physicians such as Liu Wansu and his insistence on cooling therapies. The heat factor approach to the treatment of the epidemic diseases of south China became more accepted during the late Ming. In one case, the physician Yu Chang noted that misdiagnosing a heat factor illness as a cold factor one had aggravated a patient's condition because warming rather than cooling medications were used.[21]

Writing in the late Ming, for example, Wu Youxing (1582?–1652) experienced the late Ming epidemics and recognized the inadequacy of cold damage doctrine. Classical doctrines saw pathogenic agents as types of *qi* that invaded primarily through the skin and disrupted the body's own processes. These types were in no sense species, and what part of the skin they invaded was of no interest. In his *Treatise on Heat Factor Epidemic Disorders* (*Wenyi lun*), completed circa 1642, Wu conjectured that heat factor disorders, unlike the rest, were due to "deviant *qi*" (*liqi*) specific to each disorder—not just unseasonable weather—and that these *qi* entered through the nose and mouth. Although not based on persuasive empirical evidence, this argument opened up a new set of alternative pathologies that some physicians took seriously.[22]

Due to high mortality rates of up to 70 percent in late Ming epidemics in the Yangzi delta, Fang Youzhi (b. 1523) and Yu Chang questioned Song and Jin-Yuan views of medical therapy. They dramatically demonstrated that important parts of the medical classics had been improperly adapted to the Cheng-Zhu framework and thus no longer represented the ancient diagnostic and therapeutic procedures advocated in the *Treatise on Cold Damage Disorders*.[23]

Ancients versus Moderns

Medical learning from antiquity did not present itself to late Ming and Qing scholar-physicians as a finished product. It had to be rediscovered and reconstructed. Qing scholar-physicians sought to reverse the adulteration of ancient medical practice in the Song-Ming period. Their appeal to the ancient wisdom in the authentic medical classics added to the growing eighteenth-century denunciations of the Cheng-Zhu orthodoxy in classical, political, and social matters. Moreover, Ming physicians such as Wang Ji (1463–1539) and Yu Chang increasingly referred to case histories (although they were not the first to do so) instead of the medical classics to advertise their therapeutic successes and explain them to students and amateurs.[24]

Qing debates between antiquarians and modernists concerning early medicine paralleled those between Han Learning (*Hanxue*) and Song Learning (*Songxue*) classical scholars. Evidential scholars of the Five Classics, for example, focused on the distant past to overcome the failures of the recent Cheng-Zhu tradition. The rhetoric entered the discussion of medical texts. Like Han Learning scholars, for example, Qing scholar-physicians began their studies with Han dynasty medical texts and the earliest classical interpretations, because the latter were closer in time to the composition of the classics and thereby more likely to reveal their authentic meaning. They rejected Song dynasty sources, which Song Learning scholar-physicians had relied on, because of their questionable authority and greater separation from antiquity.

The overlap between evidential research methods and textual study of the medical classics began in the late Ming and early Qing. Such textual approaches to the *Basic Questions* and *Divine Pivot* portions of the *Inner Canon* also led to investigations of the original content of the *Treatise on Cold Damage Disorders* and cold damage theory. Editing and collating the variants for the current editions of the ancient medical texts enabled scholar-physicians to reexamine the original import of the medical classics. New works appeared on the *Inner Canon*, Zhang Ji's *Treatise*, the *Essentials of the Golden Casket*, and the *Canonical Materia Medica of Emperor Shennong* (*Shennong bencao jing*). A proliferation of new annotations emerged, with some 289 such works during the Qing compared to only 59 during the Ming.[25]

Because the *Inner Canon* had been compiled in antiquity, it may have represented the therapeutic experience in the ancient Yellow River region in the northwest. The *Inner Canon* had inherited a web of correlations among the *qi* circulation tracts, the five phases, and the six yin-yang modalities worked out during the Han dynasty, which linked the four seasons, climatic elements, and all sources of pathology. The *Treatise on Cold Damage Disorders*, on the other hand, represented southern medical traditions, which also used yin-yang to delineate six stages of disease. Zhang Ji had likely used the *Inner Canon* and adapted its yin-yang and five phases perspectives. Wang Shuhe's commentary then read the entire theoretical structure of the *Inner Canon*—itself full of contradictions—into the *Treatise*, which became the orthodox view.[26]

Qing medical scholars demonstrated that Tang and Song medical works depended on Later Han texts that later interpreters had refracted. The tense interplay of an admired antiquity with a discredited Song medical orthodoxy suggests that medical studies in late imperial China were an adaptation of classical antiquity. Like evidential scholars of the Classics, Qing scholar-physicians thought their rediscoveries would improve contemporary medical therapies.[27]

In his exposition of Zhang Ji's *Treatise*, printed in 1648, Yu Chang reconstituted what he considered the 397 prescriptions of the original version, whose therapeutic efficacy he explained differently from Wang Shuhe's readings, which were based on the circulation tracts and five phases theory in the *Inner Canon*. The title of Yu's influential work was abbreviated as *A Respectful Discourse on the Treatise (Shanglun pian)*, and in it he criticized Wang Shuhe for reading his own views into Zhang Ji's writings. Yu Chang's efforts to overcome eight centuries of misinterpretation was highlighted by the editors of the most valued medical works included in the Qianlong Imperial Library in the 1780s. They were critical of the excessive reliance on Wang Shuhe in medieval and Song accounts of the *Treatise* and added that Yu had "rediscovered the principles of cold damage illnesses."[28]

The restoration of Zhang Ji's *Treatise* in the late Ming and Qing began as a revival of ancient medicine and thus overlapped with the call to return to antiquity (*fugu*) so prominent in classical studies and in mathematical learning after 1600. Heat factor therapy was reinvented by Qing literati-physicians who perceived that its medical efficacy complemented the cold damage tradition. Among the leaders of the Qing revaluation of the medical classics, Xu Dachun (1693–1771) represented a Yangzi delta physician who came from a literati family but did not take the civil examinations. He studied astrology, music, geography, and medicine and was also interested in popular religion. Aware of the parallel changes in the fields of classical learning, Xu compared the medical dispute over the *Treatise* to the scholarly debate concerning the textual provenance of the ancient version of the Great Learning and the authenticity of certain chapters of the *Documents Classic*. As a scholar-physician, Xu also advocated returning to the early medical classics such as the *Inner Canon* and the *Treatise on Cold Damage Disorders*.[29]

In 1713, Xu Dachun prepared a discussion of 113 prescriptions for fevers in the *Treatise*. In 1727, he prepared commentaries on the *Canon of Eighty-one Problems of the Yellow Emperor (Huangdi bashiyi nanjing)* and the *Canonical Materia Medica of Emperor Shennong* (1736). The classifications Xu used in his 1757 *On the Development of Medical Studies (Yixue yuanliu lun)* clearly reflected an ancient versus modern (*gujin*) split in medical traditions. In their account of Xu's works, the editors of the Imperial Library accused Song-Yuan physicians of losing the ancient methods (*gufa*) that Xu and others had restored. Several other literati-physicians in the early Qing, such as Ke

Qin, also thought that Wang Shuhe's editing of the *Treatise* was not true to Zhang Ji's original and joined Xu to restore ancient medicine. Ke argued that because much of original had been lost, Wang as the Jin palace physician had added his own materials.[30]

Qing followers of Jin-Yuan medical traditions remained prominent, however, and they opposed the efforts by Xu Dachun and others to restore ancient medicine. Called the modernists (*jin*) by the editors of the Qianlong Library, Jin-Yuan enthusiasts relied minimally on the medical classics. Among this group, Ye Gui (1666–1745) came from a Suzhou family of physicians. In addition to writing a work on heat factor disorders, Ye became a pioneer in prescribing aromatic stimulants for epidemic fevers and clarified Li Shizhen's materia medica. Xue Xue, a townsman, also followed Ye's form of treatment, and their prescriptions were subsequently published together.[31]

The editors of the Imperial Library took note of these contending medical traditions and described them using the traditional designations of schools as lineages of transmitted learning. Indeed, in 1739 the Qianlong emperor had already authorized compilation of the *Mirror of the Gold Casket for Orthodox Physicians* (*Yuzuan Yizong jinjian*), which collected annotations of Zhang Ji's *Treatise on Cold Damage Disorders* in southern medical editions by Yu Chang and others. Published in 1743, it became the standard textbook for students in the Palace Medical Service. The Imperial Library catalog adduced it as evidence of the reconstitution in the eighteenth century of ancient meanings (*guyi*) from the orthodox school of medicine.[32]

School designations were taken for granted as evidence for the filiations of scholars and physicians, as well as artists and poets, who through personal or geographical association, doctrinal or intellectual agreement, or master-disciple relations were linked into identifiable groups. These retroactive genealogies reflected real personal relations but not permanent institutions that could be called schools. Hence, when the editors prepared introductory remarks for the section on medicine (*yijia lei*) in the official catalog for the Imperial Library, they contended that the split in contemporary lineages of classical learning went back to the Song dynasty, while the separation of medical lineages went back only to the Jin-Yuan period.[33]

In this manner, the editors associated Ming-Qing literati-physicians such as Yu Chang and Xu Dachun with ancient learning, while the upholders of Jin-Yuan medical traditions were pejoratively labeled as modernists. Of course, neither group divided up so neatly, nor did they refer to themselves in such terms. Because Xu Dachun thought that the Song-Jin-Yuan medical tradition had falsified ancient medicine, he and his cohort were described as representatives of the Han school of medicine, which paralleled the classical revival of Han Learning by evidential research scholars. Ye Gui and others were tarred with the feathers of Song Learning.

In the midst of these eighteenth-century controversies, however, the heat factor tradition grew increasingly prominent among all the medical traditions. Yangzi delta scholar-physicians and hereditary doctors turned away from cold damage treatments. A fully independent medical tradition associated with heat factor therapies—a nineteenth-century invention—assembled a heat factor canon from scattered ancient writings and those of early Qing physicians, mainly in Suzhou. The final shift from a universal medical doctrine based on Jin-Yuan cold damage therapy to regional medical traditions dealing with heat factor epidemic diseases in the southeast or northern cold disorders began in the seventeenth and eighteenth centuries. Not until the late Qing, however, did the battle over the ancients and moderns end. By then, a collision with modern Euro-American medicine, especially tropical diseases, was nascent.[34]

Chen Yuanlong and the *Mirror of Origins* Encyclopedia

For the most part, High Qing evidential scholars preferred not to use the terminology for investigating things and extending knowledge (*gezhi*) to describe their research agenda, because in the mid-eighteenth century such phrases were associated with Song Learning and particularly with Zhu Xi's "studies of principle" (*lixue*), which they opposed. In the seventeenth century, however, investigating things and evidential studies had been allied. The Ming loyalist Lu Shiyi (1611–1672), for example, equated inquiry and proof (*kaoju*) with investigating things to extend knowledge and bemoaned the fact that contemporary scholars who followed Wang Yangming's teachings had forgotten this connection.[35]

Under Kangxi, the dynasty reaffirmed Cheng-Zhu learning. Song Learning views of investigating things and natural studies remained perennial issues in the eighteenth century. The emperor's own concerns for investigating things, for instance, included odd creatures, Turfan melons, stone salt, the speed of thunder, stone fish, the earth as a globe, earthquakes, and the geographical origins of the Yangzi River. For every thing (*wu*) integrated into classical works, the emperor expected to find proofs (*zhengju*) based on actual experience. After meeting Mei Wending, moreover, the emperor became convinced that European learning was derivative and that ancient learning in China was the source of all reliable knowledge.[36]

Politics and Encyclopedias

During the early eighteenth-century Rites Controversy, the Kangxi emperor encouraged several massive encyclopedias that would gather together all accepted knowledge of his age, an approach that he hoped would revive the comprehensive quest for knowledge and information that had characterized Song and Ming encyclopedias. These ranged from collections of classical quotations and phrases to compendia of historical writings and the works of the ancient masters:

- *Classified Repository of Profound Appraisals* (*Yuanjian leihan*, 1710): presents classical quotations from antiquity to 1556.
- *Thesaurus Arranged by Rhymes* (*Peiwen yunfu*, 1711; supplement in 1720): presents classical phrases, synonyms, and antonyms arranged by rhymes common then.
- *Essential Meanings of Works on Nature and Principles* (*Xingli jingyi*, 1715): compendium of Cheng-Zhu teachings on the Learning of the Way.
- *Compendium of Parallel-prose Phrases* (*Pianzi leibian*, 1719): index to first characters of important classical compounds.
- *Essence of the Masters and Histories* (*Zishi jinghua*, completed in late Kangxi period but not published until 1727): collection of essential quotations from the Dynastic Histories and the ancient masters.[37]

As the second Qing emperor, Kangxi was mindful of early Ming precedents. He recognized how the Yongle emperor had authorized the compilation of the *Great Compendium of the Yongle Reign* (*Yongle dadian*) in 1404–1407 to glorify his reign by incorporating the entire classical legacy in an imperial project. For this cultural cum imperial purpose, the Kangxi emperor early on chose Chen Yuanlong (1652–1736), who came from a renowned family lineage in Haining, Zhejiang, to compile a comprehensive encyclopedia that would surpass its Song-Ming predecessors.[38]

During the Ming and Qing dynasties, the Chen lineage produced some 31 palace graduates and 103 provincial graduates on the civil examinations, three of whom entered the Grand Secretariat directly in the service of the emperor. Chen Yuanlong was one of the three when he became a grand secretary in 1729 under Yongzheng. He had received the Kangxi emperor's orders, in 1704, to prepare a wide-ranging collectanea, ranging from astrology, geography, and human affairs to plants, trees, and insects, which he titled *Mirror of Origins Based on the Investigation of Things and Extending Knowledge* (*Gezhi jingyuan*).

By his own admission, Chen peculiarly held on to the huge manuscript for almost twenty years before publishing it in 1735 during the Yongzheng reign. He had worked on it a great deal from 1707 to 1708, and by 1717, while Kangxi still reigned, it was complete. A repository of detailed information culled from a variety of sources, the *Mirror of Origins* represented a collection of all knowledge by a well-placed scholar in both the Kangxi and

Yongzheng courts. He chose "the investigation of things and extending knowledge" for the title to refer to the Song and Ming encyclopedias that he was now superseding, such as Luo Qi's 1474 *Origins of Things* (see Chapter 1). The editors of the Qianlong Imperial Library noted that Chen had purposely cited all his sources in contrast to Ming encyclopedias, which often failed to mention theirs.[39]

Simultaneous with Chen Yuanlong's project, Chen Menglei (b. 1651), a former friend of Li Guangdi, had returned from exile in 1698 and was then serving in the household of the Kangxi emperor's third son, Yinzhi (1677–1732). Under the prince, he was compiling a classified encyclopedia that became the *Synthesis of Books and Illustrations Past and Present*. Chen Menglei's first draft, called the *Compendium* (*Huibian*), was done by 1706. He then sought the approval and support of the Kangxi emperor to add sources from the Imperial Library to complete it.

Between 1706 and 1722, roughly the same time he was supporting Chen Yuanlong, the Kangxi emperor also supported Chen Menglei's project and gave it its final name as the *Synthesis*. Prince Yinzhi was concurrently working with the Jesuits in the Academy of Mathematics (see Chapter 4) and was appointed by his father as the head of the Studio for the Cultivation of Youth (*Mengyang zhai*). Printing of the *Synthesis* was underway when the Kangxi emperor died in 1722, but early in 1723 the Yongzheng emperor, who had been opposed in the succession crisis by Yinzhi and the Jesuits, exiled Chen Menglei for his links to the prince and appropriated the *Synthesis* project in the name of his deceased father. He obliterated all mention of Yinzhi's and Chen Menglei's role in the project. The revised manuscript version of the *Synthesis* was published in 1728, which despite the new editor's claims differed little from Chen's draft.[40]

Following the lead of the Kangxi emperor, his son, the Yongzheng emperor, and grandson, the Qianlong emperor, each appropriated the classical legacy to establish their dynastic prestige and political legitimacy. The Yongzheng emperor co-opted both the *Mirror of Origins Based on the Investigation of Things and Extending Knowledge* and the *Synthesis of Books and Illustrations Past and Present* to glorify his reign. The appropriation of these virtually completed projects drew attention away from the succession crisis in which the Yongzheng emperor, according to later rumors, murdered his father, eliminated several of his brothers, and humiliated Prince Yinzhi.[41]

Given this political context, it is interesting that the *Synthesis* prepared by Chen Menglei under the auspices of Prince Yinzhi included a significant amount of information about Jesuit mathematics, Europe, and the world, while Chen Yuanlong practically ignored the Jesuits and European learning in the *Mirror*. Yinzhi's ties to the Jesuits and the Academy of Mathematics raised the profile of European learning. The fusion of knowledge in Chen Menglei's *Synthesis* acknowledged the Jesuits and their contributions since the late Ming. Because of the Kangxi emperor's increasing stress on the Chinese origins of all learning, and the Yongzheng emperor's rigorous opposition to the Jesuits from 1723, however, Chen Yuanlong did not decorate his *Mirror* with European trappings when it was finally published in 1735.

A similar fate befell Western Learning in the compilation of the *Ming History*. We have seen that Mei Wending prepared a draft of the calendrical treatise for the official *Ming History* project when it was initiated in 1679, which described late Ming efforts to amalgamate European and Chinese expertise in mathematics and astronomy to reform the calendar. Later Mei's draft for the treatise passed through many hands involved in the project, before the committee presented a *Draft Ming History* to the Kangxi emperor in 1723. Wan Sitong (1638–1702), whose draft version of the *Ming History* was passed on to Wang Hongxu (1645–1723) in 1702 when Wang became director of the project, was the first to ignore Jesuit contributions in astronomy. Wan and others involved in the project had sympathized with Yang Guangxian's political charges against the Jesuits.[42]

Zhang Tingyu (1672–1755), who served both the Kangxi and Yongzheng emperors, finished the *Ming History*. As a grand secretary, Zhang rigorously excluded the Jesuits and their religious and scientific teachings from the *Ming History*. In the 1739 published edition of the *Ming History*, Ming literati who had associated with the Jesuits, such as Xu Guangqi, and the Jesuits themselves were referred to exclusively in connection with the Chinese origins of Western learning. The contemporary *Ming History* and the *Mirror of Origins Based on the Investigation of Things and Extending Knowledge* were both censored as a result of the Rites Controversy and the Yongzheng emperor's anti-Jesuit policies.[43]

When Korean envoys entered Beijing for a visit early in the Qianlong reign, for instance, they noted how much more open the intellectual atmosphere now was when compared to their previous visit during the Yongzheng reign, when tensions over Zeng Jing's (1679–1736) famous defamation of the emperor had run high. Printed in 1735 in an anti-Jesuit atmosphere, Chen Yuanlong's encyclopedia reverted to a Jesuit-free account of investigating things and extending knowledge that superseded Hu Wenhuan's controversial late Ming *Collectanea of Works Investigating Things and Extending Knowledge*.[44]

Content of the Mirror of Origins

In his 1735 preface, written when he was sixty-four years (*sui*) old and had long since completed the project, Chen Yuanlong paraphrased the 1448

"Preface" prepared by Yan Jing for Gao Cheng's Northern Song *Record of the Origins of Things and Affairs* (see Chapter 1):

> Things have myriad functions. Heaven and earth are both things. We know the heavens fill our minds with knowledge. A gentleman would be embarrassed by failing to know even one thing. Therefore, we prioritize the investigation of things and extension of knowledge. When we appropriate things we call them matters. When we refer to them we name them. When we gather things together we classify them. Hence, things depend on the three powers [heaven, earth, humankind] and change according to the five phases.

We will see that Protestant missionaries and their Chinese collaborators borrowed similar language from Ming-Qing encyclopedias to introduce modern science to a Chinese reading public in the 1880s.[45]

The "Outline" that Chen prepared for the encyclopedia made it clear that the recovery of ancient knowledge in nativist terms was the goal of this summation concerning heaven, earth, and humankind:

> The main purpose of this work is to correct and emend our studies for investigating things and extending knowledge. For every entry about a thing, it is necessary first to explore its origins, carefully delineate its name and title, annotate its substantive classification, and look at its makeup to contribute to understanding its uses. A comparative phraseology of things is not employed here. Hence, what has been included about things in poetry and rhyme-prose from antiquity has not been recorded here at all so as to distinguish it from other encyclopedias. I have included only about one in a hundred of materials about the different names for things and anomalies.

Chen's decision to focus on the origins of things and their proper classification, left out all allegedly fictional material that Qing scholars charged permeated late Ming encyclopedias and Hu Wenhuan's collectanea. In the 1780s, the editors of the Imperial Library catalog explicitly preferred Chen Yuanlong's *Mirror of Origins* over Hu Wenhuan's disorderly *Collectanea*, which they considered unreliable and poorly organized. As a post-Jesuit compilation, the *Mirror of Origins* was included in the Imperial Library while Hu's *Collectanea* was only reviewed in the imperial catalog. To enhance its reliability, Chen Yuanlong included sources for each item, which came primarily from the Classics and Dynastic Histories but also included records of things from other detailed sources.[46]

Chen's efforts to demarcate the familiar from the strange set limits between what he considered verifiable natural phenomena and those wonders that popular lore misconstrued or wrongly explained. His encyclopedia narrowed the scope of the genre to cover mainly the arts and natural studies. The *Mirror of Origins* gave special attention to the evolution of printing and the importance of stone rubbings, in addition to topics dealing with geography, anatomy, flora and fauna, tools, vehicles, weapons, tools for writing, clothing, and architecture.

Although Chen never mentioned the reasons for his elision of European learning, at times he did refer to the Jesuits. For example, among the entries on timekeeping in the section on heavenly correspondences, where he discussed the chime clock as a curiosity, Chen presented Matteo Ricci as a clock maker, which drew on the eighteenth-century materialization of the Jesuits as artisanal experts. Pointedly deploying a Buddhist term, Chen referred to Ricci as a "Western priest" (*Xiseng*). Chen cited Xie Zhaozhe's late Ming miscellany for his source. In his discussion of astronomy and the manufacture of instruments such as the armillary sphere, however, Chen made no mention of any Jesuit contributions or of the Tychonic cosmology then in use in the Astro-calendric Bureau.[47]

The accounts of natural phenomena that Chen presented via a chronological listing of quotations from reliable sources were systematic but conservative. He made no effort to evaluate or build on the listings he prepared. On the increasingly suspect tradition of "waiting for the *qi*" (*houqi*) to date the onset of spring, for instance, the *Mirror of Origins* perfunctorily cited a few sources, without discussing the early Qing controversies over the technique, which had embroiled the Kangxi emperor in the Yang Guangxian affair. Nor did it mention contemporary critiques of the approach, such as those by the Kangxi emperor. The descriptions Chen cited assumed it was appropriate and indicated that the practice was still in force as a technique for measuring the earth's emanations of *qi*. Indeed, the Qianlong later reaffirmed it.[48]

The discussion of magnets (*cishi*) in the *Mirror of Origins* also reveals the conventional nature of Chen's encyclopedia. Early classical sources already cited in the Northern Song encyclopedia *Materials of the Taiping Xing Guo Era for the Emperor to Read* had appealed to some sort of reciprocity (*shu*) to explain the impact of a stone-magnet and the reliability of the compass needle. Other works dating since at least the Yuan dynasty had placed the action of the magnet within the scope of natural principles (*ziran zhi li*), while Li Shizhen explained how the concatenation of similar *qi* caused magnetism.[49]

Qing literati did not perceive the problem of periodicity as Europeans did. In the seventeenth century, Kepler sought to pin down the force that made planets move in periodic, elliptical orbits. His focus on the sun's attraction of the planets, however, neglected the mutual attraction of the latter, which Lagrange later corrected for by computing additional variables for the elliptical paths of the planets. Medieval accounts of the natural affinity between iron and the magnet depended on occult qualities. This view remained standard until the middle of seventeenth century, when Descartes explained magnetic attraction through a hierarchic system of vortices that provided a one-way passage for particles of suitable shape. In this view, magnetic attraction was simply a mechanical impulse within corpuscular matter, a view dissimilar from Li Shizhen's organic concatenation of fluid *qi*.[50]

Amber (*hubo*) was a transparent yellowish mineral that like the magnet also drew attention in China. Literati considered the phenomenon of a magnet attracting iron analogous to amber attracting small, light plant materials when rubbed. Zhang Hua's *Treatise on Broad Learning of Things* had described a process of millennial change (*qiannian hua*) that produced amber from tuckahoe when pine seeds entered the earth. This finding was taken up many times thereafter, and the phenomenon was discussed by Li Shizhen in his *Systematic Materia Medica*, who cited the *Progress Toward Elegance* dictionary and elaborated the reason for amber's yellowish transparency: "When the lion dies, its pure soul enters the earth and changes into a stone resembling it."[51]

In *De Magnete* (*On Magnets*, 1600), William Gilbert had also analyzed the amber effect, which distinguished its promiscuous behavior from the pure bond of sympathy uniting iron and loadstones. Gilbert's catalog of differences between static electric and magnetic phenomena represented the essential step in the emergence of electricity and electrostatics as an object of study. Gilbert concluded that electricity was not occult or due to electrical sympathy. Rather it arose from the direct action of matter through corpuscles. The alleged affinity between amber and the magnet, challenged by Gilbert, remained unquestioned in Chinese natural history until the nineteenth century, an indication of the conservative context within which Chen Yuanlong compiled his mid-Qing encyclopedias.[52]

Similarly, in the sections that chronologically cited writings on lightning and thunder, Chen Yuanlong simply repeated the information that had been included in earlier encyclopedias and in Han compendia such as the *Master of Huainan* (*Huainan zi*). For example, Chen referred to Lu Dian's Song dynasty *Adding to Elegance* dictionary, which had explained thunder in terms of the concatenation of yin and yang that produced lightning. Although not mentioned by Chen, Zhu Xi had described thunder as "righteous *qi*" (*yiqi*), which when released produced bright light.[53]

Unlike their Protestant contemporaries in northern Europe, the Jesuits who preached the advantages of Christianity in China often succumbed to a traditional, theological view of thunder as an instrument used by God to reveal his power. Jesuits such as Vagnoni had stated that all things, including rain, dew, thunder, and lightning, were all composites of the four elements. Nevertheless, Jesuit research in physics had made contributions to optics, mechanics, magnetism, and electricity, which even the fellows of the Royal Society in London cited in the seventeenth century. On the continent, Jesuits still monopolized chairs in the arts at still-conservative—anti-Descartes and anti-Newton—Catholic colleges in German-speaking lands.[54]

When physics became an independent discipline in the eighteenth century, French Jesuits such as Amiot also adopted the new mechanical philosophy despite their misgivings about Descartes's natural philosophy and Newtonian mechanics. Jesuit physics moved from an Aristotelian scheme to one stressing experiment. Amiot and the Jesuits in Beijing, for example, received an electrical machine from St. Petersburg, and the Berlin physicist Georg Wilhelm Richmann (1711–1753), who died after he caught lightning on his insulated pole, had arranged to have his papers sent to Beijing. Additionally, a Jesuit finding for an experiment on a compass needle in Beijing led to the realization in Europe that an electric force could polarize glass.[55]

In eighteenth-century Europe, the study of electricity was developing well beyond explanations based on the malevolence of inanimate objects or the attractions effected by occult qualities. Although there was no law of magnetic force in the early eighteenth century to parallel Newton's universal gravitation, the forces of attraction and repulsion, under his influence, began to take precedence when weightless substances entered physics as carriers of the forces associated with heat, light, fire, electricity, and magnetism. When Newton saw Francis Hauksbee (1670–1713) perform experiments on electricity at the Royal Society, he thought it might explain universal gravitation. Electricity was elastic enough, he thought, to fill the void of space and so weightless that it offered no resistance to the planets. By 1745, most British electricians assumed there was a special matter for gravity that mediated as an electrical force from a distance.[56]

The first spark Europeans intentionally drew from the sky occurred in the 1752 Marly experiment in France, which confirmed the efficacy of a lightning rod grounded in the earth. By the last third of the eighteenth century, the best physics texts in Europe assumed an unfettered instrumentalism that explained electrical phenomena by attraction and repulsion as if God had attributed such forces to it. D'Alembert, for instance, thought that electricity and magnetism seemed to arise from an invisible fluid that caused an attraction between terrestrial bodies.[57]

The account of amber, the magnet, lightning, and thunder in the *Mirror of Origins* literally mirrored earlier traditions of learning in China. Chen Yuanlong was unaware of the changing views of such phenomena in Europe in part because the Jesuits had not presented any substantial account of new studies of electricity. A disjunction existed between the changing scholarly world of Europe and the conservative teachings spread by missionaries in the field. As a result, Chen Yuanlong was never presented with a new understanding of lightning and thunder that would deviate from explanations based on the grinding of yin and yang *qi* against each other. Because of Amiot's caution about letting Chinese in Beijing know about his experiments in electricity, the parallel disjunction between Chinese and European natural studies remained intact.

We will see that the Newtonian revolution in mechanics in France and Great Britain was not transmitted to China until the Taiping Rebellion (1850–1864) via French engineers at the Fuzhou Shipyard and British machine workers at the Jiangnan Arsenal. In addition, the domestication of European Learning during the Kangxi reign was ensnared in political whirlwinds that neither the *Mirror of Origins* nor the *Synthesis of Books and Illustrations Past and Present* could avoid.

In the midst of the inward turn in evidential studies during the eighteenth century, which we have documented in geography, medicine, and encyclopedias, we will see below that a similar turn occurred in the study of mathematics and astronomy. The impact of evidential research made itself felt in the attention classical scholars gave to the European fields of mathematics and astronomy introduced by the Jesuits in the seventeenth century. Such interest had built upon the earlier findings of Mei Wending, a favorite of the Manchu court, once the emperor recognized his expertise in mathematical astronomy.[58]

Revival of Ancient Chinese Mathematics

Mei Wending had linked mathematical astronomy to the classical investigation of things. Earlier accounts have stressed Mei's efforts to merge classical studies with astronomy, but Mei Wending was not a proponent of evidential studies. His classical learning remained focused on Song orthodoxy and the Cheng-Zhu elaboration of moral and natural principles. Nevertheless, Mei's stress on comprehensiveness (*tong*) in mastering both Chinese and European mathematics became the conceptual underpinning for the late eighteenth-century call by evidential scholars such as Ruan Yuan (1764–1849) for comprehensive literati (*tongru*). They would master both traditions of learning within the framework of the "Chinese origins of Western Learning."[59]

Mei Juecheng, who served in the Studio for the Cultivation of Youth from 1712 to 1722 and became a Hanlin academician after attaining the palace degree in 1715, was the key figure in transmitting his grandfather's classical ecumenism to the emerging mainstream of evidential scholars in the mid-eighteenth century. In the midst of his official duties from 1737 to 1746, Mei worked on supplements to the *Sources of Musical Harmonics and Mathematical Astronomy,* such as the *Supplement to the Compendium of Observational and Computational Astronomy* and *Supplement to Exact Meaning of the Pitchpipes (Lülü zhengyi houbian).* He also served on the *Ming History* commission and reviewed the chapter on mathematical astronomy for errors that had accrued since his grandfather had contributed to its compilation.

In retirement, Mei Juecheng edited Cheng Dawei's *Systematic Treatise on Computational Methods,* which was reprinted in 1771. Ricci and Li Zhizao had adapted some of Cheng's problems in general arithmetic for the abacus in their *Translations of Guidelines for Practical Arithmetic.* Cheng's work had also included Chinese numerology based on magic squares, which Bouvet introduced to the Kangxi emperor. In addition, Mei compiled the *Surviving Treasures from the Vermilion Water (Chishui yizhen),* in which he reproduced three of nine mathematical formulas dealing with circles expressed in a series introduced by the French Jesuit Peter Jartoux. Minggatu and others elaborated the formulas.[60]

Mei Juecheng lamented the destruction of the astronomical instruments in use in the Astro-calendric Bureau until 1672, when they were replaced by Verbiest's new instruments. Mei had seen the originals in storage, but in 1715 Bernard-Kilian Stumpf, then in charge of the bureau, had several melted down to build a bronze quadrant. By 1744, only the armillary sphere, simplified sphere, and a celestial globe were left of the older instruments. Such material losses of the traditional astronomical heritage influenced Mei Juecheng's efforts to recover in their original form the single-unknown (*tianyuan shu*) techniques to solve polynomial equations with a single variable. This enterprise became a major mathematical aspect of evidential research.[61]

Under the Ming, the tradition of mathematical calculations associated with the *Computational Methods in Nine Chapters* were continued. But works such as Wu Jing's 1450 *Comprehensive Collection of Classified Problems to the Computational Methods in Nine Chapters (Jiuzhang suanfa bilei daquan),* did not understand the pioneering methods for solving polynomial equations developed by Qin Jiushao (1202–1261), Li Ye, and Zhu Shijie (fl. end of the thirteenth century). Mei focused on the Yuan minor official Li Ye's (1192–1279) *Sea Mirror of Circular Measurement (Ceyuan haijing)* of 1248, which was the oldest extant work on the single-unknown technique.[62]

Recovery and Collation of Ancient Chinese Mathematical Works

During the Kangxi revival of interest in mathematics, Mei Juecheng and others could not find many of the works originally included in the medieval Ten Mathematical Classics (*Shibu suanjing*). Moreover, in addition to Li Ye's *Sea*

Mirror of Circular Measurement, the seminal works of Qin Jiushao on polynomial algebra and other important topics were not widely available and were perhaps lost during the Manchu conquest of the Ming. In the midst of the unsupportive policies of the Yongzheng emperor and his successors a large-scale effort to recover and collate the treasures of ancient Chinese mathematics became an important part of the late eighteenth- and early nineteenth-century upsurge in evidential studies.[63]

In addition to famous evidential scholars, such as Dai Zhen, Qian Daxin, Ruan Yuan, and Jiao Xun (1763–1820), who stressed mathematics in their research, the editing of ancient mathematical texts and the continued digesting of European mathematical knowledge was carried out by a series of literati mathematicians who were also active in evidential studies. We will cite them here and discuss them below:

Chen Shiren (1676–1722)
Minggatu (Ming Antu) (d. 1763)
Li Huang (d. 1811)
Wang Lai (1768–1813)
Li Rui (1773–1817)
Xiang Mingda (1789–1850)
Shen Qinpei (1807 provincial degree)
Luo Shilin (1789–1853)
Dong Youcheng (1791–1823)
Dai Xu (1805–1860)
Li Shanlan (1811–1882)[64]

Many of the mathematical texts were collated under imperial auspices during the last years of the Kangxi reign, when the *Synthesis of Books and Illustrations Past and Present* encyclopedia was also completed. When published in 1726, the latter included some European texts from the late Ming and early Qing *Mathematical Astronomy of the Chongzhen Reign.* Five works on Chinese mathematics were also included in the astronomy section of the encyclopedia:

1. *The Gnomon of the Zhou Dynasty and Classic of Computations* (*Zhoubi suanjing*)
2. *Notes on the Mathematical Heritage* (*Shushu jiyi*)
3. *Mathematical Manual of Xie Chawei* (*Xie Chawei suanjing*)
4. *Mathematics Part of the Brush Talks from the Dream Book* (*Mengxi bitan*)
5. *Systematic Treatise on Computational Methods* (*Suanfa tongzong*)

When the first set of the Qianlong Imperial Library collection was completed between 1773 and 1781, its compilers included several collators well versed in classical mathematics, such as Dai Zhen, Kong Jihan (1739–1784),

Chen Jixin, Guo Zhangfa, and Ni Tingmei. The astronomy and mathematics (*Tianwen suanfa*) category incorporated fifty-eight works into the collection (see below). Several older, lost mathematical texts were recopied from the early Ming *Great Compendium of the Yongle reign,* which had survived in the imperial court relatively intact. The general catalog of the Imperial Library, for example, included twenty-five notices on mathematics. Of these, nine were on Tang classics, three were for Song-Yuan works, four on works from the Ming period, including the Ricci-Xu translation of Euclid's *Elements,* and nine on works from the Qing, most importantly the *Collected Basic Principles of Mathematics* (*Shuli jingyun*) and several works by Mei Wending.[65]

The eighteenth-century recovery of ancient mathematical works extended beyond the borders of the Qing dynasty. The role of Korea and Japan in preserving lost Chinese works is generally well known. Worthy of special mention, however, was Ruan Yuan's recovery of the lost *Primer of Mathematics* (*Suanxue qimeng*) by Zhu Shijie from a 1660 Korean edition that had been reprinted as a textbook in 1433. When transmitted to Japan, the *Primer* and its single-unknown (*tianyuan shu; tengenjutsu* in Japanese) algebraic notations played a significant role in Japan in the seventeenth century. Some have claimed that Li Shanlan's nineteenth-century works on the accumulation of discrete piles as a finite series (*duoji,* lit., summing piles) was inspired by Seki Takakazu's (1642?–1708) *Compendium of Mathematical Methods* (*Katsuyō sanpō*), published in 1712.[66]

Published in 1299 in Yangzhou, Zhu Shijie's work described the rudiments of a polynomial algebra using counting rods and not written equations. The first section presented problems on additive multiplication, multiplication on conversion, the subtractive division method, equivalence and proportion, and areas and volumes. Section II gave problems on areas and volumes, differential ratios, and fair distribution. The final section furnished problems on fractions, surplus and deficiency, the tabulation of positive and negative numbers, and root extractions of numerical equations up to the fifth degree.[67]

Korean missions had regularly visited Beijing during the eighteenth century. Several Korean emissaries stayed in Beijing in the early nineteenth century, when Ruan Yuan was simultaneously a high official and a patron of Han Learning. His scholarship was also influential, particularly his work on ancient technology titled *Explications Using Diagrams of the Design of Wheeled Carriages in the "Artificer's Record"* (*Kaogong ji chezhi tujie*). Later Kim Chŏng-hui (1786–1856) visited Beijing and met Ruan Yuan in 1810. As a result, Kim sent Ruan the Korean edition of the *Primer of Mathematics,* and Ruan presented a number of his works to Kim. Because Ruan and others were interested in Zhu Shijie's role in the formation of single-unknown methods, he published the book in 1839. The Jiangnan Arsenal's Translation Department reprinted it in 1871 as a sign of its relevance for mastering modern algebra.[68]

As the Ten Mathematical Classics were partly reconstituted, the Song-Yuan works of Qin Jiushao, Zhu Shijie and Li Ye, among others, also became available. A special edition of seven of the Ten Computational Classics was reprinted by the Imperial Printing Office, including *The Gnomon of the Zhou Dynasty and Classic of Computations* and *Computational Methods in Nine Chapters*, as well as one hundred chapters (*juan*) from the *Sources of Musical Harmonics and Mathematical Astronomy* of the Kangxi era. Traditional mathematical works were also reprinted in several important collectanea.[69]

Reconstruction of the Ten Mathematical Classics

In the late Ming, Xu Guangqi had claimed that the Ten Mathematical Classics, which had provided a set of classical problems to solve, were inferior to Jesuit mathematics. Xu Guangqi's claims about the superiority of Jesuit mathematics were increasingly disparaged by evidential scholars who recovered and collated ancient mathematical texts. They appealed to the Chinese origins of Western Learning as a historical reality and not just a political tactic to justify calendrical reform, as had been the case for Xu in the last years of the Ming.[70]

Twelve works (see Appendix 1), were used to teach mathematics during the Sui and Tang dynasties at the Imperial Academy (*Guozi xue*). Emperor Gaozu of the Tang (r. 618–626) had ordered their use and decreed that like the Five Classics the Ten Mathematical Classics should also be annotated. There was no collection of the Ten Mathematical Classics until the eighteenth century. Erudites of mathematics (*suanxue boshi*) provided instruction in the Imperial Academy. They paralleled the erudites of the Five Classics who were established earlier during the Former Han dynasty.[71]

Between 644 and 648, Li Chunfeng (602–670), then director of the Tang Astro-calendrical Bureau, with the help of others collated the ancient mathematical texts for the official mathematics examinations (*mingsuan*) (see Appendix 1). From 1084 on, eight surviving pre-Song books were printed officially, but these editions were lost by the beginning of the Qing. A private publisher reprinted them in 1213. Subsequently, in 1403–1407, court scholars copied some works from the Ten Mathematical Classics into the *Great Compendium of the Yongle Era*. Because mathematics examinations had died out before the end of the Tang period, literati became less interested in such texts.[72]

Only *The Gnomon of the Zhou Dynasty and Classic of Computations* was widely available in Ming times. The sources extant in the late Ming derived from Southern Song editions or from the early Ming *Great Compendium*. In the late Ming, Mao Jin (1599–1659) and Mao Yi (b. 1640) collated seven of the mathematical classics from unique copies in the Rehe Imperial Library for their Suzhou printing house known as the Pavilion Reaching to the Ancients (*Jigu ge*):

1. *The Gnomon of the Zhou Dynasty and Classic of Computations (Zhoubi suanjing)*
2. *Sunzi's Computational Canon (Sunzi suanjing)*
3. *Computational Canon of the Five Administrative Departments (Wucao suanjing)*
4. *Zhang Qiujian's Computational Canon (Zhang Qiujian suanjing)*
5. *Computational Canon of the Continuation of Ancient Techniques (Jigu suanjing)*
6. *Xiahou Yang's Computational Canon (Xiahou Yang suanjing)*; not the original[73]
7. *Computational Methods in Nine Chapters (Jiuzhang suanshu)*; incomplete version

Although criticized for allegedly shoddy xylography, the Pavilion Reaching to the Ancients versions of the Classics and the Dynastic Histories were highly prized. Among Mao's specialties was his care in making facsimiles of Song editions by tracing every feature of the rare books he borrowed from others. The *Zhou Dynasty Canon* and *Notes on Bequeathed Mathematical Arts (Shushu jiyi)* were also published in a Wanli-era (1573–1619) collection.[74]

Later, however, the Mao family's own copies were dispersed and fell into book collectors' hands. Only five of the mathematical classics were intact in the early Qing. A manuscript copy of the *Computational Methods in Nine Chapters* was sent to the Kangxi court and kept in an imperial library. Subsequently, collating of the Ten Mathematical Classics accelerated after the 1728 publication of *The Gnomon of the Zhou Dynasty and Classic of Computations* and *Notes on Bequeathed Mathematical Arts* as part of the *Synthesis of Books and Illustrations* encyclopedia. The celebrity that Mei Wending had achieved as a mathematician, coupled with the publication of several new European mathematical works during the late Kangxi reign, brought mathematical astronomy into the mainstream of classical studies.

The influence of the *Collected Basic Principles of Mathematics (Shuli jingyun)* as a translation of what Jesuits led the Qing court to think were the latest features of European mathematics stimulated scholars associated with evidential studies to emulate Mei Wending and rediscover the Chinese origins of Western mathematics. The *Collected Basic Principles* introduced a new version of Euclid's *Elements* that had wider circulation than the Ricci-Xu version. It also introduced algebraic formulas for computational solutions and reiterated the importance of plane geometry. Logarithms and algebra (*jiegen fang*) introduced in the *Collected Basic Principles* were also more widely available now. As a result, Mei Juecheng and others increasingly equated single-unknown problem-solving techniques with more formal algebra.

While serving on the Imperial Library staff in the 1770s, Dai Zhen collated seven of the ten mathematics classics from the *Great Compendium of the Yongle Era*. In addition, he recovered two more from manuscript copies originally held by the Mao family. The Imperial Printing Office then published them as rare editions in a special collectanea. Dai's colleague Kong Jihan had them reprinted in the *Ripple Pavilion Collectanea* in 1773 under the title Ten Mathematical Classics (*Suanjing shishu*), which represented the first collectanea with this name. Subsequent editions, including Dai's annotations, were based on these versions.[75]

The reconstruction of the mathematical classics stimulated interest in them, and evidential scholars such as Dai Zhen, Li Huang, Shen Qinpei, and Gu Guanguang increasingly studied them. They produced several important works on them that echoed the follow-up scholarship known as "additions and corrections" (*buzheng*) in Qing dynasty classical book titles. Li Huang, for example, prepared works titled *Worked Out Examples with Diagrams of the Computational Methods in Nine Chapters* (*Jiuzhang suanshu xicao tusho*), *Worked Out Examples with Diagrams of the Sea Island Computational Canon* (*Haidao suanjing xicao tushuo*), and *Critical Notes on Annotations of the Ancient Mathematical Classics* (*Jigu suanjing kaozhu*). Gu Guanguang published his *Collation Notes for The Gnomon of the Zhou Dynasty and Classic of Computations* (*Zhoubi suanjing jiaokan ji*).[76]

Recovery of Song-Yuan Mathematical Works

Reconstructions of the single-unknown and four-unknowns techniques for solving polynomial equations in several unknowns and to several powers, and collations of mathematical texts produced by a long-neglected group of Song-Yuan minor officials and commoner mathematicians were particularly prominent in the late Qianlong era. Qin Jiushao's *Computational Techniques in Nine Chapters* (*Shushu jiuzhang*, 1247), for example, provided general algorithms for solving simultaneous congruences (*dayan qiuyi shu*, lit., "great expansion procedure for finding 1," a method for solving the Chinese remainder problem). He also investigated techniques similar to the Horner-Ruffini method devised in the early nineteenth century for calculating the roots of polynomial equations.[77]

Although Qin's work followed the overall structure of the *Computational*

Techniques in Nine Chapters, his algorithms were much more sophisticated. In addition, because Qin had studied in the Song Astro-calendric Bureau in Hangzhou as a youth, his work also treated certain astronomical predictions as problems in remainder theory. Finally, the *Computational Techniques* calculated the area of any triangle as a function of the lengths of its three sides, which was similar to the proto trigonometric relational features for computing the sides of a right triangle (*gougu*) revived in the late Ming. Later, this approach drew the attention of Dai Zhen and others interested in correlating such proto trigonometric relational features with Jesuit trigonometry.[78]

Dai Zhen copied Qin's *Computational Techniques* from the *Great Compendium of the Yongle Era* into the Imperial Library. Jiao Xun and Li Rui each studied Qin's findings. Later Shen Qinpei discovered a Ming manuscript of the *Computational Techniques,* which he compared to the version in the *Great Compendium.* Song Jingchang, Shen's student, prepared notes and comments for both editions. A definitive edition of the *Computational Techniques* was then included by Yu Songnian in the 1842 *Collectanea of the Yijia Hall* (*Yijia tang congshu*) and became the basis for modern versions. Hua Hengfang's (1833–1902) *Notes on Mathematical Studies* (*Xuesuan bitan*) for 1882–1888 presented a list of must-read books, which included both Chinese and Western works. Although he was well versed in modern mathematics, Hua still recommended the *Computational Techniques* for solving remainder problems.[79]

After the Mongols conquered north China in 1232, Li Ye (1192–1279) (also called Li Zhi) prepared two works: the *Sea Mirror of Circle Measurement* (*Ceyuan haijing,* 1248) and *Adding to Ancient Techniques for Computing Geometric Figures* (*Yigu yanduan,* 1259), which were copied into the eighteenth-century Imperial Library. Although Li lived in reclusion after the Mongol triumph, he was called to the Mongol court to consult on governance and earthquakes. The *Sea Mirror* survived in a book in Li Huang's private library. Scribes subsequently copied Li's *Adding to Ancient Techniques* into the *Compendium of the Yongle Reign.* It was also listed in Ming bibliographies of mathematical works. Li Rui later collated both in the early nineteenth century, and they were also printed in the *Collectanea from the Can't Know Enough Pavilion.*

The *Sea Mirror* presented 170 problems on single unknown techniques. Li based the problems in it on a single diagram, a triangle with an inscribed circle (in the formulation of the problem, he constructed a triangle around a city wall; see Figure 6.1). Using a method equivalent to polynomial equations (including negative powers of an unknown), Li calculated the diameter of the circle, which was solvable using other methods known at the time. Using his diagram, Li also calculated the lengths of segments in the figure based on the single-unknown computational techniques emerging in the Song-Yuan period. In his *Adding to Ancient Techniques,* Li dealt with the algebraic and geometric construction of sixty-four polynomial equations, which may have been designed as an introduction to the more difficult *Sea Mirror.*[80]

Although Yang Hui's (fl. in Hangzhou during the Southern Song) mathematical works were all included in the *Compendium of the Yongle Reign,* Dai Zhen did not copy them into the Imperial Library. Part of Yang's *Continuation of Ancient Mathematical Methods for Elucidating the Strange [Properties of] Numbers (Xugu zhaiqi suanfa)* was, however, edited for the *Collectanea from the Can't Know Enough Pavilion* and printed by the Hangzhou bookman Bao Tingbo (1728–1814). In 1840, Yu Songnian included Yang's commentary and supplement for the *Computational Methods in Nine Chapters* and the *Yang Hui's Calculation Methods (Yang Hui suanfa,* 1275), which was still incomplete, in the *Collectanea of the Yijia Hall.* Yang's complete work was lost in China but was rediscovered in the late Qing by Li Rui. In the early twentieth century, a 1433 Korean edition of the *Calculation Methods,* based on a 1378 Ming edition, was found. It was recopied in Japan by Seki Takakazu when Japanese scholars discovered it among the books brought from Korea by Hideyoshi's invasion forces.[81]

On the other hand, the editors had not recovered Zhu Shijie's 1303 *Jade Mirror of the Four Unknowns (Siyuan yujian)* and his 1299 *Primer of Mathematics* in time for inclusion in the Qianlong Imperial Library. Although Zhu's *Jade Mirror* purported to solve practical issues dealing with architecture, finance, military logistics, etc., it energized late Qing evidential scholars, who found in it a Chinese algebra to extract roots (*kaifang*) using counting rods that predated the Jesuit's borrowing roots (*jiegen fang*) approach. Zhu's polynomial equations went beyond the second and third degrees up to the fourteenth.[82]

During the Jiaqing reign (1796–1820), while governor of Zhejiang, Ruan Yuan acquired from a Korean envoy a version of the *Primer of Mathematics,* which he used to reconstitute the *Jade Mirror of the Four Unknowns.* He then sent the latter to Beijing to be included as an uncollected book of the Imperial Library. Ruan also gave Li Rui a copy to collate, which Li left uncompleted. Others such as Xu Youren (1800–1860) and Shen Qinpei worked on it, although Shen's commentary was never published.[83]

Luo Shilin obtained a copy of the *Jade Mirror of the Four Unknowns* in 1822, and after ten years of collation he presented a definitive reconstruction of four-unknowns techniques, which he titled *Jade Mirror of the Four Unknowns with Detailed Calculations (Siyuan yujian xicao,* 1843), and had it printed in Yangzhou. After the rediscovery of Zhu Shijie's *Primer,* Luo Shilin also had it

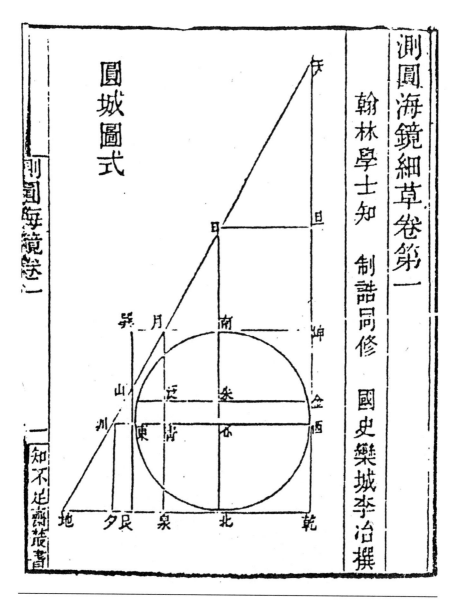

Figure 6.1. Li Ye's "diagram of a triangle with an inscribed circle" (representing a triangle around a city wall).

Source: Zhi buzu zhai edition of the *Ceyuan haijing* (*Zhu buzu zhai congshu*, 1776–1823).

reprinted in 1839 based on the Korean edition of 1660 recovered by Ruan Yuan (see above). As noted above, the *Primer* was also an important clue to the fundamentals of the single-unknown and four-unknowns polynomial algebra.[84]

Zhang Dunren in 1801 solved the problems in the *Computational Canon of the Continuation of Ancient Techniques* based on single-unknown procedures. Zhang also prepared an 1831 work titled *Technique for Finding "1"* (*Qiuyi suanshu*, i.e., indeterminate analysis based on the great expansion [*dayan*] method for solving congruencies). Li Rui, who was Zhang's personal secretary, helped complete this harvest of Song-Yuan mathematics and bring it into the mainstream of evidential studies in the first half of the nineteenth century.

Antiquarianism abetted the inward turn to native mathematics described above. Chinese classical scholars at the cutting edge of evidential studies grasped the importance of advanced algebraic techniques for solving complicated equations based on sophisticated mathematical problems. At the same time, however, they focused on the recovery of ancient texts and lost knowledge. When Alexander Wylie and John Fryer finally introduced the differential and integral calculus in the middle of the nineteenth century, Li Shanlan and others appreciated its sophistication because they had already mastered single-unknown and four-unknowns problem-solving skills. No literati-mathematician in the eighteenth century recognized such possibilities.[85]

Jesuit introduction of limited aspects of early modern European medicine, astronomy and mathematics, Tychonic cosmology, and early modern algebra had fortified the resolve of mid-Qing literati to recover classical Chinese medicine and mathematics. Such disjointed imports, however, had none of the corrosive cultural and intellectual implications that science and technology brought in the nineteenth century.

The Macartney mission may have been a harbinger, but neither Lord Macartney nor the Qianlong emperor could foresee that the Industrial Revolution in England would produce British military superiority and Protestant evangelism in China. We should not read the events of the first and second Opium Wars back into the eighteenth century. Qianlong arrogance encoded in the emperor's infamous 1793 edict to King George III and British imperialism were inseparable. Macartney's ship mathematician James Dinwiddie described the limits of their mission: "What information could we derive respecting the arts and sciences in a country where we could not converse with the inhabitants."[86]

7

Seeking the Truth and High Qing Mathematics

Since the late Ming, a growing emphasis on ancient classical learning had occasioned a revival of the Five Classics, in addition to the medical and mathematical classics, as the cornerstone of literati scholarship. Accepting the call to seek truth from actualities (*shishi qiushi*) rather than to investigate things and extend knowledge (*gewu zhizhi*), Qing scholars increasingly focused on study of the Great Learning and the Doctrine of the Mean as a way that challenged the legitimacy of the Four Books. Many connected with this debunking movement had some contact with Jesuit learning, and their circles took substantial interest in European astronomy and mathematics.

Evidential scholars also turned classical and Jesuit mathematics into an independent discipline, rather than just a tool that the court used to produce the calendar. While they incorporated mathematics into classical studies, they did not support new directions in research or experimental applications. Nevertheless, their efforts still laid the foundation for nineteenth-century developments, when Protestant missionaries introduced the calculus and Newtonian mechanics in China. In an age of evidential studies, literati prioritized mathematics as an important field of study.

High Qing Views of the Investigation of Things

Mid-Qing literati scholars took a range of positions concerning natural studies. The seventeenth-century impact of Jesuit knowledge in China was not always elided or easily domesticated in the eighteenth century. A private scholar, Jiang Yong (1681–1762), for instance, combined loyalty to Zhu Xi's teachings with knowledge of Jesuit writings obtained through his evidential studies. In his work on mathematical harmonics, Jiang drew on Zhu Zaiyu's pioneering late Ming work on musical pitches. Jiang also prepared a treatise on astronomy with equations for calculating the courses of sun, moon, and planets as well as lunar and solar eclipses.

Jiang was radical in his critique of both Han Learning scholars and Mei Wending for exalting nativist natural studies. Jiang recognized the advantages European astronomy had over native traditions, while at the same time he continued to uphold the cultural superiority of the Cheng-Zhu view of morality. Although Jiang preferred European learning for understanding the principles of natural events, he maintained a clear distinction between astronomical methods and native cultural values.[1]

Increasingly, the mid-Qing scholarly agenda of evidential scholars led them back to Han or pre-Han sources. They placed proof and verification at the center of their organization and analysis of the classical tradition, which now included aspects of natural studies and mathematical astronomy. The terminology they preferred came from the Han dynasty, and they gainsaid what they considered the misappropriation of the Great Learning by Zhu Xi and others to investigate things as the means to fathom principles. Instead, they preferred the slogan of "seeking truth from actualities" and its implied focus on empirical findings based on experience.[2]

Suzhou Traditions of Han Learning and Investigating Things
in the Eighteenth Century

In Suzhou, the home of Han Learning in the mid-eighteenth century, for example, Hui Shiqi (1671–1741) critiqued Zhu Xi's emendation of the Great Learning. A Hanlin academician and educational commissioner during his official career, Hui contended in his scholarly assessments that Han scholarship was more reliable than Tang or Song classical learning. In this light, he favored the Han commentaries for the *Spring and Autumn Annals* and the *Rituals of Zhou*. He also condemned Zhu Xi's prioritizing of the investigation of things passage in the Great Learning to elaborate Zhu's airy notion of exhaustively mastering principles (*qiongli*). Instead, Hui regarded the central message in the original Han dynasty version of the Great Learning in more concrete terms so as to achieve broad learning (*boxue*).[3]

Hui's son, Hui Dong (1697–1758), chose a scholarly career over official service and through his many writings brought Han Learning in Suzhou to broad notice among evidential scholars centered in the Yangzi delta. Hui Dong's Han Learning inspired reconstructions of the Five Classics and Four Books. In his influential *Ancient Meanings in the Nine Classics* (*Jiujing guyi*), for instance, Hui Dong interpreted the investigation of things in terms of measurement, which later commentators regarded as evidence of the importance of mathematical astronomy in the classical canon. Hui's father had produced two works on astronomy and mathematical harmonics. The Hui lineage in Suzhou initially represented a private tradition of Han Learning that became increasingly influential in the Yangzi delta overall and offered an alternative to the public-examination studies that were predicated on the mastery of Cheng-Zhu learning.[4]

The Suzhou native Qian Daxin, arguably the greatest classical scholar of his age, openly opposed Zhu Xi's emendations of the Great Learning in favor of the Old Text version. For Qian, the investigation of things required a scholar to unravel the orderly manifestations of things. Animated by a restorationist concern, Qian Daxin successfully appropriated technical aspects of European astronomy and mathematics into the classical framework for investigating things and extending knowledge. Qian acknowledged this broadening of the literati tradition, which he saw as the reversal of centuries of focus on moral and philosophical problems:

Comparing lands of the Eastern [China] seas with those of the Western [Atlantic], we note that their spoken languages are mutually unintelligible and that their written forms are different. Nonetheless, once a computation has been completed [no matter where], there will not be the most minute discrepancy when it is checked. This result can be for no other reason than the identity of human minds, the identity of patterns of phenomena, and the identity of numbers [everywhere]. It is not possible that the ingenuity of Europeans surpasses that of China. It is only that Europeans have transmitted [their findings] systematically from father to son and from master to disciple for generations. Hence, after a long period [of progress] their knowledge has become increasingly precise. Literati scholars have, on the other hand, usually denigrated those who were good mathematicians as petty technicians. . . . In ancient times, no one could be a literatus who did not know mathematics. Chinese methods [now] lag behind Europe's because literati do not know mathematics.[5]

When he retired from official life at forty-eight in 1776, Qian spent his middle years engaged in specialized historical and classical research. Later in 1789, he was appointed director of his Suzhou alma mater, the Ziyang Academy. He taught there for the final sixteen years of his life. During his tenure, over two thousand students matriculated. According to academy accounts, they mastered "ancient learning" and "sought the truth in actual facts" (shishi qiushi). Graduates of Ziyang during this time, such as Li Rui (see below), went on to distinguish themselves as specialists in mathematics, geography, paleography, technical statecraft, and historical studies.[6]

Dai Zhen and the Great Learning

Dai Zhen also challenged Zhu Xi's emendation of the Great Learning based on historical philology. For example, at the age of ten years (*sui*) Dai was studying Zhu Xi's standard commentary on the Great Learning, when he asked his teacher about one of Zhu Xi's comments, which read:

> The preceding chapter of classical text is in the words of Confucius, handed down by Master Zeng. The ten chapters of explanation that follow contain Zeng's views and were recorded by his disciples. In the old copies of the work [in the *Record of Rites*], there appeared considerable confusion from the disarray of the [original] wooden slips. But now, availing myself of the decisions of Master Cheng [Yi], and having arranged anew the classical text, I have rearranged it in order as follows.[7]

Dai asked about Zhu Xi's rearrangement of the Great Learning into a classical text by Confucius and ten commentarial chapters by Master Zeng:

> *Dia Zhen asked:* How does one know here that these are the words of Confucius recorded by Master Zeng? Moreover, how does one know that Master Zeng's intentions were recorded by his followers?
>
> *The teacher replied:* That is what the earlier literatus Zhu Xi said in his annotation.
>
> *Dai Zhen asked another question:* When did Zhu Xi live?
>
> *The teacher answered:* Southern Song [1127–1280].
>
> *Dai asked again:* When did Confucius [circa 551–479 B.C.] and Master Zeng live?
>
> *The teacher replied:* Eastern Zhou [770–221 B.C.].
>
> *Dai asked again:* How much time separates the Zhou [dynasty] from the Song?
>
> *Reply:* About two thousand years.
>
> *Dai queried again:* Then how could Zhu Xi know it was so?
>
> The teacher could not reply.[8]

Dai Zhen, like Wang Yangming, defied Zhu Xi's classical authority. Dai, however, exposed as historically undocumented Zhu's claims that the Great Learning in the *Record of Rites* was corrupt because it failed to make the investigation of things passage the starting point for classical learning. Despite such vocal criticism of Zhu's emendation of the Great Learning since the Song, it is interesting that in their late Ming attacks on Zhu Xi the Jesuits such as Matteo Ricci never raised the issue.

Dai Zhen's criticisms occurred within a sea of Ming-Qing literati opinion that considered Zhu's appreciation of the investigation of things superior to Wang Yangming's inward quest for moral principles in the mind. The more orthodox assessment of the Cheng-Zhu legacy by the editors of the Qianlong Imperial Library in the 1780s indicated that Zhu Xi's view of investigating things and extending knowledge was an acceptably balanced theory of knowledge.[9]

Yangzhou Scholars and the Late Qianlong Revaluation of the Investigation of Things

Although he had doubts as a schoolboy about Zhu Xi's emendations, Dai Zhen was more sympathetic with them in his later writings when attacking Wang Yangming for relying on the Old Text version to develop his baseless views for the investigation of things. Later in the 1760s, upon the death of Hui Dong, Dai became the most heralded opponent of Cheng-Zhu learning. His views of the investigation of things favored concrete studies. The sage's ideal had been "to investigate each event or thing that appeared and carefully master its reality to the smallest detail." Dai Zhen's views became very prominent in Yangzhou, where he first taught in the Wang household from 1756 to 1762.[10]

Dai Zhen had in mind a systematic research agenda that built on paleography and phonology to reconstruct the meaning of classical words. Later, Dai's student Wang Niansun (1744–1832), and his son Wang Yinzhi (1766–1834) extended his approach and attempted to use etymology to reconstruct the intentions of the sages. The technical phonology applied to the history of the classical language reached unprecedented precision and exactness.[11]

Describing the academic environment in Yangzhou, Wang Zhong (1744–1794) recalled:

> At this time, ancient learning [guxue] was popular. Hui Dong of Yuanhe [in Suzhou] and Dai Zhen of Xiuning [in Anhui] were admired by everyone. In the area north of the Yangzi River [that is, Yangzhou], Wang Niansun promoted ancient learning and [Li Chun (1734–1784)] did the same. Liu Taigong [1751–1805] and I came along and continued [their efforts]. We worked hard together to realize our potential, and each of us formed his own [specialty of] learning.[12]

Concerning the Great Learning debate, Wang Zhong argued that Song scholars had used the Four Books to spread Chan Buddhist ideas instead of the classical heritage.

Other Yangzhou scholars such as Ling Tingkan (1757–1809), Cheng Yaotian (1725–1814), and Jiao Xun (1763–1820) were also critical of Cheng-Zhu learning for lapsing into Chan Buddhism. Combining a stress on morality and scholarship, Ling favored the ritualization of knowledge. In this way, he thought, correct ritual behavior would become the end point of investigation of things rather than the purely contemplative ideal that he contended had penetrated Cheng-Zhu learning. Cheng Yaotian likewise stressed a literal reading of the investigation of things that would be penetrating enough to comprehend the principles of things. Again, morality and scholarship were presented as compatible. For his part, Jiao Xun presented the investigation of things as a way for the government to harmonize the desires of individuals so that the emperor's policies would conform with the desires of the people. In this manner, the conflicts among individuals would not lead to litigation.[13]

Knowledgeable about native and European mathematical astronomy, Jiao agreed with his Yangzhou patron Ruan Yuan that mathematics was one of the keys to concrete studies. Jiao prepared a detailed study of early Chinese mathematics, and he aided Li Rui, a disciple of Qian Daxin, in Li's reconstruction of the Song-Yuan single-unknown craft of algebra. Through his interests in astronomy and mathematics, Dai Zhen also realized that a keen understanding of mathematics was required to understand passages in the Classics dealing with technical, mathematical, and astronomical phenomena. In his study of the "Artificer's Record" (*Kaogong ji*) chapter of the *Rituals of Zhou*, Dai indicated that mathematical training was needed to understand the texts describing ancient technology and production techniques. Qian Daxin and other eighteenth-century scholars trained in mathematical astronomy followed Dai in this line of argument.

Dai, for example, used mathematics to estimate the size and shape of the ceremonial bronze bells mentioned in the "Artificer's Record." Dai included diagrams based on his calculations of what the bells should look like. Because he employed texts and relics as archaeological evidence, Dai's drawings proved to be highly accurate when a large bronze bell was unearthed during the Qianlong era (see Figure 7.1). Jiang Yong also had made attempts to reconstruct the appearance of the bells, but his drawings were less accurate. Like Dai Zhen's diagrams, Jiang's drawings showed how to reconstruct ancient objects. Cheng Yaotian read Dai's and Jiang's descriptions and attempted to cast the bells. On the basis of archaeological and technical data, Cheng tried to reconstruct the actual instruments used in ancient music.[14]

Ruan Yuan's first published work in 1788 was titled *Explications Using Diagrams of the Design of Wheeled Carriages in the "Artificer's Record."* There he improved on Dai Zhen's earlier research on this problem and reconstructed the dimensions of ancient vehicles, which he maintained anyone could now build if they followed his guidelines. Archaeological research was taking on a momentum of its own as a field of exact classical scholarship. Applying mathematics and using ancient relics to decipher ancient texts, Dai Zhen and Qian Daxin paved the way for a focus on artifacts and their importance as historical evidence. They were also reconsidering the scope of traditional technologies in the "Artificer's Record" in light of European achievements as documented by Schreck and Wang Zheng in their *Diagrams and Explanations of the Marvelous Devices of the Far West.*

The revaluation of the investigation of things during the Qianlong era by Han Learning scholars paralleled their domestication of Jesuit learning. Evidential scholars recognized that reinterpretation of the Cheng-Zhu maxim would enable them to reframe investigating things and extending knowledge to suit their more intellectualist agenda. European studies and Cheng-Zhu learning could be elided by appealing to an antiquity that preceded and encompassed both. At the end of the eighteenth century, the Yangzhou scholar Ruan Yuan attacked the Ming stress on the mind for investigating things in favor of more practical affairs tied to evidential research. Ruan equated the investigation of things with the Han Learning slogan of searching truth from facts.[15]

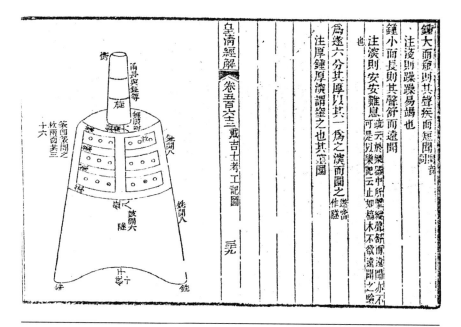

Figure 7.1. Dai Zhen's "diagram of the Zhou dynasty ceremonial bronze bells."
Source: Huang-Qing jingjie, compiled by Ruan Yuan et al. (Guangzhou: Xuehai tang, 1829).

In the early nineteenth century, scholars increasingly softened their positions on the distinction between Han and Song Learning and followed the lead of evidential scholars such as Ruan Yuan, who noted, "Many earlier literati discussed the investigation of things, but they associated it with empty meanings and apparently failed to grasp that such meanings were not the basic intent of the sages. I don't dare to be different in my discussion of the investigation of things, but [I do say] it is the search for the truth from facts and that's all."[16]

Mathematics in an Age of Evidential Research

Heavily biased in favor of Han Learning and evidential research, the catalogers of the Qianlong Imperial Library initiated a well-publicized search in the 1770s for every published work in the empire. They then engaged in a critical review of every book available to them, selected books worthy of inclusion in the collection, and carefully collated the final versions chosen for inclusion. They also prepared critical summaries of important works that were not copied into the collection. Leading classical scholars in various fields were appointed to evaluate and collate books in their specialties. More than 360 scholars officially worked at the apex of a staff of several thousand in Beijing.

The Critique of Western Learning in the Qianlong Era

The editors separated the investigation of things and extension of knowledge from its long-standing ties with Cheng-Zhu Song Learning and associated it instead with their emphasis on certain knowledge derived from empirically based research. They contended, for instance, that Chen Yuanlong's entries in the *Mirror of Origins* were all examples of broad learning and thus the work was justifiably titled "investigating things and extending knowledge."[17]

In some cases, like Chen's *Mirror of Origins,* the editors elided any mention of European learning when they could. For example, their summary of Fang Yizhi's late Ming *Notes on the Principles of Things,* which we have seen generally accepted Jesuit explanations of natural phenomena, linked Fang's material investigations to encyclopedias compiled since medieval times in China. The final version of the Imperial Library account never mentioned the notions of a spherical earth, limited heliocentrism, or human physiology that Fang had culled from Jesuit translations.[18]

Unlike Chen's encyclopedia, the editors could not completely ignore the translations and other works that had been published in Chinese since the late Ming by the Jesuits and their collaborators. They addressed the Jesuit role in the Kangxi-era *Compendium of Observational and Computational Astronomy* and its supplement. The prestigious catalog mentioned some thirty-six European works, over twenty of which were copied into the library. All were on natural studies, although the focus on Christianity was more obvious in some than others.[19]

On the other hand, the ten works from the applications (lit., "implements and utensils") section of Li Zhizao's *First Collection of Celestial Studies* were included, but not those from the principles and patterns section, except for a geographical work by Aleni. Similarly, only two of Verbiest's works were mentioned in the Imperial Library, while the library's catalogers discussed Aleni's *A Summary of Western Learning* but dismissed it as a heterodox study.[20]

The editorial account of Aleni's *Summary,* a work that was not copied into the Imperial Library, presented the six-fold classifications of the sciences that corresponded to sixteenth-century European standards of learning. Seeing some parallels with native traditions, the editors analogized the Jesuit fields of grammar, history, poetry, and the art of writing to the native category of lesser learning (*xiaoxue,* i.e., philology). They likened Jesuit logic, physics, metaphysics, mathematics, and ethics to the teachings in the Great Learning.

Moreover, the editors praised the system of knowledge in Aleni's *Summary* and compared it with the investigation of things and exhaustively mastering principles (*gewu qiongli*). Indeed, Aleni used the investigation of things to exhaustively master principles (*gewu qiongli*) as the common Chinese term for contemporary European philosophy. More or less simultaneously, Japanese students of Dutch natural studies in Nagasaki and Edo used exhaustively mastering principles (*qiongli* = *kyūri*) to translate *natuurkunde* into Japanese (*kanbun* = Chinese readings).[21]

Despite such correspondences, however, the Qianlong editors dislodged their eighteenth-century views of the investigation of things and extension of knowledge from its late Ming association with Jesuit learning as well as with Cheng-Zhu studies. Again, they linked the teachings of the Great Learning to an emphasis on verifiable knowledge derived from empirical studies. Moreover, they concluded that Aleni had forced equivalences between European and native learning in order to substantiate Christianity as an ancient teaching in China:

> They also investigate things to exhaust principles, and seek to understand substance for practical use; in that they are roughly similar to our literati. However, the things they investigate are mostly petty, and the principles they seek to exhaust are mostly esoteric and untestable. That is why this book is considered heterodox doctrine.

The Jesuit tactics that Yang Guangxian had detected, and which the Rites Controversy had exposed, demonstrated to eighteenth-century scholars such as Qian Daxin and Dai Zhen that European learning of the seventeenth century was inferior. On this basis, the editors of the Imperial Library also ridiculed Vagnoni's *Treatise on the Composition of the Universe* because it had tried to introduce a new reading of *qi* as air that favored the scholastic four elements.[22]

Overall, the editors, unaware of eighteenth-century trends in Europe, ridiculed the religious content of Western works. In particular, they noted that despite the Jesuits' attacks on Buddhism, in spite of their efforts to seek a middle ground between Chinese classical learning and Christian theories of life and death, their conclusions were as nebulous as Buddhist views. According to the editors, Christianity and Buddhism shared more similarities than differences. What interested the Imperial Library reviewers was the natural philosophy, mathematics, astronomy, geography, agronomy, and ingenious machines that the Jesuits and their collaborators had introduced since 1600. Their accounts of European works in these categories were by and large very positive, although such praise was usually muted by efforts to discredit—and thereby ignore—European innovations as derivatives of ancient Chinese learning.

European Astronomy and Mathematics in the Qing Imperial Library

The *General Catalog*, completed in 1782, was the most comprehensive collection of abstracts compiled by classical scholars in imperial China. The editors attempted to present an accurate and concise idea of the nature and importance of each work so that readers could choose whether to read it. Even though the editors chose to elide or ridicule those Jesuit works that were included in the *General Catalog*, they did innovate when they presented abstracts on works dealing with astronomy and mathematics.

As a frequent beneficiary of the director Ji Yun's (1724–1805) patronage, Dai Zhen worked for the Imperial Library commission on the mathematical astronomy included in the "Astronomy and Mathematics" section, which we have seen enabled him to recover several of the Ten Mathematical Classics. Of the ten thousand titles reviewed by the editors, about a third finally were copied into the imperial manuscript collection, including fifty-eight works on astronomy and mathematics. Forty of the latter were recovered from the *Great Compendium of the Yongle Era* by Dai Zhen and his colleagues Chen Jixin, Guo Changfa, and Ni Tingmei. Some six Jesuit translations in this category were positively evaluated by the editors.[23]

As an indication of the emerging disciplines that comprised Qing classical scholarship, the late eighteenth-century classification of knowledge revealed the manner in which the variety of learning was perceived and the nature and structure of the concepts used to order that variety. The bibliographic clustering of subjects in the Qianlong Imperial Library enables us to deduce the culturally conditioned biases in Qing Han and Song Learning. Moreover, the eighteenth-century conception of the structure of knowledge shaped evidential studies and influenced how new research on mathematical astronomy would be understood.

Representing the classical scheme of disciplines in the late eighteenth century, the Imperial Library was based on the four classifications system (*sibu*), which incorporated astronomy and mathematics (*tianwen suanfa*), as well as medicine, as subcategories under the early and later masters (*zibu*) category (see Table 7.1). Similarly the mathematical aspects of music were subsumed under the Classics, while chronography and geography were listed under the category of history. There was no single unified category for natural studies in the earliest imperially authorized bibliographies. The ancient *Seven Summaries* (*Qilue*) and medieval *Seven Records* (*Qilu*) begun in 523 by Ruan Xiaoxu (479–535) organized such studies under the general category of calculating skills (*shushu*) or skills and techniques (*shuji*).[24]

The editors, rather than placing mathematics in the lesser studies (*xiaoxue*) section as the late seventeenth-century compilers of the *Ming History* had done, linked it commonsensically to astronomy. They separated the calculating arts (*shushu*, i.e., numerology) from their traditional association with mathematics and granted them an independent status because of their association with exotic fields such as astrology, chronomancy, the five phases, milfoil divination, prognostication, and geomancy.[25]

Although they gave little or no credit to the Jesuits for their innovation, the Qianlong compilers of the Imperial Library catalog broke new bibliographic ground by linking mathematics and astronomy. The editorial overview that opened the Astronomy and Mathematics section of the *General Catalog* represented a native response to the Jesuit's impact on new methods (*xinfa*) for reforming the mathematical calculations associated with the Qing calendar. Moreover the pride the editors took in acknowledging how Qing scholars had balanced and unified Western and Chinese mathematics in the process of recovering the single-unknown techniques for solving quadratic and higher polynomial equations made it clear that they regarded the Chinese origins of Western Learning as the key rationale for overcoming the less-informed mathematics of their Ming predecessors.[26]

Ruan Yuan and the Biographies of Mathematical Astronomers

Overall, Ruan Yuan's compilation of the *Biographies of Astronomers and Mathematicians* (*Chouren zhuan*) while serving as governor of Zhejiang province in Hangzhou from 1797 to 1799 climaxed the celebration of nat-

Table 7.1. Forty-four subdivisions of the Imperial Library (*Siku quanshu*). Fields associated with natural studies have been printed in bold face.

Classics	*History*
Change	Dynastic Histories
Documents	Annals
Poetry	Topical Records
Rituals	Unofficial Histories
Spring and Autumn Annals	Miscellaneous Histories
Filial Piety	Official Documents
General Works	Biographies
Four Books	Historical Records
Music	Contemporary Records
Philology	**Chronography**
	Geography
	Official Registers
Early and Later Masters	Institutions
Literati	Bibliographies and Epigraphy
Military Strategists	Historical Criticism
Legalists	
Agriculturalists	
Medicine	*Literature*
Astronomy and Mathematics	Elegies of Chu
Calculating Arts	Individual Collections
Arts	General Anthologies
Repertories of Formulas, Recipes	Literary Criticism
Miscellaneous Writers	Songs and Drama
Encyclopedias	
Novels	
Buddhism	
Taoism	

Benjamin A. Elman, *From Philosophy to Philology: Social and Intellectual Aspects of Change in Late Imperial China*. Second edition. Los Angeles, UCLA Asia Institute Monograph Series, 2001.

ural studies within the Yangzi delta literati world of the late eighteenth century. Containing summaries of the works of 280 mathematicians and astronomers, including thirty-seven Europeans, the *Biographies* was followed by four supplements in the nineteenth century. The collection was reissued in 1829 with only Qing biographies, and it was later enlarged and reprinted in 1849. In 1840, for example, Luo Shilin added forty-three sections on Song-Qing mathematical astronomers based on new sources for the four-unknowns techniques recovered in 1822 from the Song and Yuan, such as *Yang Hui's Calculation Methods* and the *Jade Mirror of the Four Unknowns*. In 1857, Alexander Wylie worked with Wang Tao (1828–1897) to improve on the views presented in the collection, particularly critiquing the Chinese origins rationale.[27]

Ruan had become famous among his peers in 1791 when his prose-poem for the Hanlin Academy special examination topic on Zhang Heng's early Yuan astronomy was singled out by the Qianlong emperor for special praise. Ruan Yuan's technical interests were also influential because of his status as a patron of Yangzi delta scholarship, particularly evidential scholars from Yangzhou. In addition, Ruan in 1799 served as director of the mathematics section of the Imperial Academy (*Guozi jian*) in Beijing.[28]

Ruan was aided in his Hangzhou *Biographies* project by many of the leading evidential scholars of the late Qianlong period: Li Rui, Qian Daxin, Jiao Xun, Ling Tingkan, Tan Tai, and Zhou Zhiping. Their efforts in astronomy have been described by Sivin as "a programmatic synthesis of traditional and Western astronomy designed to encourage the study of the latter in order to improve the former." Their efforts culminated an ongoing process reaffirming the value of mathematics and astronomy as part of a classical education.

Ruan did not include fortune-telling or numerology in the collection, and he opposed connecting mathematical astronomy with mathematical harmonics or studies of the *Change Classic*. Moreover, he was critical of three new findings introduced by Michel Benoist: (1) heaven and earth are round; (2) planets follow elliptical paths; and (3) the sun is stationary. Although Benoist had by then finally presented the Copernican system to China as the European norm, Ruan Yuan found such views unacceptable, in part because they contradicted earlier Jesuit presentations of Copernicus, which had denied that his cosmology differed from the Tychonic system. Ruan sought a fusion of European and Chinese mathematics based on shared common conceptions. For astronomy, he sought an accurate, predictive computational system that would be based on improved techniques, not cosmology.[29]

Interest in mathematical astronomy among literati, which had developed steadily since Mei Wending, grew in importance outside the imperial court by the late eighteenth century. This growth was tied to the popularity of evidential studies among literati outside the patronage networks of the Manchu court, which had focused on Manchu and Mongol bannermen to control such knowledge. By connecting mathematics and astronomy to classical studies, Ruan Yuan successfully integrated mathematical astronomy with evidential studies. Because mathematics and natural studies remained dependent on classical studies, Ruan Yuan and his colleagues revived the ancient category of metrologists (*chouren*, i.e., those associated with computational astronomy, mensuration, and surveying), whom he now considered mathematical astronomers.

In the mid-eighteenth century, the official *Ming History* had reiterated the ancient dispersion of classical metrologists (*chouren*) to the "Western region," specifically to the Islamic world. Moreover, the term for classical metrologist (*chouren*) had been used in the canonical *Artificer's Record* (*Kaogong ji*) and Sima Qian's *Records of the Official Historian* during the Han dynasty. Earlier in the eighteenth century, the Kangxi emperor had singled out palace graduates in mathematical astronomy (*chouren jinshi*) such as Wang Lansheng, Minggatu, and Mei Juecheng for special honors.

Such usage was now reworked by Ruan Yuan and Tan Tai in their lead accounts of the meaning and scope of mathematical astronomers in the *Biographies*. They employed the term as a classical sanction for a new intellectual and social category of contemporary scholar-literati such as Mei Wending, Dai Zhen, and Qian Daxin, implying a genealogy of professionalized skills in mathematics and astronomy going back to antiquity. This orthodox term was the first of several used in the eighteenth and nineteenth centuries to describe European scientists in classical Chinese.[30]

The category for mathematical astronomer (*chouren*) also had its conceptual roots in mathematics as one of "six arts" (*liuyi*) of ancient, aristocratic scholars. Beginning in the late Ming, as literati increasingly engaged in the study of mathematics and astronomy, two types of experts emerged: (1) specialists in computational astronomy; (2) literati with an academic interest in mathematics. These two categories were most evident during the Kangxi era when mathematical study was a tool needed for calendar reform. The academic climate among evidential scholars, along with imperial patronage, helped make mathematics and astronomy a collateral branch of classical learning.[31]

Despite their distance from the court, literati who favored mathematical astronomy, such as Ruan Yuan, never thought it should gain an independent position from other fields of classical learning. As calendrical difficulties declined in importance during the eighteenth century due to the success of the Ming-Qing reforms, mathematics emerged as an independent field of inquiry in evidential research and among Han Learning scholars, particularly in Yangzhou. As the cultural and political disputes over the calendar receded from view, literati debates shifted first in the early nineteenth century to the achievements of native mathematics, and then to the superiority of traditional Chinese medicine over modern European medicine.[32]

Looked at from the angle of the cultural hierarchy then in place, which paralleled the social and political hierarchies, we see that natural studies were justified by Suzhou and Yangzhou literati as the proper concern of the scholar-official precisely because they could be included within the orthodox system. Experts, as long as they were subordinate to dynastic orthodoxy and its official representatives, were necessary parts of the cultural, political, and social hierarchies. The literatus-official coexisted with the astronomer, imperial physician, or Jesuit artisan in the bureaucratic apparatus but at higher levels of political status, cultural prominence, and social prestige.

Dai Zhen, for example, insisted on deferring to the native tradition when using foreign knowledge. Dai maintained that the essential elements of astronomy and mathematics could be located in ancient classical texts. If studied properly, according to Dai, the Classics would prove themselves repositories of mathematical and astronomical knowledge that had been lost due to neglect and lack of understanding. He set out, for instance, to show—mistakenly as it turned out—that a cryptic passage in the *Documents Classic* revealed that the ancients had plotted the different paths of the sun on the celestial sphere. In the process, Dai concluded that this and other examples "are clear proof that Western methods were derived from *The Gnomon of the Zhou Dynasty and Classic of Computations* (*Zhoubi suanjing*)," which he had recovered from the Ming archives (see Figure 7.2).[33]

Similarly, other evidential scholars at times valued the recovery of ancient mathematical texts over mathematical problem-solving. Li Rui's biography of

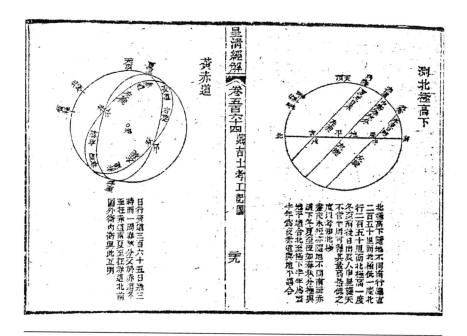

Figure 7.2. Dai Zhen's "drawing of the sun's paths along the celestial sphere." The drawing on the right is titled "Altitude of the North Pole" from Dia's drawings for the "Artificer's Record" *(Kaogong ji)*. The drawing on the left shows the ecliptic and equatorial. Their intersections represent the vernal and autumnal equinoxes. Also shown are the ecliptic and equatorial poles.

Source: Huang-Qing jingjie, compiled by Ruan Yuan et al. (Guangzhou: Xuehai tang, 1829).

Qian Tang (1735–1790) in the *Biographies of Astronomers and Mathematicians*, for instance, cited Qian's confirmation that an earlier value of π (*pi*) had been "3.16," and that this was more acceptable than a later, medieval finding of between "3.1415926" and "3.1415927" (see below). That Li Rui and Qian Daxin both upheld the more ancient finding indicates that many evidential scholars skewed their search for the truth in light of ancient learning and not mathematics per se.[34]

On the other hand, Wang Lai (1768–1813) preferred the European mathematical notation used in the Imperial Observatory, for which he was condemned. Dong Youcheng (1791–1823) criticized Qian Tang's reaffirmation of π = 3.16 and supported the medieval value by referring to the Jesuit-inspired *Collected Basic Principles of Mathematics* (*Shuli jingyun*). Wang Lai sardonically noted, "Contemporary philologists always focus on what their predecessors did, and do nothing but copy what has already been written. They are never able to discover what their predecessors had not yet discovered."[35]

Nativism and Early Nineteenth-Century Mathematics

Due to Jiao Xun's, Wang Lai's, Li Rui's, and Luo Shilin's work in mathematics, Qing dynasty literati were increasingly conversant with mathematics before the Opium War. Moreover, due to their mastery of Jesuit algebra (*ji-gen fang*) and native techniques (*tianyuan*), they generally appreciated both. Until Luo Shilin and others fully reconstructed the four-unknowns (*siyuan*) techniques from 1823, they considered the Western approach the equivalent of single-unknown procedures. After that, however, they favored the four-unknowns technique.[36]

Minggatu, for instance, elaborated the geometric power series for the expansion of trigonometric functions. Literati mathematicians were still few in number, however, and they lacked a Newtonian mechanics to find practical applications outside the domains of astronomy and cartography. Alexander Wylie (1815–1887) and Li Shanlan's 1859 preface to *Step by Step in Algebra and the Differential and Integral Calculus* (*Dai weiji shiji*), which was translated from Elias Loomis's (1811–1889) *Elements of Analytical Geometry and of the Integral Calculus* (New York: Harper and Brothers, 1851), noted that such reasoning on trigonometric series was close to infinitesimals and represented the starting point for studies of trigonometric expansions in the nineteenth century.[37]

Luo Shilin collated and edited Minggatu's work for publication, which despite its links to Western methods (*Xifa*) also received praise from Ruan Yuan. Luo later added Minggatu to the *Biographies of Astronomers and Mathematicians*. Luo thought that the new approach Minggatu employed was similar to Qin Jiushao's general algorithms for solving simultaneous equations and remainder problems (*dayan qiuyi*). Hence, Luo referred to Minggatu's approach as a "new method" (*xin fa*) rather than as a Western method, which echoed the early Qing critique of Adam Schall's Jesuit astronomy as new rather than Western.[38]

Wang Lai worked on a theory of equations and the determination of the nature of roots in his *Mathematical Studies* (*Hengzhai suanxue*), published in 1802. In his collations of Song dynasty works by Qin Jiushao and Li Ye, Wang was the first Chinese mathematician to note whether equations had negative as well as positive roots. He listed the various types of quadratic and cubic equations and noted what roots they have. Subsequently, Li Rui simplified Wang's ninety-six types of quadratic and cubic equations into three rules. Wang and Li also worked on sums of finite series, which followed up earlier work by Chen Shiren, and their work was then continued by Luo Shilin and Li Shanlan. Wang's student Xia Xie later collected and edited Wang Lai's posthumous works for publication in 1834.[39]

Although Jiao Xun had introduced Wang's European-oriented *Mathematical Studies* to Liu Rui in 1801, Jiao concentrated on the laws of mathematical operations in the *Computational Methods in Nine Chapters* classic and created a symbolic science of algebra of his own using the ten "celestial stems" (*tiangan*) for arithmetic operations. He was the first in China to concern himself with the functional aspects of addition, subtraction, multiplication, and division, which raised new problems in light of traditional Chinese problem solving. Like Wang Lai, however, Jiao's new findings did not receive a positive response from mathematicians with ties to evidential studies. Jiao's efforts were better received when, like Li Rui, he applied single-unknown methods to solve problems using ancient mathematics.[40]

Xiang Mingda (1789–1850), like Wang Lai, discredited the priority of textual criticism in Chinese mathematics. He continued to work on trigonometric expansions after Minggatu developed his own method to find the circumference of an ellipse. On the other hand, Zhang Dunren focused on indeterminate analysis to continue traditional expansion procedures for remainder problems. Again, Qin Jiushao's Song dynasty procedures for indeterminate problems were the model.[41]

Perhaps the most influential mathematician in China until his death in 1818, Li Rui was a disciple of Qian Daxin at the Ziyang Academy in Suzhou. The *Posthumous Works of Li Rui* (*Lishi yishu*), published in 1823 by Ruan Yuan, included eleven mathematical treatises, with seven on ancient astronomical systems and four on equations and traditional methods for computing the sides of a right triangle. The last clarified Gu Yingxiang's mid-Ming *Calculations of Segments of Circles* (*Hushi suanshu*), which had been based on Guo Shoujing's Yuan-dynasty *Draft for the Season-Granting Calendar* (*Shoushi li cao*). The goal of Li Rui's work was to reverse the procedures in Gu's approach to circle segments.

Ruan Yuan's account in the *Biographies of Mathematical Astronomers* had noted that single-unknown procedures had fallen into disuse in Ming times. Gu had also missed the trigonometrical aspects of Guo Shoujing's work. Moreover, he had eliminated single-unknown procedures in Li Ye's *Sea Mirror of Circular Measurement* of 1248. As a corrective, Li Rui worked out all of Gu's problems using single-unknown procedures rediscovered in the eighteenth century, which restored the mathematical power of pre-Ming mathematical texts.[42]

Literati Natural Studies: Classics and Mathematics

Literati in the nineteenth century were not doomed by the premises of Qing classicism to ignore natural studies and mathematics before the Opium War. Lord Macartney had recognized the Chinese interest in astronomy, but he had erred when he reduced it simply to "astrological trifling, the goal of which is the calculation of auspicious times." Moreover, he was unaware of Chinese expertise in mathematical astronomy. He mistakenly thought that the Chinese had no notion of even the craft of algebra and possessed only a limited understanding of geometry and trigonometry. Englishmen such as Macartney were continuing a tradition of denigrating Chinese natural studies by Europeans that the Jesuits had begun.[43]

Eighteenth-century literati associated with evidential studies had successfully restored a place for mathematics within the framework of the Chinese origins of Western learning. Even the massive *Qing Exegesis of the Classics* (*Huang Qing jingjie*), a scholarly collection published in the early nineteenth century and devoted exclusively to evidential scholarship, included a significant number of works on natural studies and mathematical astronomy. Because it was the first comprehensive collection of Qing contributions to classical scholarship, the 1829 publication of the *Qing Exegesis of the Classics* in Guangzhou, after four years of compiling and editing at the Sea of Learning Hall under the auspices of Ruan Yuan, then governor-general there, was greeted with great acclaim in China, Korea, and Japan.[44]

An imposing anthology of some 180 diverse works by 75 seventeenth- and eighteenth-century authors in more than 360 volumes, the *Qing Exegesis of the Classics* was a major repository of the research carried out by evidential scholars. It exemplifies the works of Yangzi delta literati in the seventeenth and eighteenth centuries. Although a continuation of earlier commentaries and annotations of the Five Classics and Four Books, the *Qing Exegesis* was also a response to early Qing collectanea (*congshu*) and encyclopedias of Song-Ming classicism.[45]

What was noteworthy about Ruan Yuan's summation of Qing classical scholarship was not only its spotlight on the Han Learning and evidential studies produced by scholars from Yangzhou, Suzhou, and other intellectual centers in the Yangzi delta but that such partisan focus also allowed Ruan to expand the scope for imperial classical studies beyond the domain of Cheng-Zhu Learning. Via the accolades for Han Learning, Ruan and his staff incorporated the works of many of the Qing authors he had discussed in the *Biographies of Mathematical Astronomers*.[46]

For example, Ruan Yuan included verbatim major sections from the *Biographies* dealing with Qing mathematical astronomers, beginning with two chapters extensively quoting from Wang Xichan's important works, followed by sections drawing on the writings of Mei Wending, Hui Shiqi, Jiang Yong, and an extensive chapter on Dai Zhen's publications. Remarkably, this was followed by four chapters from the *Biographies* on Westerners, which included mention of ancient Greek scholars, including Aristotle and Ptolemy, and early modern Europeans such as Copernicus, Brahe, Ricci, Schall, Verbiest, and Smogolecki. Newton was also included in a brief mention of his revision of Tycho Brahe's length of the solar year, which was taken from the 1742 *Supplement to the Compendium of Observational and Computational Astronomy*. Inclusion of foreigners to this degree was unprecedented in a repository of Chinese classical learning. The era of eliding European scholarship, dominant since the Yongzheng reign, was over.[47]

In addition, three works on the "Artificer's Record" (*Kaogong ji*) chapter of the *Rituals of Zhou* were incorporated in the *Qing Exegesis,* including Dai Zhen's and Cheng Yaotian's illustrated studies of the ceremonial bronze bells. Ruan also included his own *Explications Using Diagrams of the Design of Wheeled Carriages in the "Artificer's Record."* Similarly, Ruan printed two important studies of the historical geography in the "Tributes of Yu" chapter of the *Documents Classic.* The lead work was Hu Wei's (1633–1714) highly praised *A Modest Approach to the Tributes of Yu* (*Yugong zhuizhi*), which corrected many mistakes in earlier commentaries. Cheng Yaotian's work on the subject was also included. In addition, numerous geographical accounts and the chronology of events in the classics were published as part of the anthology.[48]

Sheng Bai'er's (fl. ca. 1756) study of astronomy in the *Documents Classic* was included in the *Qing Exegesis.* Sheng was not known for any other significant works, although he had taught in academies for over a decade. His expertise in astronomy and trigonometry influenced many of his students. Sheng's work was filled with illustrations based on the Tychonic geoheliocentric system from the late Kangxi *Compendium of Observational and Computational Astronomy*. It also had many references to Jesuit scholarship on star maps, eclipses, and the motion of the planets.[49]

Ruan Yuan added Chen Maoling's 1797 *Examination of Mathematics and Astronomy in Classical Works* (*Jingshu suanxue tianwen kao*) to the collection. Chen had become interested in Mei Wending's work in 1793 and subsequently learned about Western learning from a European in Guangzhou who was on his way to Beijing to serve in the Astro-calendric Bureau. Chen's own inquiry stressed that the *Computational Methods in Nine Chapters* classic had originated remainder problems, single-unknown procedures, and other calculation techniques. Based on Ruan Yuan's findings, for example, Chen affirmed the theory of a round earth. He then claimed that the geoheliocentric position, which he argued was first enunciated in the *Rituals of Zhou*, proved that the earth was round. Chen's *Examination* served as a repository of native mathematical astronomy that had been reinvigorated by the impact of Jesuit studies.[50]

Many other straightforward classical works that also incorporated natural studies were included, but their technical content was hard to locate. To make the technical subjects in the *Qing Exegesis* more accessible, the Hanlin academician Yu Yue (1821–1907) later prepared a topical index to Ruan Yuan's collection as part of late Qing efforts to document the Chinese priority in scientific knowledge at a time when modern science was being introduced by Protestant missionaries. Yu deemed this index important for his students at the Refined Study for the Explication of the Classics (*Gujing jingshe*) Academy in Hangzhou where he taught for thirty years beginning in 1867. While governor of Zhejiang in 1801, Ruan Yuan had established the "Refined Study" in Hangzhou to honor Later Han classicists and to link a classical education with a commitment to "concrete studies" (*shixue*). Ruan had seen to it that students there were examined in astronomy, mathematics, and geography, in addition to their literary and textual studies.[51]

Yu Yue's index of topics in the *Qing Exegesis* at first sight looked like the table of contents for a Ming-Qing encyclopedia of the sort I described in earlier chapters. It opened with the category of astronomy (*tianwen*), which was tied to the use of mathematics to measure the heavens. The index went on to include forty-two other categories ranging from human relations, morality, and political matters to rituals, food and drink, and things. Things were delimited to animals (with feathers or hair) and plants (*zhiwu*) from grasses and vegetables to crops, trees, and bamboo. Despite the similarity to earlier encyclopedias, however, the terminology Yu and others employed was drawn from a new era. What Yu called astronomy was *tianwen*, the earlier term for astrology. Instead of *bencao* (materia medica), the index used *zhiwu*, the new term for plants derived from the study of horticulture, which had been introduced through Protestant translations.[52]

General Knowledge of Mathematics: Evidence from Novels

In addition to the classics, we discern a more general interest in mathematics and natural studies in some late Qing novels, which however fictionalized were repositories of literati interests and daily life. Casual references to some aspects of Chinese mathematics had been included in the late eighteenth-century novel *A Country Codger's Words of Exposure* (*Yesou puyan*, ca. 1780) by Xia Jingqu (1705–87), which sinified Europe within the Qing orbit. Certain topics of mathematics and natural studies were presented in far more detail, however, in Li Ruzhen's (ca. 1763–1830) early nineteenth-century novel, *Flowers in the Mirror* (*Jinghua yuan*, ca. 1821–1828). His fantasy about the adventures of one hundred talented women visiting imaginary overseas kingdoms included mention of chats and quizzes by women at parties held to celebrate their success in imaginary civil examinations. Seven of the female characters had some knowledge of mathematics.[53]

Li Ruzhen's fascinating story included inversions of gender in which men suffered the mortification of having their ears pierced and feet bound in a "Women's Kingdom" where women were in charge. The fantasy of a realm of women knowledgeable in the classics and mathematics was prescient in the early nineteenth century, although Li Ruzhen's ending for the novel reaffirmed the literati world of his day. That women in the novel understood mathematics parodied an academic world in which most literati were still not well informed about natural studies. The larger questions *Flowers in the Mirror* raised about conventional concepts and practices, such as concubinage and foot-binding, made Li's inclusion of mathematical puzzles in the novel both entertaining and enigmatic.[54]

Failing to attain a civil examination degree because he refused to master the despised eight-legged essay, Li in the time-honored manner of Ming-Qing literary critics used his novel as an allegory to spoof the civil examinations via a description of gifted women. Upon returning to China, the women were appointed to government posts by Empress Wu Zetian (r. 684–704). Empress Wu herself was an acknowledged pioneer in establishing medieval civil examinations. Moreover, examination questions on mathematics had been required during the Tang dynasty, but they were no longer in use during the Ming and Qing dynasties. This brief excursion out of his own society allowed Li Ruzhen to explore a different world, in which women were well informed about mathematics and could take on male roles.[55]

Lu Xun (1881–1936) saw Li's 1820s novel as an encyclopedia tied to the many specialized fields inspired by evidential studies. Based on his classical expertise, for instance, Li Ruzhen used the novel to display his achievements in phonology and etymology, two of the three key tools of evidential scholarship. In fact, contemporaries criticized Li for his lack of concern with the third, paleography. The overlap between evidential studies and mathematics, which we have outlined, suggests why difficult mathematical problems turn up in a popular novel.[56]

Moreover, Li Ruzhen was tied to a circle of Yangzhou scholars who combined evidential studies and the mathematics that he ascribed to women in the novel. Li's teacher, Ling Tingkan, was an acknowledged expert in mathematics and astronomy who helped Ruan Yuan compile the *Biographies of Astronomers and Mathematicians*. Ling studied under Dai Zhen and was influenced by Dai's works on mathematics, such as *Calculations Using Counting Rods (Cesuan)* and the *Record of Measuring Segments of a Circle by Computing the Sides of a Right Triangle (Gougu geyuan ji)*, as well as the mathematical classics Dai had retrieved from the Imperial Library.

Another Yangzhou native, Cheng Yaotian, was linked to both Ling Tingkan and Dai Zhen and had written two works on *The Gnomon of the Zhou Dynasty and Classic of Computations* dealing with properties of the square. Jiao Xun, another of Dai Zhen's Yangzhou partisans, prepared several mathematical works on squares and explanations of mathematical processes. He had also completed a textbook titled *General Explanations of Extracting Roots (Kaifang tongshi)* and the more technical *An Explanation of Single Unknown Procedures (Tianyuan yishi)*. In addition, Li Ruzhen's brother-in-law, Xu Guilin (1778–1821), had prepared instructional manuals for using counting rods for single-unknown procedures.[57]

If Li Ruzhen's *Flowers in the Mirror* reflected literati interest in the mathematics of his time, the problems he chose to present came from the classical mathematical canon. One dealing with the rule of false position (*yingnü suanfa*) asked:

> A tray of fruit is distributed to a number of persons. If everyone gets 7 [pieces of] fruit, there is 1 left over; if everyone gets 8, there are 16 short. How many persons and [pieces of] fruit are there?

The solutions were not diagrammed in the novel, any more than they would have been in the mathematical canon, because they were reached using counting rods and not via formal presentation of a series of written equations that could be described analytically. The heroine, after presumably working out the numerical steps in her head, simply gave the correct answer. Such problems could be traced to the chapter on excess and deficit (*ying buzu*) in the *Computational Methods in Nine Chapters* and also appeared in *Sunzi's Computational Canon* and Cheng Dawei's *Systematic Treatise on Computational Methods* (see Appendix 1).[58]

Similarly, problems in arithmetical progression (*chafen fa*) were presented:

A set of 9 gold cups is made from gold weighing 126 taels. What is the weight of each cup?

This poorly presented problem exemplified an arithmetical progression from the smallest (2.8 taels) to the largest cup (25.2 taels), which the heroine again worked out in her head. Related problems were included in the *Computational Methods in Nine Chapters, Zhang Qiujian's Computational Canon*, and the *Systematic Treatise*. The problem in the novel was solved using horizontal counting rods instead of earlier vertical rods. Mei Wending had changed such counters from vertical to horizontal, with the numbers written vertically, and Dai Zhen had also presented this new technique in his 1744 *Calculations Using Counting Rods*.[59]

Another of the problems raised in the *Flowers in the Mirror* asked about two different types of lanterns, but it followed the same form of a famous problem about pheasants and rabbits enclosed in the same cage, which derived from a question included in *Sunzi's Computational Canon*, the *Systematic Treatise*, and *Yang Hui's Calculation Methods* of 1275:

There are pheasants and rabbits of unknown numbers in a cage. 35 heads can be seen on the top, and 94 feet on the bottom. How many pheasants and rabbits are there? Answer: 12 rabbits and 23 pheasants.

Moreover, the character in the novel who asked the question indicated that she was following the "pheasants and rabbits caged together" approach to solve the problem.[60]

Measurements of the circumference of a circle in Li Ruzhen's novel were based on the common approximation of 3 for π. Evidential scholars chose between several values for π when discussing the ratio of a circle's radius to its circumference. In addition to the two mentioned above, some drew on the archaic value of 3 drawn from *The Gnomon of the Zhou Dynasty and Classic of Computations*. This choice was far from obsolete because the Season Granting Calendar of 1280, representing the high point of Chinese computational astronomy, adopted it for technical reasons. Some used diverse values, extremely close to the modern one (3.14159+), worked out by Zu Chongzhi (429–500) and his successors. Unlike evidential scholars such as Qian Daxin and Li Rui who settled for a value of $\pi = 3.16$, Li Ruzhen chose 3.14, approximate but adequate for the novel's purposes.[61]

Aspects of the natural world were also discussed by women in *Flowers in the Mirror*. To determine the weight of matter and the speed of sound, for instance, Li Ruzhen presented the problem in light of the velocity of sound through the medium of lightning and thunder. The speed the novel gave was

1,285.7 Manchu feet (*chi* = 1,303.5 English feet) per second. Newton had given an experimental value of 1,142 English feet (= 1,043.9 *chi*) per second for the speed of sound. It is likely that Li Ruzhen's figure was derived from the *Collected Basic Principles of Mathematics* (*Shuli jingyun*), where the solution to the velocity of sound of a canon blast was given as 1,285.7 Manchu feet (*chi*) per second. Converted to meters, the velocity was about 397.3 meters (= 1,303.5 feet) per second, about 17 percent higher than the modern value (340.29 meters per second = 1,116.4 English feet per second).[62]

The level of knowledge of mathematics and natural studies in Li Ruzhen's novel reflected the academic climate among evidential scholars, when mathematics became an essential subject. A special civil examination in mathematics was promoted as early as 1867; this was not due solely to the impact of modern science via the Protestants after the Opium War. The receptivity of Li Ruzhen and others to both evidential studies and mathematics was a preliminary step in the process of acceptance of modern mathematics.[63]

The novel *Flowers in the Mirror* satirically appraised many topics that interested evidential scholars in the early nineteenth century. It reflected Chinese scholarship, however, that involved a small part of the elite. On the eve of the Protestant infiltration of treaty ports along the south China coast, the popularization of mathematics and natural studies in China was not nearly as pervasive as in late eighteenth-century Europe. Interest in the new fields of natural learning had developed, but the full impact of modern science and technology was not felt until well into the twentieth century. By the 1880s, however, late Qing novelists increasingly used scientific information gleaned from Protestant translations to imagine technological utopias.[64]

The Usefulness of Recovering Ancient Mathematics

Scholars such as Dai Zhen and Qian Daxin took up mathematics and astronomy as a means toward the reconstruction of antiquity. Because they valued ancient knowledge of mathematics and astronomy over new, they were not tempted to develop natural studies as an independent field of inquiry. They also lacked foreign stimuli. There was little contact with Europeans until the 1793 Macartney mission, and after that no Europeans during the Napoleonic wars transmitted the new sciences and mechanics to China. Exaggerations about Qing interests in mathematics and astronomy before the Opium War based on a long account of a single novel, however precocious, would be ill-advised.

The concern with documents that pervaded the evidential studies movement restricted eighteenth-century scholars to a textual focus, even if they occasionally used archaeological findings or computed astronomical phenomena. Once astronomical prediction ceased to be problematic, scholars valued its techniques mainly for application to classical studies. Such matters were rarely more than ancillary to classical and historical concerns. Inquiries into natural phenomena, for the most part, depended on textual evidence, not experimentation.

Evidential scholars were curious about the natural world and mathematics, but the philological biases that dominated their scholarship did not independently support the research and experimentation that might have led to a step-by-step quantification of the natural world. In light of the important place mathematics and astronomy occupied in evidential research, however, we cannot simply dismiss the eighteenth-century recovery of traditional Chinese mathematics. In succeeding chapters, we will see how—not overnight to be sure—Manchu and literati elites adapted to the needs of science and technology in the nineteenth century. After the Opium War of 1839–1842 and Sino-Japanese War of 1894–1895, the politics of defeat in China, not science or technology, made it appear as if little progress had been made.

The triumph of modern science in China, as in Europe, required an intellectual transformation that would break the boundaries of textual scholarship, and social, economic, and political changes that would challenge the paramount status of classical studies. Such intellectual and technological transformations began during the seventeenth century in Europe, but they were not fully understood in China until after 1865. European science in eighteenth-century China could not have developed past the limits sketched in this and earlier chapters partly because of flawed, inadequate Jesuit transmission that failed to challenge native classicism or provide an academic alternative. The preeminent position of classical studies, oriented toward antiquity, remained intact. That preeminence was shaken only after the Taiping Rebellion.[65]

Nevertheless, important mathematical research did take place, and the institutions required for precise scholarship existed when Protestant missionaries made their way to the China coast after the Opium War. Moreover, the successful reconstruction of Song-Yuan-Jin mathematical texts made it easier for a significant number of Chinese literati in the nineteenth century to recognize the precise significance of and the need to master the new developments in advanced algebra, analytic geometry, and the calculus that Protestant missionaries introduced via translation. These translations were possible because of the cooperation of traditional Chinese mathematicians who fully understood the single-unknown and four-unknowns techniques that had been successfully restored.[66]

In the early nineteenth century, for example, Luo Shilin commented on the relative strengths of traditional and European mathematics before the impact of the calculus in China. Luo was versed in European mathematics after he studied in the Astro-calendric Bureau as a stipend student for seven years. His early work followed the bureau's European tradition, but he changed his mind when he went to the capital in 1822 for the provincial examination, which he never passed. While in Beijing he finally read Zhu Shijie's *Jade Mirror of the Four Unknowns*. In Zhu's work he discovered a method powerful enough to solve sophisticated mathematical problems.[67]

Luo, who perished in the Taiping assault on Yangzhou in 1853, pointed out that European mathematics—trigonometry, logarithms, and its method of borrowing roots and powers—was not as powerful as Song-Yuan single-unknown and four-unknowns techniques, by now completely reconstructed. Luo successfully explored the geometrical properties of the cone using single-unknown operations, and his 1840 *Supplement to the Calculations of Segments of Circles* (*Hushi suanshu bu*) extended Li Rui's work on arcs by adding many problems solved using single-unknown procedures. Not yet aware of the calculus, Luo urged Chinese scholars not to follow European mathematics too closely.

Via evidential studies, philology became one of the key tools that literati scholars such as Xu Shou and Li Shanlan employed to adapt European learning to traditional Chinese natural studies. They introduced the modern sciences and mathematics as compatible with native classical and technical learning. Works designed to retrieve the Chinese origins of mathematics served as a useful transition to the differential and integral calculus in the mid-nineteenth century. Li Shanlan, for instance, considered Li Ye's *Sea Mirror of Circular Measurement* (*Ceyuan haijing*) the most influential native text he had studied before learning Western mathematics. Later, when he taught at the Beijing School of Foreign Languages (*Tongwen guan*), he used the *Sea Mirror* to prepare students to master modern algebra (*daishu*).[68]

While introducing these branches of mathematics, Li Shanlan and other Chinese mathematicians reminded their readers that the four-unknowns notation was perhaps superior to Jesuit algebra. But Wylie acknowledged that the Chinese had never developed the calculus. These most influential Chinese mathematicians were no longer devoted exclusively to the revival of ancient Chinese mathematics. They aimed to merge European and Chinese mathematics. We will stress in Part IV the usefulness of High Qing mathematics as a foundation for mastery of the calculus.[69]

IV

Modern Science and the Protestants

8

Protestants, Education, and Modern Science to 1880

After the French Revolution and Napoleonic Wars, Christian missionaries again took the lead in Sino-European interactions. The victories over Napoleon at sea in Egypt and on land at Waterloo validated Great Britain's global importance. By 1820, the British Empire controlled one quarter of the world's population, many incorporated between the 1756 Seven Year's War and 1815. Moreover, due to the American Revolution, Britain's expansionist aspirations were redirected toward south and southeast Asia and China.[1]

British Protestants increasingly perceived the Qing empire as an obstacle to open commerce and Christian evangelism. English politicians, merchants, and evangelicals articulated through religion and science their goal of opening China to the international community of nations and enlightening its peoples. When Manchus and Chinese resisted the goals of the Macartney mission in 1793 and the Amherst mission in 1816, Britain grew anxious over the usefulness of diplomacy to gain access to the most important market in the world. Moreover, between 1828 and 1836 more than $38 million flowed out of Qing China to pay for illegal opium imports sponsored principally by the British East India Company. The call for free trade in England climaxed in 1833, when after four years of politicking free traders and their evangelical supporters convinced Parliament to abolish the East India Company monopoly in China.[2]

Protestant Missionaries in China

Once free traders forced the East India Company to permit missionaries into its territories, the Methodists, Evangelical Anglicans, and other denominations quickly organized as missionaries. Similar groups organized in mainland Europe, and the Jesuit order regrouped to enter China again. Officially non-denominational, the London Missionary Society (LMS) played an important role in China throughout the nineteenth century after its founding in 1795 by a coalition of Anglican, Methodist, Presbyterian, and Independent ministers. Similarly, the Church Missionary Society was organized in 1799 as an independent voluntary organization within the Church of England. The "Evangelical Revival" in Britain and the "Great Awakening" in the United States spawned new denominations such as the Methodists and new organizations such as the Salvation Army, the Sunday school, as well as the YMCA and YWCA.[3]

Other Protestant missionary societies were also organized throughout Europe, although the English presence was felt most in China: the English Baptists in 1792; the Church Missionary Society in 1799; and the British and Foreign Bible Society in 1804. Similarly, the American Board of Commissioners for Foreign Missions organized in 1810. Extraordinary wealth and energy generated in the aftermath of the Industrial Revolution in England, the United States, and in Europe was in part channeled overseas toward the missionary movement. Although Britain and the United States remained political antagonists, the religious controversies of the nineteenth century among English and American Protestants were not isolated events, as the career in China of the Englishman become American, John Fryer, will demonstrate. The religious revival that swept through Great Britain in this period also swept through the churches and universities of New England.[4]

The Protestant mission in China began when Robert Morrison (1782–1834) labored for the LMS, which was then restricted to Guangzhou (Canton) and Portuguese Macao. Morrison had served as an East India Company translator from 1809 to 1815 and in 1816 was an interpreter for the Amherst mission. He founded an Anglo-Chinese College in 1818, initially in Malacca, which led to establishment of the Morrison Education Society after his death. The first Christian works to reach China were published in Malacca circa 1815, in Dutch Batavia in 1828–1829, and in Singapore in 1830s. The first English paper, the *Canton Register,* was published from 1828 until 1843 by James Matheson with the help of Morrison, but it was mainly an expression of East India Company interests. In 1833, Christian publication began in Guangzhou.[5]

Such efforts laid the foundation for future work by arranging a corpus of Christian literature for printing. It also provided groundwork for gathering books in Chinese by the missionaries. Morrison, for example, put together an impressive library collection that included the Kangxi-era *Collected Basic Principles of Mathematics,* Yu Chang's revaluations of the *Treatise on Cold Damage Disorders,* and Mei Wending's complete works in mathematical astronomy. Established at Malacca in 1818 at the Anglo-Chinese College, the London Missionary Press (also founded in 1795) of the LMS later moved to

Hong Kong and then to Shanghai in 1842, where it was called the Inkstone Press (*Mohai shuguan*). When Alexander Wylie began its supervision after 1847, the Inkstone Press in Shanghai became the publishing mecca of missionary activities.[6]

In contrast to the Jesuits at court, who thought the Bible alien to Chinese literati and focused instead on using Scholastic materials from Coimbra, Morrison with the help of William Milne (1785–1822) prepared the Old and New Testaments for printing in 1819. Milne later compiled the first Chinese-English dictionary. There were also missionary journals, such as the first Chinese magazine titled *A General Monthly Record, Containing an Investigation of the Opinions and Practices of Society* (*Cha shisu meiyue tongji zhuan*), which was issued between 1815 and 1821 in Malacca and included some introductory materials on modern science. Both Walter Henry Medhurst (1796–1857) of LMS and Samuel Wells Williams (1812–1884) of the American Board were trained printers, and by 1840 there were twenty missionaries in China representing half a dozen different societies from several European countries.[7]

In the 1830s, missionaries and their Protestant presses penetrated south China. Their publications presented journalistic portraits of the European powers, especially of Britain and the United States, which through translation initiated a new Chinese vocabulary for political institutions, economic prosperity, and national aspirations. The first leading missionary publication in China was *The Chinese Repository*, a monthly begun in Guangzhou in 1832 by the American Elijah Coleman Bridgman (1801–1861). He was joined on the journal in 1833 by Samuel Williams. It was the main outlet for serious Western scholarship in China until the 1850s.[8]

Among medical missionaries, the American Peter Parker (1804–1888) organized the Medical Missionary Society in Guangzhou in 1838. The German Karl Friedrich August Gützlaff (1803–1851) and Bridgman were active in Singapore and Guangzhou and served as secretaries when the joint missionary-merchant publication *Society for the Diffusion of Useful Knowledge in China* was founded in November 1834. In 1837, the society initiated an ambitious plan to present in Chinese some twenty-four works on history, geography, natural history, medicine, mechanics, natural theology, and other subjects.

For this series, Bridgman produced a treatise on the United States that was printed in 1838, revised in 1846, and expanded for publication in 1862. The Guangzhou scholar Liang Tingnan (1796–1861) later included information from it in his 1846 *Four Talks on the Maritime Countries* (*Haiguo sishuo*). Liang had in 1835 helped compile a gazetteer on coastal defense in Guangdong province, and from 1840 he taught at the Sea of Learning Hall (*Xuehai tang*) founded by Ruan Yuan. Because he knew about Euro-American affairs, he advised numerous Qing officials.[9]

In 1838, Bridgman also produced a dictionary based on Cantonese, titled *Chinese Chrestomathy,* for the Society for the Diffusion of Useful Knowledge, but it was not printed until 1841. In the work, Bridgman presented English translations for Chinese phrases, with Cantonese romanization of the Chinese characters. Among the seventeen sections, he included scientific materials on the human body, mechanics, architecture, agriculture, mathematics, geography, mineralogy, botany, zoology, and medicine. Despite the initial influence of the *Chinese Chrestomathy,* however, its limited circulation meant that few of Bridgman's terms became normative.[10]

Gützlaff published the *Eastern-Western Monthly Magazine* (*Dong xiyang kao meiyue tongji zhuan*) in the 1830s in Singapore and Guangzhou, which Lin Zexu's (1785–1850) staff later monitored and translated when Lin arrived in Guangzhou in March 1839 as the imperial plenipotentiary to solve the opium crisis in south China. Lin's *Gazetteer of the Four Continents* (*Sizhou zhi*) included portions of Gützlaff's magazine, which was one of the major sources for Wei Yuan's (1794–1856) influential *Treatise on the Maritime Countries* (*Haiguo tuzhi*), initiated in 1841, published in 1844, enlarged in 1847, and doubled in size in 1852. Xu Jiyu's (1795–1873) *Brief Survey of the Maritime Circuit* (*Yinghuan zhilue,* 1848) also drew on Bridgman's and Gützlaff's publications.[11]

After the Opium War (1839–1842), the treaty between Britain and the Qing dynasty opened the south China coast to European and American trade. As a result of the first treaty settlement, the ports of Guangzhou, Shanghai, Fuzhou, Amoy, and Ningbo were also opened to foreign residence. Hong Kong island became a British colony. The Christian missions from several European states in Batavia, Malacca, Singapore, and Macao quickly moved their missionaries and printing equipment to the south China coast. After the Sino-French Agreements of 1858 and 1860, France assumed the role of protector of Roman Catholic missions in China.[12]

Early Translators of Modern Science and Medicine in South China

Wei Yuan's *Treatise on the Maritime Countries* contained a plan for Qing defense that included construction of a navy yard and arsenal near Guangzhou where Westerners could teach Chinese how to build ships and manufacture arms. The Manchu governor-general in Guangzhou, Qigong (1777–1844), called for including mathematics and manufacturing in the civil examinations in the early 1840s. Because such requests fell on deaf ears in the court in Beijing, the early introduction of modern science and technology was left to the Protestants and their converts in south China.

Dr. Benjamin Hobson (1816–1873) and Dr. Daniel Jerome Macgowan (1814–1893) were the key translating pioneers in the late 1840s and early 1850s. The American Macgowan initially served as a medical missionary in Ningbo and later became a freelance lecturer and writer, as well as a member of the Qing Maritime Customs service. After moving from Macao to Hong Kong, Hobson, an English medical missionary, pioneered a series of medical and science translations coauthored by Guan Maocai and Chen Xiutang for his premedical classes in Guangzhou. Wylie came to Shanghai from England to join the LMS Inkstone Press. He had been selected to join the Press as its printer by James Legge (1815–1897), who pioneered the translation of the Chinese Classics at Oxford with the help of Wang Tao (1828–1897).

While in Ningbo, Macgowan published a science almanac titled the *Philosophical Almanac* (*Bowu tongshu*, 1851), which was devoted mainly to electricity and electrotherapy. Focusing on glass devices for storing electricity and metal objects to transmit electricity, its discussion of magnetism made Chinese encyclopedias obsolete. Macgowen used the electrolysis of water to demonstrate that it was a composite substance and not a single, unified agent. He also introduced what many thought was the curative power of electricity via electrodes attached to the body. Both he and Hobson regarded electricity as a newly discovered key to understanding the nervous system.[13]

Similarly, Hobson used the broad learning of things (*bowu*) in his title, which in English was *Treatise of Natural Philosophy* (*Bowu xinbian*, 1851). He also used it to refer to scientists (*bowu zhe*). Initially published by the Guangzhou Hospital, Hobson's translation drew on Macgowan's Ningbo *Almanac* but became the more influential work when reissued in Shanghai. Wylie and Li Shanlan, on the other hand, preferred investigating things and extending knowledge (*gezhi*) in their translations for the Inkstone Press. Thus, as with Jesuit translations of Aristotelian natural studies, Chinese coworkers, such as Li Shanlan and Guan Maocai, were still influencing the choice of terms in Protestant scientific translations.[14] From the 1850s, terms for the sciences selected by the Europeans and their Chinese collaborators were quickly transmitted from China to Japan.[15]

Hobson's Anatomy and Traditional Chinese Medicine

Before the LMS translations in Shanghai became prominent, Hobson also produced a series of other works to educate his students. His *Summary of Astronomy* (*Tianwen lunlue*, 1849), the *Treatise on Physiology* (*Quanti xinlun*, 1851), in addition to the *Treatise of Natural Philosophy*, were designed for medical students. In addition, Hobson published the *First Lines of the Prac-*

tice of Surgery in the West (*Xiyi luelun,* 1857), the *Treatise on Midwifery and Diseases of Children* (*Fuying xinshuo,* 1858), and the *Practice of Medicine and Materia Medica* (*Neike xinshuo,* 1858). The *Treatise on Physiology* reintroduced the centrality of the brain and the nervous system, which Schreck and Verbiest had tried unsuccessfully to promote in the late Ming and early Qing.[16]

This unprecedented series of modern medical works remained standard in China until the late nineteenth century, and each volume was reproduced widely in Japan. Later Dr. John G. Kerr (1824–1890), Dr. John Dudgeon (1837–1901), and John Fryer translated new texts on medicine that superseded Hobson's works. The Medical Missionary Association was formed in 1886 and printed its own medical journal in Chinese. The Scientific Book Department associated with the Shanghai Polytechnic (*Gezhi shuyuan*) had twenty-four Western medical works for sale in Chinese translation by 1896.[17]

The missionaries believed that medicine was at a low ebb in China. Yet when Hobson was translating Western medical works into classical Chinese, the heat factor tradition grew increasingly prominent among traditional physicians in south China, where the missionaries were often assigned. Regional traditions dealing with southern infectious diseases and northern cold damage disorders continued to evolve in the nineteenth century. Some late Qing figures, such as the Suzhou physician Wang Shixiong (1808–1890), rejected the universality of cold damage therapies in favor of a local class of heat factor therapies in the Yangzi delta. In the process, heat factor illnesses became a new category of their own.[18]

The mid-nineteenth century emergence of a medical tradition stressing heat factor therapies coincided with the introduction of Western medicine in the treaty ports, particularly Guangzhou, Ningbo, and Shanghai. The newly systematized canon for heat factor disorders was constructed from many scattered writings on heat factor illnesses in the ancient medical canon and from more recent works produced mainly by early Qing physicians in Suzhou. The latter remained the cultural center of the Yangzi delta until the aftermath of the Taiping Rebellion in the 1860s, when the treaty port of Shanghai became the center of the region.[19]

Wang Shixiong's *Warp and Weft of Warm and Hot Factor Diseases* (*Wenre jingwei*), for example, was published in 1852, a year after Hobson's *Treatise on Physiology* appeared. Earlier in 1838, Wang finished a treatise on cholera based on cases he treated during an epidemic. He synthesized heat factor materials in the *Inner Canon of the Yellow Emperor* and the *Treatise on Cold Damage Disorders* with new medical texts from the Qing. In *Warp and Weft,* he acknowledged the usefulness of Hobson's anatomical depictions of female reproductive organs and human physiology to improve diagnosis, but he subordinated such new information to the traditional therapeutic regime, which was based on a system of circulation tracts that mapped acupuncture and moxibustion on the main and branch conduits of *qi* in the body.[20]

Chinese accepted anatomy when they could assimilate it within the Chinese focus on internal conduits of *qi*. Moreover, acupuncture and moxibustion therapy had by Song times been mapped onto the skeletal body, and the internal organs had also been drawn and modeled. An ancient dynastic history records a complete human dissection during Wang Mang's brief reign (A.D. 9–23). The abdomens of rebels, for example, were opened in Chinese dissections of the human body in 1045. Illustrations of internal organs, which were drawn from a systematically dissected cadaver after an execution, were widely available in 1113. Forensic medicine later applied this knowledge of internal organs. The 1113 Chinese drawings were the source of Persian drawings of the body circa 1313–1314 in Ilkhan Tabriz.[21]

In 1878, Song Zhaoqi's *Discriminating Examination of Southern Diseases* (*Nanbing biejian*) presented Qing medical authors writing on heat factor disorders as a distinctive southern tradition, separate from the northern tradition associated with cold factor disorders. Song drew on the works of eighteenth-century Suzhou physicians to invent retroactively a new medical tradition (see Chapter 6). In this intellectual environment, both traditional natural studies and Chinese medicine faced challenges.[22]

Emergence of a Suzhou medical tradition for treating infectious illnesses paralleled the gradual emergence of tropical medicine during the late nineteenth century when European settlers considered the tropics the "White Man's Grave." We do not fully understand these parallel developments, nor whether there was any interaction between them, but the focus on the infectious diseases was obvious in both. The British Empire in the tropics was increasingly populated with its own physicians. These networks of doctors and their medical reporting system from Africa to India and south China in turn addressed interregional infectious diseases such as malaria. Colonial physicians cumulatively sent back information about epidemics and infectious illnesses to London, now the metropole of global medicine.[23]

A port surgeon and medical officer in the Imperial Chinese Customs Office since 1866, Patrick Manson (1844–1922), for example, helped establish the London School of Tropical Medicine in 1898, which focused on diseases such as malaria in light of parasitology and bacteriology. Assigned to semicolonial treaty-ports in Taiwan initially and Amoy for over two decades, Manson studied tinea, Calabar swelling, and blackwater fever before he grew concerned about tropical hygiene. He distinguished himself for his filariasis research (*The Filaria sanguinis hominis and Certain New Forms of Parasitic Diseases in India, China and Warm Climate Countries*, London, 1883), a disease endemic in South China for which neither Chinese nor European medicine had a remedy. In particular, he observed in 1878 that the filariae worms causing elephantiasis passed part of their natural life cycle in the *Culex* mosquito, thus demonstrating transmission by parasites and explaining their natural history.[24]

Until unseated by the germ-parasite theory of disease in the late 1890s, Europeans regarded malaria as a miasma defined by its symptoms of fever. Hobson had earlier associated malaria with putrid air rather than parasites. In the latter half of the nineteenth century, Western physicians such as the Brigade Surgeon of the Indian Army, C. F. Oldham (*What Is Malaria?* London, 1871), tried to explain such extreme fevers by using a chill theory that described tropical illnesses according to the degree of change in an individual's physiology. Hot days and cold nights produced such fevers, Oldham thought.

Such views were not dissimilar with Chinese notions of cold and heat factor illnesses, within which southern miasmas were also identified medically as the causes for infectious diseases. Moreover, the focus on variations in local *qi* to explain southern medical disorders entailed demoting the cold damage tradition to a northern medical tradition. Like Europeans, Chinese were turning from a focus on diseases of the tropics as a subset of all illnesses to specific illnesses such as leprosy and malaria associated with south China.[25]

This demarcation of illnesses unique to north or south China meant that cold damage therapies were inappropriate for southern illnesses. These medical currents represented a critical test for a therapeutic tradition cobbled together from the *Inner Canon of the Yellow Emperor* and the *Treatise on Cold Damage Disorders* since the Song dynasty, which represented ancient ideals. The challenge of Western medicine in the nineteenth century and its increasing focus on parasitic causes of tropical diseases added to native debates that involved late Qing medical practitioners. Many of them had rejected the universality of the Jin-Yuan framework of cold damage therapies for heat factor illnesses. In Chapter 9, we will address the prize essay contests on medicine that were administered in Shanghai from 1886 to 1894. In those essays, the Chinese increasingly acknowledged the need to synthesize Chinese and Western medicine.[26]

In the twentieth century, late Qing radicals such as Zhang Binglin (1868–1936) rejected the *Inner Canon* in favor of heat factor therapies, which Zhang maintained until his death were superior to Western medicine in dealing with febrile illnesses. Others such as Ding Fubao linked cold damage disorders to the specific illness that Western physicians identified as typhoid fever. Germ theory was added to the discussion of warm versus cold factor illnesses. Chinese physicians increasingly explained the wasting of the body's natural vitality in terms of tuberculosis (= wasting disease) and gonorrhea (= depletion illness).[27]

Unlike the Ming-Qing astronomical computus, which was completely reworked in the seventeenth and eighteenth century by the introduction of Western techniques, traditional Chinese medicine did not face a challenge from Europe until the middle of the nineteenth century. George Thomas Staunton (1781–1859), a case in point, had learned Chinese while accompanying his father on the 1793 Macartney mission. Later, his book on vaccines against smallpox was imported to China and Korea. It had been based on Alexander Pearson's 1805 pamphlet on Jennerian vaccination, which he prepared in collaboration with a Chinese physician while Pearson served the East India Company.[28]

Except for smallpox inoculations, quinine therapy for malaria, and a number of herbal medicines unknown in China, the European medicine brought by Protestant missionary physicians was not superior in therapeutic results until a relatively safe procedure for surgery combining anesthesia and asepsis was developed at the turn of the twentieth century. The translations Hobson prepared led to some literati questioning of traditional Chinese medicine in the nineteenth century, however. Xu Shou was one of the first scholars to attack the application of Cheng-Zhu theory to medicine—particularly the dualism of form-principle (*li*) versus *qi*—that had favored medical prescriptions based on the correlation between cooling therapies and moral cultivation to calm bodily excesses.[29]

In an article published in April 1876 (see below), Xu also attacked the yin-yang dichotomy and the five phases correspondence system when compared to the medical theories introduced by Hobson and Guan Maocai. Moreover, Xu Shou complained that while literati had integrated Western and Chinese mathematics, they paid little attention to the strengths of Western medicine. Xu called for a similar synthesis of Western experimental procedures, linking chemistry and Chinese strengths in materia medica.[30]

Xu Shou's 1876 plea for more attention to Western medicine indicated that outside the missionary hospitals and clinics established in the treaty ports, Hobson's translations were not attractive due to the Chinese distaste for surgery. Minor surgical procedures such as cutting warts, lancing boils, cauterizing wounds, removing cataracts, and castration for eunuchs were the preserve of the nonliterati majority of physicians. Hobson's *Treatise on Physiology* and his *Treatise on Midwifery and Diseases of Children* introduced invasive surgery for childbirth drawn from the anatomical sciences that had evolved in Europe since the sixteenth century. Although anatomy could pinpoint childbirth dysfunctions in utero, such procedures were dangerous even by Western standards until modern surgery integrated sterilization techniques with anesthetization procedures to make local interventions secure.[31]

Nevertheless, Hobson's anatomical procedures were unprecedented by the standards of medicine for women (*fuke*) in imperial China. Chinese procedures did not address anomalies of childbirth, such as anatomical abnormalities or uterine dysfunction. Hence, Hobson's "anatomy-based, surgical midwifery" was not easily integrated with traditional procedures or therapies. Rather than invasive surgery, Chinese physicians preferred practical therapies for women based on the holistic, interactive model of the human body. Wang Tao, then Hobson's associate, noted that Western treatises for women's medicine did not prescribe therapy for dealing with illnesses, which was the strength of procedures in the Chinese tool kit. To reply to this complaint, Hobson prepared a more practical guide in 1857, titled *First Lines of the Practice of Surgery in the West* (*Xiyi luelun*), in which he presented the medical prescriptions left out in his earlier *Treatise on Physiology*.[32]

The struggle to understand Hobson's work was also apparent in the Chinese reprints of the *Treatise*. The first edition in 1851 contained illustrations in lithographic plates for the twelve thousand copies published. Another edition by Hobson soon appeared using woodblock illustrations because of their greater accuracy. The governor of Guangdong province had woodblock illustrations made from Hobson's lithographic versions, which were included in a reprint. This version was included by Pan Shicheng (1832 provincial degree), from a Cohong merchant family, in an updated 1852 anthology of fifty-six works on science and medicine by Chinese and Western authors from Ricci and Schall to Hobson.

In the various editions, Hobson's detailed original illustration for the "Frontal and Rear View of the Human Skeleton" (see Figure 8.1), for example, was replaced in Pan's edition with a Chinese version that did not resemble the original. As in the eighteenth century, the refraction of European anatomy via Chinese lenses for viewing the dead body yielded grotesques rather than verisimilitude (see Figure 8.2). We are reminded here of Luo Ping's intriguing "Fascination of Ghosts" handscroll, which had portrayed ghosts in the mid-Qianlong era anatomically as human skeletons (see Figure 5.4, page 219). An anatomically exact depiction of the body was not required in Pan's renderings. Medical authors such as Hobson clearly couldn't control the illustrations of their works.[33]

Because Chinese physicians who accessed Hobson's translations were not interested in surgery or obstetrics, they tailored Hobson's anatomy to suit their interests. Moreover, in his preface to Hobson's *Treatise on Physiology* Pan Shicheng rejected the theological underpinnings that Hobson used to link Western medicine to a natural theology common in Britain at that time. This teleology explained the complexity of the human body in light of the "Lord of Creation," an approach that echoed the Jesuits and which we will see below informed most missionary presentations of modern science in China.[34]

In addition, Pan eliminated Hobson's discussion of the human soul, which had paralleled religiously Hobson's presentation anatomically of the human brain in the *Treatise on Physiology* as the headquarters of the human body. We noted in Chapter 3 that many Chinese literati believed that the heart-mind was the sovereign of all intelligence. Hence, they rejected the Jesuit claim that knowledge and memory was located in the brain. Now they refuted the existence of the soul.

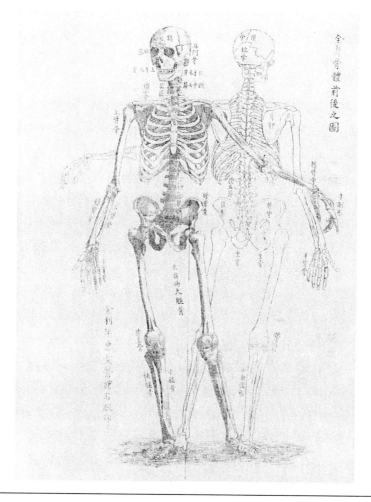

Figure 8.1. Hobson's "Frontal and Rear View of the Human Skeleton," in the *Treatise on Physiology* (*Quanti xinlun*).

Source: Hobson, *Quanti xinlun* (Guangzhou: Hum-le-fow Hospital, 1851).

The physician Wang Qingren (1768–1831), for example, was one of the few Chinese to take anatomy seriously. Wang seems to have had unwitting access to an indirect transmission of Jesuit anatomy via the work of a late Ming Chinese convert. Based on his own examination of human corpses, Wang

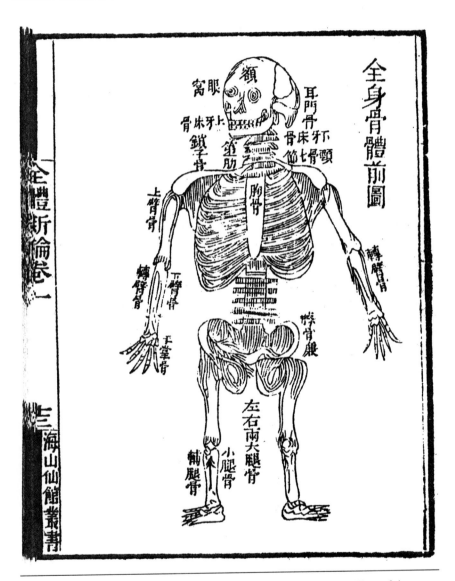

Figure 8.2. Pan Shicheng's substitution for the "Frontal and Rear View of the Human Skeleton."

Source: Haishan xian'guan congshu, edited by Pan Shicheng (Shanghai, 1852; reprint. Taipei: Yiwen yinshu guan, 1967), Volume 119.

contended that the bodily depictions in the medical classics were all inaccurate. His *Corrections of Errors in the Forest of Medicine* (*Yilin gaicuo*) also maintained that the brain was the central organ of the body, a view that became more prominent after Protestant medical texts such as Hobson's were translated into Chinese.[35]

Wang's work was more widely available after 1850, and his anatomical inquiries were quickly linked to Hobson's work. Publishers reprinted Wang's *Corrections of Errors* when the *Treatise on Physiology* first came out. Pan Shicheng and others also attacked Wang Qingren because of his affinity with Hobson's views. Many traditional physicians argued that examining a dead body did not enhance understanding a living person. Pan contended that Hobson's anatomy of a dead body left out the circulation of the vital essences of *qi* in the living. His goal was to show the limits of anatomical medicine. Nevertheless, he and others acknowledged that a better map of the living human body enhanced diagnosis. But he only required a rough rendering of the body and not a precise drawing.[36]

From Western Medicine to Modern Science

We should not underestimate Hobson's contributions as a missionary-physician. His work represented the first sustained introduction of the modern European sciences and medicine in the first half of the nineteenth century. His 1849 digest of modern astronomy, for instance, presented the Copernican solar system in terms of Newtonian gravitation and pointed to God as the author of the works of creation. Thereafter, Newtonian celestial mechanics based on gravitational pull (*yinli*) was increasingly presented in Protestant accounts of modern science—with God central.

A natural theology also informed Hobson's *Treatise of Natural Philosophy*, which was the first work to introduce modern Western chemistry (see Appendix 2), and which established the names of gases in Chinese. The textbook also presented the fifty-six elements, but Hobson still presented God as ultimate creator behind all the myriad changes in things. Although later changed, Hobson's chemical terminology presented the chemical makeup of the world, which supplanted the four elements theory of the Jesuits and challenged the Chinese notion of the five phases.[37]

The English missionary John Fryer described a group of Chinese literati investigators who earlier had met informally in Wuxi, northwest of Shanghai, to go over Jesuit works on mathematics and astronomy and had used the 1855 Shanghai edition of Hobson's *Treatise* to catch up with findings since the Jesuits had transmitted Aristotelian natural studies and the Tychonic world system to China. This group, which included Xu Shou and Hua Hengfang, also carried out experiments. After fleeing the Taipings in the early 1860s, they were invited by the leader of the Qing armies, Zeng Guofan (1811–1872), to work in the newly established Anqing Arsenal. Hua began translation projects with Alexander Wylie and Joseph Edkins (1823–1905), while Xu worked on constructing a steamboat based on Hobson's diagrams.[38]

Hobson had compiled his *Treatise* for his premedical classes in the Guangzhou Hospital to provide Chinese students with a general survey of modern science in preparation for their medical studies. By including sections on physics, chemistry, astronomy, geography, and zoology, Hobson unexpectedly attracted the interest of dissatisfied literati, most unsuccessful in the civil examinations. When Medhurst published *China Serial* (*Xia'er guanzhen*) from 1853 to 1856 in Hong Kong, for example, he drew much of his material on science and medicine from Hobson's treatises and the Rev. William Muirhead's (1822–1900) *Complete Gazetteer of Geography* (*Dili quanzhi*), which was also published in the early 1850s.[39]

Protestants and Modern Science in Shanghai

Although their enthusiasm for China was often fired by evangelicalism back home, British and American Protestants in China swiftly recognized that Chinese literati were interested in the sciences that the missionaries accepted as part of their Christian heritage. Like their Jesuit predecessors, British and American missionaries viewed science as emblematic of their superior knowledge systems. To introduce modern science and medicine to China was not only a missionary tactic. The missionaries presented the wealth and power of Western nations to the Chinese as inseparable from the Christian gospel they preached.

Among treaty ports, Shanghai by 1860 was the main center of foreign trade, international business, and missionary activity. Initially, the *New Paper for China and Abroad* (*Zhongwai xinbao*) was published in Ningbo from 1854 to 1861, before moving to Shanghai. Until then, the Ningbo paper had been the chief rival for the main LMS publications, namely the *China Serial* published in Hong Kong and the *Shanghae Serial*.

The London Missionary Press in Shanghai became the most influential publisher after 1850. It published translations from a distinguished missionary community, which included Alexander Wylie and Joseph Edkins. They worked with outstanding Chinese scholars such as Li Shanlan and Wang Tao, who had both moved to Shanghai after failing to gain a place in the imperial civil examinations. Wang went to Shanghai in 1849 and was invited by Walter Henry Medhurst to serve as the Chinese editor at the Inkstone Press.[40]

Alexander Wylie and the Shanghae Serial

In the 1850s, other Protestant journals, such as the *Shanghae Serial* (*Liuhe congtan*), introduced new fields in the Western sciences in Chinese, rigorously categorizing them under the general rubric of "learning based on the investigation of things and the extension of knowledge" (*gezhi xue*). Beginning with the *Shanghae Serial,* the traditional literati notion of investigating things (*gewu*) moved from representing a methodological universal for classical learning, encompassing natural studies, to designating a specific domain of knowledge within the natural sciences. Protestant use of these terms differed from their use in Ming-Qing literati writings and in Jesuit translations.[41]

The Jesuits had not challenged the diffuse methodological focus of the native view of the investigation of things. They simply added the fields of *scientia* to it. Through the translation work of Wylie, Li Shanlan, and others involved in the *Shanghae Serial,* however, the investigation of things demarcated the new Western natural sciences. A scientist was now called "someone who investigates things and extends knowledge" (*gezhi shi* or *gewu jia*). The British Royal Society was translated in the *Shanghae Serial* as the "Public Association for Investigating things and Extending Knowledge" (*Gezhi gonghui*), while the French Academy of Science was called the "French State School for Investigating Things and Extending Knowledge" (*Falangxi gezhi taixue*) in Chinese. During the Kangxi reign the Paris academy was called the "Academy for Investigating Things and Exhaustively Mastering Principles" (*Gewu qiongli yuan*).[42]

An LMS-sponsored publication, Medhurst's Hong Kong *China Serial* was the immediate precursor to the *Shanghae Serial.* The North China Branch of the Royal Asiatic Society was also formed in the 1850s for scholarly research. Unlike the *North China Herald and Supreme Court and Consular Gazette,* founded in 1850 in Shanghai, which began as a weekly and eventually became the leading foreign language newspaper in China, the *Shanghae Serial* took as its mission the deeper presentation in classical Chinese of Western affairs and culture, new scientific fields, and Christian theology.[43]

In the early days of the Inkstone Press, which began publishing the *Shanghae Serial* in January 1857, Wang Tao and Li Shanlan, who joined the bookstore in 1852, were the chief Chinese translators. It soon was a meeting place for Chinese literati living in or passing through Shanghai who were interested in Western learning, such as Guo Songtao (1818–1891), Xu Shou, Hua Hengfang, and many others. Many, like Guo, would go on to distinguished diplomatic and educational careers serving the Qing dynasty after the imperial government embarked on a moderate program of military and scientific reform in the aftermath of the Taiping Rebellion.[44]

A talented missionary printer and translator, Alexander Wylie produced the *Shanghae Serial* for LMS as a monthly for a total of thirteen issues in 1857 (there was an intercalary fifth month in the Chinese lunisolar calendar) and for two more issues in 1858, before it suddenly stopped on June 11. After arriving in China, Wylie made some remarkable inquiries about Chinese science and mathematics with the help of Li Shanlan. Through this interaction, Li successfully completed the transition from the traditional craft of algebra to the modern calculus. Like his English contemporaries in mathematics, Wylie looked back to the pioneering work of the French in China, such as the Jesuit Antoine Gaubil, who in the eighteenth century had written authoritatively on the history of Chinese astronomy, and Edouard-Constant Biot (1803–1850).

Around 1860, Wylie and Li also started but left unfinished a complete translation of Newton's *Principia*. It presented for the first time Newton's laws of motion. About 1868, Li Shanlan completed the translation in the Jiangnan Arsenal's Translation Department with the help of John Fryer, but the work was never published because it was too hard to understand. After 1899 but before a revision could be completed by Hua Hengfang, the manuscript was lost. The title for the Li-Fryer version of the *Principia* was "Investigating and Extending Knowledge of Mathematical Principles" (*Shuli gezhi*).[45]

Wylie's and Li's 1859 translation of John Herschel's (1792–1871) *The Outline of Astronomy* (1851, *Tantian*) grew out of their early collaboration. Cambridge-educated, Herschel drew for his astronomy on Laplace's *Celestial Mechanics*, which stressed the harmony and simplicity of planetary orbits, and Lagrange's disparagement of geometrical intuition. Herschel moved away from the late eighteenth-century Newtonians in England, who had stressed geometrical demonstrations over algebraic processes. This new generation gainsaid the static model of the solar system, such as the planetarium that Lord Macartney had thought a grand gift for the Qianlong emperor in 1793, and moved toward a more powerful mathematical physics by frequent and precise observations, applying partial differential equations. In effect, Wylie and Li's translation of *The Outline of Astronomy* presented a British version of the French Newtonians who had informed Napoleon's École Polytechnique.[46]

Wylie's focus on advanced mathematics and astronomy was unprecedented by Jesuit standards. He and Li Shanlan quickly translated Books 7–15 of Euclid's *Elements of Geometry* in 1855, which had been left out of the Jesuit versions. The initial printings were destroyed, however, when Songjiang was attacked by the Taipings. Zeng Guofan rescued the project in 1865, when he printed it in his Nanjing military headquarters. When compared to Newtonian mechanics and the calculus, however, it had less relevance for industrial enterprises in the arsenals (see Chapter 11). In 1853, Wylie prepared his

Compendium of Arithmetic (*Shuxue qimeng*) with the help of his student Jin Xianfu. This primer contained the rudimentary rules of arithmetic, the theory of proportion, and logarithms, but Wylie solved traditional Chinese mathematical problems to make the translation accessible.[47]

Wylie also used the *Compendium of Arithmetic* to refute Mei Juecheng's eighteenth-century claim that Jesuit algebra was the same as Song-Yuan single-unknown methods. Wylie and Li Shanlan also published through Inkstone Press a translation titled *Elements of Algebra* (London: John Taylor, 1836, *Daishu xue*, 1859), from the original by Augustus De Morgan (1806–1871), who had graduated from Trinity College in Cambridge in 1827 and taught at University College, London. Part of the first generation of mathematical reformers at Cambridge in the 1820s, De Morgan sought to apply the calculus to mechanics, physical astronomy, and other branches of natural philosophy. He saw algebra as the universal logic of meaning underlying empirical observations. In Scotland, William Thomson (1824–1907) followed De Morgan and sought mechanical models of quantitative phenomena. Wylie's Chinese preface for the *Elements of Algebra* made clear that algebra had advanced beyond the version introduced by the Jesuits two centuries earlier.[48]

Similarly, Wylie and Li stressed modern algebra as a mathematical language for the natural sciences. They related it to traditional Chinese mathematics by substituting it for single-unknown and four-unknowns procedures, which were the pinnacle of traditional Chinese mathematics. Wylie emphasized that Chinese "quadrilateral algebra" (*siyuan shu*, i.e., four-unknowns procedures) had been superior to the Jesuits' elementary algebra (*jiegen fang*) and that Western scholars had not studied the two traditional methods carefully. Nevertheless, Li and Wylie also refuted the theory that the science of algebra had originated in China.[49]

The Religious and Scientific Content in the Shanghae Serial

Five thousand copies of the first issue of the *Shanghae Serial* were printed in January 1857 at the cost of ninety thousand cash (eighteen cash per issue). In 1858, however, there were only about 378 foreigners in Shanghai, including 17 consuls, 32 teachers, 46 women, 5 doctors, 200 merchants, 15 bankers and gold traders, 6 printers, 9 shipbuilders, 25 harbor officials, and 23 others. Most of the subscriptions were ordered by Shanghai foreign residents, while subscriptions from Shanghai Chinese comprised only 8 percent.

The journal was also available in five treaty ports and Hong Kong. Another 11 percent of the subscribers were Chinese and foreigners from Fuzhou. Sold mainly as single copies, the *Serial* was not widely available outside of these

Western enclaves, because the print run for issues declined gradually from a peak of 5,190 copies in March 1857 to 2,500 copies for the last two issues in 1858. It was widely read in Japan, however, when prefectural libraries ordered it and other translations from China (see Chapter 11).

To attract Chinese readers, each issue from the second on opened with an ephemeris, usually followed by a monthly account of worldwide current events, and data of Sino-British trade history, new books of note, market prices and imports, and other general topics. As print runs diminished, the editors added a series on mechanics and later on popular mechanics to interest a less exclusively literati audience. The rest dealt with a balance of theological and religious issues, the core message of the LMS. The scientific coverage represented Wylie's influence, and political-commercial events drew readers from the business community.[50]

The first issue opened with a "Brief Introduction" in which Wylie stressed the value of science for understanding the world and introduced the Chinese terminology for its most important fields: (1) chemistry (*huaxue*), (2) geography, (3) animals and plants, (4) astronomy, (5) calculus (*weifen fa*), (6) electricity (*dianqi zhi xue*), (7) mechanics (*zhongxue*), (8) fluid mechanics (*liuzhi*), and (9) optics and sound (*shi zhuxue*). Wylie also mentioned Chinese traditional single-unknown mathematics and then discussed the importance of algebra and the calculus for astronomical calculations.[51]

Because the *Shanghae Serial* was an LMS publication, Wylie quickly turned in his "Brief Introduction" to natural theology to elaborate the providential design of the "maker" in the universe: "All things that are born and mature on the earth are made by the Lord on High, and when we examine the reasons, these also result from the Lord on High." In this way, the *Shanghae Serial* celebrated God, creation, and salvation. For example, an article in the second issue titled, "Natural Theology—Existence of God," explained that "Heaven and earth were not formed naturally; rather everything had been made by the Lord on High." Furthermore, God had created sixty-four (up from fifty-six) chemical elements, which were not a creation of natural forces. We will see similar adjustments to Darwinian evolution.[52]

Protestant theology in Shanghai bore the imprint of William Paley (1743–1805) and his *Natural Theology*, published in 1802. Paley's combination of theology and utilitarianism also influenced Charles Darwin, who studied under Paley at Cambridge. Paley's work was subtitled "Evidences of the Existence and Attributes of the Deity, Collected from the Appearances of Nature," which was a major theme that informed most scientific texts translated by missionaries in China after 1850. The *Shanghai Daily* (*Shen bao*) favorably mentioned Darwin and his 1871 book *The Descent of Man, and*

Selection in Relation to Sex (called *Renben,* lit., "Origin of humankind") as early as 1873, but until 1895 Protestants and Catholics ignored Darwin's theory of biological evolution based on natural selection.[53]

The articles included in the first issue of *Shanghae Serial* ranged from one on geography by William Muirhead, which became a regular series in later issues, to another on physical characteristics of the globe. These were balanced by Joseph Edkins's piece on "Greek the Stem of Western Literature" and one on "The Bible" by Alexander Williamson (1829–1890), who wrote many of the theological articles in subsequent issues. Edkins played a special role in transmitting his knowledge of Greek and Latin literature in later issues, as did Williamson who specialized in biblical topics.[54]

Beginning in the fifth issue of 1857, Wylie and Wang Tao translated the article titled "Progress of Astronomical Discovery in the West," which was published serially in the *Shanghae Serial* until June 1858. It opened with a theological refrain concerning God's powers and the smallness of humankind but then quickly addressed ancient astronomical traditions in China, Egypt, India, and Mesopotamia. According to Wylie, Chinese observances of solar eclipses recorded during the Shang were the most ancient. In Chinese, Wylie and Wang presented the premodern Western man of science in light of the ancient Chinese category of a computist (*chouren;* see Chapter 7), which was borrowed from the equation of "*chouren* = mathematical astronomers" by Ruan Yuan and evidential scholars circa 1800.[55]

The article gave more historical information for Western astronomy than did Wylie's and Li Shanlan's 1859 translation of Herschel's *The Outline of Astronomy.* The sequels, which resumed with issue nine in September 1857, carried the history of Western astronomy from ancient times to the Greeks and Ptolemy; from the Middle Ages to Copernicus, Brahe, and Kepler; and from Galileo and the telescope to Newton and the Royal Society. Wylie and Wang Tao refuted the still-popular notion in China that Western studies originated in China.[56]

An article by Williamson titled "Advantage of Science" ("Gewu qiongli lun," lit., "On investigating things and exhaustively mastering principles"), appeared in the sixth issue in June 1857 and equated national strength with science: "A country's strength derives from the people. The people's strength derives from the heart-mind. The mind's strength drives from science [lit., investigating and things and fathoming principles]." Williamson noted that the Chinese were very skilled, but he was critical of them for wasting themselves on the eight-legged civil examination essay, later a common theme among missionaries and reformers. He also disdained Chinese literature and poetry. Williamson was most responsible for the theological pieces in the *Shanghae*

Serial and was the first to proclaim that China required scientists (*gezhi shi*) and science schools (*gezhi yuan*) to revive the country.

Protestant stress on weaknesses in Chinese culture when compared to Western power in the nineteenth century influenced Li Shanlan, Wang Tao, and later reformers such as Feng Guifen (1809–1874). Feng, who had studied under Li Rui, stressed "selectively using Western studies." He prepared a mathematical primer in 1847 that introduced Li Rui's traditional methods for measuring segments of a circle, and in 1865 another on the 1859 calculus textbook translated by Wylie and Li Shanlan.[57]

By appealing to political and economic power, however, Williamson and the LMS missionaries were treading a treacherous path, both for them as Christians preaching the gospel of science and for the Chinese who sought military power in that gospel. Like the Jesuits, who failed to convince Ming literati that their *scientia* was inherently Christian, so too the Protestants would find that Chinese reformers could disentangle their rhetoric of science and state power from Protestant theological musings. In the end, China would Westernize without Christianity.

The abrupt end of the *Shanghae Serial* may not have been due simply to subscriber limits or because of the lack of articles to publish. The link between science and Christianity that Wylie championed had perplexed the LMS. Consequently, the *Shanghae Serial* was probably stopped because the LMS wished to refocus Inkstone Press on missionary activities.

This tension between science and religion in the Protestant movement in China mirrored earlier divisions in the Catholic mission during the late Ming. Dissenting orders had convinced the pope that the Jesuit focus on mathematical astronomy had compromised the transmission of true Christianity. Similarly, Protestant dissatisfaction with too much science in the *Shanghae Serial* created tensions within the LMS and with other such organizations. The LMS and James Legge chose a more religious focus.[58]

Although the issue of religion versus science played a part—whether large or small—in closing the journal, Inkstone Press was the first after the Opium War to introduce modern science in China beyond the treaty ports. Wylie and his LMS colleagues working on the *Shanghae Serial* were instrumental in introducing modern science in Shanghai, the most vital of the five treaty ports. Through the collaboration of Li Shanlan, Xu Shou, Wang Tao, and Hua Hengfang, these new trends influenced the late Qing reformist initiatives of Zeng Guofan and Li Hongzhang (1823–1901). The earliest use of contemporary terms for science, such as chemistry (*huaxue*), for example, occurred in this monthly.[59]

Introduction of Modern Mathematics and the Calculus

While collaborating on the *Shanghae Serial,* Li Shanlan and Alexander Wylie also translated a work that they titled, *Step by Step in Algebra, the Differential and Integral Calculus* (*Dai weiji shiji,* 1859), which reintroduced, this time successfully, the Cartesian algebraic symbols rejected in 1712 by the Kangxi emperor. More importantly, this path-breaking translation of Elias Loomis's (1811–1889) *Elements of Analytical Geometry and of the Differential and Integral Calculus* (New York: Harper and Brothers, 1851) presented for the first time an important improvement developed in the late seventeenth century, but which the Jesuits failed to transmit to China.[60]

Wylie noted in his English preface the impact the calculus would have:

> [T]here is little doubt that this branch of the science will commend itself to native mathematicians, in consideration of its obvious utility. . . . A spirit of inquiry is abroad among the Chinese, and there is a class of students in the empire, by no means small in number, who receive with avidity instruction on scientific matters from the West. Mere superficial essays and popular digests are far from adequate to satisfy such applicants.[61]

Wylie also hinted at the early nineteenth-century changes that the English sources of his translations represented. Loomis's *Elements* followed a series of works, such as William Whewell's (1794–1866) *An Elementary Treatise on Mechanics* (1819). Whewell represented an earlier generation when British mathematicians at Trinity College in Cambridge University, the intellectual home of Bacon and Newton, began to overtake the French. Although British mathematicians had contributed to the development of geometry, algebra, physical astronomy, pure and applied calculus, and probability, Newton's calculus of fluxions was not comparable in utility to the continental differential and integral calculus because it was too clumsy in notation and awkward in methodology.[62]

Because of Newtonian notation, the British began losing ground in mathematics in the 1740s. It took another century for them to catch up. Cambridge reformers in the 1820s laughed at Newton's dot notation and British adherence to geometrical methods as signs of the "dot-age" of academic mathematics. They blamed its decline on a chauvinistic attachment to Newton's inconsistent notation, which obscured the coherence of the calculus. Using Leibniz's notational forms, the French had successfully developed a series of rule-like differential and integral equations that France's engineers need not reconceptualize each time.[63]

James Thomson's *Calculus* (1831), for instance, blamed the inferiority of the fluxional notation for the lack of significant progress in British mathematics after Newton. De Morgan was committed to the Lagrangian school of calculus and the elimination of Newtonian fluxions. Similarly, Whewell's mechanics built on Lagrange's *Mecanique Analytique* at the same time that he preserved Newton's geometric proofs in the *Principia*. Joseph Edkins and Li Shanlan finished a translation of Whewell's *Treatise* (*Zhongxue*) in 1859, but its publication was delayed until 1867.[64]

The British continued to eschew the latest trends in continental mathematics. Any comparison of Chinese efforts to master modern Western science should consider the variations in European stages of scientific learning country by country. The predicament the British faced in the 1820s was repeated in the 1850s when Chinese mathematicians such as Li Shanlan and Hua Hengfang worked with Wylie and Fryer to translate the calculus into Chinese.[65]

Li Shanlan had contended that the four-unknowns notation (*siyuan shu*) was equivalent as a calculating craft to modern Western algebra, but he could find no native precedent for the dynamic calculus, which affirmed Wylie's claim in his 1859 preface to the *Step by Step in Algebra* that there were no Chinese origins for the calculus, "so far as the translator is aware." Hence, while Wylie was editing and publishing the *Shanghae Serial* in 1857–1858, he and Li Shanlan were working on translations of several English mathematical works and carrying out sophisticated research into the strengths and limits of native mathematics. Later translators of Western mathematics into Chinese, such as John Fryer, could not appreciate these investigations.[66]

In 1874, for example, John Fryer and Hua Hengfang together completed the *Origins of the Differential and Integral Calculus* (*Weiji suyuan*) for the translation department in the Jiangnan Arsenal. This curious version was translated from William Wallace's (1768–1843) article on Newton's fluxions published posthumously in the *Encyclopedia Britannica* (eighth edition, 1853). When Wallace's "Fluxions" had appeared in the fourth edition of the *Encyclopedia Britannica* (Edinburgh, 1810), for instance, it was one of leading treatises on the calculus before the introduction of Lacroix's *Traité Elémentaire* in 1816.

Although the Cambridge reformers had gone beyond fluxion notation, as had Wylie and Li Shanlan in 1859, Wallace's article attributed the discovery of the calculus to Newton's inverse method of fluxions and Leibniz's differential and integral calculus, which he represented as two distinct origins. Seeking a compromise, Wallace mixed the continental, Lagrangian approach with Newton's limited definition of the fluxion. Wallace had published a longer article titled "Fluxions" in the *Edinburgh Encyclopedia* in 1815 using differential notation, which was the first complete mathematical text in England to do so and helped establish the foundations for the revival of British mathematics in the early nineteenth century.

At this time encyclopedias such as the *Britannica* served as unique venues for the popularization of French mathematics in Britain. Fryer's use of the 1853 *Encyclopedia Britannica* version of Wallace's 1810 article in 1874, however, meant that as an Englishman he was still loyal to old-fashioned British efforts to maintain Newton's fluxions, however modified, rather than the continental calculus produced at Cambridge and introduced to China by the Wylie-Li Shanlan translation of Loomis's more up-to-date textbook.[67]

Neither Li nor Hua likely knew about the British-French rivalries that affected Fryer's introduction of fluxions. There were three potential groups of Chinese readers after 1865: scholar-literati, technicians needing Western mathematics for work in arsenals and state schools, and students at various missionary schools. The major audience for Fryer's and Hua's British version of the calculus were in the arsenals and technical schools, particularly the Jiangnan Arsenal in Shanghai. In the Fuzhou Navy Yard, however, French engineering influence was paramount.[68]

Wylie and Li ingeniously translated mathematical notation, concepts, and theorems for Loomis's calculus. They made the symbols look Chinese, enough to be compatible with native four-unknowns notation. New terms had to be convincing if the Chinese were to grasp the suitability of the calculus. Traditional mathematics thus contributed to shaping the translation of the calculus. The translations for the differential and integral calculus reflected Li Shanlan's and Hua Hengfang's training in traditional mathematics. Li Shanlan invented the Chinese notations that would become standard for these terms. These traditional forms were replaced, however, by Japanese expressions in the early twentieth century, ironically after Li Shanlan's and Hua Hengfang's notations had first been introduced in Japan.[69]

That is why the modern Chinese terms for the differential and integral first appeared in Liu Hui's circa 263 commentary on the *Computational Methods in Nine Chapters*. The differential (*weifen*) was employed in Liu's comments on the section dealing with methods for measuring circular fields, which established the first theory of limits and the first method of approximation in Chinese mathematics. When Liu Hui obtained the value of $\pi = 3.146$, the differential as a term was directly related to this value. Liu used the method of circular fields to obtain the area of a circle by inscribing a series of regular polygons to measure the tiny parts (*weifen*) between the circle and the final 3,072-sided polygon.

The result was not only a tiny fraction but also incorporated an increment between two related values from the successively inscribed polygons. That is why Li Shanlan chose this term for the differential. Loomis had defined the differential of a function as its rate of variation through successive increments. The object of the differential calculus was to make such increments eventually approach zero. This conception of the differential reminded Li Shanlan of the tiny parts in traditional mathematics.

The modern term for integral, literally, "accumulated parts" (*jifen*), appeared twice in Liu Hui's commentary on the *Computational Methods*. The integral (*ji*) literally meant "to store up, to amass, or to accumulate." In its first appearance in the *Computational Methods,* the compound *jifen* referred chiefly to the area of a figure, the volume of a solid, or the product of numbers. In the second instance, Liu used it in comments on the method for extracting the square root to designate the number whose square root was to be extracted. The common feature it had with the integral was that it represented a product of several parts. Liu also used it for the result of an inverse operation, i.e., moving from the square back to the root and vice versa. Loomis defined the integral as a function that was the reverse of differentiation. Li Shanlan understood this to mean that by "dividing an integral we get countless differentials[;] combining these countless differentials we get back the integral." Li Hui's second instance for the integral fit in well with Li Shanlan's understanding of integral calculus.

On the other hand, Hua Hengfang in his collaboration with Fryer translated the concept of a differential in Newton's fluxions as an exceeding ratio (*yilü*) in a flow, which also showed traditional influence. As a compound, an exceeding ratio depended on the use of ratio (*lü*) in traditional mathematics. For Liu Hui, a ratio also meant numbers in a certain relation, as in the problem of exchanging fifty units of millet for thirty units of inferior rice and twenty-seven units of superior rice. In Hua's and Fryer's translation of the *Origins of the Differential and Integral Calculus,* ratio defined the concept of the differential as a series of numbers where the differential was not the Leibnizian difference but a ratio of Newtonian fluxions.

The Wylie-Li and Fryer-Hua translations emphasized the application of the calculus as a powerful tool to solve new, dynamic mathematical problems. Despite the adaptation of Chinese terminology, the introduction of the calculus shocked Chinese literati interested in mathematics because, unlike the craft of algebra, nothing similar was found in traditional mathematics. Initially, the power of the calculus was seen as a practical technique for calculations (*shu*), which paralleled its use in Europe by engineers. Chinese mathematicians perceived calculus as a tool. It did not become a separate field of mathematical theory in China until A. P. Parker's new translation of Loomis's work appeared in 1905.

Chinese mathematicians regarded the calculus as a development from algebra, which provided the general language, but they realized that algebra could only deal with constants and static variables while the calculus addressed dynamic variables. This approach enabled mathematicians to deal with certain curves, surfaces derived from curves, and solids derived from surfaces, which they could solve only when they related them to motion. In 1899 the first *Mathematical Periodical* (*Suanxue bao*) in China presented the calculus as the highest achievement of Western mathematics. Prefaces for the first issue noted that "algebra and the calculus are more powerful than any Chinese methods."

Although Chinese such as Li Shanlan could not find a precedent for the calculus, a few still argued that the West had borrowed from China various ideas essential to its invention. Lin Chuanjia in 1900 tried to show that many essential ideas of the calculus came from the mathematical classics by building on earlier claims about the Chinese origins of algebra. Lin saw the Western focus on curves in terms of Chinese analysis of the relational features for computing the sides of a right triangle (*gougu*), and consequently he did not think Newton had gone beyond Liu Hui.

The initial science and natural history translations by Inkstone Press and the LMS group of Chinese collaborators represented a small but promising scholarly community of scientific educators in the treaty ports in the 1850s and 1860s. Due to civil war, however, such new developments had little influence in the areas of south China affected by the Taiping Rebellion until Chinese armies loyal to the Qing dynasty recaptured Nanjing in 1864. The deaths during the rebellion of traditional mathematicians such as Luo Shilin, Shen Qinpei, and Dai Xu, who had played a key role in reviving traditional algebraic techniques, also weakened the literati community in the Yangzi delta that had championed the link between evidential studies and mathematics.

After Qing armies defeated the Taipings, the calculus spread through the new arsenals first established in the 1860s. The wide practical applications of the differential and integral calculus for arms manufacture and shipbuilding made it essential for the Chinese working in the new arsenals and technical schools under British or French advisors. A new understanding of European mathematics among Chinese literati emerged, which emphasized integrating Chinese and Western mathematics. This emerging hybrid no longer favored Chinese origins. Until the nineteenth century, literati had understood mathematics as a tool. With the introduction of advanced algebra and calculus after 1850, Chinese began to move to a view of mathematics as a field of learning with its own principles.[70]

The Shanghai Polytechnic and Reading Room

After the Taipings were defeated, the weakened Qing dynasty and its literati-officials faced up to the new technological requirements to survive in a world filled with menacing, quickly industrializing nations. Literati such as Xu Shou and Li Shanlan, who had worked with the Protestant missions in the 1850s, began in the 1860s to move from their LMS positions to teaching and translating the new natural sciences for the government's new westernizing institutions, such as the Jiangnan Arsenal in Shanghai. They built bridges between post–Industrial Revolution science and traditional Chinese natural studies.

The *North China Herald,* for example, editorialized on December 26, 1872, calling for the creation of a reading room for Chinese in Shanghai that would stress secular learning. The British Consul in Shanghai, Walter Medhurst, later submitted a detailed plan in a letter to the paper on March 12, 1874. An earlier version of Medhurst's proposal had appeared in the Western-owned Chinese newspaper *Shanghai Journal* (*Shenbao*) on March 25, 1873. He focused on the "intellectual and commercial development of the Chinese" to foster "appreciation for all kinds of Western knowledge" and "encourage an interest in polytechnic and general information, rather than in mere literary research."

Medhurst's proposal included a prospectus. He suggested calling the Reading Room the "School for Search after Knowledge" (*Gezhi shuyuan,* lit., "Academy for investigating things and extending knowledge"). Public subscriptions by foreigners and Chinese in the foreign settlements would provide the funding. It would house Chinese periodicals, newspapers, and translations of standard foreign works. Periodic lectures and exhibitions would be organized to promote Western manufactures, a model based on similar exhibitions then prominent in Europe and the United States.[71]

A committee of foreigners was formed including Medhurst, Muirhead, Edkins, Wylie, and Fryer. They named the school in English "The Chinese Polytechnic Institution and Reading Room." In Chinese it was still called the "Academy for Investigating Things and Extending Knowledge." At the organizational meeting on March 24, 1874, after some debate over the inclusion of religious publications in the reading room, Alexander Wylie drew up a list of appropriate books and periodicals. More secular Europeans wished to privilege science over religion, while others still clung to the Protestant agenda for linking Christianity to science.

The debate reflected the increasing secularization of scientific training in Europe and the United States after the Darwinian revolution challenged the natural theology of prominent Protestants such as William Paley at Cambridge. For example, Charles Lyell's *Elements of Geology,* translated by Macgowen and Hua Hengfang at the Jiangnan Arsenal in 1873, was premised on

Lyell's determination to "free the science from Moses" and the biblical chronology of creation. Lyell revised his *Elements* several times after 1830, but in each edition he contended that the geological record revealed that flora and fauna had been created as fixed rather than evolving forms in response to a constantly changing environment. Although he remained a Christian, Lyell challenged scriptural geologists. We will see in Chapter 9 that although the earliest missionary translations of Darwin dated to 1895, Chinese had already begun to show interest in Darwin in the 1880s.[72]

The committee met again on January 8, 1875. Through Xu Shou's efforts, generous donations from Chinese high officials such as Shen Baozhen (1820–1879) and Li Hongzhang, then the governor-generals of Jiangnan (in the Yangzi delta) and Zhili (the capital region) respectively, allowed the Polytechnic to open formally on June 22, 1876. Students started in 1879. Even so, the lecture room and classes for publicizing science were neglected until the 1890s. More than 80 percent of the funds were provided by Chinese officials and merchants.[73]

Although Xu Shou was an able fund-raiser when deficits in the Polytechnic deepened in 1877, Western subscribers attributed its failure to draw Chinese to its reading room to the Xus and their bureaucratic ties to Li Hongzhang. Medhurst's letter to *The North-China Herald*, which was cited in a March 15, 1877, editorial, contrasted the disappointment of the Polytechnic in Shanghai to the success of the Hong Kong Museum. The editorial attributed the difference to the lack of a tradition of public reading rooms in China.

The Xus treated the Polytechnic like their private library. The directors also noted that they did not encourage visitors and ignored the wishes of the foreign committee. After Xu Shou's death in 1884, the foreign committee led by Fryer reasserted control, and by Chinese New Year in 1886 the reading room and exhibition rooms were back in working order. Wang Tao, a veteran translator since the 1850s, took over Xu Shou's responsibilities. In 1886 Fryer invited Wang to serve as the curator of the Polytechnic and its reading room.

When Liu Xihong, the envoy to Germany from 1877 to 1878, visited the academy, however, he criticized what he considered its distasteful practical focus and wrote that he thought it ought to be called the *Hall of Artisans* (*Yilin zhi tang*) rather than a high-minded "Academy for the Investigation and Extension of Knowledge." He insinuated that the education there was unsuitable for literati elites. Classical studies for the civil examinations remained a superior calling. Fryer's goal of popularizing the sciences to address "the practical requirements and affairs of an ordinary Chinaman's every day life" ignored the gulf between ordinary Chinese and literati. The latter would no more automatically take to the gospel of science than would Oxbridge gentlemen.[74]

In 1885, classes and public lectures were finally implemented, and the scientific essay contest was proposed (see Chapter 9). Instructional plans included an appointment for a foreign professorship of science, which did not materialize. By 1894, the directors authorized a program of free lectures on scientific and technological subjects in Chinese, which was tied to the curriculum of six fields detailed in the Western curriculum of the Polytechnic: mining, electricity, surveying, construction engineering, steam engines, and manufacturing. In 1895, Fryer compiled a *Mathematical Problems Textbook* (*Xixue kecheng shuxue keti*). Regular, free classes on Saturday evenings were also started in 1895. Because students were unqualified, he started a preparatory class in arithmetic on May 11, 1895.[75]

In January 1894, just before the Sino-Japanese War, Fryer applied for a teaching position at Berkeley University, which he accepted in November 1895. The renewed success of the Polytechnic in 1894–1897 was the result of China's 1895 defeat in the war, which had shocked many Chinese and led to new respect for Western studies. Fryer's departure for Berkeley in 1896 left only Rev. F. H. Jones to carry on with the evening lectures and class instruction. Wang Tao's death in 1897 left the Polytechnic plagued with administrative problems.

In the wake of the Boxer Rebellion, new government schools (*xuetang*) were established, where the instruction drew on foreign language schools and the official schools already established in the arsenals. Chinese increasingly went abroad to study in Japan or the West. The Polytechnic thus lost its unique position, and by 1907 science classes there ceased. In 1917, a new state school was built on the site. The scientific raison d'être of the Protestant missionary schools empirewide had been superseded by Japan's success (see Chapter 11).[76]

From The Peking Magazine *to* The Chinese Scientific Magazine

Because the Polytechnic did not attract the interest that Fryer and the oversight committee in Shanghai had hoped for, Fryer and Xu also created the earliest scientific journal in China to reach Chinese in the treaty ports. Indirectly affiliated with the school, the journal was known in English initially in 1876 as *The Chinese Scientific Magazine* and later in 1877 as *The Chinese Scientific and Industrial Magazine,* with the subtitle of *A Monthly Journal of Popular Information Relating to the Sciences, Arts, and Manufactures of the West* with which is incorporated *The Peking Magazine.* In Chinese the name remained "Compendium for Investigating Things and Extending Knowledge" (*Gezhi huibian*). It first ran monthly issues in 1876–1877 and 1880–1881 in Shanghai before turning into a quarterly from 1890 to 1892. Fryer took leave in England after 1877, and the journal resumed in 1879 when he returned.

Fryer's new journal immediately drew support from the Society for Diffusion of Useful Knowledge in China (*Guangxue hui*) in Beijing, which was closing down in 1875 and ending its illustrated monthly magazine known as *The Peking Magazine* (*Zhongxi wenjian lu*). *The Peking Magazine* was edited by William Martin (1827–1916) and Joseph Edkins, among others. It was printed starting in 1872 as a monthly for thirty-six issues before closing in August 1875. Supported by the Society for the Diffusion of Useful Knowledge, the magazine was devoted to Western and international news, but it also included articles on astronomy, geography, and general science (*gewu*). Martin had worked on the journal while heading and teaching at the Beijing School of Foreign Languages (*Tongwen guan*). The society's members in Beijing transferred their subscriptions to the Polytechnic.[77]

Beijing remained central for missionaries because of the official School of Foreign Languages established by the Qing dynasty in 1861, which Martin called the "University of Peking." Qing officials, foreign scholars and Chinese students at the School promoted foreign affairs (*yangwu*) in Beijing in the 1860s and early 1870s. In addition, the school's two tier system of a five-year and eight-year program of study, established in 1870, prioritized engineering in the longer program, which was complemented by focus on science and mathematics in the shorter course. Hence, the School of Foreign Languages used *The Peking Magazine* to promote science and missionary concerns. The missionaries supported Li Hongzhang and the Self-Strengthening (*ziqiang*) partisans in their efforts to reform the Qing regime.[78]

Moreover, the journal's free monthlies promoting science and technology were usually reprinted in the *Review of the Times* (*Wan'guo gongbao*), which had originally been called the *Chinese Missionary News* (*Jiaohui xinbao*) in 1868. With Young J. Allen (1836–1907) as editor, the *Review of the Times* was published weekly in Beijing from 1874 and monthly after 1889. Both the *Magazine* and the *Review* addressed issues that concerned those in the emerging arsenals and shipyards (see Chapter 10).

Of the 361 essays in *The Peking Magazine* by fifty-four foreign missionaries, traders, and diplomats, 166 (46 percent) dealt with science and technology. The topics included astronomy, geography, physics, chemistry, medicine, as well as the technology of railroads, mining, and the telegraph. Biographies of Western scientists were added. Many other articles dealt with political economy, current affairs, etc.[79]

William Martin's and Joseph Edkins's efforts to use *The Peking Magazine* to promote science in Beijing was complemented by the numerous contributions of Li Shanlan and his mathematics students in the School of Foreign

Languages to the journal. Often the examination papers in the sciences and the mathematics homework of Li's students were included in the *Magazine* as a reminder that it was the school's journal. The March and June 1875 issues, for instance, carried articles on Martin's reaction to examination papers debating whether the earth or sun was at the center of the universe.[80]

Some 199 articles (55 percent) came from the students or teachers. They often replied to "Difficult Questions" (*Nanti zhengda*) and established the precedent for "Answers to readers' queries" (*Duzhe laixin huida*), which *The Chinese Scientific and Industrial Magazine* in Shanghai later made a regular feature. *The Peking Magazine* was an important model for Fryer's scientific journal, which became the public voice simultaneously of the Shanghai Polytechnic and the Jiangnan Arsenal.[81]

Although the Shanghai journal was published and sold through the Polytechnic, it was a separate enterprise that Fryer was responsible for. Fryer's classical Chinese was not good enough to produce the journal by himself. Hence, he employed Luan Xueqian as his private secretary to help translate the unattributed Chinese articles in the journal. Luan trained at Calvin Mateer's (1836–1908) Hall of the Culture Society (*Wenhui guan*), a school in Shandong, where science and advanced mathematics courses were taught in Chinese. Fryer set up the Chinese Scientific Book Depot (*Gezhi shushi*, lit., "Bookshop for investigating things and extending knowledge") in 1884 as a retail outlet for Jiangnan Arsenal and outside translations, which were distributed by Ernest Major's (1842–1908) *Shanghai Journal* (*Shen bao*) throughout China.[82]

Fryer worked with Luan Xueqian since at least 1877 when *The Chinese Scientific and Industrial Magazine* commenced. Luan, for example, prepared an account of a chemistry class at Shanghai Polytechnic before he collaborated with Fryer. Moreover, Luan was probably also involved in the Science Outline Series (*Gezhi xuzhi*) and Science Handbook Series (*Gezhi tushuo*) that were published from 1882 to 1898 (see Appendix 3). In addition, Luan managed the book depot for Fryer after 1885, until Fryer turned it over to him in 1911.[83]

Copies of the Polytechnic's journal were available first at twenty-four and then twenty-seven of the most important trading centers in East China and Japan. From thirty agents in early 1880, these increased to seventy by the end of the year. Although *The Chinese Scientific and Industrial Magazine* picked up from *The Peking Magazine,* it went beyond the latter by focusing on the natural sciences and technology in Europe and the United States. With more Chinese participants in its production, the translations in the Polytechnic's journal improved. It was regularly mentioned in treaty port newspapers such as the influential *Shanghai Journal*. Fryer relied on the empirewide Chinese audience of the latter, which Ernest Major created for the *Journal,* to reach a broader audience outside the treaty ports.[84]

The journal initially printed three thousand copies and usually sold out in several months. Nine months later, the first nine issues were reprinted in a second edition to meet the demand. With four thousand copies printed per issue at its peak in the 1880s and 1890s, it reached some two thousand readers in the treaty ports. Fryer hoped that popular presentations of mathematics and the industrial sciences would become acceptable among literati and merchants. He expected that *The Chinese Scientific and Industrial Magazine* would also compensate for the limited scope of the Jiangnan Arsenal's translations, which were usually printed in runs of only a few hundred copies. Later in 1891, reprints of previous issues were sold. Missionary reviewers in China effusively praised the civilizing mission of the issues. A review in the winter 1891 issue presented the contents of the issue, ranging from photography to steam engines and mathematics, and concluded, "Priestcraft and superstition cannot long maintain their hold under the search-light of modern progress."[85]

Altogether, sixty issues were published intermittently over seven years. After 1880, the magazine shifted its emphasis from introductory essays on science to accounts of its specialized fields. Fryer increasingly paid attention to mathematics as the foundation of scientific knowledge. The Polytechnic's teaching program of mathematics and science was most effective after 1885. According to a General Affairs Office (*Zongli yamen*) memorial of May 18, 1887, which advocated modifying the civil examinations to allow candidates to be examined in mathematics, the Shanghai Polytechnic was training most of the top students of mathematics in the empire.[86]

In addition *The Chinese Scientific and Industrial Magazine* published some 317 inquiries sent to the journal as "Letters to the editor" (*Huxiang wenda*), which San-pao Li has divided roughly into two groups: (1) questions that were practically motivated, and (2) those that were chiefly motivated by curiosity. One hundred and seventy-six of these letters (55 percent) came from the treaty ports. One letter in the November 1880 issue, for example, challenged the claim that the theory of a spherical earth was Western in origin. True to the "Chinese origins of Western learning" (*Xixue Zhongyuan*) position articulated by Chinese since the late Ming, the letter writer cited *The Gnomon of the Zhou Dynasty and Classic of Computations* as proof of the Chinese origin of the notion of a spherical earth.

The letters stressed the practical value of technology and showed less interest in what Li essentializes as "pure science." About 123 letters (38 percent) showed some interest in or knowledge of scientific theories or abstract scientific models, which is a relatively high percentage for a journal oriented to popular science or popular mechanics. The queries paralleled the technological interests of those involved in the Self-Strengthening movement. The large volume of letters to Fryer anticipated a more widespread awakening of interest in science after 1895.[87]

After publication stopped in 1892, *The Chinese Scientific and Industrial Magazine* was reprinted in 1893, 1896, and 1897. In the 1890s the first four volumes were reissued to meet demand. Such reissues sold very well in the Chinese Scientific Book Depot after the Sino-Japanese War in 1895. Later in 1901–1902, the past issues of the journal were reorganized topically and edited by Xu Jianyin under the title, *Collectanea of Science* (*Gezhi congshu*, lit., "Collectanea of works investigating things and extending knowledge") and reprinted in Shanghai. This compendium of articles from *The Chinese Scientific and Industrial Magazine* echoed in its title the much more amorphous collectanea published three centuries earlier by Hu Wenhuan, which had preceded both Jesuit *scientia* and the Protestant version of science in China. Xu made no mention of this precedent.[88]

After 1904, the Polytechnic gave up its teaching program, but the library and reading room were reopened on March 6, 1905. By 1906–1907, the library had some fifty thousand volumes. In 1911, the Polytechnic offered to lease the property to the Shanghai municipality to build a secondary school. The deed was registered in 1914, and the municipality took a permanent lease on the property. A new polytechnic public school opened in 1917.[89]

The Content of The Chinese Scientific and Industrial Magazine

The first issue of the *Scientific and Industrial Magazine* appeared in February 1876. There were three breaks between the three series of publications before it finally ceased in 1892. Volumes 1–2 appeared in 1876 and 1877, volumes 3–4 in 1880 and 1881, both as monthlies, and volumes 5–7 in 1890 and 1892, published as a quarterly. The third series moved from the Polytechnic to the Chinese Scientific Book Depot. Advertisements were mainly from British manufacturers of machinery, tools, and scientific and industrial equipment. Some advertisements for banks, a hospital, and for consumer goods also appeared.

Fryer presented the *Scientific and Industrial Magazine* as a Chinese journal within the world community of science. He sent the first issue to the editorial committee of the *Scientific American* with a note that it was his "ambition to make this magazine to China what the *Scientific American* is to every country where the English language is known." Seeing this as a sign of progress and that in another generation the West would begin to "feel the industrial competition of China," the editorial response to Fryer's initiative was very favorable.[90]

On the other hand, Xu Shou placed the scientific articles within the classical tradition of the investigation of things. Xu's classical gloss for the Chinese name for the Polytechnic prioritized the "extension of knowledge and the investigation of things" (*zhizhi gewu zhi xue*), which he noted drew from the Great Learning as the foundation for moral cultivation, ordering the family, governing the dynasty, and bringing peace to the empire. Xu then elaborated Zhu Xi's orthodox system of knowledge that had encompassed the principles of affairs and things (see Chapter 1).

In the English translation, which was likely prepared by Fryer, Xu's classical references were flattened into a more literal modern rendering: "Science constitutes the primary basis of personal cultivation and state governing." Zhu Xi, it was argued, had broadened human knowledge by stressing "a thorough study of the rationale of things." Xu referred to Fryer as a comprehensive scholar (*tongru*), which we have seen in Chapter 7 was a standard reference to a classical scholar in the late eighteenth century who had incorporated mathematics into his broad learning. Fryer translated this reference to him as "an English learned scholar" who had learned the Chinese language.

Xu then reviewed the new fields of science that Fryer had introduced in his arsenal translations and which would comprise the majority of articles in the *Scientific and Industrial Magazine*. He stressed that the fields of mathematics, such as algebra, calculus, trigonometry, and Newton's mathematical physics (*shuli*, lit., "principles of numbers"), were the foundation for the sciences. At the same time, however, Xu emphasized that the new journal would introduce practical learning in metallurgy, geology, coastal defense, navigation, and astronomy.[91]

In addition, every issue during the first years included a final section titled, "Miscellaneous Aspects of Science" ("Gewu za shuo"). This loosely organized part mentioned new theories for explaining earthquakes, for example, and science fairs and industrial exhibitions in Europe and the United States, as well as new products and books. Beginning in 1880, a new illustrated series, in lead articles through spring 1892, was called "Explanations of Scientific Instruments" ("Gezhi shiqi") for scientific and medical applications.

The 1880 issues also began a multipart article titled "On Chemistry in Public Health" ("Huaxue weisheng lun"). Fryer used this series to present a Euro-American model of "hygienic modernity" (*weisheng*), which redirected the Chinese medical traditions of "nourishing life" (*yangsheng*) away from individual meditative, sexual, and dietary regimes for enhancing the body toward an abstract, modern understanding of the chemical processes of life. In subsequent works on this topic in the 1890s, Fryer added temperance as a precondition for health.[92]

The lead article in the inaugural issue, February 1876, was titled "Introduction to Modern Science" ("Gezhi luelun") and appeared serially with copious illustrations in the first twelve issues ending in January 1877. The series derived from *Chamber's Introduction to Science* and covered major fields or topics of modern science: astronomy, gravitational laws of matter and motion, geology, geography, heat, light, electricity, chemistry, botany, and physical anthropology. Fryer acknowledged that the source was an English primer for children, which contained some three hundred items organized into simple explanations.

The opening section addressed the limitations humans faced in accumulating knowledge of the world from their villages. To gain an understanding of the stars and the puzzling spherical shape of the earth, for instance, one had to break away from one's mundane view of things and entertain broader views of the universe. Scientists (*gezhi jia*), for example, had discovered through the use of the telescope that the sun and fixed stars were comparable because they each emitted their own light, while planets and the moon only reflected light.[93]

If astronomy was the focus in the opening section of the "Introduction to Modern Science," gravitation and mass were the heart of the second section. Nature (*zaohua zhu*, lit., "the supreme maker") had allocated a fixed rule for all things with mass, which enabled them to attract other objects of mass, depending on their distance from each other. Pulleys could be devised to take advantage of objects with mass to move them easily by distributing the weight of the object more efficiently. Mining the earth was the chief topic of the third section, and this important discussion of geology carried over to the fourth, fifth, and sixth parts. The latter section also initiated a discussion of the varieties of plant and animal life and the role of combustion in maintaining life.[94]

Chemistry and optics were the next major fields that introduction took up. The importance of sunlight for photosynthesis in plants and of electromagnetism in things was traced back to the thermodynamics of heat. The role of electricity in the operation of the brain was also noted. Atmospheric gases, their pressure, and the transmission of sound were discussed in the context of weather and its changing atmospheric makeup through the action of the wind. Meteorology was the focus of the eighth part, which stressed the importance of hydraulics.

This part then presented the sixty-four basic chemical elements (*yuanzhi*), which through combining and separating made up all things, whether in pure form or as compounds. The ninth part detailed the new field of botany (*zhiwu xue*), while the tenth and eleventh together addressed zoology (*dongwu xue*). The final portion of the article, which appeared in January 1877, emphasized anthropology, human origins, and human physiology. The last items in this regard were related to the human soul (*lingxing*) and the nature of human intelligence.[95]

In addition to this elementary foray into popular science, the early issues of the *Scientific and Industrial Magazine* included more technical articles, such as Xu Shou's "Discussion of Medical Studies" ("Yixue lun"), which, as we have seen above, attacked traditional Chinese medical concepts in favor of Western medicine. While Western remedies were appropriate for external medicine, according to Xu, Chinese physicians felt that they were not as good as Chinese therapeutics for internal medicine. Western physicians were as adept in using metals and minerals in medicine as the Chinese were, but they were deficient in applying herbs and plants. Nevertheless, Western medicine was able to pinpoint the causes of an illness before delineating its cure. Xu noted that through his collaborations with John Fryer he had learned to appreciate an odd grouping of chemistry, acoustics, manufacture, and medicine.

Xu added that Fryer had collaborated with Zhao Yuanyi (1840–1902) to translate a general work by Frederick Headland titled, *A Medical Handbook* (London, 1861; *Rumen yixue*, Shanghai, 1876), which taught clergy and laypeople the routines to "nurture life" (*yangsheng*), methods of treatment, the characteristics of drugs, and the key aspects of illnesses. Xu appealed to a synthesis of Western and Chinese medicine, which would parallel earlier efforts in the eighteenth century to synthesize Chinese and Western mathematics. In this way, Xu Shou believed Chinese medicine would make breakthroughs that would go beyond the ancients.[96]

William Muirhead's three-part "Principles of Science" ("Gezhi lilun"), and his "General Discussion of the New Methods of Science" ("Gezhi xinfa zonglun"), in five parts, addressed the methodology of science. The "Principles of Science" first presented examples based on annual and periodic occurrences in astronomy, physics, and botany. His scientific insight that "all things have an appropriate principle to explain them" resonated with the Cheng-Zhu natural philosophy that was the generally accepted view among literati. Muirhead redirected the moral aspects of such natural principles away from the Chinese notion of an uncreated and eternal cosmos, which had bedeviled the Jesuits, to an affirmation of an omniscient and omnipotent prime mover.

In the second part of the essay, Muirhead focused on daily and periodic aspects of the earth's rotation and its impact on plants and living things. Muirhead elaborated this point in 1877, eighteen years after Darwin's *The Origins of Species* appeared, to prove that a supreme power had created a world of living things appropriate to each habitat. The daily realities of humans, animals, and plants also encoded a natural way of life informed by a greater plan, which could be grasped through a natural theology.

The third part of Muirhead's essay focused on the biological impact of changes in gravitation on plants and animals and their ability to move. The ecological constraints perceptible in living things on high mountains or in the depths of the ocean actually held for all life forms on the earth. Each organism was placed in its proper niche by a higher power. Religion, although Muirhead did not mention it, overlapped with science in the narrative of life forms, with the former justifying the latter. Like the *Shanghae Serial* of the 1850s but less explicitly, Fryer's and Xu's *Scientific and Industrial Magazine* drew on missionary strategies that had been in use since the Jesuits to highlight Christianity as a theological narrative informing science.[97]

Similarly, Muirhead's series titled "General Discussion of the New Methods of Science," first published in April 1877, represented an elaborate, if cursory, history of Western science from Aristotle to Bacon, Newton, and Lavoisier. The narrative was a morality tale explaining how the true principles of nature were discovered. Francis Bacon, for example, had unraveled the mysteries in the natural design of the prime mover. Others built on Bacon's scientific studies to overcome the Aristotelian natural studies of the Jesuits, just as Copernicus gainsaid the Ptolemaic geocentric cosmos. Newtonian mechanics and the chemistry of Lavoisier also made significant strides in grasping the true way of the world.

Muirhead cleverly pointed the Chinese reader to the superior, Protestant provenance of modern science, when compared to its deficient, Catholic predecessor. In Muirhead's triumphal Anglican pedigree of science, Bacon "was considered the founding father of the experimental sciences." Newton then appeared as an intellectual star whose prodigious discoveries in optics, mechanics, mathematics, and astronomy grew out of Bacon's contributions to the "establishment of the methods of science."

Muirhead encouraged the Chinese to join this march toward truth through the application of the "new methods" of science. His English narrative of science remained influential in twentieth-century English-language accounts of the history of science. The "General Discussion" served as an overview for the series of articles on Bacon's contributions titled "New Methods of Science" ("Gezhi xinfa," i.e, the *Novum Organon*), which Williamson authored in April 1877.[98]

Muirhead did not mention Darwin in any of these articles or the implications of natural selection for botany, zoology, or biology. We will see in the next chapter that this strategic silence in the face of the Darwinian brouhaha in England was not atypical. The omission supported a continued faith in natural theology, which also informed more technical accounts of the sciences prepared for the Qing state by missionaries in the 1880s.

Late Qing literati who associated with a form of classical studies that stressed philology, mathematics, and astronomy, namely, evidential studies (*kaozheng xue*), helped create the intellectual space needed to legitimate study of natural studies and mathematics in China after the Taiping Rebellion. Chinese literati and Christian missionaries stressed different aspects of their joint translation of science into Chinese (*gezhi*), a more native trope for investigating things among literati and a notion of modern science informed by Christian natural theology for missionaries. Such compromises reproduced the Sino-Jesuit term for natural studies (*scientia*), but it now referred to modern science (*gezhi xue*), not early modern science (*gezhi*). In this way, mathematics and the other sciences such as chemistry slowly became acceptable, if still less popular, than the civil service careers for literati.[99]

9

The Construction of Modern Science in Late Qing China

The missionaries presented refracted versions of Euro-American science and mathematics to the Chinese. They encoded late Qing science translations with the cultural imperatives of their Christian beliefs. Qing officials and Chinese literati often failed to realize that the Western sciences were packaged via translation in a natural theology that favored a Protestant affinity with science. Particularly in the case of Darwin, we will see below that Christians delayed communicating a new set of scientific ideas until the 1890s. The absence of knowledge in China about Darwin from the 1870s parallels the nonappearance of Copernicus in seventeenth- and eighteenth-century China. Darwin's replacement by natural theology reminds us of how Jesuits substituted a Tychonic cosmos for the heliocentric system of Copernicus and Galileo.

Although the Protestant missionaries often did not present the Chinese with the most up-to-date aspects of Euro-American science and technology, the translations that Fryer, Martin, and others produced in collaboration with their Chinese colleagues influenced a broad range of Chinese people in Qing society. In this chapter, we will focus first on the science primers used in missionary and state schools and then describe the science essay contests that promoted writing about science in ways that mimicked the way literati mastered the classical curriculum required in the civil examinations.

In the following chapter, we will shift from these promising educational developments to describe the role of the many arsenals, navy yards, and factories built before the Sino-Japanese War of 1894–1895, where Chinese literati, artisans, and workers applied the science they learned to the technological production of modern ships, armaments, and chemicals. The arsenals the Qing government established in China in the 1860s gave it a head start in industrialization over Meiji Japan, which after 1868 initiated similar plans. Only after the Iwakura Mission (1871–1873) returned to Tokyo did Japanese industrialization based on Euro-American models begin in earnest.[1]

Early Science Primers

William Martin's *Elements of Natural Philosophy and Chemistry* (*Gewu ru-men*) was one of the first texts widely used for science instruction at missionary schools and in the Beijing School of Foreign Languages, where Martin was the director. The school published the *Elements* in 1868 and reprinted it in many later editions. Martin initially prepared it serially for the *Chinese Missionary News* (*Jiaohui xinbao*), published by the Presbyterian Mission Press. It covered the fields of hydraulics, meteorology, heat and light, electricity, mechanics, chemistry, and mathematics. Anatole Billequin's *Guide to Chemistry* (*Huaxue zhinan*) was also a popular textbook.[2]

Joseph Edkins prepared an article titled "Outline of New Studies in the Sciences" ("Gezhi xinxue tigang"), which was published in the *Chinese Missionary News* from 1872. It presented a history of modern science. Similarly, a series of articles in three chapters published from 1873 to 1876 by Alexander Williamson, with the likely help of Li Jiyuan, was called in English *Natural Theology and the Method of Divine Government* (*Gewu tanyuan*, lit., "Inquiry based on the investigation of things"). The series expanded into a textbook of six chapters for use in the Beijing School of Foreign Languages in 1880 after the final part appeared in the weekly *Review of the Times* (*Wan'guo gong-bao*). In 1874, the latter succeeded the *Chinese Missionary News* and focused on international news and scientific knowledge. Williamson's *Natural Theology* stressed that religion and science were complementary, with "religion as fundamental and science as useful" (*zongjiao wei ti gezhi wei yong*). Using the late Qing formula for separating substance from function (*tiyong*), which Zhang Zhidong later championed, Williamson tried to demonstrate the providential design of the creator through science.[3]

In the late 1870s, Fryer and the translation department staff in the Jiangnan Arsenal also emphasized translating a series of introductory textbooks to prepare students for more-advanced studies in science and technology. More ambitious than Hobson's outdated *Treatise of Natural Philosophy*, the "Science Outline Series" (*Gezhi xuzhi*, lit., "What to know about the investigation of things and the extension of knowledge") was conceived as a multivolume set of texts. A "Science Handbook Series" (*Gezhi tushuo*, lit., "Illustrations of the investigation of things and the extension of knowledge"), which would match each "Outline Series" title, complemented the textbook series.

Fryer and his staff at the Jiangnan Arsenal translated and published the sources for the Science Outline Series. Fryer had traveled to England to gather more materials on science before returning to Shanghai in 1873. He chose the science primers compiled by Henry Roscoe (1833–1915) and others, which represented the current scientific fields of learning in Great Britain.

The translation was titled *Primers for Science Studies* (*Gezhi qimeng*, lit., "Primers for the investigation of things and the extension of knowledge"). When completed around 1875 with the collaboration of the American missionary Young J. Allen and Zheng Changyan, the arsenal printed it in 1879–1880 in four volumes, one each dealing with the fields of chemistry, physics (*gewu xue*), astronomy, and geography.[4]

Macmillan and Company had published Roscoe's *Science Primers* in England in 1872. The series represented the collaboration of the Darwin supporter Thomas Huxley (1825–1895), Sir Archibald Geikie (1835–1914), and Balfour Stewart (1828–1887), along with Roscoe. Roscoe did the volume on chemistry; Geikie prepared the volume on physical geography and another on geology. Stewart, a contemporary of William Thomson (Lord Kelvin, 1824–1907), specialized in thermodynamics and physics, and Huxley prepared the general introduction. J. Norman Lockyer (1836–1920) completed the volume on astronomy.

From 1857 to 1870, for instance, Roscoe had with the support of Huxley, then at Oxford, remade Owens College (later part of Manchester University) into a scientific college and medical school with a focus on a scientific education, which broke ranks with the Oxbridge focus on an education for gentleman, although Cambridge had had a reputation for mathematics and physics since Newton. Stewart became a professor of natural of philosophy, that is, physics, at Owens. We will discuss below why Huxley's views on Darwinism were not prioritized in Fryer's translations after he returned to China with the *Science Primers* in hand.[5]

Roscoe's struggles to make science respectable among elites in England in the 1870s paralleled Fryer's efforts in contemporary China. After joining the British Royal Society in 1863, Roscoe had been instrumental in organizing the journal *Nature*, which began publishing in November 1869. Although London was the home of the Royal Society, and Manchester had been at the center of the British Industrial Revolution, the gentlemanly practice of science and the grimy production of factories and arsenals had not yet merged to encourage an ethos favoring science and technology among men of science and industrialists.[6]

Such technical fields were also less acceptable socially and intellectually for England's middle class and elites than an Oxbridge classical education. Stewart noted in his physics primer that "thoroughly competent teachers" in scientific subjects "are not yet numerous." Consequently, the late Qing Chinese reaction to modern science must be measured by British elite resistance to a science education, such as the 1880s debate between Huxley and Matthew Arnold (1822–1888) over the significance of science in relation to literature and culture, and not by twenty-first-century hindsight.[7]

As a civilian, Roscoe promoted modern science in the 1860s through a public lecture series and by organizing cultural events as part of the Manchester Literary and Philosophical Society. His activities were the model for Fryer's science translations inside the Jiangnan arsenal and the latter's secular programs outside the arsenal among Shanghai literati and merchants to heighten appreciation for Western learning. Since 1872, a group of foreigners in Beijing had formed the Society for Diffusion of Useful Knowledge in China (*Guangxue hui*). Their goal was to introduce modern science and liberal thought as a means to overthrow ancient superstitions and to prepare the way for inevitable innovations in China.[8]

A sample of the twenty-seven works in the Science Outline Series published from 1882 to 1898 shows the range and scope of the translation project (see Appendix 3). In May 1889, for example, texts for the series were available in algebra, trigonometry, calculus, mensuration, conic sections, drawing and mathematical instruments, and electricity. Nineteen volumes were published from 1882 to 1896; twenty-one volumes appeared between 1886–1898. A volume in the "Science Handbook Series" accompanied most of them.[9]

Edkins's *Primers for Science* and the Problem of Darwin in China

By the 1880s, Henry Roscoe's original *Science Primers* were out of date and limited to just four fields. While on leave in England, the Qing customs director Robert Hart (1835–1911) discovered an updated and more complete 1880 version of the science series published by Macmillan and Company (London) and by Appleton (New York), which included the basic works by Huxley (introductory), Roscoe (chemistry), Geikie (geology and geography), Lockyer (astronomy), and Stewart (physics) that Allen and Zhang had translated, as well as new works on botany by Joseph Dalton Hooker (1817–1911), physiology by Michael Foster (1836–1907), and logic by W. Stanley Jevons (1835–1882).[10]

On his return to China, Hart published the entire series as a new textbook for science instruction in the Beijing School of Foreign Languages and other state schools. In its summer 1891 quarterly the *Scientific and Industrial Magazine* ballyhooed the publication in 1886 of Joseph Edkins's *Primers of Western Studies* (*Xixue qimeng*). This new series solicited by Hart was also called *Primers on Science Studies* (*Gezhi qimeng*, lit., "Primers on the investigation of things and extension of knowledge). Li Hongzhang, then the northern superintendent of trade, actively supported it and wrote an enthusiastic preface. The Imperial Maritime Customs Office published it in Beijing under the auspices of the General Affairs Office.[11]

When Hart asked Edkins, with the encouragement of Li Hongzhang, to undertake the translation of this more-detailed primer series, this was an important step in missionary collaboration with Qing officials to advance science in the dynasty's schools and academies. Zeng Guofan's elder son Zeng Jize (1839–1890) graced the publication with a note echoing Ming encyclopedias in which he appealed to the ideals of classical erudition when "a scholar [*Ru*] would be embarrassed by failing to know even one thing" (see Chapter 1). Zeng contended that the new textbook would build on the specialized studies pioneered in the ancient *Progress Toward Elegance* (*Erya*) dictionary. More than a decade before Yan Fu's translation of Huxley's "Evolution and Ethics" appeared in 1898 under the title of *On Evolution* (*Tianyan lun*), Edkins introduced Huxley's general views on science, although he modified Huxley's Darwinian position.[12]

A noted translator and regular contributor to the *Shanghae Serial*, Edkins had earlier worked in Shanghai with Li Shanlan to translate Whewell's *An Elementary Treatise on Mechanics* (*Zhongxue*, 1859). After a Taiping attack on Songjiang destroyed the first edition, however, it was published in Nanjing in 1867 and widely read by those involved in the Foreign Affairs Movement. Regarded by Fryer as the "greatest living authority on the Chinese language and Chinese literature," Edkins produced a notable corpus of English works on Chinese religiosity and the prospects for their conversion (1859), a grammar of colloquial Chinese (1864), an introduction to Chinese characters (1876), a historical sketch of Chinese Buddhism (1880), a lesson in spoken Chinese (1886), the evolution of the Chinese language (1888), and an account of Chinese money (1901), in addition to his many translations and articles in Chinese.[13]

Fields of Science

According to his preface, Edkins began to translate the *Primers* by himself in 1880 and completed them in 1885 just as Li Hongzhang was founding the Broad Learning of Things Academy (*Bowu shuyuan*), which also used the series as textbooks. Li's preface described Edkins as a scholar (*Ru*) who in sixteen detailed volumes had presented an encyclopedic introduction to Western science. Edkins also added materials on Greece, Rome, logic, and European history, from the Macmillan introductory textbook on history and literature (see Appendix 5).

The opening volume in Edkins's *Primers* presented an "Overview of Western Learning," with the seventh chapter organized around a larger group of twenty-three fields that formally made up the sciences (see Appendix 6). Each of these categories was initially presented in very general terms to define the scope and delimit the content of the modern sciences for beginners. Designed as introductory materials for students, when mastered they allowed students to move on to the more specialized volumes that followed.[14]

The theoretical perspective informing this survey and the later specialized volumes combined the experimental method (*shiyan zhi fa*) and natural theology. The fields of science were presented with great factuality, but the complexity of nature was ultimately informed by a prime mover whose plans for creation were illuminated through the study of science. Because things in the world were infinite, they were "all conceived by the supreme maker."[15]

Consequently, the intellectual vision informing Edkins's translation was essentially as conservative from the viewpoint of Christianity as it was for Qing literati seeking classical parallels to modern science. In neither case was science a revolutionary break with past knowledge. Rather, it was a more precise, cumulative outgrowth of earlier learning, which it supported rather than made obsolete. A curious overlap existed between the Protestants' efforts to domesticate modern science within a Christian worldview and the Qing literati's attempt to tame Western studies within the framework of classical learning.

For both Edkins and his Chinese colleagues, the scientist "exhaustively mastered principles through things themselves (*ji qi wu er qiong qi li*). This mind-set [*xinfa*, lit., "methods of mind," i.e., "moral cultivation"] was transmitted for generations by scientists. The Chinese translation Edkins and his Chinese collaborators chose thus echoed the mind-centered methodology for moral cultivation of the Cheng-Zhu tradition that the Jesuits had earlier adopted to render *scientia*.[16]

Edkins translated Huxley's "Introductory" for the *Science Primers* as the *Primer for the General Learning of Science* (*Gezhi zongxue qimeng*). Edkins's complete translation of Huxley's work was paralleled by a briefer translation of Huxley's work for the Jiangnan Arsenal Translation Department by Henry Loch (1827–1900), who joined the arsenal in 1880 and worked on the translation with Qu Anglai, which was titled *Brief Introduction to Science* (*Gezhi xiaoyin*). The latter was often too brief, however, to get across Huxley's detailed presentation on nature and science.[17]

All sixteen volumes of the 1886 edition of the *Primers for Science Studies* were slightly reorganized and later reprinted in Shanghai in 1896 under the title of *Primers of Western Learning* (*Xixue qimeng*). Edkins's translation of Huxley's "Introductory," included in the 1896 reprint, was also made part of the Shanghai *Complete Collection of Books and Materials* (*Tushu jicheng*) collectanea in 1898. The preface for the 1896 edition prepared by Zhang Yuan-fang noted that the press thought a reprint of Edkins's complete translation was necessary because the original edition had not circulated widely enough and was no longer available. Zhang added that although Zhu Xi had advanced the principle of the investigation of things, Zhu had not recognized its wider efficacy for promoting technical knowledge. Hence, the sciences Edkins introduced were a valuable boon to literati seeking to master political economy (*fuqiang zhi shu,* lit., "techniques of wealth and power").[18]

Because Edkins provided explanatory material, his complete version was more understandable to the Chinese. Edkins, however, prepared his translation principally by himself. While his written Chinese was probably the best of any scholar-missionary of his day, it was still grammatically odd. Liang Qichao (1873–1929) described it in 1897 as a flawed translation but essential to understanding science. Moreover, the terminology Edkins employed was not widely used in Chinese. It often differed from the translations Loch and Qu used, for example, which were based on the arsenal's more systematic approach. Moreover, Edkins injected into his translations of Huxley, a promoter of Darwin's theory of evolution, and into the works of the other authors in the Science Primer Series, a more Christian view of biological evolution.[19]

Modern Geography and Geology

Volumes 3–5 in the *Primers* dealt with the earth's topography (*dizhi*), physical geography (*dili zhixue*), and geology (*dixue*). These parts advanced significantly beyond the late Ming Jesuit *mappa mundi* and Kangxi-era mensuration of the Qing empire. In essence the new, scientific view of the earth presented the inner core and outer surface of the planet. These primers on the study of the continents and oceans of the earth effectively superseded the inward turn toward local geography, which had marked literati interest in historical geography during the age of evidential studies from 1750 to 1850. Zou Yan's Warring States theory of the nine continents, which literati in the late eighteenth century had used to attack Ricci's division of the world into five continents, was discarded to accord with the new science of geography.[20]

The volume that discussed topography paired complete maps of the physical earth with chapters on the earth's land and sea areas. The advantages and disadvantages of each ocean and continent were delineated in light of climate, ecology, tides, and temperature. The physical makeup of the earth was presented as a physical explanation of how favorable climates and rivers made China and India so productive in Asia and why Europe's unique land and ocean orientations had provided it advantages in dominating the globe.[21]

In addition to exploring the oceans' depths and their variable temperature and tides, the primer on topography explained the natural phenomena of volcanoes, the formation of canyons and plateaus, as well the importance of the continental shelf for rivers, and the sediments carried by rivers to the deltas and seas. The account also explained the homology between physical characteristics and the names that were associated with such physical phenomena after human habitation. The interaction between continents and the oceans was a major theme for understanding how the physical conditions of the earth had evolved.[22]

The second primer addressed the physical geography of the globe and noted how the earth's position in its solar system and its relations to the moon and sun explained the alternation of day and night and the atmospheric winds that were produced. The role of the winds and precipitation in the formation of streams, lakes, and rivers on the continents was presented as an explanation for the variations and changeability in the earth's characteristics on its surface and in its inner core. The volume on geology continued this discussion, which examined the earth's crust in light of how stones and water had interacted with plants, some now fossils, to provide a favorable ecology for life on earth.[23]

The rhetoric of such geographical and geological findings appealed to the Chinese reader to master the real aspects (*shixue*) of the natural world. Such mastery would yield an abiding understanding of the physical conditions informing life on earth. The goal of scholarship and research was to clarify the actual conditions that had determined the formation of the earth and the life-forms on it. This account of the inanimate, natural contours of our earth reached the borders between science and natural theology but stopped just short of the latter. Geography and geology were valuable as fields of knowledge that enhanced our ability to live in the world. The accounts of plant, human, and animal life in Volumes 6–8 of the primers, however, would cross that border.[24]

Botany, Physiology, and Evolution

The natural theology Edkins strategically placed in his translations was clear in the volume on botany by Hooker. To be sure, botanists such as Hooker and physicists such as Balfour Stewart had not relinquished notions of a blueprint of nature. In vigorously opposing Darwin, William Thomson maintained that physical laws upheld nature's design. Factually, however, Hooker's analysis of the biology of plants in his primer stressed the importance of chemistry and physics for explaining the makeup of plants and the functioning of their parts. Edkins's translation accordingly accentuated the importance of testing (*ceyan*) for ascertaining the chemical aspects of botany. His account also addressed the Greek and Roman nomenclature underlying all biological taxonomy.[25]

In 1858, Edkins along with Li Shanlan and Alexander Williamson completed an important translation titled *Botany* (*Zhiwu xue*), based on John

Lindley's *Elements of Botany* (London, Bradbury and Evans, 1849), which they prepared before the Darwinian challenge. This volume was one of the first to develop Chinese botanical terminology within a modern classificatory framework that would go beyond the traditional focus on plants (in addition to animal and mineral drugs) as medicine, which subsumed botany to materia medica (*bencao*) in Chinese natural studies.[26]

To explain the evolutionary changes in plants, Edkins took great pains in his 1886 *Primers* to avoid mention of Darwinian natural selection. Instead, he explained that "there were two possible reasons why plants were different in kind": first, the creator (*zaowu zhu*) simply followed his original plan to cause plants to take their myriad forms; or second, the creator might have first created a pure type or several pure types, which had evolved over time into many different forms of plants. Of the two explanations, Edkins explained that the first was "impossible to verify" and thus was an "unhelpful form of empty speculation," implying that the second was not.[27]

When we look closely at the original English, however, we find that Edkins massaged Hooker's *Botany* textbook to insinuate a natural theology into what Hooker called the "origin of species," a direct reference to Darwin's work. Edkins presented the notion of origins in Chinese as "why plants that grew on earth were different in kind." Moreover, Hooker had articulated the two opinions about plant differences as a choice between "independent creation" and "evolution," the latter again a direct reference to Darwin, which Edkins refracted into a choice about when exactly the creator intervened in creation.

What Edkins did not translate from Hooker's *Botany* was also significant. Although he translated Hooker's chief reservation about evolution (that the "apparent fixity of a species" survived with only limited changes), Edkins failed to tell his Chinese audience that Hooker praised the theory of evolution for having "taught much" and that Hooker accepted the Darwinian notion that only the "fittest survive." There was no natural theology in Hooker's account of the origin of species.[28]

Edkins's translation favored an evolutionary perspective, which enabled scientists to explain the myriad changes in plants over time. Unlike Hooker, however, his presentation depended on God's creation of life forms. A sort of Christian Darwinism was taking shape in Great Britain and the United States in the late nineteenth century. This compromise position in biological evolution resembled the Tychonic system as a Catholic compromise between the Ptolemaic and Copernican world systems in the seventeenth century. Edkins's translations in the 1886 *Primers* reflected an Anglo-American modification of the Darwinian model. Instead of natural selection, Edkins preferred to define evolution in light of the creator's purposes.

In this manner, Edkins could embrace a Christianized version of biological change that explained the slow, evolutionary changes in plants from the original pure forms the creator had sanctioned to the many forms of plant life today. In addition, he anticipated an objection from someone who might question whether or not plant forms really changed over hundreds and thousands of years, when it was clear that the peaches and plums we have today are no different from those in ancient times. Going beyond Hooker's more circumscribed critique of Darwin, Edkins replied to this query by explaining that because the weather and fertility of the earth in a place did not change drastically, a new plant would appear very rarely and could not be very different from its immediate predecessor. Between the lines of his translation of Hooker, Edkins was silencing Darwin by presenting evolution in Christian terms.[29]

Never mentioning Darwin's view of the accumulation of biological changes that enabled some species to compete with other life forms for survival, Edkins revised Hooker's views of botany into a diluted form of evolution within a post-Darwinian natural theology. God remained the ultimate creator of all life forms. Changes over time did occur in the makeup of a particular species, although such changes were always gradual and never drastic. Survival of the fittest in an amoral world of natural selection was as anathema for Protestants as the unseating of earth as cosmic center had been for Catholics.[30]

A similar, revised natural theology permeated Edkins's volume on human physiology, which was translated from a work by Michael Foster (1836–1907). Although he detailed Foster's version of the complex process of chemical oxidation that sustained life through food consumption, Edkins ultimately attributed the physiological processes that an organism used for digestion to the heat the creator had intended for the human body to survive. We are reminded here of the heat factor and the agency of fire that had informed traditional Chinese etiology based on reinterpretations of the *Treatise on Cold Damage Disorders* during the Qing dynasty.

The theological argument Edkins adopted, although included in his description of the importance of oxygen (*yangqi*, lit., "nourishing *qi*") in the bloodstream for energy, was clearly articulated in his translation of Foster's summation of human physiology. Human perception of the outer world depended on a remarkable internal network of organs of perception tied to the brain (*naotui*). Within the brain, however, the "soul remained the master of the body."

Again, Edkins's spiritual vision trumped Foster's biological and chemical account of the body. On the other hand, Edkins's account of the brain, which Hobson's anatomy had emphasized in the 1850s, did play a role in the eventual Chinese rejection of the priority of the heart-mind complex as the venue for mental operations, that is, knowledge and memory. The Kangxi court had ridiculed such anatomical claims when they were introduced by Ferdinand Verbiest in 1683.[31]

As up to date as the 1886 version of the *Primers* was on the sixty-seven fundamental elements, and as pioneering as Edkins's accounts for gases and substances were, Edkins still maintained the creator had endowed the planet with life and spirit. Theology still took precedence in his account of modern chemistry and physics. Hence, the 1886 version of the *Primers* was a highly mediated mirror of the natural sciences in the aftermath of the Darwinian revolution. As a result, the full Chinese account Edkins prepared as a version of Christian evolution was never as popular as the *Brief Introduction to Science*, which the Jiangnan Arsenal published as an alternate version of Huxley's "Introductory." In Edkins's hands, a Christian version of Huxley's agnosticism materialized instead.[32]

Political Economy as a Science

In addition to the natural sciences, Edkins included in the *Primers* a volume titled, *Policies for Enriching the Dynasty and Nourishing the People* (*Fuguo yangmin ce*). Derived from a textbook titled *Political Economy*, produced by the London University professor W. Stanley Jevons in 1878, this volume made the case that science must carry over to government policies concerning the formation of wealth for both rich and poor based on the encouragement of trade. Capitalism as an economic system for maximizing financial resources was presented as the means to enrich the dynasty.[33]

William Martin and his collaborators in the Beijing School of Foreign Languages had translated a British textbook by Henry Fawcett (1833–1884) titled *A Manual of Political Economy* (London, Macmillan, 1874–1876), which they called in Chinese *Policies for Enriching the Dynasty* (*Fuguo ce*, 1883). A Japanese edition appeared in 1881. As in Meiji Japan, the slogan of "strengthening the dynasty" (*fuguo*) among reformers such as Yan Fu in the 1890s became a prerequisite for restoring the people to wealth and the dynasty to power (*fuqiang*).[34]

Edkins's volume contended a government must treat politics and economy as technical fields of specialized learning comparable to the natural sciences in order to move from poverty to prosperity. Rather than remaining tied to a vision of a moral economy, which gave priority to ethical factors in determining how things should be used for human purposes, an enlightened state would employ a rational economic policy to maximize its resources for the material benefits of all its citizenry. Such practical fields of learning, when mastered and effected, would eventually prove as useful as the natural sciences in enriching the dynasty and nourishing the people: "Those who contend that political economy is not worthy as a field of study do not yet understand its benefit and efficacy. . . . Those who maintain that advocates of political economy merely seek to enrich the already rich and that it has no benefit for the poor are greatly misguided."[35]

We see here in China the beginnings of the disciplinary matrix that would yield the formation of the social sciences in Europe and the United States. The first period of that disciplinary formation from 1870 to 1914 placed the fields of politics, society, and economy increasingly within an evolutionary and comparative perspective that drew on Darwinian biology, European historicism, physical mechanics, and a notion of social progress to enunciate the new research agendas. To study society, the economy, or politics became just as scientific, its practitioners claimed, as the study of astronomy, chemistry, or physics.

If the sciences were the tools of modernity, then they naturally carried over to all aspects of modern life and culture. In effect, a revised form of Enlightenment positivism, described in Chapter 4, emerged in late nineteenth-century Europe and the United States that heralded the gospel of science as a new secular version of modernity encompassing the natural, political, and social worlds within a single research agenda. Applying mathematics and science to public policy became de rigueur. The later publication in 1896 of a collectanea of some forty works in political economy, titled *Collectanea of Western Studies on Wealth and Power* (*Xixue fuqiang congshu*), highlighted the growing importance of economic and political issues in the mastery of Western learning.[36]

According to Edkins's account, the formation of capital (*ziben*), which drew on the legacy of Adam Smith's *Wealth of Nations* (*Fuguo tanyuan*), depended on removing restrictions placed on trade and labor. In addition, the American Indians had remained poor because they hadn't produced any industry to make things and had only relied on their natural environment for survival. Only through industry and science (*jiyi gezhi*) could a state ensure the political and economic well-being of its people.[37]

This inclusive vision of science enabled Edkins to include volumes on the history of Hellenic Greece to Rome and Europe as part of the *Primers* series. In addition, Edkins added a volume from the Primer Series titled *Logic Primer* (Macmillan, 1876) by W. Stanley Jevons, immediately following the account of Greek history to stress the importance of deductive and inductive reasoning for the sciences. Edkins saw this as an effort to continue Ricci's efforts in the late Ming, and his own in the *Shanghae Serial* and *The Peking Magazine,* to educate the Chinese in proper forms of inference and proof. Edkins's example for the Aristotelian syllogism, for instance, had a Qing dynasty flavor to it:

> Jiangnan people all are the emperor's sons and people.
> Suzhou people are all Jiangnan people.
> Therefore, we know that all Suzhou people are the emperor's sons and people.[38]

The history of Greece, Rome, and Europe allowed Edkins to conclude with a late nineteenth-century assessment of world affairs. The experience of Europe, according to Edkins, revealed that weaker nations invariably lost out to stronger nations. This admonition for China of the importance of self-strengthening (*ziqiang*) was remarkably Darwinian in its realpolitik. Edkins and the Protestants could accept an amoral vision of natural selection in political economy, presumably as long as Europe was dominant, but they could not extend this competitive model to the biological evolution of the species, where moral considerations intervened and trumped Darwin's notion. Edkins's account of contemporary Europe ended with the recent linkage between self-government and the rise of technology.

These dual aspects of recent European historical development augured well for the future, Edkins thought. Such confidence also explained why Christians such as Edkins were convinced that the gospel of science they were transmitting to China would eventually refashion the Qing empire into a Western-style nation and people. Until the Sino-Japanese War of 1894–1895, such Christian hope for China was the rule, not the exception. As we will see below, however, the Protestant silence about Darwin would not prevail among the Chinese literati who turned to Western studies and modern science by reading the *Primers for Science Studies*, translations from the Jiangnan Arsenal, and the other books sold at Fryer's Scientific Book Depot.[39]

From the Scientific Book Depot to the China Prize Essay Contest

In his "Report of the Chinese Scientific Book Depot, Shanghai, 1887," John Fryer summarized the results of its first three years of operations, which had aimed at "facilitating the spread of all useful knowledge literature in the native language throughout China, and especially of books, maps and other publications of a scientific or technical character." Branch depots were established in Tianjin, Hangzhou, and Shantou in 1886, and four more in Beijing, Hankou, Fuzhou, and Xiamen were added in 1887. Fryer noted that seventeen thousand dollars' worth of books and maps were sold since 1885. That translated into approximately 150,000 volumes of scientific and educational literature that in two years had "found their way to the most distant parts of China as well as to Japan and Corea [Korea]."[40]

The 1886 catalog of the Book Depot, for instance, listed 371 works for sale. Fifty-nine titles were on science, forty-nine on Chinese studies (to attract Chinese customers), and thirty-five on mathematics. The catalog also included forty-four works that Fryer had translated, twenty-eight of them for the Jiang-nan Arsenal. By 1888, there were 650 titles on Western topics at the depot. Of these, 228 were original Chinese works. Books on the Sino-Japanese War were popular after 1895 when the depot became a mecca for young Chinese students of the sciences and mathematics.[41]

Another hopeful sign Fryer included in his report was that the Qing civil service examinations at times required knowledge of mathematics and science. W. T. A. Barber from the Wuchang High School later reported in July 1888, for example, "At the recent examination for the Sin Ts'ai [*xiucai*; local graduate] degree held in Han-yang-fu, the district [prefecture] containing the mart of Hankow [Hankou, Hubei], three candidates presented themselves in mathematics." Two passed by answering questions dealing with four questions on right triangles to find the sides and two questions on volumes to find the dimensions:

1. In a right triangle the hypotenuse is greater than the base by 16 feet, and than the perpendicular by 2. Find the sides.
2. In a right triangle the sum of the hypotenuse and base is 98 feet; the sum of the hypotenuse and perpendicular is 100. Find the sides.
3. The solid content of a regular parallelopipedon is 3,724 cubic feet. Its length and breadth are equal, its height 5 feet more. Find its dimensions.
4. The solid content of a regular parallelopipedon is 2,448 cubic feet. Its length and breadth are equal; its length and breadth together are 29 feet. Find its dimensions.
5. In a field the shape of a right triangle, the area is 30 square paces; the difference between hypotenuse and base = 8 paces. Find the sides.
6. In a field the shape of a right triangle, the area is 210 square paces; the sum of the hypotenuse and base = 49 paces. Find the sides.

Barber overoptimistically regarded the answers as equivalent to the "Pass B. A." at Cambridge or Oxford. He added that three years earlier three of the sixty-six graduates had received their provincial degrees for mathematics in Hunan. Candidates for the local degree, he thought, were learning sufficient mathematics to pass. He was apparently unaware that these problems had long been solvable using traditional Chinese four-unknowns crafts instead of algebraic equations. Indeed, they were not even high-school-level questions in England.[42]

In addition to reporting on the success of the Book Depot, Fryer's piece in *The North-China Herald* appealed for help to establish further branch depots throughout China and provide books to encourage the "demand for Western learning." Despite the considerable efforts Fryer, Xu Shou, and Wang Tao expended on maintaining the viability of the Polytechnic and its affiliated if irregular *Chinese Scientific and Industrial Magazine,* as well as the Book Depot, they were aware of the limits they faced in reaching educated Chinese outside the literati in Shanghai and the other treaty ports. The mainstream of literati were still drawn to what the Protestants increasingly belittled as the useless focus on literature and poetry in the more prestigious civil examinations.[43]

Contest Procedures and Official Patronage

To attract the interest of the mainstream, Fryer and Wang Tao devised the China Prize Essay Contest (*Gezhi shuyuan keyi*) in 1886. Since 1873, Ernest Major had solicited Chinese civil service examination essays for publication in Chinese in his widely read *Shanghai Journal* (*Shen bao*). Upwards of twenty thousand original eight-legged essays had been submitted to Major by 1879. Trying to tap into this large audience of literati writers and readers, Fryer conceived of the essay-writing contest to attract the many literati proficient in examination essay writing to write about foreign subjects, including science and technology, which the *Shanghai Journal* essays had not addressed. Fryer linked the Shanghai Polytechnic's educational focus to the Chinese examination culture around it, which paralleled the activities of prestigious Ming-Qing private academies such as Ruan Yuan's Sea of Learning Hall in Guangzhou:

> To popularise Western knowledge among the *literati* it is necessary to take advantage of all such existing national characteristics; and hence it was conceived that in essay writing there existed a most powerful means for inducing the better classes of Chinese to read, think, and write on foreign subjects of practical utility, and thus carry out one of the main objects for which the Polytechnic Institution was founded.

Fryer failed to add that numerous prize essays solicited by the royal academies in London, Paris, and Berlin had similarly been of strategic importance to focus interest on specific issues at the cutting edge of the sciences and mathematics during the seventeenth and eighteenth centuries in Europe. Moreover, school examinations in the sciences in Britain had been instrumental in promoting such fields of learning.[44]

In time this experiment became one of the most successful undertakings of

Shanghai Polytechnic to spread Western learning beyond the treaty ports. Fryer described the selection process as follows:

> A high official is asked to give a subject on which prize essays are invited, and to promise not only to look over the essays himself but to bestow certain sums of money upon some of the more successful essayists in addition to the regular quarterly amount of Tls. 25 [35 silver dollars] voted from the funds of the Institution for this special purpose. Every quarter a fresh subject is advertised in the native newspapers, and a date is fixed after which no essay will be received. The bundle of essays is then forwarded to the co-operating official who reads them carefully over and adjudicates their order of merit, affixing a criticism of greater or less length in his own handwriting, pointing out the features of excellence or defect in each. . . . Three receive the highest awards and ten or more receive smaller sums. . . . At the end of each year the three highest names for each quarter are honoured by having their essays, with the criticisms, printed in the form of a book, complimentary copies of which are sent to co-operating high officials and successful essay writers.[45]

The Polytechnic's essay contest closely paralleled the civil examination process of reward and fame for prized essays. Mimicking the palace examination, the three major and ten minor prizes were also announced in the Chinese and Western press. The best essays were released to newspapers such as the influential *Shanghai Journal*. Wang Tao also edited the special prize science essay volumes published by the Polytechnic, which paralleled reprinted collections of the infamous Ming-Qing civil examination essays (known as eight-legged essays, that is, composed in eight parts). This tactic also mimicked the official publication of the civil policy essays of the top three finishers on the regular palace examination. Although Fryer never indicated that the formalistic style of the dreaded eight-legged essay was required in the contest, the Qing officials charged with formulating the questions assumed that they were grading the equivalent of policy essays for the palace examination, which normally used parallel prose styles.[46]

When Fryer prepared his "Second Report of the Chinese Prize Essay Scheme" in 1889, he proudly announced:

> By its means the existence of the Polytechnic Institution has become known far and wide; the cooperation of some of the highest officials in the Empire secured; and an interest in western ideas has been created in some of the most influential quarters. By the annual expenditure of only a hundred Taels (139 silver dollars) or thereabouts, and by working in harmony with the Chinese methods of thought, and time-honoured sys-

tems of literary competition, a result has been obtained which the use of large sums of money in other ways would have failed to produce.[47]

Qing officials quickly saw the efficacy of applying the examination ethos, so well entrenched among the literati, to Western learning. Li Hongzhang and Liu Kunyi (1830–1902) as the northern and southern superintendents of trade respectively, each consented to give an extra theme every year during the spring and fall, in addition to the quarterly arrangement in place. Fryer noted that when Li Hongzhang's extra theme for the first half of 1889 was issued, thirty essays were forwarded to him for ranking. Li's list of the twenty-seven successful writers, and the extra awards of $204 given them, together with his personal criticisms, were published in the local Chinese papers.[48]

Literati Participation

Literati who sent in essays for the competition between 1886 and 1893 were getting their information for the topics mainly from Jiangnan Arsenal translations, materials prepared at the Beijing School of Foreign Languages, and articles on the sciences that had appeared in *The Chinese Scientific and Industrial Magazine*, *The Peking Magazine*, as well as the *Shanghae Serial*. The popularity of the "Answers to readers' queries" in *The Peking Magazine* and "Letters to the editors" in *Scientific and Industrial Magazine* also indicates the popularity these journals had among literati in and outside the treaty ports, although the real boom in reprinting such publications came after the Sino-Japanese War.

Following these pioneering compilations, other compendia on the sciences were also widely available (see Appendix 7). Compiled from 1877 to 1903, they reprinted many science works from the Jiangnan Arsenal Translation Department and from Fryer's Science Outline Series. Literati found out about these new compilations via book advertisements in the emerging Western and modern Chinese press and the catalogs of Western books mentioned earlier. Advertisements in the 1897 issues of the *Shanghai Journal*, for instance, included mention of new works on chemistry and public affairs directly aimed at the civil examination market. One advertisement, for example, claimed that a particular book on Western history was "absolutely essential for success" on the policy essays required in the civil examinations.[49]

The essay competition was enthusiastically received, and many of the essays were not only printed in newspapers but also included in reformist encyclopedias such as the *Collected Writings on Political Economy from the August [Qing] Dynasty (Huangchao jingji wenbian)*. In a "Table Showing the Results of the Chinese Prize Essay Scheme," Fryer presented a breakdown of the competition (see Table 9.1). Fryer's table presented figures through spring 1889, and I have updated these to include contests through 1893. Based on the figures for forty-two contests held from 1886 through 1893, in which 2,236 total essays were sent in, an average of 46.1 percent of the essays received prizes. Overall, each contest drew about fifty-three essays.[50]

Seventeen Chinese officials presented a total of eighty-six questions, with several presenting questions a number of times:

Gong Zhaoyuan	4 times	4 questions	(Shanghai Circuit)
Li Hongzhang	5 times	15 questions	
Liu Kunyi	3 times	7 questions	
Sheng Xuanhuai	6 times	6 questions	
Wu Yinsun	6 times	12 questions	(Ningbo Circuit)
Xue Fucheng	3 times	3 questions[51]	

Altogether 1,878 Chinese submitted essays over the eight years of the competition. Among the ninety-two essayists who were honored (4.9 percent), several produced five or more essays that were awarded prizes:

Yang Minhui	Licentiate	14 essays
Li Dingyi	2nd class provincial student	7 essays
Xu Keqin	Supplementary student	7 essays
Wang Zuocai	Senior tribute student	6 essays
Yin Zhilu	Polytechnic graduate	6 essays
Li Jingbang	2nd class provincial student	6 essays
Zhong Tianwei	Expectant official	5 essays
Ye Han	Purchased county degree	5 essays[52]

Despite such success and patronage, Fryer was still troubled by the relatively small number of essays Chinese sent in for the extra spring 1889 competition: "The fact that only thirty essayists dared to tackle all three subjects is an evidence of the general ignorance of the literati on everything outside the ordinary curriculum of Chinese study; while at the same time it shows how effectively this prize essay scheme is doing its work." Fryer also saw some limits in the sorts of questions that other officials such as the Shanghai circuit intendant or governor-general in Nanjing prepared: "[A]lthough their questions relate, perhaps, more to political economy and commerce than to the severer branches of science, it is still gratifying to see how patriotic they are, and how they regard knowledge from the practical, utilitarian point of view, rather than from the theoretical alone."[53]

In time, the Shanghai Polytechnic prize essays themselves, discussed below, became sources of information, as indicated in Appendix 7 by the publication of the *Compendium of Prize Essays on Science* (*Gezhi keyi huibian*) in 1897

Table 9.1. "Table showing the results of the Chinese prize essay scheme"

Year	Quarter	No.	Total Essays	Total Prizes	% with Prizes
1886	Spring	1	77	15	19.5
	Summer	2	49	18	36.7
	Autumn	3	81	23	28.4
	Winter	4	46	24	52.2
1887	Spring	5	49	15	30.6
	Summer	6	38	20	52.6
	Autumn	7	35	22	62.9
	Winter	8	58	11	19.0
1888	Spring	9	52	12	23.0
	Summer	10	49	25	51.0
	Autumn	11	26	20	77.0
	Winter	12	47	15	31.9
1889	Spring	13	37	20	54.1
	Extra	14	30	27	90.0 (Fryer Table stops)
	Summer	15	74	40	54.1
	Fall	16	52?	52	100.0
	Extra	17	66	32	48.5
	Winter	18	30?	30	100.0
1890	Spring	19	41	10	24.4
	Extra	20	61	35	57.4
	Summer	21	64	36	56.3
	Fall	22	126	30	23.8
	Extra	23	53	28	52.8
	Winter	24	47	28	59.6

Year	Quarter	No.	Total Essays	Total Prizes	% with Prizes
1891	Spring	25	57	30	52.6
	Extra	26	36	21	58.3
	Summer	27	53	28	52.8
	Fall	28	73	28	38.4
	Extra	29	60	30	50.0
	Winter	30	72	8	11.1
1892	Spring	31	36	24	66.7
	Extra	32	41	24	58.5
	Summer	33	100	36	36.0
	Fall	34	52	28	53.8
	Extra	35	64	20	31.3
	Winter	36	30	20	66.7
1893	Spring	37	33	20	60.6
	Extra	38	56	33	58.9
	Summer	39	56	28	50.0
	Fall	40	32	16	50.0
	Extra	41	50	25	50.0
	Winter	42	47	24	51.1
Total			2,236	1,031	46.1

and the *Shanghai Polytechnic Prize Essay Competition* (*Gezhi shuyuan keyi*) again in 1898. Key bibliographies of Western learning were also compiled in the 1890s by Liang Qichao and Xu Bianze. Xu's *Books on Eastern and Western Learning* (*Dong Xixue shulu*), when published in 1899, praised *The Chinese Scientific and Industrial Magazine* published by the Polytechnic for introducing the sciences to a generation of literati since the 1870s.

These essay competitions had a great deal of impact as model essays on the reformed civil examinations of 1901–1904, promulgated after the Boxer Rebellion. The reforms required all candidates to master the new fields in the sciences and world affairs. Often, candidates who had prepared essays for the Polytechnic Prize Essay Competition, such as Zhong Tianwei (1840–1900), went on to take the reformed civil examinations. In addition, many of the policy questions used in the reformed civil examinations were derived from the topics chosen by officials such as Li Hongzhang for the Prize Essay Competition. Although the civil examinations were abrogated in 1904, these scientific texts and science topics remained important in the new schools (*xuetang*) formed after 1905.[54]

Hence, from the 1880s we can document a significant increase in numbers of literati, such as Du Yaquan (1873–1933), who were educated in the modern sciences. Du taught himself mathematics by studying the works of Li Shanlan and Hua Hengfang. The future chancellor of Beijing University, Cai Yuanpei (1867–1940), invited him to teach at a new school focusing on Chinese and Western studies in Shaoxing, Zhejiang, in 1898. Du built on his knowledge of science at the end of the nineteenth century by teaching science. He founded the first modern chemistry journal in Shanghai in 1900. Subsequently he took charge of the science publications section of the Shanghai Commercial Press in 1904. In 1911, Du became the editor of *Eastern Miscellany* (*Dongfang zazhi*), a scholarly journal published by Commercial Press. The growth of a significant community of scholar-officials conversant with science, which accompanied the growth of thousands of technicians, engineers, and skilled artisans in the empirewide arsenals, began in the 1880s and 1890s—before the Sino-Japanese War.[55]

Prize Essay Topics and Their Scientific Content

From 1886 through 1894, forty-six prize essay contests were held on Western learning. A total of ninety-two essay topics were selected by the Polytechnic and Qing officials. Of these, thirty-three (36 percent) dealt with political economy and industry. The sciences (*gezhi xue*) were next with twenty-four topics (26 percent). Fryer noted the importance of the latter in his first report for 1886 and 1887.[56] (Several of the queries on science can be found in Appendix 8.)

The questions that Li Hongzhang and his Qing colleagues prepared also influenced the reformed civil examinations initiated after 1900 as part of the New Governance (*Xinzheng*) policies initiated in the last decade of the Qing dynasty. The revamping of the civil examinations held empirewide for some 150,000 provincial candidates meant that science questions were now regularly introduced under the category of science (*gezhi xue*). For example, five of the eight post-1900 essay topics on the natural sciences were phrased as follows:

1. Much of European science originates from China; we need to stress what became a lost learning as the basis for wealth and power.
2. In the sciences, China and the West are different; use Chinese learning to critique Western learning.
3. Substantiate in detail the theory that Western methods all originate from China.
4. Prove in detail that Western science studies mainly were based on the theories of China's pre-Han masters.
5. Explain why Western science studies are progressively refined and precise.
6. Itemize and demonstrate using commentaries that theories from the Mohist Canon preceded Western theories of mathematical astronomy, optics, and mechanics.[57]

Such questions and their answers revealed that among imperial examiners the wedding of the traditional Chinese sciences and Western science was still intact.

Xu Xingtai's spring 1887 theme comparing the sciences of China and the West, and Nie Jigui's spring 1894 question on locating the Western principles of mathematical astronomy, optics, and mechanics in the ancient text of *Master Mo's Teachings* (*Mozi*) became model policy questions that Qing examiners might use for civil examination policy questions between 1902 and 1903 (see Appendix 8).

The Polytechnic's prize essays thus give us a vantage point from which to evaluate the conventional understanding of modern science among Qing literati. Normally we focus on the perspectives of a few leading classical scholars such as Kang Youwei (1858–1927), Liang Qichao, or Tan Sitong when we consider the impact of science in the late Qing, favoring a few prominent reformers who were more interested in the political importance of science than the study of science itself. The ninety-two questions prepared by eighteen Qing officials for the Polytechnic's forty-six regular and special essay competitions from 1886 to 1894 provide better balanced examples.[58]

For example, Zhong Tianwei prepared an essay for Li Hongzhang's spring 1889 "Extra Theme" on native and Western science. Parts of his policy essay, presented below, were written following the exact parallelism of an eight-legged civil examination essay:

> With regard to the sciences, China and the West are different.
>
> Speaking of it from the angle of what is above form, then earlier literati scholars clarified everything leaving nothing out.
>
> Speaking of it from the angle of what is below form, then the new principles of the West daily emerge without end.
>
> It is likely that China has emphasized the Way while undervaluing the arts. Therefore, Chinese science valued meaning and principles.
>
> The West has emphasized the arts and undervalued the Way. Therefore, Western science focused on principles of things.
>
> This is why China and the West diverged.[59]

In a spring 1889 essay prepared for Li Hongzhang's "Extra Theme" on the development of Western science since Aristotle, Wang Zuocai, a student at Shanghai Polytechnic, argued:

> Therefore, Master Zhu [Xi] appended a chapter to the [Great Learning] commentary . . . , but what he elucidated was a form of science [*gezhi*] that stressed meanings and principles and not the *gezhi* that emphasized the patterns of things. China has stressed the Way and undervalued the arts. Anything to do with statutes and institutions, or rituals and music for moral edification and civilizing, were always stressed without leaving anything undone. If a sage were to reappear, he would have nothing to add. Only the makeup of the principles of things, which have forms that can be investigated, have been discarded.[60]

This perspective stressed that since the Greeks the West had focused on things themselves as objects of analysis in contrast to the political, moral, and institutional focus of classical scholars in imperial China.

Medical Missionaries since 1872 and Medical Questions as Prize Essay Topics

Since Dr. Benjamin Hobson's pioneering translations of modern medicine, Chinese literati and physicians increasingly noted the anatomical and surgical strengths of Western medicine, when compared to the therapeutic efficacy of classical Chinese medicine. The missionary vocation that informed Hobson's work and that of his successors had focused on medical education and hospitals in China. For example, the English missionary and physician John Dudgeon introduced courses in anatomy and physiology at the School of Foreign Languages in Beijing in 1872. Dudgeon also published a six-volume work in Chinese on anatomy, which the Qing government subsidized in 1887. A companion volume on physiology also appeared.[61]

Between 1874 and 1905, the number of professional medical missionaries rose from ten to about three hundred. In 1876, forty missionary hospitals and dispensaries treated 41,281 patients. Three decades later, 250 treated approximately 2 million annually. Dr. John G. Kerr, who associated with the American Presbyterian Mission, took over Peter Parker's Guangzhou hospital and supervised the treatment of over a million patients during his half century of service to his Cantonese patients. Other large missionary hospitals were established in Hangzhou and Tianjin. Li Hongzhang's wife endowed the latter in Tianjin to repay Dr. John K. Mackenzie for saving her life.

A generation of Chinese trained as modern physicians also emerged after 1870. By 1897, about 300 Chinese doctors had graduated from missionary medical schools, with an additional 250–300 in training. Many, such as Sun Yat-sen (1866–1925), were trained in missionary hospitals in China or Hong Kong. Numerous fully trained native assistants made up the staffs of most such hospitals and dispensaries. Between 1886 and 1929, some 3,816 Chinese graduated from twenty-four modern private or public medical schools.[62]

Several Chinese revolutionaries studied Western medicine early in their careers. Sun Yat-sen, for example, was a member of the first graduating class of the Hong Kong College of Medicine in 1892. Both Guo Moruo (1892–1978) and Lu Xun traveled to Japan to study modern medicine, and Guo completed his premedical and medical courses there. Lu Xun turned to literature after the Russo-Japanese War, but he described his youthful zeal for Western medicine as a reaction against the "unwitting or deliberate charlatans" who posed as classical Chinese physicians. Lu also gleaned from existing translations that the Japanese revival had been based in part on the "introduction of Western medical science to Japan."[63]

Not triumphant until the twentieth century, Western medicine still faced a sizeable opposition among traditional physicians in the late Qing, although many Chinese had joined Xu Shou in his critique of the conceptual weaknesses in the etiology provided by Chinese medical theories. Traditional Chinese physicians gradually integrated the Western anatomy of blood vessels and the nervous system. Accordingly, when the Polytechnic included medical topics for the Prize Essay Contest, the threat to classical medicine was still not noticeable.[64]

In spring 1891, for instance, the topic chosen was the medieval traditions of materia medica that had linked particular foods to human health. This query was directed at "students who researched the principles of things" (*wuli*), a term that had not yet come to mean physics. Moreover, it was framed in light of Ji Kang's (Xi Kang, 233–262) medieval advocacy of medical and spiritual techniques for "nourishing life," which had informed many late Ming encyclopedias.

The prize essays appealed to the methods for investigating things that had informed medieval lexicographies such as Yang Quan's *On the Principles of Things* (*Wuli lun*) and Zhang Hua's *Treatise on Broad Learning of Things* (*Bowu zhi;* see Chapter 1). In addition, the essays compared the healthful aspects of Chinese and Western foods. Most importantly, however, the essays pointed to the strength of modern chemistry to elucidate the alchemical findings of medieval adepts. Citing Martin's *Elements of Natural Philosophy and Chemistry,* one essay noted that native traditions for "nourishing life" and Western chemistry together could reveal the efficacy of materia medica. When Wang Tao added his evaluations for the published version, he agreed that alchemy had paralleled Western chemistry in important ways.[65]

Similarly, when Liu Kunyi prepared an essay topic on medicine for the special fall 1892 Prize Essay Contest, he asked authors to address which medical tradition, Chinese or Western, was superior conceptually (see also Appendix 8). The Zhejiang literatus Xu Keqin, who altogether submitted seven prize essays, stressed the achievements of ancient Chinese physicians, especially Zhang Ji, whose *Treatise on Cold Damage Disorders* had by early Ming times reached canonical status. Xu added that Western physicians emphasized the nervous system, but they were unaware of the twelve circulation tracts (*jingluo*) important to understand the body's susceptibility to illness. In particular, Xu's and the other essays still focused on the strength of heat factor and cold damage therapies in Chinese medicine, while pointing to the dangers in the surgical techniques employed by Western physicians. Neither the Chinese nor the missionaries were aware that Patrick Manson was unraveling the natural history of infectious diseases, based on his experiences in Amoy.[66]

In his prize essay for the summer 1893 contest, Xu Keqin addressed a query by the Ningbo circuit intendant Wu Yinsun on the comparative strengths and weaknesses of Chinese and Western medicine. The essayists were asked to demonstrate their knowledge of the history of Western medicine and its most famous physicians. Xu provided such a summary, but he added that Chinese therapies were superior to the invasive, surgical techniques used in Western countries and now in China. He admitted, however, that each tradition had certain strengths that should be selected or combined.

The Anhui literatus Li Jingbang's 1893 prize essay on medicine described the place of the brain in Western medicine. He also noted the institutional importance of the hospital for the success of Western medicine. Li's observation on hospitals was interesting because, except for the poor, they were not that prominent by the 1890s. Li claimed that traditional medicine had also stressed the brain, but his apologetics were a transparent effort to modify the still-prominent claim of the Chinese origins of Western learning to medical studies. In Li's view early Europeans had come to China, not vice versa.

Li claimed, for example, that during the Roman empire Westerners had come to China and that they had taken back the *Basic Questions* part of the *Inner Canon of the Yellow Emperor*, among other canonical medical texts. Over time, Western medicine had built on these Chinese texts to produce new findings drawn from chemistry. We will see in Chapter 11 that such traditionalist claims were increasingly suspect by 1900 as more Chinese were trained as modern physicians. Moreover, Li placed the blame for the decline of Chinese medicine on its recent practitioners, who had failed to live up to the comprehensive understanding achieved by ancient physicians.

Yang Minhui's prize essay for the summer 1893 medicine theme betrayed the fact that Western medicine was increasingly respected in official and popular circles and feared by native physicians. Yang tackled this change by appealing to the theoretical superiority of classical medical principles while at the same time admitting the advantages of Western medical practices. Western knowledge of electricity and chemistry, according to Yang, were two areas that when applied therapeutically superseded native medicine. Altogether Yang presented ten areas in which Western medicine was superior but only four for native medicine. To salvage the strengths of Chinese medical principles, Yang proposed that a "great mediation" (*dazhong*) could be achieved if new medical procedures from the West informed ancient principles.[67]

The Polytechnic's Prize Essay topics on medicine also influenced the reformed civil examinations after 1901. Essays from both the 1891 Polytechnic query on the medieval materia medica and the 1892 theme of "which medical tradition was superior theoretically" were presented as model policy essays. The Shandong licentiate Sun Weixin's 1891 prize essay on the materia medica and Li Jingbang's 1893 essay on the history of Western medicine served the same purpose in a 1903 collection of New Governance–era policy answers. More significantly, civil service examiners now regarded medicine as one of the modern sciences.[68]

Natural Theology, Darwin, and Evolution

Li Hongzhang's spring 1889 "Extra Theme" straightforwardly asked contestants to explain Darwin's and Spencer's writings (see Appendix 8). The essays on this topic reveal a remarkable combination of ignorance and knowledge concerning Darwin's famous claims about the evolution of all life forms. For example, Jiang Tongyin's prize essay simply bluffed its way through the complexity of Western learning. The judges, including Li Hongzhang, apparently did not know the difference. Jiang's essay received the first prize, but he identified Darwin as a famous geographer who also wrote on chemistry and Spencer as an expert in mathematics. That may seem odd for someone who came from the suburbs of Shanghai, where Western learning was fairly common. Jiang added, "Today the works by both gentlemen are widely prevalent abroad, . . . but I regret not yet seeing any translations." Jiang relied on missionary translations, which were silent about Darwin and Spencer, for his information.[69]

Wang Zuocai, because he had studied at Shanghai Polytechnic, was more discerning in his second-place essay. Wang noted that Darwin had argued that organisms adapted to the world in different ways and thus evolved according to different levels of complexity. Without adapting, such life forms could never survive. Although Wang recognized that Darwin had discovered a principle that no one had ever understood before, his account missed the key role of natural selection. Moreover, when he came to Spencer, Wang bluffed his way through by linking Spencer to Darwin and simply stressed the popularity of the latter's writings.[70]

The third place finisher, Zhu Dengxu, like the first-place essayist, was totally ignorant of Darwin's and Spencer's views. A supplementary student in Shanghai county, Zhu described Darwin as a geographer who became famous for his travels. Darwin's research, Zhu added, was based on science and chemistry, but Zhu also added a clue explaining his limited knowledge: none of Darwin's major works were translated into Chinese. According to Zhu, Spencer was skilled in practical applications of mathematics to science.

Such shallow characterizations of Darwin and Spencer in 1889 by Chinese literati in the Shanghai area are at first sight surprising, given John Fryer's ties to British science since his early 1870s home visit. After returning to Shanghai in 1873, Fryer clearly avoided the Darwinian controversy, although favorable mention of Darwin's *The Descent of Man* (London, 1871) had appeared in the *Shanghai Journal* on August 21, 1873. No works of Darwin or Spencer were translated at the arsenal, although the arsenal did translate a very general piece by Huxley introducing the *Primers for Science Series* in 1875.[71]

Fryer's and the other missionary translators' silence about Darwin becomes more disquieting when we examine more carefully the essay that won fourth place, which we have briefly looked at above. Zhong Tianwei had received a local degree in the civil examinations at the age of twenty-six *sui*, and he was awaiting appointment as a magistrate's aide in Guangdong province. Failing to advance through the more competitive provincial examinations in Nanjing, Zhong at age thirty-three (circa 1873) entered the Foreign Language School located in the Jiangnan Arsenal, where he studied Western learning for three years. Subsequently, Zhong traveled abroad in Europe for two years, and then he worked with Fryer and others in the Jiangnan Arsenal's Translation Department. Because of his ties to the Arsenal, both as a student and translator, and his travels abroad, he likely had access to accounts of Darwin and Spencer unavailable to others in Shanghai.

After 1901, Zhong's answer to a question on "comparing the vicissitudes in Chinese and Western mathematical astronomy" was chosen by examiners as a model essay for the reformed civil examinations. The question, and Zhong's reply, cited information that Wylie and Li Shanlan had included in their 1857 article "Progress of Astronomical Discovery in the West," which had appeared in the *Shanghae Serial*. Zhong's prize essay also drew literally on Edkins's *Primers for Science Studies*.[72]

Among the works Fryer brought back with him in 1873 after his leave in England was Roscoe's series of *Science Primers* published in 1872. Thomas Huxley, a champion of Darwin since 1860 when Huxley debated the bishop of Oxford, wrote one volume for Roscoe's project. Fryer must have heard of Darwin's *The Origin of Species* while in England. He had the arsenal's translation department translate Roscoe's series, including Huxley's volume. Hence, we find in Zhong's prize essay a remarkably accurate account of Darwin's theory of evolution. Why was it ranked fourth, then?[73]

David Wright has explained that the essays submitted for the Chinese Prize Essay Contest contain some of the earliest references in Qing China to Charles Darwin and his theory of evolution. The first, brief documented reference to Darwin came in the Jiangnan Arsenal translation of Lyell's *Elements of Geology* (sixth edition; *Dixue qianshi*, 1873) by Hua Hengfang and Macgowan, and again that year in articles in the *Shanghai Journal*.

A vague mention of human evolution also appeared in a September 1877 article in *The Chinese Scientific and Industrial Magazine*, titled "The Theory of Chaos" (*Hundun shuo*), in which an anonymous author discussed human evolution from apes in an account of how the world might end. This article argued that it was more important to consider the end of life forms rather than their origin, thereby cleverly sidestepping Darwin. Fryer and Luan usually prepared the anonymous articles in the journal.

In other words, in the 1870s it was best for an unnamed Christian—no matter how secular—and his Chinese aide to critique Darwin obliquely in Chinese. Jesuits had introduced Copernicus in a similar, misleading manner—though they signed their work—and later were chided by eighteenth-century literati for the contradictions in their presentation.[74]

Zhong Tianwei's 1889 fourth-place essay opened with a well-informed account of Greek science. Then it described the evolution of science from Aristotle to Bacon, Darwin, and Spencer. Zhong, however, went well beyond the natural theology that essays on science in the *Shanghae Serial* and the *Scientific and Industrial Magazine* had added to their translations of botany and biology. After surveying the development of Greek science from natural philosophy (*gezhi lixue*) to metaphysics (*xingli xue*) and dialectics (*bianli xue*), Zhong summarized the modern contributions of Francis Bacon, Charles Darwin, and Herbert Spencer:

> Two thousand and three years later, the Englishman Bacon first appeared and transformed Aristotle's theories . . . At the age of thirteen he entered the state school to study, but he dismissed old learning, which demonstrated his independent stance . . . Then he focused on the study of science . . . The main point of his study was that in all scientific matters it was necessary to provide substantiation through demonstrable proof so that each principle can be exhaustively grasped without enunciating that principle a priori. In this manner, through an evidential analysis of the nature of things, its principle will become manifest. . . .
>
> Darwin was born in 1809. . . . the grandson of a physician and the son of a scientist . . . Growing up he was selected to attend Edinburgh University in Scotland. Later he traveled around the globe on an English naval vessel carrying out surveys and preparing drawings while investigating each plant and animal in its ecological setting . . . In 1859 he prepared his magnum opus "on the origin of the species of all things." He also declared the "principle of the survival of the fittest" [*wanwu qiangcun ruomie zhi li*, lit., "the principle that the strong survive and the weak perish"] in Spencerian terms.
>
> All species of plants and animals undergo changes over time and have never remained unchanged. Those plants and animals that are not successful in adapting slowly perish. Those that successfully adapt survive for the long term. This is the natural principle of the heavenly way [*tiandao ziran zhi li*, i.e., close but not quite "natural selection"]. His theory, however, contradicted the teachings of Jesus, and thus scholars from every country refused to follow his words. At first he was greatly attacked, but today those who honor him have gradually increased. Hence, science underwent a great change, and Darwin can be called a superior man who arises once in a thousand autumns.
>
> As for Herbert Spencer, . . . he was with Darwin for eleven years in his youth. His life works mainly expanded on Darwin's theories, enabling people to grasp the principles of psychology . . . He claimed that only the external appearance of all things was knowable. The inner subtleties of all things were in fact unknowable . . . Comparing it to what Christianity has called God [*shangdi*] and what science calls an element

[*yuanzhi*], although the human intellect does not have the power to know or measure them, yet the point is that without any doubt such things actually exist. Moreover the changes that all things go through go back in origin to one thing. This one thing is the root, and all other things are its branches.[75]

Although couched in the rhetoric of evidential studies and the investigation of things, Zhong Tianwei's remarkable essay for the Polytechnic's essay contest represented a succinct summary of Darwin's theories and introduced Spencer's methodology almost a decade before Yan Fu's Chinese translation of Huxley's 1893 Romanes Lectures on "Evolution and Ethics" appeared in 1898 under the title of *On Evolution* (*Tianyan lun*). Yan, for example, translated natural selection as "heavenly selection" (*tianze*), which like the missionary introduction of evolution in botany prepared by Edkins avoided mentioning the key issues of variation and natural selection.

Zhong circumvented the natural theology that informed the missionaries' account of evolution. Zhong noted in his essay that he had seen a recent translation of Spencer's first work, which was called *Essential Guides for Study* (*Siye yaolan*). Zhong's student account thus presented Darwin through Spencer's slogan of the survival of the fittest. Zhong tied his account of species variation to the "natural principle of the heavenly way," which contrasted with the more classically domesticated notion of "heavenly selection" that Yan Fu presented a decade later in his translation.[76]

The lower rank that Zhong's essay received, despite its more informed account of Darwin and Spencer, becomes understandable, however, when we review the comments by Wang Tao, who was the overall supervisor of the essay contests. In the edition of the 1886–1893 essays that he published together as an annual group, Wang Tao's comments were often included in the top margins of the page, as were the comments of Ming and Qing examiners and private or academy teachers in the collections of model literary and policy essays for the civil examinations. In his comments on Zhong's explication of the principle of the survival of the fittest, Wang wrote of Darwinism:

This essay describes the flourishing of all living things whereby those most suitable survive the longest. What is referred to as "those most suitable" means "those most benefited." The theory that "under heaven the strong survive and the weak perish" has no basis in fact.[77]

Wang's comments, which exposed Spencer's—not Darwin's—efforts to justify the social order of his time, indicate why the Darwinian view of evolution was unacceptable in the 1880s and 1890s for Protestant missionaries and their converts. Opposition carried over to their longtime collaborators, such as Wang Tao, who earlier had worked with the Scottish missionary James Legge to translate the Chinese Classics into English. Hence, the lower ranking of Zhong's 1889 essay was an indication of the antagonism Darwin's views provoked among the contest judges, even though Zhong's essay was the only one to describe clearly the controversy of how Darwin's views "contradicted the teachings of Jesus."

Nevertheless, Zhong's essay was awarded fourth place, published in 1889, and reprinted in later collections. As a low-level translator and educator, Zhong supported educational reform in China within the balanced framework of adapting Western science and technology to Chinese learning, thus unifying the practical arts (*yi*) and the Way. His precocious analysis of Darwin revealed the potential for Chinese literati to reject the Christian packaging of the modern sciences, a process that began in earnest after the Sino-Japanese War, at the same time that Zhong continued to award Chinese learning prominence in theoretical matters.[78]

Just as the Chinese eventually learned about Copernicus despite the Jesuits' efforts to conceal his theory of a heliocentric cosmos, so too they learned about Darwin in spite of missionary attempts to replace the theory of natural selection with a natural theology of Christian evolution. The Chinese Prize Essay Contest reveals that the Protestant enterprise, once its religious agenda was exposed, was no more convincing to many Chinese in the late nineteenth century than the Jesuit translation agenda had been in the seventeenth and eighteenth. Hence, after the Sino-Japanese War, Chinese literati quickly turned to Meiji Japan for the latest currents in modern science.

In the late 1880s, most Chinese scholar-officials, such as Li Hongzhang, who posed the spring 1889 question, were still ignorant of Darwin's theory and followed instead the revised natural theology of Edkins, Williamson, and others. Ironically, the essay contest reveals that some Chinese such as Zhong Tianwei were more in tune with the Darwinian turn in biology, zoology, and botany than their missionary teachers or their older Chinese predecessors in science studies such as Wang Tao.

After Wang Tao's death and Fryer's departure for Berkeley University, their successors at Shanghai Polytechnic no longer enthusiastically promoted the essay contests, although they were still held irregularly, sometimes monthly, sometimes quarterly in 1901, 1904, 1906, and 1907. We will see in the next chapter that the Sino-Japanese War heightened the disjunction between events before the war and those after. In particular, many of the events from 1865 to 1894 leading up to the establishment of modern science in China were forgotten as the public paid more attention to reformers, iconoclasts, and revolutionaries after the 1895 debacle. The later reformers soon designated themselves as decisive, and not those who had been part of the Protestant era or worked in the arsenals and schools associated with the dispersed foreign affairs movement.

Liang Qichao, for instance, overlooked the remarkable expansion of newspapers after 1850 in his own self-serving accounts of a new and critical journalism in the late nineteenth century. In the process, others, such as Tan Sitong and Kang Youwei, who also rose to prominence after the 1894–1895 war, received the credit for many of the contributions that Li Shanlan, Hua Hengfang, Wang Tao, and others including Xu Shou and Zhong Tianwei made in breaking new intellectual ground in the 1870s and 1880s. As we will see in Chapter 11, both Tan Sitong and Kang Youwei as publicists appropriated science to legitimate their millennial visions without understanding much of it.[79]

Another example of this displacement was the meteoric rise of Yan Fu as a public figure. His reputation in relation to his predecessors as an iconoclast, the pioneer translator of Spencer, and the introducer of Darwin's theories replaced his earlier career as a Fuzhou Navy Yard school teacher and administrator, which we will describe in Chapter 10. The spotlight on Yan Fu after 1895 has overshadowed earlier aspects of the rise of modern science and Western learning in China before the Sino-Japanese War.

We have described above and in Chapter 8 many of the strengths and weaknesses of nineteenth-century science in China. Nevertheless, the next chapter may alarm some readers when we explain that from the vantage point of science and technology, Qing China could have won the Sino-Japanese War of 1894–1895. We will see that the many arsenals, shipyards, and factories established in the treaty ports and provinces beginning in the 1860s were important venues for technological experiments to produce weapons, ammunition, and navies, which were in many ways comparable to those of Meiji Japan.

V

Qing Reformism and Modern Science

10

Government Arsenals, Science, and Technology in China after 1860

In Chapters 8 and 9, we explained the relative success of traditional Chinese natural studies and Western science in developing together among literati elites from the 1860s under the rubric of investigating things and extending knowledge (*gezhi xue*). Most historical accounts of the post-1865 era portray those who were drawn to scholarly and technical work in the new industrial arsenals or who accepted translation positions in the official Foreign Language Schools, as marginal literati because they usually had failed the more prestigious civil examinations. In this chapter, we will discover that a decade before anything comparable in Meiji Japan many literati and artisans saw in Western learning and the modern sciences an alternative route to fame and fortune.[1]

This chapter will also reassess the naval wars that the Qing dynasty lost in the late nineteenth century. Historians have used the Sino-French naval battles of 1884–1885 and the 1894–1895 Sino-Japanese War to prove the failure of the Self-Strengthening reforms after the Taiping Rebellion (1850–1864). In particular, my account below will address how Chinese naval defeats contributed to the transformation of official, elite, and popular perceptions in the Chinese and missionary press, which shaped the emerging national sense of crisis among Han Chinese.

Chinese elites increasingly opposed the Manchu dynasty in power. Their disappointment with the military losses convinced many Chinese that more radical political, educational, and cultural changes were required to follow Japan's lead in coping with foreign imperialism. Chinese quickly forgot or repressed the earlier adaptation of new technological and scientific learning that had begun before the Meiji government's industrial programs of the 1870s. Many Protestant missionaries and experts who had aided in the Qing dynasty's scientific translation projects concluded that the Qing dynasty, the Chinese language, and traditional culture were doomed. We will first problematize and then refute the received wisdom concerning late nineteenth-century developments in science and technology.[2]

From Chinese Working for Missionaries to Missionaries Working for the Dynasty

After the treaty ports were opened in 1842, Protestant missionaries quickly established links with literati and common people in their assigned areas, as the Jesuits, Franciscans, and Dominicans had in the seventeenth century. Not until the devastations of the Taiping Rebellion did literati who were trained in the sciences by working with the Protestants, such as Xu Shou and Li Shanlan, establish links with the ruling dynasty by serving as official advisors and translators.

Not only Chinese who had worked for Inkstone Press in Shanghai, for example, moved from Protestant missions to the dynasty's arsenals and new schools. Just as the Jesuits had changed their focus from proselytizing among Chinese literati to working in the Qing bureaucracy to gain access to the imperial court, Protestant missionaries such as Fryer, Wylie, Allen, Dr. John Macgowan, and Carl Kreyer came to work in the translation department of the Jiangnan Arsenal after it was established in Shanghai. They remained committed to the gospel of science in China because they also thought its success in the precincts of the Qing government would redound to Christianity.

In the 1850s, Xu Shou, Li Shanlan, and Hua Hengfang had been hired by LMS to work as translators with Wylie and Macgowan in the Inkstone Press. Now in the 1860s, Wylie and Macgowan were employed by the Qing government as translators to work with Xu, Li, and Hua in the Jiangnan Arsenal. A small coterie of exceptional Chinese literati also joined the translation project as editors and proofreaders. In this milieu, Zhong Tianwei (1840–1901) grasped modern evolution long before Yan Fu (1853–1921) did in the 1890s, and Zhao Yuanyi (1840–1902), a cousin of Hua Hengfang, became a pioneering translator of Western medical works.[3]

Like the Jesuits in the late Ming, many leading Protestants in China moved from serving their missions in the 1850s to becoming minions of the Qing dynasty in the 1860s and 1870s. This transition troubled many Protestant missionaries, as it had their Jesuit predecessors, because their medical and scientific work soon outweighed their missionary activities. From the very beginning, Christians faced the quandary of presenting the progressive aspects of Western Europe and the United States to the Chinese primarily via medicine and science or via religious instruction. These endeavors were theoretically complementary, but in practice each took up a sizable part of a missionary's daily work.[4]

Wylie, for example, worked with Xu Shou on translating Mains's *Manual of the Steam Engine* for designing and building a steam engine, which was published by the Jiangnan Arsenal in 1871. Nevertheless, both Wylie and Allen returned to missionary work after working in the arsenal for several years because their translation and teaching work for the Qing dynasty distanced them from their calling as missionaries. On the other hand, the more secular Fryer and William Martin, the latter teaching English, political economy, and international law in Beijing from 1864, spent their careers as well-paid and hardworking servants of the Qing state. Martin publicly announced his teaching position in the Beijing Foreign Language School as "Professor of Hermeneutics, Political Economy and International Law in the University of Peking."[5]

During this era, conservative Manchu officials, such as Woren (d. 1871), and traditionalist literati echoing earlier critics of the Jesuits such as Yang Guangxian attempted to derail the movement to enhance foreign learning in official schools such as the Beijing School of Foreign Languages. Literati who feared that Western learning would again subvert state orthodoxy produced three major nineteenth-century anti-Christian tracts, each of which contained substantial sections from Yang Guangxian's early Qing work: (1) an edition of Xu Changzhi's work titled, the *Collection of the Sacred Qing Exposing Heterodoxy* (*Shengchao boxie ji*), published in 1855; (2) *A Veritable Record to Ward Off Heterodoxy* (*Bixie shilu*), printed in 1870; and (3) *A Record of Truths to Ward Off Heterodoxy* (*Bixie jishi*), published in 1871. Although never as successful as early Qing xenophobes had been, Woren and others affirmed classical learning and ancient models.[6]

As a result of the 1860 Conventions of Beijing, when the capital was occupied by British and French forces, the northern city of Tianjin became a treaty port, and Beijing became a site for Western embassies and Protestant missions. Portions of the capital, especially the imperial residence and gardens of the Lofty Pavilion (*Yuanming yuan*) in the northwestern suburbs, were sacked. Under duress, the dynasty began an era of reform known as Self-Strengthening (*ziqiang*) to cope with the external threats the dynasty faced from the superior military firepower of Western navies. In this period the technological competition between Britain and France to influence China via technology surpassed the eighteenth-century disputes between the French Jesuits, Germans, and the Portuguese in China over astronomy and influence in the Astro-calendric Bureau.[7]

Post-Taiping Reformers and Late Qing Science

The court leaders of the 1860s innovations in Beijing were Prince Gong (Yixin, 1833–1898), who signed the 1860 Conventions on the behalf of the dynasty, and Wenxiang (1818–1876), who assisted in the negotiations. Zeng Guofan, Li Hongzhang, and Zuo Zongtang took the lead in the provinces after Nanjing and the south were recovered from Taiping rebels. Prince Gong and Wenxiang together led the Grand Council and the newly founded General Affairs Office (*Zongli yamen*) in 1861. Until Wenxiang died and Prince Gong lost power in the 1870s, these capital-based and provincial groups were largely responsible for the Qing appropriation of the Protestant missionaries as westernizing servants of the dynasty.[8]

The 1861 proposal by Prince Gong and Wenxiang to establish the General Affairs Office had also included a proposal for a School of Foreign Languages in Beijing. Like the 1712 Kangxi emperor's Academy of Mathematics, students for this school were drawn from the eight Manchu banners and not from Han literati. Li Hongzhang advocated similar schools in Guangzhou and Shanghai in 1863. His proposal was based on Feng Guifen's 1861 recommendation for establishing an arsenal and shipyard in each Chinese port for better arms and ships to defend the coast. Feng also stressed establishing schools in Guangzhou and Shanghai for instruction in Western languages.[9]

The dynasty's pursuit of Western technology began in earnest when Zeng Guofan established the Anjing Arsenal in Anhui province in 1862. Two former LMS translators, Xu Shou and Hua Hengfang, served as directors. Yung Wing (Rong Hong, 1828–1912), a Cantonese who graduated from Yale in 1854, represented Zeng in buying all-purpose machinery in Europe in 1864. Yung had advised Zeng in 1863 to launch an ironworks in Shanghai. Li Hongzhang initially established two small arsenals in 1863, the first in Shanghai under Ding Richang (1823–1882) and another in Songjiang under Halliday Macartney, a former British army surgeon who had commanded an army under Li against the Taipings.[10]

The Songjiang arsenal moved to Suzhou in 1864 when the Qing court arranged for a machine shop to be brought to China as part of the Lay-Osborn Flotilla engaged in 1862. Zeng moved the arsenal to Nanjing, and it was then called the Nanjing Manufacturing Bureau (*Jinling zhizao ju*). Although formally under a Chinese director, Macartney actually managed the Nanjing Arsenal. He was granted an annual appropriation of 50,000 taels (69,500 silver dollars) to produce fuses, shells, friction tubes for firing cannon, and small cannon for the Anhui Army. New machinery was added in 1867–1868 along with some British mechanists. By 1869, Nanjing was producing rockets and was trying to forge larger guns.[11]

In 1866, Zuo Zongtang suggested creating a modern navy yard in Fuzhou to build and operate Western-style warships. Prince Gong and the regents of the Tongzhi emperor (r. 1862–1874) quickly authorized the proposal. Although Zuo had initiated the project, Shen Baozhen (1820–1879) became director-general of the Fuzhou Navy Yard in 1867 when Zuo was sent on military campaigns to the far northwest in Chinese Turkestan (Xinjiang) to put down secessionist rebellions. Depending on French engineering and technological know-how, Fuzhou quickly became the largest and most modern of all the Chinese military defense industries established in the 1860s and 1870s. It also had the largest gathering of foreign employees. Until the Sino-French War of 1884–1885, Fuzhou remained a major center of French interests.[12]

Subsequently, in 1866–1867 the court approved a proposal to add a Department of Mathematics and Astronomy to the Beijing School of Foreign Languages. The goal was to teach students about modern science through instruction in chemistry, physics, and mechanics. When William Martin returned to Beijing in 1869 to teach physics after more advanced study in the United States, he assumed the leadership of the School of Foreign Languages.

The addition of mathematics and astronomy in particular was vigorously opposed by Woren in an 1866 memorial sent while he was a Hanlin academician and imperial tutor. Although not against the teaching of mathematics per se, Woren—like Yang Guangxian before him—was incensed that foreigners such as Martin would be the teachers. When he was asked by the court to recommend native mathematicians and astronomers for service in the new department, however, Woren—again like Yang—pleaded for time. Subsequently he used an illness as an excuse to avoid an assignment in the General Affairs Office to supervise foreign affairs. He was soon relieved of all duties. Later some followers of his student, a prince in the court, openly sponsored the Boxers during their 1900 uprising.[13]

When Woren failed to stem the reformist tide in the Manchu court, Li Shanlan left Shanghai and the Jiangnan Arsenal in 1869 and accepted an appointment as professor of mathematics in the Beijing School of Foreign Languages. Unsympathetic with Woren's agenda, Li moved only after the Beijing School was upgraded to a college and a department of mathematics and astronomy was secure. Earlier Zou Boqi (1819–1869) had turned down an invitation to teach mathematics there in 1864 and 1867–1868. Li taught mathematics at the school for thirteen years. A special civil examination in mathematics, however, was opposed in the 1870s, although Li's mathematics examinations at the School of Foreign Languages were influential.[14]

The Jiangnan Arsenal in Shanghai

In the summer of 1865, Li Hongzhang, then Jiangsu governor, and the Shanghai customs intendant Ding Richang rented a machine shop in the Hongkew section of Shanghai from Thomas Hunt and Company, an American firm in the Shanghai Foreign Settlement that was the largest foreign machine shop in China. Li also approved the purchase of the machine shop and the shipyard of Hunt and Company for the Suzhou Foreign Arms Office. Additional machinery was imported, and subsequently the Jiangnan Machine Manufacturing General Bureau (*Jiangnan jiqi zhizao zongju*), usually called the Jiangnan Arsenal, was established to administer the industrial works and educational offices.[15]

Initially, the Jiangnan Arsenal used 250,000 taels (348,000 silver dollars) for production facilities, drawn mainly from maritime customs funds collected at Shanghai. Ding Richang was appointed director in 1865, and Ying Baoshi (b. 1821) was appointed 1866–1868. The arsenal moved outside the old Chinese part of Shanghai in the summer of 1867. By 1870, according to Mary Wright, the arsenal had arguably become the greatest manufacturing center of modern arms in East Asia.[16]

Zeng Guofan, Li Hongzhang, and their advisors considered building machines to be the fundamental building blocks for industry. In their view the three basic ingredients for constructing such new industry were: (1) manufacturing machines; (2) creating a new institutional category of engineers (*zhiqi zhi ren*, lit., "machine workers"); and (3) the translation of scientific and technical texts. Via armaments manufacture, the Qing state would master contemporary useful knowledge and break the Western monopoly of warships and cannons.[17]

Technical work at the arsenal was left in hands of foreigners such as Hunt's chief engineer, the American T. F. Falls, who was the superintendent. Eight of Hunt's machinists were retained, and six hundred workers from Hunt and Company were transferred directly to the Jiangnan Arsenal. Many others were later added. They produced serviceable muskets and small howitzers after initial failures in rifle production. By mid-1867 the arsenal was producing fifteen muskets and a hundred twelve-pound shrapnel daily. Twelve-pound howitzers were produced at rate of eighteen per month and used as munitions in the northern Nian wars of the 1860s. In 1871 the arsenal finally produced breech-loading rifles of the Remington type. By the end of 1873, 4,200 were produced, but they were more costly than and inferior to imported Remingtons. In 1874–1875, Li Hongzhang advised establishing a branch to produce powder and cartridges instead.[18]

John Fryer and the Translation Department

In 1867, Xu Shou, Hua Hengfang, and Xu Jianyin initiated the translation department at the Jiangnan Arsenal, which Zeng Guofan enlarged in 1868 to include a school to train translators (see Figure 10.1). In addition to relying on foreign manufacture, Zeng and Li Hongzhang regarded translation as the foundation for learning the techniques of modern manufacture and the engineering mathematics, that is, the calculus, on which it was based. Their precedent was the late Ming and early Qing translation projects that had enabled successful reform of the imperial calendar in the Astro-calendric Bureau based on new techniques and models introduced by the Jesuits.[19]

John Fryer had worked at the Jiangnan Arsenal since he left the Anglo-Chinese School in Shanghai in 1867, but he was not officially hired until November 1868. Before that, Fryer was professor of English in the School of Foreign Languages in Beijing from 1863 to 1865, but he was forced to resign when his fiancé, who had been seduced by the captain on the ship from England, arrived in Beijing in 1864 apparently pregnant. When Fryer officially joined the arsenal as a translator of scientific books, he indicated that he welcomed the appointment over teaching, which he did not enjoy, and missionizing, for which he was considered too secular. The scandal concerning his wife also made his religious ordination impossible.[20]

Figure 10.1. The Translation Office in the Jiangnan Arsenal: Xu Shou, Hua Hengfang, and Xu Jianyin.
Source: Chuanshi yanjiu 8 (1985).

In Shanghai, Fryer also edited the *Chinese Missionary News* (*Jiaohui xin-bao*) from 1866 to 1868. The paper may have reached some five thousand readers because Fryer's articles were reproduced in three other Chinese newspapers. When Fryer officially moved to the arsenal in 1868, he turned the editorship over to Young J. Allen to avoid a conflict of interest with the Qing government. Allen worked as translator and teacher in Shanghai for ten years.[21]

Others whom the Jiangnan Arsenal hired in the Translation Bureau included Alexander Wylie (who stayed eight years), Macgowan, and Rev. Carl Kreyer (who stayed for nine years). They were free to choose books for translation without direction from the imperial government. Wylie, for instance, was contracted to work with Xu Shou on Mains's *Manual of the Steam Engine*. Fryer worked with Xu Jianyin to translate William Burchett's *Practical Geometry* (1855 edition), and with Li Shanlan he began a translation of Newton's *Principia*. Macgowan joined Hua Hengfang to translate Charles Lyell's (1797–1875) *Elements of Geology* (sixth edition), published in 1873.

By renewing his contract with the Jiangnan Arsenal in 1871 and becoming the head of the new school there, Fryer chose to work for the reform of China. He stayed for twenty-eight years before accepting the Aggasiz Chair in Oriental Languages at Berkeley University and leaving for California in the summer of 1896. The total of Fryer's contributions at the arsenal came to 129 translations, with 77 published by the arsenal. Fourteen were released between 1896 and 1909 when Fryer was in California. The missionary-sponsored School and Textbook Committee and the Chinese Scientific Book Depot published the remaining thirty-eight.

Fifty-seven of Fryer's seventy-seven arsenal translations dealt with the natural sciences. His translations of physics were concentrated in the years between 1885 and 1894. He also completed five works on mathematics in 1887–1888, compared to seven from 1871 to 1879. Forty-eight works dealt with applied science, with eighteen on manufacturing. Because the Chinese government prioritized machinery, Fryer early on stressed Chinese adaptations from *The Engineer and Machinist's Drawing Book*, published in Britain in 1855. Aided by Xu Jianyin, the Arsenal published Fryer's translation in 1872.[22]

Enlightenment through Translation and Terminology

Writing in May 1886 at a symposium titled "The Advisability, or the Reverse, of Endeavoring to Convey Western Knowledge to the Chinese through the Medium of Their Own Language" Fryer noted:

Next if we examine Chinese secular literature we find astronomy and mathematics with kindred subjects have always been popular among the

Chinese. The most highly prized books on these subjects are the translations or compilations made by the Jesuits two or three centuries ago. These are found in the library of every Chinaman who has any pretensions to general scholarship. Coming down to more recent times . . . [t]here is strong demand for whatever useful knowledge foreigners have to impart. The cry on all sides is for more books.[23]

In 1880, Fryer rejected the popular view that Chinese was inadequate for scientific discourse. Moreover, Fryer discarded the notion that English would become a universal language or that China would ever be ruled by foreign powers. He then described the method of translating he and his co-workers employed and noted the translators' efforts to establish a systematic nomenclature based on three linguistic choices: using existing nomenclature by searching through native works on the arts and sciences; coining new terms by creating a new character, inventing a new descriptive term, or phoneticizing a Western term according to the Mandarin dialect; or constructing a general vocabulary of terms and list of proper names.

Earlier plans had not been very successful, according to Fryer, because translators had not appreciated the need to use the same terms throughout a series of publications. Fryer noted that Hobson's pioneering terms, for example, had unfortunately not been followed, which created confusion among Chinese readers. Hence, standardized terms became a key goal of the Jiangnan Arsenal's Translation Bureau. When translating an English work, the foreign employee and Chinese writer collaborated via dictation in Chinese sentence by sentence, which the Chinese then revised. The resulting translation was then printed using traditional woodblocks because of its efficiency for reprints.[24]

To promote modern science beyond the arsenal, Fryer cooperated with other missionaries, such as Martin, Williamson, Calvin Mateer (1836–1908), Allen, and Rudolph Lechler (1824–1908), in the School and Text-Book Series Committee (*Yizhi shuhui*) beginning in 1877. The committee met regularly in Shanghai to approve textbooks on science and other subjects for use in missionary schools. It drew up a plan to prepare a series of elementary and advanced texts covering ten subjects: mathematics, surveying, astronomy, geology, chemistry, zoology, geography, history, language, and music. Although Fryer resigned from the committee for a time, he became the general editor of the series in 1879. At the 1880 meeting, Fryer presented the terms he had used for translation at the Jiangnan Arsenal and compiled the *Translater's Vade Mecum* (*Yizhe zhinan*) based on his experience. By 1886, 104 books were published and Fryer contributed one-quarter of them.[25]

The need for a consensus for systemizing terminology that would become standard was critical for the success of the science textbooks and primers that were produced in the 1870s and 1880s. Justus Doolittle's 1872 *Vocabulary and Handbook of the Chinese Language, Romanized in the Vernacular* (*Ying-Hua cuilin yunfu*) tried to reach a consensus by incorporating translations from the leading translators of the day for specialized vocabularies in eighty-five fields covering over nine thousand terms. Wylie, for instance, prepared a section with 1,016 entries on "Terms Used in Mechanics, With Special Reference to the Steam Engine." Martin presented a list of "Terms Used in Natural Philosophy," and chemical terms were provided by John Kerr (1824–1901).

Doolittle included a section titled "Elements of Natural Science" ("Bowu zhi li"), which borrowed 322 terms from Benjamin Hobson's more complete 1858 *Medical Vocabulary in English and Chinese* (*Yixue Ying-Hua zishi*). Even so, the terminology that Doolittle deployed never became normative. For the terms Doolittle listed in his *Vocabulary,* he rarely took into account earlier translations, many of which later became standard.[26]

Chemical terminology in particular became contentious among the translators. Considerable variations in Chinese terms for the same chemical elements, for example, occurred in five early and influential chemistry translations by Benjamin Hobson in 1855, William Martin in 1868, John Fryer in 1869, John Kerr in 1870–1871, and Anatole Billequin (1837–1894) in 1873, among others (see Appendix 2). Like Martin, Billequin was then teaching chemistry at the Beijing School of Foreign Languages. Fryer and Xu collaborated in the Jiangnan Arsenal's Translation Department, while Hobson and Kerr initially published their translations in Guangzhou. The German missionary Wilhelm Lobscheid developed an earlier nomenclature for chemical elements from 1866 to 1869 for his *English and Chinese Dictionary* (*Ying-Hua zidian*) in Hong Kong, but his translations neglected the phonetic element in Chinese characters and were problematic.[27]

Even the Chinese term for the notion of a chemical element varied among translators, from "original material" (*yuanzhi*) to "original agent" (*yuanxing*). Rival translations created animosities. Fryer and John Kerr began corresponding in November 1869 about their different translations of the same chemical text, namely David Wells's *Principles and Applications of Chemistry* (New York and Chicago, 1858, 1862). In 1895, their enmity over the issue still lingered on when Kerr was criticized by Fryer in the April issue of *The Chinese Recorder.*[28]

Because their translations of fourteen of the chemical elements differed, Fryer charged that Kerr should have delayed publication of his translation on chemistry to take into account Fryer's terms. Kerr replied in the June 1895 issue that before his translation was published in 1870–1871, he had consulted with Fryer. Those exchanges "cannot be called 'negotiations for a compromise,' because the only arrangement that could be made was adoption by me of the terms used by Mr. Fryer." Fryer's self-assurance often stood in the way of the goals he advocated.[29]

When the missionaries created the School and Text-Book Series Committee in 1877 (later reorganized in 1890 as the Educational Association of China) to prepare textbooks for elementary and advanced training in mathematics and science, they made a concerted effort to unify terminology, but the results were still mixed until 1890. By then, they adopted some thirty-six thousand terms. At the second national Missionary Conference of the Series Committee, which convened in Shanghai in 1890, Fryer reiterated his view that rules for terminology should be strenuously applied. Curiously, he blamed his Chinese collaborators for the inadequacies that others pointed to in his translations.[30]

When the 1890 Missionary Committee decided to act, they acknowledged the confusion in the terminology for chemistry. By that time, the Fryer-Xu Shou translation was the most successful in coining Chinese terms for the chemical elements because they had adopted new Chinese characters using the appropriate organizing radicals, whereas the Kerr-He Liaoran version had not used fixed principles in their translations. In 1884, moreover, Fryer published through the Jiangnan Arsenal a *Vocabulary List of Names of Chemical Substances* (*Huaxue cailiao Zhong-Xi mingmu biao*) based on the Fryer-Xu *Mirror of the Origins of Chemistry* (*Huaxue jianyuan*) and its follow-ups. Fryer's terms for organic and inorganic compounds and chemical concepts, however, were less successful. Two of the committee members, Calvin Mateer and W. M. Hayes, complained about Fryer's choices for Chinese terminology:

> If Dr. Fryer's chemical names and terms had been all that the occasion demanded they would no doubt have vindicated their place in the public estimation. They have had ample time and opportunity, and the fact that they have not done so is the patent proof that changes are needed.

Fryer replied to the committee with his usual confidence:

> Your committee ought not to change my terms unless they are radically wrong and impossible to be used. Should any terms of mine be shown to be erroneous, absurd, or otherwise unserviceable and another be . . . without defects I will gladly yield to it and not otherwise.[31]

Similarly, Fryer complained in summer 1891 that Joseph Edkins had not compared the scientific terminology in other works when Edkins prepared his 1886 science primers, which duplicated four of the volumes that Young J. Allen and Zheng Changyan had prepared in 1875 for the arsenal. The latter had used the scientific terminology that Fryer had unified for the Jiangnan Arsenal's Translation Department, but Edkins had ignored those efforts.[32]

Competition among translators was not the only problem, however. In the case of the Chinese term for mechanics, for example, two legitimate translations competed in the 1860s. These terms could be traced back to the Swiss Jesuit Johann Schreck's *Diagrams and Explanations of the Marvelous Devices of the Far West* (*Yuanxi qiqi tushuo luzui*). Schreck's 1627 compilation discussed the principles of weight (*qingzhong zhi litui*) and force (*liyi zhi litui*). Although Schreck's account coined the term for the "study of weight" (*zhongxue*), that is, a pre-Newtonian notion of mechanics, his account also stressed the role of force (*liyi*). The former term likely influenced nineteenth-century translations of mechanics by Wylie and Fryer in Shanghai. The first volume of the *Shanghae Serial* in January 1857, for instance, emphasized that its issues would introduce the Newtonian science of mechanics (translated as *zhongxue*) to its readers.

Wylie and Wang Tao prepared a two-part article in 1858 titled "Popular Mechanics" (*Zhongxue qianshuo*), which presented diagrams of levers for moving things and described the mechanical operation of the wheel and axle, pulleys, the inclined plane, the wedge, and the screw. It appeared in the first volumes of the *Shanghae Serial*. Likewise, Joseph Edkins and Li Shanlan published a translation of William Whewell's *An Elementary Treatise on Mechanics* (Cambridge 1819) in Chinese as the *Study of Weight* in 1867. Justus Doolittle's 1872 *Vocabulary and Hand-book of the Chinese Language* included Wylie's piece titled "Terms used in mechanics, with special reference to the steam engine," which also used "study of weight" as the standard translation for mechanics.[33]

William Martin, however, challenged this translation in 1868 when he published his *Elements of Natural Philosophy and Chemistry* (*Gewu rumen*) for instruction at the Beijing Foreign Language School. Martin included a section on mechanics, which he titled "Introduction to Force" ("Lixue rumen"). His translation of mechanics as "the study of force" replaced the stress on weight. Later in 1883, when challenged about his translation, Martin justified his choice by subsuming the study of weight under the study of force, thus making weight secondary in the study of mechanics.

Fryer countered in 1889, when he used the term for the study of force (*lixue*) as the title of his primer *Dynamics* (*Lixue xuzhi*). Martin's version remained influential enough in China and Japan, however, that the prominent scholar, reformer, and publicist Liang Qichao (1873–1929) introduced Newtonian mechanics as the study of force in his widely read essay on the history of science composed while in exile in Japan in 1902, although at the same time he reserved the study of weight (*zhongxue*) for the mechanics of Archimedes. A similar confusion of terms for mechanics prevailed in Meiji Japan.[34]

Given the potential market for textbooks in the sciences for missionary and dynastic schools and in the arsenal training programs, a missionary-translator could achieve fame and fortune, as Fryer and Martin both did, through the publication of their works. Hence, the competition over translation terminology since the 1850s was also a claim for priority in the lucrative textbook market. The School and Text-Book Series Committee's 1890s response to the continuing controversy over chemical terminology, for instance, chastised Fryer for not fulfilling his own call for unified rules for nomenclature:

> Conformity to such a diction as this would make it impossible for us to do anything but adopt Dr. Fryer's system *in toto*. The same principle carried out in Mathematics, Physics, Astronomy, Medicine . . . would make short work of the whole business and leave our committee without any reason for their existence. The spirit of Dr. Fryer's remarks is in fact, just what has stopped all progress in the matter of terminology for the past twenty years.

Through the intervention of colleagues, Fryer relented. He became an energetic member of the committee once the controversy had been aired and resolved. In 1898 Mateer issued his *Revised List of the Chemical Elements,* which resolved the differences among the three remaining systems of nomenclature developed by Fryer, Kerr, and Billequin.[35]

Fryer also described the enlightenment project of the Translation Department. His goal was to break up intellectual stagnation in China, which Fryer felt he had in part achieved by 1880. The works the Jiangnan Arsenal had translated were well received, and they were used as textbooks in Beijing at the School of Foreign Languages and in higher-level mission schools. Moreover, bookstores sold translations to provincial civil examination candidates in Nanjing. Fryer also regarded the recent creation of Chinese professorships at Oxford, London, Paris, and Harvard as emblematic of the increasing importance of Chinese language study abroad.

Despite his series of specialized and elementary translations, Fryer remained cautionary in 1880 about the future of the enlightenment project and the popularization of science in China. The "work of supplying useful knowledge to the Chinese by means of their own language" was still needed, Fryer noted, to overturn "the system of ignoring everything but the Four Books and the Five Classics at the Government examinations." In effect, Fryer, Wylie, Macgowan and Kreyer were "spreading intellectual light" within the precincts of the Qing bureaucracy.

Fryer's translations served the gospel of science. Missionaries working for the Qing dynasty interpreted the daily drudgery of translation in light of "bringing about the intellectual and moral regeneration of this great country." Like the Jesuits before them, the Protestants saw the translation of science as a way to spread Western knowledge and Christianity. While visiting the Jiangnan Arsenal in 1877, Zeng Guofan presented Fryer with a fan on which Zeng composed a poem comparing Fryer favorably to Verbiest and Schall, which delighted Fryer.[36]

One major weakness in the School and Text-Book Series Committee's efforts to unify terminology was its deprecation of the contributions made by the Chinese collaborators Fryer and the missionaries worked with. Moreover, Wylie pointed out in 1867 that because the Jesuits had unwittingly translated many things as new, particularly in mathematics, the new nomenclature they used for algebra (i.e., *jiegen fang*), for example, initially replaced an older established terminology (i.e., *tianyuan*). When Qing scholars successfully restored the single-unknown and four-unknowns (*siyuan*) terminology for the native craft of algebra in the eighteenth and nineteenth centuries, however, there were now two systems of terms, which according to Wylie "introduced a looseness and inaccuracy of phraseology, little to the advantage of mathematical studies." This looseness continued in the parallel curriculums for modern and traditional mathematics that Chinese students in the post-Taiping imperial arsenals empirewide were expected to master.[37]

Similarly, other terms, such as tool or implement (*qi*), controlling device (*ji*), and device or weapon (*xie*), were applied to technical machinery in the nineteenth century. The technical name for the Jiangnan Arsenal, for example, was the "Machine Manufacturing General Bureau" (*jiqi zhizao zongju*). When Li Hongzhang proposed establishing new categories for the civil examinations 1867, he included "mathematics and science" (*suanshu gezhi*) and "technology and manufacturing" (*jiqi zhizuo*) as two of the eight categories.[38]

When the Protestant missionary influence peaked in the 1890s, Chinese translators took matters into their own hands and independently translated works in the natural and social sciences from Japanese into Chinese. Unaffected by the natural theology encoded in Christian-inspired textbooks, the Japanese textbooks in the sciences became models after the Sino-Japanese War. Thereafter, the Chinese translators no longer relied on missionary informants. In chemistry, for instance, Japanese books on the subject continued to use earlier Chinese terms for the elements and compounds, but the terms and concepts informing chemical theory had new translations. The same changes occurred in physics, biology, and geology.[39]

Technical Learning in the Jiangnan Arsenal and Fuzhou Navy Yard

Before Fryer joined it, the translation project for the Jiangnan Arsenal was initially very modest. The Chinese and their collaborators planned to produce an encyclopedia that would resemble the *Encyclopedia Britannica*, but this goal was quickly seen as too elementary and perhaps too traditional, that is, a mimicking of the Ming-Qing encyclopedia tradition. Instead, the core group of Chinese and Western translators hired at the Translation Bureau began producing a series of industrial treatises focused on technology and machinery rather than mathematics and the natural sciences.[40]

From 1863, when the Imperial Court approved its creation, the Shanghai School of Foreign Languages remained an independent school of translation. In 1869, however, the school was moved to the Jiangnan Arsenal and renamed the School for the Diffusion of Languages (*Guang fangyan guan*). The Shanghai Maritime Customs also paid for its new buildings. Fryer's work was narrowly defined to translate Western books on manufacturing for the new Arsenal school. He initially focused on the fields of engineering, navigation, military technology, and naval affairs.[41]

The poetic couplets Feng Guifen and Li Hongzhang wrote to celebrate the unification of the Shanghai school with the arsenal presented the name of the Foreign Language School as the "Hall for Investigating Things and Extending Knowledge" (*Gezhi tang*). Feng and Li were referring to the new industrial sciences, but classical learning and traditional mathematics were also continued in the Jiangnan Arsenal after the Foreign Language School moved in. Teachers in the Foreign Language School stressed study of the Four Books and Five Classics in the hope that the graduates would pass the more prestigious civil examinations. Hence, the school attracted the sons of Shanghai merchants and Christian converts in a more foreign environment, who saw the new learning as an alternate means to access the civil service.

The School made classical and historical studies part of the lower-division curriculum. Texts drawing on Zhu Xi's legacy, such as the *Essential Meanings of Works on Nature and Principles* (*Xingli jingyi*), which the Kangxi emperor had authorized in 1715 as an official reiteration of Cheng-Zhu orthodoxy under Manchu rule, were required. Teachers also drilled students in the eight-legged essay at the same time that mathematics was also given priority. For the latter, the teaching staff used the Ten Mathematical Classics, which had been reconstituted by Qing scholars, and the four-unknowns notational form to tutor the students in traditional Chinese mathematics.[42]

Following the Anglo-American model for training engineers, students also studied Western algebra, geometry, trigonometry, astronomy, and mechanics in the lower-division curriculum. Teachers provided training in international law, geography, and mechanical drawing. The upper-division curriculum for students emphasized seven fields:

1. mineralogy and metallurgy
2. metal casting and forging
3. wood and iron manufacturing
4. machinery design and operation

5. navigation
6. naval and land warfare
7. foreign languages, customs, institutions

It took students three years to complete the two divisions. Outstanding graduates, it was hoped, would then take special provincial examinations in Beijing.[43]

At its crest, the Jiangnan Arsenal contained four institutions: (1) the Translation Department; (2) the Foreign Language School; (3) the School for training skilled workmen; and (4) the Machine Shop. In addition, the Jiangnan Arsenal had thirteen branch factories. By 1892, it occupied seventy-three acres of land, with 1,974 workshops and a total of 2,982 workers. The arsenal acquired 1,037 sets of machines and produced forty-seven kinds of machinery under the watch of foreign technicians, who supervised production.[44]

Shipbuilding in the Jiangnan Arsenal

From 1868 to 1876, according to Meng Yue, shipbuilding in the Jiangnan Arsenal was highly productive. It built eleven ships in eight years. Ten were warships. Five of these had wooden hulls; the other five, iron hulls. All parts of each ship, including the engine, were built at the arsenal. The arsenal also experimented with different designs, from single- to double-screw engines, wooden and iron hulls, and simple warships to turreted vessels. When compared to the warships built following French models at the leading Japanese dockyard in Yokosuka in the 1870s, the level of shipbuilding technology at the Jiangnan Arsenal was actually more advanced.[45]

Founded in 1867 to replace the late Tokugawa regional dockyards in Nagasaki and elsewhere, Yokosuka's dock was designed by the French architect-engineer François-Léonce Verny (1837–1908), who had graduated from the L'École Polytechnique in 1858. Verny worked on shipbuilding projects in Ningbo and Shanghai for two years before coming to Japan in 1865. In the midst of the Meiji Restoration in 1868, his Japanese superiors at Yokosuka thought he should take a short leave to study the Fuzhou Navy Yard, which the Qing government was then building with French technical advice. In 1871, the construction director at Fuzhou visited Verny in Yokosuka, now under the Meiji Public Works Ministry. Verny covetously noted that the Fuzhou Navy Yard's budget was three times his own.[46]

The Yokosuka Dockyard did not produce its largest wooden warships until 1887–1888. They were each armed with twelve guns and boasted 1,622 horsepower. Neither was the match for the largest warship built at the Jiangnan Arsenal in 1872, which had 1,800 horsepower and was armed with twenty-six guns. The arsenal produced five iron-hulled warships before 1875, while the Japanese did not complete their first iron gunboats until after 1887. In terms of armaments, those manufactured at the Jiangnan Arsenal also were generally superior to those of Japan.[47]

The Chinese fleet of iron and wooden ships quickly fell behind the new ironclad ships of Europe, however. Moreover, the compound engine in Europe had replaced the outmoded single- or double-screw engines in Chinese vessels. The Chinese began to build compound engines only in 1877. An earlier proposal was turned down because of the lack of funds. Hence, China's ships overall were still behind Europe's in the 1870s. Moreover, because Chinese shipyards could not produce enough ships, more warships were built in Europe for the Chinese navy. Although the Qing continued to employ foreign technicians to build large modern warships, Chinese ships were still outmoded by the 1890s because Chinese training did not keep pace with continued Western technological progress. Japanese officers and sailors, in contrast, were better trained to manage their ships and guns by 1894.[48]

Shipbuilding in the Jiangnan Arsenal dramatically slowed after 1876. In 1885, after the arsenal completed its first steel gunboat, it ceased to be a military shipyard. The technological switch toward steel and armored warships in Europe highlighted the difficulty of transporting iron and coal to make steel in coastal China. At the same time, imported steel remained prohibitively expensive. Nevertheless, shipbuilding technology in Jiangnan and the Fuzhou Navy Yard probably remained slightly better than in Japan until 1889, when a French engineer designed new steel and iron warships for the Yokosuka Dockyard. Its first modern warship had more horsepower and a higher top speed than the same type of warship built at the Jiangnan Arsenal.[49]

Once shipbuilding was no longer its major task, the Jiangnan Arsenal adapted its machinery to produce the most advanced foreign guns and small arms for military use. As of 1874, the arsenal had produced a total of 110 cannons and a variety of guns modeled after products from the Armstrong factory in Britain. Three types of large 120 mm, 175 mm, and 200 mm caliber muzzle-loading guns made by the arsenal were deployed at the Wusong fort guarding the mouth of the Yangzi River. In the late 1880s, the arsenal produced large breech-loading guns that initially used black and later brown gunpowder. By 1885, Li Hongzhang favored the German arms industry over the British or French, and the scale of Krupp arms sales to China increased.

Before the Sino-Japanese War commenced in 1894, the Jiangnan Arsenal was producing large breech-loading Armstrong guns with a range from seven thousand to eleven thousand yards. They were capable of firing projectiles from eighty to eight hundred pounds. The arsenal also became known after 1890 for its success in producing the rapid-firing machine gun, which was important in enhancing sea power and coastal defense forts. By 1892 the Jiangnan Arsenal had manufactured ten forty-pound rapid-firing guns. Two years later, the arsenal finished making rapid-firing machine guns capable of launching forty-pound and one-hundred-pound shells. Because annual production in the arsenal was insufficient to supply the Chinese army, the Qing military still had to purchase such arms from abroad. Japan by comparison did not begin its ambitious artillery program until 1905, during the Russo-Japanese War.[50]

The Fuzhou Navy Yard and French Technology in China

Besides the Jiangnan Arsenal in Shanghai, the second major industrial site for shipbuilding and training in the Western sciences, engineering, and technology was the Fuzhou Naval Yard. When Zuo Zongtang submitted his 1866 memorial to establish a complete navy yard at Fuzhou, the expectation was that after five years the need for foreign experts would be eliminated. The estimated start-up costs of 300,000 taels (417,000 silver dollars) and the 600,000 taels (834,000 silver dollars) for annual operations were to come from maritime customs duties and the interprovincial taxes (*lijin*) collected in Fujian, Zhejiang, and Guangdong provinces. In return, those provinces would receive naval protection from the Southern Fleet based at Fuzhou.[51]

In contrast to the hybrid British and French influence at the Jiangnan Arsenal, Zuo and his successor Shen Baozhen from the start relied on French expertise. Once the Navy Yard was established, however, the Fujian maritime customs turned over only 400,000 taels (556,000 silver dollars), with another 50,000 (69,500 silver dollars) per month for operations, leaving the venture in a perpetual financial bind. At its peak the shipyard employed 3,000 workers. When later construction was completed the force dropped to 1,900, with 600 in the dockyard, 800 in workshops, and 500 coolies. Some 500 soldiers guarded the premises. The Navy Yard had more than forty-five buildings on 118 acres set aside for administrative, educational, and production purposes. By comparison, the Jiangnan Arsenal as the largest ordnance enterprise in 1875 had thirty-two such buildings on seventy-three acres.[52]

In terms of scale, the Fuzhou Navy Yard was the leading industrial venture in late Qing China. Designed as a Westernized enterprise based on technical machinery and organizational efficiency, the whole plant was served by a modern tramway with turntables at important workshops and intersections. The Navy Yard's goal was to build a modern Chinese flotilla between 1868 and 1875. Nineteen ships were planned with 80- to 250-horsepower engines. Of these, thirteen would be transport ships with 150-horsepower engines. Sixteen ships were finished during this time. Ten transports with 100-horsepower engines, and one corvette as a showpiece with a 250-horsepower engine, were realized in 1869–1875 while Shen Baozhen was in charge. Nine of the 150-horsepower transports cost over 161,000 taels (224,000 silver dollars) each; five of the 80-horsepower ships cost over 106,000 taels (147,000 silver dollars), with the Yangwu corvette alone requiring 254,000 taels (353,000 silver dollars).[53]

Like the Jiangnan Arsenal, the Fuzhou Navy Yard also compared favorably with the Yokosuka Dockyard. The latter had a budget in 1865 of 1.3 million taels (1.8 million silver dollars) for a four-year period, compared to 4 million taels (5.6 million silver dollars) allotted to Fuzhou over five years. Actual expenditures at Yokosuka doubled the budget, while the Fuzhou Navy Yard expended 5.36 million taels (7.5 million silver dollars) from 1866 to 1874. By 1868, Yokosuka had completed eight ships with eleven more on the way. In comparison, Fuzhou was also at the forefront of naval and technological development. With two major industrial sites in the Yangzi delta and in Fujian province, the Qing was ahead of Japanese modernization efforts in the 1860s and 1870s, but such aggregate advantages in relation to Japan did not translate into organizational superiority when the Fuzhou naval fleet faced the French flotilla alone and unaided in 1884.[54]

The industrial results in Fuzhou were at first gratifying for the Qing dynasty and praised in the December 10, 1875, *North-China Daily News*. Like ships built at the Jiangnan Arsenal, however, those in the Fuzhou Southern Fleet were mainly wooden ships, which were still common in the French navy until 1873, and thus vulnerable to European ironclads after the British began producing them in 1865. They were not equipped with the latest compound engines. When faced with war with France in the 1880s and Japan in the 1890s, some Qing officials unfairly blamed the French, particularly Prosper Giquel (1835–86), a French naval officer, for purposely dumping obsolete equipment and designs on the Chinese navy.[55]

Giquel had joined the Chinese Imperial Maritime Customs as a commissioner of customs at Ningbo in 1861 and later in Hankou until 1866. He signed a contract in 1866 to be the foreign director of the Fuzhou Navy Yard. Zuo Zongtang had also suggested opening a school for technical training called the Hall for the Search for Truth (*Qiushi tang*), an evidential research slogan, which served as the School for Naval Administration. Between 1866 and 1874, almost five hundred Chinese received technical training according to French standards. Twenty became engineers, and another 348 were highly skilled laborers. An additional one thousand men trained as machine and tool workers.[56]

Foreigners taught English, French, mathematics, and drafting. At the same time, just like candidates for the local civil examinations, students were expected to master the *Classic of Filial Piety* and the "Sacred Edict" and "Amplified Instructions" of the Kangxi and Yongzheng emperors. The Qing dynasty's long-term goal for the training that the French engineers and skilled workmen provided in Fuzhou was to create Chinese naval architects and engineers and to generate modern workmen, carpenters, ironworkers, brass workers, ship construction workers, etc., based on the French industrial model.

Giquel reported in 1873 that the Navy Yard set up two divisions of French and English schools. The French division included departments of naval construction, design, and apprentices. A naval academy with departments of theoretical navigation, practical navigation, and engine-room training were in the English division. The naval construction department opened in February 1867 with a curriculum that included French, arithmetic, algebra, descriptive and analytic geometry, trigonometry, calculus, physics, and mechanics. The five-year program suffered a high rate of attrition, however. In the first group of 105 beginning students, only 39 remained at the end of 1873.[57]

To train Chinese officers to operate warships, the English division, headed by John Carroll from England, created a department of theoretical navigation with a curriculum that paralleled that for English and French civil engineers:

Arithmetic: for knowledge of fractions, proportions, interest, etc.
Algebra: for quadratic equations of second degree, ratios, proportions, progressions, etc.
Geography: used Anderson's General Features of the Globe.
Trigonometry: plane and spherical, for solutions of triangles in navigation and nautical astronomy.
Geometry: used Todhunter's edition of Euclid's *Elements* (three books and part of sixth).
Navigation: used Raper's Correction of Compasses, the Sailings, as usually taught, and the Day's Work manuals.
Nautical Astronomy: finding latitude and longitude methods and errors of the compass.

Besides building the dockyard and training personnel, the Navy Yard launched fifteen ships between June 1869 and February 1874. However, only nineteen were completed between 1874 and 1897 due to the lower caliber of administration after Prosper Giquel's departure. During this era the Qing was expected to manage the shipyard on its own. The Navy Yard also faced a curtailment of operating funds due to the decline of interest by Beijing and provincial officials.[58]

A period of Qing self-management commenced in 1874 when operations in the Navy Yard were carried on without foreign technicians, and ended when five new Frenchmen arrived in 1897. The schools attracted native students until the late 1880s. Students in the French division were usually from Fujian; those in the English division were from Guangdong or Hong Kong. After 1874 the Navy Yard sent graduates to Europe, especially England and France, for advanced training to keep up with new technological developments. In 1877 Giquel led a party of twenty-six students. Twelve students from the English division went to England with five attending the Royal Naval College at Greenwich. Nine of the fourteen students from the French division studied hull construction and engine principles in France; the other five studied mining and metallurgy.

A second group of eight graduates was sent out in late 1882 for three years of advanced training. Five studied fortifications, defenses, and gunpowder explosives in France; two studied navigation and naval command in England; and one went to Germany for training in naval mines and torpedoes. A third group of thirty-three graduates was sent in 1886, with ten from the English division, fourteen from the French division, and nine from the Tianjin yard. Thirty completed their training; eighteen studied hydrography, ironclad warship navigation, naval artillery and small arms in England; twelve studied hulls and engines, mathematics and ship construction, river control, bridge and railway construction, and international law in France. A fourth group was scheduled to go to Europe in 1894, but the war with Japan interrupted that.

In 1874, as a twenty-one year-old graduate, Yan Fu, for instance, was acting captain of a small steamer owned by the Fujian-Zhejiang administration but not built by the Fuzhou Navy Yard. As a graduate of the Fuzhou naval division, Yan was eligible to receive advanced training in Europe. On his return to China he became a dean and professor of navigation and mathematics for many years at the Fuzhou Navy Yard. In the early 1880s he became professor of navigation and mathematics in the Tianjin Naval Academy, where he was a teacher and administrator for nearly twenty years.

After the overwhelming defeat by Japan in the Sino-Japanese War, the Navy Yard considered an 1896 recommendation to hire foreign teachers in China rather than sending students to Europe, but the Zongli yamen still wished to send the best naval students to Europe for advanced training. Ten were sent in 1897 for six years of training, but only six went to France. They were recalled in 1900 after three years due to insufficient funds.[59]

Industrial decline at the Fuzhou Navy Yard due to financial troubles had set in by 1876–1877. Expenditures totaled 5.35 million taels (7.4 million silver dollars) for the six and a half years to July 1874. This amount significantly exceeded original estimates, partly due to the high monthly wages per person (371.1 silver dollars) for the forty-five foreign advisors and workmen, which used up 15 to 24 percent of operation costs. By contrast, the total wages for two thousand Chinese craftsmen and nine hundred laborers amounted to only from 12.5 to 20 percent of total operation costs per month, or 4.8 silver dollars monthly per Chinese. Corruption and nepotism ate away at the rest.

The Chinese staff under Shen Baozhen had to work together with Giquel and his Europeans for construction to remain on schedule. Because the Navy Yard was financed as a traditional enterprise with numerous sources of income, traditional Qing budgetary practices did not take into account inflation, growth, or retooling. Long-term planning became impossible. After 1880, the Fujian Maritime Customs failed to turn over regularly the full annual allocation of 600,000 taels (834,000 silver dollars). By the 1890s, the allocation fell to between 200,000 (278,000 silver dollars) and 300,000 (417,000 silver dollars) and under 200,000 taels by 1895. As a result, the schools and dockyard were less active in the 1890s.[60]

Naval Warfare and the Refraction of Qing Reforms into Failure

Europeans and Japanese generally acknowledged that the Jiangnan Arsenal and the Fuzhou Navy Yard were more advanced technically than their chief competitor in Meiji Japan, the Yokosuka Dockyard, until the 1880s. Leaving out the issues of Japanese skills and personal motivations, which were decisive in trumping China's superior numbers in 1895 (see below), David Pong has contended that had the Qing navy engaged the Japanese in a naval battle over Taiwan in 1874–1875, after the Japanese threatened the island in April 1874, Chinese maritime defense preparations would have gained greater support. Perhaps, but due to a failure to reach consensus, the Qing government wavered and sued for peace to avoid hostilities. Subsequently, the budget for the two modern naval fleets in north and south China was cut to 4 million taels (5.56 million silver dollars), much less than was needed. By the late 1870s China's armament industries were mainly producing ammunition. Besides financial difficulties, corruption was also rife among leading officials who competed with each other for the remaining funds.[61]

According to John Rawlinson, only three Japanese ships with about 3,600 hundred men were in the 1874 Japanese expedition to Taiwan. The Japanese naval ministry was established in 1872, and by 1874 it had just seventeen ordinary ships with an aggregate of about fourteen thousand tons. Foreign observers thought China's twenty-one steamers in the one-thousand-ton class could repel the Japanese threat, but, as in 1894–1895, the Chinese ships were not organized into a unified fleet.

Since it would take time to gather a fleet in Taiwan, and because he feared—wrongly—that Japan had two ironclad warships, Shen Baozhen as the director-general of the Fuzhou Navy Yard ended the crisis with a financial payment to Japan. The Qing government also recognized Japanese de facto control of the Liuqiu (Ryūkyū) Islands. By 1879, China had two ironclad steamships, which had been ordered from the Vulcan factory in the Baltic for the Northern Fleet and were more advanced than anything the Japanese navy had at the time. They were both later sunk in the Sino-Japanese War when the Japanese had caught up technologically. In gunpowder manufacture, moreover, the machinery used in Germany, interestingly, was not as advanced as that in Shanghai at the Jiangnan Arsenal. Japan subsequently caught up with China technically in the 1880s.[62]

The Impact of the Sino-French War

The lack of coordination between the northern and southern navies became the chief weakness of the Qing fleets compared to their counterpart in Japan, which was a unified fleet stationed in Yokosuka under a central command. This disadvantage became clearer after 1874 when the French claimed Vietnam as protectorate, leading to conflict with Qing China in the upper Red River border in northern Vietnam. France began a naval buildup on the China coast that provoked several naval engagements. France did not win all the battles of the Sino-French War in 1884–1885, but it did win the war because of the lack of coordination between the vulnerable Chinese fleet based at the Fuzhou Navy Yard and the Northern Fleet under Li Hongzhang's control. The irony that a French-sponsored navy at anchorage in Fuzhou would be destroyed in a preemptive attack by an invading French flotilla based in Vietnam points to the dangers of relying on European aid in an age of imperialism.[63]

The Qing had over fifty modern naval ships in 1884, with more than half built in China. Among the others, thirteen were British-built Armstrong gunboats, two were Armstrong cruisers, and two more were German-built ships with two eight-inch guns. The latter two pairs were divided equally between the northern commissioner's Beiyang fleet and the southern commissioner's Nanyang fleet. The Qing navy, however, was divided into four fleets: the Northern Fleet at Weihaiwei and Port Arthur, one in Shanghai, another in Fuzhou, and the smallest in Guangzhou. In the 1884–1885 war, the Fuzhou flotilla fought the French nearly alone in the climatic battle.

At its home port of Mawei, the Fuzhou fleet was destroyed in fifteen minutes when French commanders sailed their war vessels uncontested past the Min River defenses protecting the Navy Yard. Because the French had not yet declared war, Qing forces naively allowed the French to exploit the situation. The French fleet thus approached the Fuzhou dockyard on August 23, 1884, unchallenged. The Qing government had not prepared contingencies for the impending war.

The Qing fleet at the Mawei anchorage numbered eleven ships. All were at least nine years old and made of wood. Eight French vessels were anchored near by and were superior, but the Chinese ships had respectable if nonstandard armaments. The Chinese were unable to use the tides to outmaneuver the heavier French vessels, which suggests that the Fuzhou captains were inexperienced. Li Hongzhang sent only two of the ships that the Qing court requested from his Northern Fleet, and he quickly withdrew them by asserting that the Japanese threat in Korea mandated their return north.[64]

The French fleet withdrew to Taiwan, but after a failed landing French forces threw a blockade around the west coast of the island. Negotiations resumed after a Chinese land victory at Lang Són. China's loss, then, was not simply due to French military superiority. French technological superiority in the 1880s was not as great as England's during the Opium War and the Second China War of 1856–1860. China had closed the gap somewhat with Europe technologically. The actual problems were (1) the political and regional disorganization of the empire; and (2) naval personnel were insufficiently trained and had a poor grasp of modern naval strategy.[65]

Except for the lost vessels, the Fuzhou dockyard had survived with little damage. In the postwar period, however, progress at the Fuzhou Navy Yard was limited in scope after its founder, Zuo Zongtang, died in 1885. Li Hongzhang thereafter purchased naval vessels for his Northern Fleet rather than building them at home. Li also had to supply his Anhui land army. After most of the Fuzhou squadron was destroyed by the French in 1884, a foreign-built ship was purchased and used as a training vessel. The Fuzhou Navy Yard also reduced its number of engineers and skilled workmen, but it continued to operate in the 1890s despite neglect. One ship each year was launched in 1891, 1892, and 1895. The war damaged or destroyed books and logistical supplies for the schools, but officials quickly restored them by 1886.[66]

The rise of Beiyang as China's chief fleet after 1885 resulted from what the Chinese called the "Disaster in the South." Although demanded by the court, subsequent efforts to create a single command for a unified naval fleet never succeeded. The new Navy Board and Li's Beiyang fleet competed for financial resources, which were declining due to further naval budget cuts between 1885–1894. The empress dowager's infamous efforts to garner funds to expand the Summer Palace did not bankrupt the imperial treasury, but they didn't help support the Chinese navy either. Inadequate funding and misplaced priorities set limits on Li Hongzhang's plans to expand the Northern Fleet.[67]

The strength of the Beiyang fleet was clear to the Japanese because of stops the Chinese fleet made in Japan in the 1880s after cruises to Vladivostok. Moreover, the inconclusiveness of the Sino-French War, which reporters conveyed to Japan, restored some Chinese prestige in Japanese eyes from the low it had reached after the Opium War. In the Nagasaki Incident of 1886, for instance, four warships of the Northern Fleet anchored in Nagasaki on their re-turn trip from the Russia. Li Hongzhang sought to impress the Japa-nese that China's naval equipment, reinforced by new ships purchased from Ger-many, was superior to Japan's. Fights between Chinese sailors, who claimed the right of extraterritoriality, and Nagasaki police, who viewed it differently, broke out during the port call, and each side blamed the other.[68]

China's flaunting of its naval superiority aroused Japanese hostility. Simi-larly, Japanese-Chinese fights provoked the Kobe Incident of 1889, which became a diplomatic dispute after a Chinese port stop there. A reporter for the newspaper *Citizen's Press* (*Kokumin shimbun*) reported another visit by the Chinese fleet in July 1890, presenting it as an instance of Chinese show-ing off their new ships. Toyama Masakazu (1848–1900), an educator and former president of Tokyo University, visited the flagship of the Chinese fleet and came away impressed with its large-caliber guns and thick steel armor. The Sino-Japanese War put an end to these diplomatic controversies when Japan exploded the notion of Chinese superiority and ended Chinese claims of extraterritoriality.[69]

The Sino-Japanese War and Its Aftermath

When the Sino-Japanese War unexpectedly began on July 24, 1894, the for-eign press generally predicted a Chinese victory even after reports of initial losses. At the time, the Qing navy (sixty-five ships) ranked eighth in the world, compared to Japan's (thirty-two ships), which ranked eleventh. China's navy was superior in armor, armaments, and tonnage. Some thought that China's two German-built battleships were more powerful than the *Maine* and the *Texas,* the United States Navy's largest warships. G. A. Ballard, vice admiral in the British Royal Navy, believed the Beiyang fleet in the 1890s was in ser-viceable condition and ready for action. Some later comparisons of the Qing and Meiji naval fleets have suggested that China could have won the sea war. On land, however, the sixty battalions of the Chinese army in the north had serious organizational weaknesses. Only twenty thousand frontline troops faced Japan's fifty-thousand-man army.[70]

Almost fourteen thousand men manned Japan's naval fleet of thirty-two warships and twenty-three torpedo boats. Ten ships were built in Britain, and two in France. The Yoshino from Britain's Armstrong's shipyard was ar-guably the fastest vessel of its time when it was timed at twenty-three knots in the 1893 trials. China's navy was still divided into the Beiyang, Nanyang, Fujian, and Guangdong fleets. In 1894, these four combined had about sixty-five large ships and forty-three torpedo boats. The strongest, the Beiyang fleet, more or less equaled Japan's entire fleet. Chinese ships were equipped with more-modern guns, but the navy lacked an adequate supply and transport system to take the offensive. The fleets took a defensive posture, which had contributed to defeat in the Sino-French War a decade earlier.[71]

If general opinion among foreigners favored Li Hongzhang's fleet over Japan's, then Japanese newspapers, magazines, and fiction were marked by exhilaration at the prospect of war with China. Some Japanese were not overly confident of victory, however. The publicist Fukuzawa Yukichi (1835–1901) warned against overconfidence, for instance, although he agreed with Japan's just cause in spreading independence and enlightenment to Korea. The Meiji emperor was reluctant to begin hostilities. He had refused to send messengers to the imperial shrines at Ise or to his father's grave to announce the war until the news of the initial Japanese victories was communicated to him. Japanese Diet members were also surprised at the easy victory.[72]

Another British observer noted that Chinese crews engaged in the war were at half strength, but salaries for full crews were paid. The greatest contrast lay in the fact that Japan's navy was unified. In the end Li Hongzhang's Beiyang navy fought the Japanese principally unaided. Li had kept his fleet out of the Fuzhou battle in 1884, and the Nanyang officers now got their revenge on the Northern Fleet by keeping their fleet for the most part out of war with Japan. So much for a united fleet.

Located between Japan and north China, Korea had historically tilted toward China rather than Japan, particularly after Hideyoshi's disastrous invasions of the peninsula in the late sixteenth century. Two hundred years later, Korea's Chŏson rulers still acknowledged their tributary status in relation to the Qing empire. Korean affairs, particularly Japan's dissatisfaction with Qing influence over Korea's closed-door foreign policies, became a flashpoint between Tokyo and Beijing. With the political and economic opening of Korea as the key dispute, hostilities commenced when Japan seized the Korean king. Li Hongzhang shortly before, in July 1894, had sent Qing troops into Korea to preserve Korea as a Qing ally. The Korean king's regent, an ally of Japan, then declared war on China. The first encounter between Chinese and Japanese ships occurred on July 24 at Fengdao, and China's two warships proved no match against an unprovoked attack. After that sea battle, the Qing Northern Fleet tried to defend the Chinese coast from Weihaiwei and Port Arthur to the mouth of Yalu River and finally declared war on August 1 (see Map 10.1).[73]

Subsequently, the Japanese naval raid at Weihaiwei on August 10 stunned the Qing court, while Li Hongzhang stalled and made excuses. Weihaiwei and Port Arthur controlled the entrance to the Bohai Bay and sea approaches to Beijing. The main Beiyang fleet gathered at the mouth of the Yalu where

Map 10.1. The Chinese coast from Weihaiwei to the mouth of Yalu River during the Sino-Japanese War.

a major naval battle with Japan commenced on September 17. It was arguably the first great naval battle employing steam-powered fleets. Each side had twelve ships in the clash. China had the advantage in armor and weight in a single salvo, while Japan had a decided advantage in speed of ships and total amount of metal thrown in a sustained exchange of salvos. Japan had more quick-firing guns, which could fire three times more weight in shells than China's six- to twelve-inch guns could.[74]

Technology alone was not the key determinant. Japan, for example, could not match China's two major battleships. Japan, however, proved to be superior in naval leadership, ship maneuverability, and the availability of explosive shells. Some observers rushed to scapegoat the Fuzhou-trained officers as cowards because they were the dominant Chinese group and had more experience and training than the Tianjin-trained officers. In 1892, for example, most engine-room appointments still went to Fuzhou graduates. Nine of the twelve captains of the Beiyang ships that fought Japan at the mouth of the Yalu were Fuzhou graduates. Rawlinson, however, has contended that cowardice was not the decisive factor. China fired 197 twelve-inch projectiles at the decisive naval battle of Yalu, with half of them being only solid shot rather than explosive shell. They scored ten hits, only four filled with explosives.[75]

From smaller guns, Chinese fired 482 shots and registered fifty-eight hits, twenty-two on one ship. They also launched five torpedoes without hits. China scored about 10 percent of her tries. The Japanese, on the other hand, with their rapid-firers scored about 15 percent of their tries. In addition, the Chinese were hampered by shortages of ammunition, especially for their bigger guns. Some were filled with cement, for example, the one that struck the Matsushima and the two that passed through the Saikyo. This suggests that there were serious corruption problems in Li Hongzhang's supply command. With hindsight, assuming that Qing and Meiji strategic decisions remained the same, it was clear that the speed and rapidity of fire of Japan's ships were more important at Yalu than the weight of the Qing vessels and their superior armor.

Shore engagements continued after the battle at the Yalu as the Japanese took advantage of their unexpectedly decisive victory at sea to launch a land war that allowed the Japanese First Army to occupy Pyongyang and then cross the Yalu to enter China at the Manchurian border. In addition, Japanese cryptographers had since June 1894 decoded Li Hongzhang's military communications. The Japanese Second Army, formed in September 1894, landed on the Liaodong Peninsula and took Port Arthur. The poor command structure of the Beiyang Fleet and the lack of a court-martial system made it impossible to place blame on any Qing officers or allocate reward properly. Moreover, the Qing personnel system was unpredictable and far from impartial. Many Chinese captains and officers simply committed suicide. No one dared to question the command structure or demand a board of review independent of the navy.[76]

The Sino-Japanese War generated intense Japanese self-confidence after 1895. The Japanese navy was enhanced by the capture of twelve Chinese warships and seven torpedo boats during hostilities, which added significant tonnage to the Meiji fleet. Moreover, Japanese industrialization accelerated after the Qing dynasty was forced to pay a considerable indemnity to the Meiji regime. The Japanese government used the indemnity as a windfall to bankroll a massive rearmament program to address the Russian expansion on the borders of northeast China. Korea and Taiwan were ceded to Japan and became colonies.

The indemnity also meant that China's huge payments to Japan could not be used to augment the Qing dynasty's reconstruction projects. The Jiangnan Arsenal and Fuzhou Shipyard in particular never recovered from the indemnities. If the Qing government was unable to integrate development so that innovative institutions reinforced each other before this, the added weight of Japanese and European imperialism after 1895 tipped the scales. The Qing reforms initiated in 1865 had even less chance of success under such political conditions.[77]

Wider Western notice of the small island kingdom that had defeated the Qing empire also ensued. The Japanese victory, however, angered the Russians, who feared Japanese expansion on the Asian continent. In concert with Germany and France, the Russians joined in a Triple Intervention after the Treaty of Shimonoseki was signed in April 1895, which forced the Japanese to withdraw from the key Liaodong Peninsula in exchange for an additional payment from the Qing government. Subsequently, in June 1896 Russia and the Qing government signed a secret alliance against Japan in which Russia was granted a railroad concession in the northeastern provinces that were increasingly referred to as Manchuria.[78]

For the Japanese public, the victory developed into the key event that energized the newly emergent Meiji press, and drowned out editorial debate over Japan's military role in Korea (see Figure 10.2). Public rage was also directed at the European powers for intervening on the side of China. When Russia later forced the Qing to lease the Liaodong Peninsula to them, the Japanese were primed for war with Moscow over the fate of China. Public enthusiasm for military adventures became a common feature when the dissemination of the national news became a central feature of the Japanese press after 1895. There were by then 600,000 newspaper subscribers in Tokyo and Osaka alone. The Japanese victory over China reverberated throughout the country and demonstrated the preeminence of Meiji Japan in east Asia. The Japanese naval victory over Russia in 1904–1905 cemented such national exuberance.

The shift from a controlled to an information press in Meiji Japan grew out of news accounts of the Sino-Japanese War, which stimulated the demand for news and information in a new, unified Japanese language. The Hakubunkai Publishing House, for example, took advantage of the outbreak of war and quickly published a trimonthly illustrated record in September 1894 titled the *Diary of the War Between Japan and Qing China* (*Nisshin sensō jikki*), which was enormously popular and helped create a cult of Japanese war heroes. Other publishers quickly followed suit, and novels, plays, and woodblock posters about the war became best sellers. The *Yomiuri shimbun* newspaper initiated a prize competition for the "best" anti-Chinese war songs.[79]

In a completely opposite way, the naval disaster at the Yalu River and the decisive Qing defeat in the Sino-Japanese War energized public criticism of the dynasty's inadequate policies and enervated the staunch conservatives at court and reformers in the provinces who had opposed westernization. The unexpected naval disaster at the hands of Japan and the way it was presented as a technological victory had shocked many literati and officials and now led to a new respect for Western studies in literati circles (see Figure 10.3). The

Figure 10.2. Nakamura Shūkō, "The Great Victory of Japanese Warships off Haiyang Island, 1894."

Source: Reprinted from *Japan at the Dawn of the Modern Age: Woodblock Prints from the Meiji Period*, Catalogue by Louise E. Virgin, with the permission of the publishers, Museum of Fine Arts, Boston. Copyright 2001.

renewed success of the Shanghai Polytechnic in 1896, for example, was tied to this event. John Fryer now reported, "The book business is advancing with rapid strides all over China, and the printers cannot keep pace with it. China is awakening at last."[80]

Unfortunately for Fryer and the missionaries, China increasingly imported science books that were translated or edited in Japan after the Sino-Japanese War. Accordingly, there was a sea change from missionary-based Chinese terminology for science from the 1840s to Japan-based Chinese terminology from 1900 with the Sino-Japanese War as key point of change. During the last decade of the Qing dynasty, the terminology in science journals in China shifted dramatically away from terminology associated with investigating things and extending knowledge (*gezhi xue*) to Japanese terms in science texts such as *World of Science* (*Kexue shijie*, 1903–1904); *First Level in Science* (*Kexue yiban*, 1907); *Science Journal* (*Lixue zazhi*, 1906–1907); and *Science* (*Kexue*, 1908–1910).[81]

The period after the Sino-Japanese War in Shanghai also saw the more-active involvement of Catholic missionaries in a new enlightenment project (*weixin*) that drew on the image of Meiji Japan as a model for the 1898 reform movement in China. It was ironic that as Japanese influence in the sci-

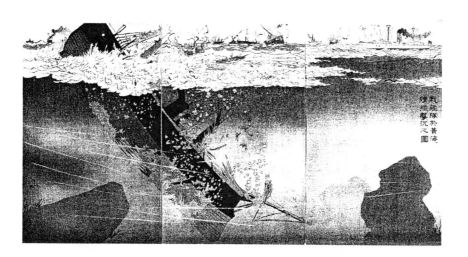

Figure 10.3. Kobayashi Kiyochika, "Illustration of our Naval Forces in the Yellow Sea Firing at and Sinking Chinese Warships, 1894."

Source: Reprinted from *Japan at the Dawn of the Modern Age: Woodblock Prints from the Meiji Period*, Catalogue by Louise E. Virgin, with the permission of the publishers, Museum of Fine Arts, Boston. Copyright 2001.

ences replaced the Protestant era of translations, the French Catholic community renewed the Jesuit effort to enlighten China through science.[82]

For example, the *Revue Scientifique* (*Gezhi xinbao,* also called the *Scientific Review*) was published monthly by the Commercial Press from March to August 1898 and was tied to the French Catholic community in Shanghai. Founded by Zhu Zhiyao (1873–1955) and Zhu Yunzuo (d. 1898), both converts, with the editorial assistance of Wang Xianli and French missionaries, it represented a clone of *The Chinese Scientific and Industrial Magazine.* On the same model, the *I-Wen-Lou et Revue Scientifique* (*Gezhi yiwen huibao*) was published as a Catholic journal in Shanghai from 1898 to 1899, which combined two earlier periodicals into a single journal. The stated goal was to introduce Western learning and practical affairs to the Chinese, as if they had not mastered their earlier lessons properly. Chinese awareness of modern science was deepened by these new journals, but they were soon superseded.[83]

Reconsidering the Foreign Affairs Movement

Chinese naval defeats contributed to the transformation of official, elite, and popular perceptions of the Self-Strengthening era. New public opinions appeared in the Chinese and missionary press that shaped the emerging national identity and sense of crisis among Han Chinese, who increasingly opposed the Manchu regime in power. Disappointment with the military losses convinced many Chinese that the foreign affairs movement had failed and that more radical political, educational, and cultural changes were required to follow Japan's lead in modernizing and coping with foreign imperialism. Earlier adaptation of new technological and scientific learning before 1895 was forgotten and repressed. Euro-American missionaries and experts who had aided in the Qing dynasty's scientific translation projects, which were used as textbooks in the arsenals and technical schools, now also thought that the Chinese nation, language, and culture were doomed.[84]

Allen Fung has recently reconsidered the "witch-hunt for the inadequacies of the Chinese army and navy" that ensued after 1895. Fung focuses on the defeat of the Chinese army in the Sino-Japanese War because Japanese land victories gave them a clear path to march on Beijing. This threat to the capital forced the Qing court to seek an immediate settlement of the war. In contrast to accounts in China that still accuse the key Qing minister, Li Hongzhang, of cowardice for his peace-at-any-cost policy, Fung maintains that Chinese armies were well equipped during the early stage of the war with Japan and that the Chinese field commanders were not incompetent. He refutes earlier claims that China's land defeats in the Sino-Japanese War were due to the failure of the Chinese ordnance industry.

Fung concludes that the primary explanations for China's losses in the land war are (1) the better military training Japanese troops and officers received compared to their Chinese counterparts; and (2) the fact that Qing troops were outnumbered by the Japanese at the major battles. I would add that the Qing court and its regional leaders underestimated the dangers of relying on European aid in an age of imperialism.[85]

We have assessed above the naval wars that the Qing dynasty lost in the late nineteenth century. Along with the Sino-Japanese War, the Sino-French naval battles have also been used as a litmus test for measuring the failure of the Self-Strengthening reforms initiated after the Taiping Rebellion. The rise of the new arsenals, shipyards, technical schools, and translation bureaus, which are usually undervalued in such "failure narratives," should also be reconsidered in light of the increased training in military technology and education in Western science available to the Chinese after 1865. Long-standing claims made by contemporaries of the Sino-Japanese War that China's defeat demonstrated the failure of the foreign affairs movement to introduce Western science and technology successfully will be reconsidered below.[86]

From "Computists" (Chouren) to "Scientists" (Gewu zhe)

We have focused above on the schools and factories launched within the Jiangnan Arsenal in Shanghai and the Fuzhou Navy Yard. John Fryer wrote in 1880, for instance, that the Jiangnan Arsenal had commenced publishing translations of Western works in 1871. By June 30, 1879, some ninety-eight works were published in 235 volumes (*juan*). Of these, twenty-two dealt with mathematics, fifteen were on naval and military science, thirteen covered the arts and manufactures. Fryer reported that another forty-five works in 142 volumes were translated but not yet published, and thirteen other works were in process with 34 volumes already completed.

Altogether, the Translation Office in the Jiangnan Arsenal had by 1880 sold 31,111 copies representing 83,454 volumes, and this had been accomplished without advertisements or postal arrangements. A work on the German Krupp guns translated in 1872 sold 904 copies in eight years, for example. Another work on coastal defense published in 1871 sold 1,114 copies in nine years. *A Treatise on Practical Geometry* (1871) sold 1,000 copies in eight years; *A Treatise on Algebra* (1873) sold 781 copies in seven years. Fryer's work on coal mining published in 1871 sold 840 copies in nine years. Publicizing these works beyond Shanghai, Beijing, and the treaty ports was difficult, but even for the latter venues such numbers were disappointing.[87]

Nevertheless, the controversial political reformer cum New Text iconoclast Kang Youwei (1858–1927) purchased all the arsenal works when he was in

Shanghai in 1882. Between 1890 and 1892, his disciple Liang Qichao purchased many of the arsenal's translations and *The Chinese Scientific Magazine*. Liang developed an influential reading list based on these materials known as the "Bibliography of Western Learning" (*Xixue shumu biao*), which was revised and published in 1896. Of these 329 published works in twenty-eight categories, 119 (36 percent) were translated by Fryer and his Chinese collaborators.

Tan Sitong (1865–1898) wrote in 1894 on scientific topics and mentioned *The Chinese Scientific Magazine* as one of his sources of scientific learning. Tan had visited Shanghai in 1893 and bought many of the arsenal's science translations, as well as works on history, politics, geography, and religion that were published by the Society for the Study of National Strengthening (*Qiangxue hui*).[88]

Besides their use as textbooks in the increasing number of missionary schools, a regional matrix of arsenals, factories, and technical schools that formed the nineteenth-century roots of the twentieth-century industrial revolution in China also used them (see Map 10.2). Hence, we should also acknowledge the scope and scale of scientific translation and military arsenals elsewhere in China after 1860. Not all of them were based on British or French models, although our two examples of the arsenal in Shanghai and navy yard in Fuzhou largely were. A list of these empirewide venues is included in Appendix 4.[89]

Once the attack on the Fuzhou Shipyard during the Sino-French War demonstrated the vulnerability of the factories and fleets to foreign naval blockade, Zhang Zhidong (1837–1909), then governor-general in Hubei and Hunan provinces in the middle Yangzi region, recognized the need for the Hanyang Ironworks (1890) and Hanyang Arsenal (1892) as protected, inland industrial sites. Not funded until 1891–1895, however, and then subject to the competing interests of Li Hongzhang's Northern Fleet and the military threat from the Japanese in Korea, the Hanyang Arsenal found that its funds were inadequate for simultaneous development of the ironworks and the arsenal. This problem lead to a slowdown in the arsenal, which failed to produce weapons or ordnance in time for the Sino-Japanese War.

Other delays in plant building and a damaging fire in summer 1894 kept the Hanyang project from achieving success before the twentieth century. Zhang wrestled with the twin goals of strategic industrialization and modern military production in the midst of the court's emergency diversion of funds and resources to deal with the Russian and Japanese threats. He chose to fund the ironworks for raw material rather than the arsenal for military arms. Hence, the Hanyang Ironworks became the hub of China's iron and steel industry during the first half of the twentieth century.[90]

If we repopulate this impressive list of factories with the human lives and literati careers they contained, then we can trace more clearly the post-Taiping successors to the native mathematical astronomers (*chouren*) that Ruan Yuan's compilation of the *Biographies of Astronomers and Mathematicians* had adumbrated circa 1800. A new group of artisans, technicians, and engineers emerged between 1865 and 1895 whose expertise no longer depended on

Partial Chronological List of Arsenals, etc., in China, 1861–1892

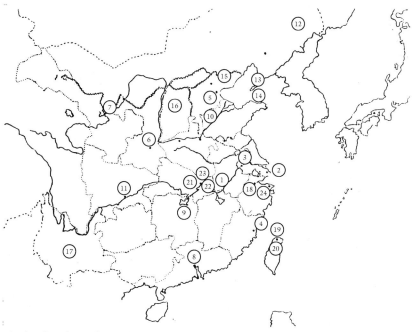

1. Anqing Arsenal, 1861
2. Jiangnan Arsenal, 1865
3. Jinling Arsenal, 1865
4. Fuzhou Shipyard, 1866
5. Tianjin Arsenal, 1867
6. Xi'an Arsenal, 1869
7. Lanzhou Arsenal, 1871–1872
8. Guangzhou Arsenal, 1874
9. Hunan Arsenal, 1875
10. Shandong Arsenal, 1875
11. Sichuan Arsenal, 1877
12. Jilin Arsenal, 1881

13. Lüshun, Port Arthur Naval Station, 1881–1882
14. Weihaiwei Shipyard, 1882
15. Beijing Field Force Arsenal, 1883
16. Shanxi Machine Shop, 1884
17. Yunnan Arsenal, 1884
18. Hangzhou Arsenal, 1885
19. Taiwan Machine Shop, 1885
20. Taiwan Arsenal, 1885
21. Daye Iron Mine, 1890
22. Hanyang Ironworks, in Hubei, 1890
23. Hanyang Arsenal, 1892
24. Zhejiang Machine Shop, 1893

Map 10.2. Location of key arsenals, shipyards, and factories during the Self-Strengthening period. (See also Appendix 4).

the fields of classical learning monopolized by the customary scholar-officials. Increasingly, they were no longer subsidiary to the dynastic orthodoxy or its official representatives.[91]

Still a necessary part of the cultural, political, and social hierarchies, the new students of the sciences in the arsenals and missionary schools emerged from the older categories of the myriad elite aspirants for official status. The scientist (*gewu zhe*) was "one who investigated things," and he now coexisted with the orthodox classical scholar in the bureaucratic apparatus but at lower levels of political rank, cultural distinction, and social esteem. The self-taught students of modern science and technology in the 1850s, such as Xu Shou, Hua Hengfang, Xu Jianyin, and Li Shanlan, were successors of the Yangzi delta mathematical astronomers who had emerged during the rise of mathematics in the age of evidential research.

Before he began work in the Jiangnan Arsenal, for instance, Xu Shou was expert in mathematics and physics, but he remained unsuccessful in the civil service competition. His classical and mathematical expertise enabled him to master the technical knowledge needed for shipbuilding and other industrial projects at the arsenal. In one curious overlap between evidential studies and modern physics, for example, Xu Shou, who was a skilled maker of replicas of ancient musical instruments, revised the findings in the English physicist John Tyndall's (1820–1893) book *Sound*, whose second edition (London, 1869) was translated by Fryer and Xu Jianyin.

Seeking to restore a mathematically accurate musical pitch pipe series for ancient harmonics, Xu Shou published an article titled "Evidential Investigation into Mathematical Harmonics" ("Kaozheng lülü shuo") in an 1880 issue of *The Chinese Scientific Magazine*, in which he noted that the ratio of the length of a pitch pipe to another one an octave higher was 9:4 and not 2:1 as Tyndall had claimed. Fryer then submitted a letter to *Nature* in 1881 that demonstrated why Tyndall's finding was in error. Seeing a scientific basis in Xu Shou's experimental critique of Tyndall's error, the editor of *Nature* noted, "It will be seen that a really scientific modern correction of an old law has most singularly turned up from China and has been substantiated with the most primitive apparatus." Tyndall initially corresponded with Fryer about the translation of his book into Chinese, but after the Xu critique of his work appeared in *Nature*, Tyndall made no effort to alter the mistake in later editions of his book.[92]

Xu Shou's training in evidential research was coupled to a rejection of the key concepts of traditional Chinese natural studies. We have seen in Chapter 8 that he also attacked traditional Chinese medicine because of its use of yin-yang and the five phases. In Xu's case, evidential research also prepared him for mastering Western science. Xu Shou, Li Shanlan, and others were in turn succeeded by those like Yan Fu and Lu Xun who were drawn to the Fuzhou Navy Yard and the Jiangnan Arsenal for formal training in science, mathematics, and engineering.[93]

By going outside the orthodox curriculum of the civil service examination, the newly educated in science, mathematics, and engineering inhabited the unprecedented arsenals, shipyards, and factories that offered non-degree-oriented engineering, mathematical, and science training. By linking science and technology, late Qing reformers produced an early version of twentieth-century Chinese techno-science (*keji*). The regional leaders of the foreign affairs movement emphasized technical expertise in engineering and mechanics and specialized knowledge of the modern sciences.[94]

Those who were drawn to scholarly work in the new industrial arsenals in Fuzhou, Shanghai, and elsewhere or translation positions in the three Foreign Language Schools in Beijing, Shanghai, and Guangzhou still tended to be Manchu bannermen or Han Chinese literati such as Xu Shou and Li Shan-lan, men who failed the civil examinations several times and saw Western learning and the sciences as an alternative route to an official career. Yan Fu and Lu Xun were famous examples of this group of outsiders from the civil examinations that initially served as the pool of highly educated men who filled the promising world of late Qing institutions oriented toward science.[95]

Lu Xun's grandfather, Zhou Fuqing, a Hanlin academician and the first important scholar in the Zhou family from Shaoxing in Zhejiang province, was jailed for attempting to bribe an examiner assigned to the 1893 Zhejiang provincial examination. The scandal forced Lu to leave his lineage school where he had prepared for the civil examinations. Before turning to literature, Lu Xun was first trained at the Jiangnan Arsenal, and he later traveled to Japan to study modern medicine at Sendai just before the 1904–1905 Russo-Japanese War.[96] Better known as a translator and publicist who was critical of late Qing reform efforts, Yan Fu, we have seen, was a graduate of the Fuzhou naval division and later received advanced training in Europe. In 1902 he was appointed chief editor for the new official Translation Bureau in Beijing after the fame he received for his translations of John Stuart Mills's *On Liberty* and Herbert Spencer's account of social Darwinism.[97]

Eventually, thousands of administrative experts, translators, and advisors—including hundreds of foreigners—served in provincial schools and arsenals under the chief provincial ministers of the late Qing. Zeng, Li, Zuo, and Zhang were the leaders of the post-Taiping turn toward foreign studies focusing on science and industry. Literati associated with statecraft and evidential studies after the Taiping Rebellion legitimated literati study of natural studies and mathematics within the framework of "Chinese studies as fundamental, Western learning as useful" (*Zhongxue wei ti Xixue wei yong*).[98]

The promising start made in missionary schools and the empirewide arsenals accelerated in the 1880s when Shanghai and Beijing took the lead in promoting the new fields associated with the foreign studies movement. In particular, the shipbuilding industry played an indispensable role in the emergence of late Qing industrial enterprises. At Jiangnan and Fuzhou the first lathes and furnaces in China to produce molten steel emerged. During the last years of the First World War, more than two hundred skilled workers at the Fuzhou Shipyard manufactured the first Chinese airplane. The arsenals, machine shops, and shipyards provided the institutional venues for an education in science and engineering. They also trained the architects, engineers, and technicians in the shipyards and arsenals who later provided the manpower for China's increasing number of public and private industries in the early twentieth century.[99]

China's defeats in the Sino-French and Sino-Japanese wars produced a pessimistic intellectual climate among both foreigners and reformers in China. Chinese literati increasingly believed that China was doomed unless more radical political initiatives were carried out. In the process, rhetoric favoring modern science became a key theme of revolutionaries in their political discourses. Earlier efforts to complement traditional Chinese natural studies and mathematics with modern science were gainsaid and deemed ludicrous. This perspective informed elite discourse about premodern natural studies and traditional Chinese medicine for much of the twentieth century.

The Construction of China's Backwardness after the Sino-Japanese War

The Sino-Japanese War provoked a striking switch in Protestant confidence about the future of Qing China. Chinese frequently pirated Young J. Allen's account of the Chinese defeat, after it was translated into Chinese. It was required reading for the 1896 Hunan provincial examination in Changsha. One of the leading Protestant missionaries and translators in Beijing, Allen outlined needed reforms in China. Allen had published an extended essay titled "Precis of Sino-Western Relations" (*Zhongxi guanxi luelun*), in the September 1875 to April 1876 issues of the *Review of the Times* (*Wan'guo gongbao*). The *Review*'s accounts of the war with Japan were republished in 1898 as a massive tome and immediately sold out its three thousand copies.[100]

Such missionary assessments were no longer gradualist, however. In the essay, Allen traced China's backwardness to three root causes: (1) superstition (*mixin*); (2) opium; and (3) civil examinations. In this series, he also stressed the importance of science (*gezhi*) as a corrective for the causes of China's backwardness. Native studies had, according to Allen, failed to grasp the universal lessons of modern science. In particular, China's assimilation of Western science missed the importance of the "study of the principles of things" (*wuli zhi xue*), that is, natural science. After 1900 this term would increasingly refer to physics, based on Japanese translations of Western scientific texts. Moreover, Allen used superstition as a modern cultural category to pigeonhole the entire Chinese classical tradition, a reduction that would become de rigueur among many Chinese radicals in the twentieth century.[101]

One of the institutional products of the political iconoclasm in China after the Sino-Japanese War, which survived the empress dowager's counter-coup against the Reform Movement in 1898, was the Imperial University of Beijing (*Jingshi daxue*). The Qing government established it at the pinnacle of an empirewide network of schools that would expand on the Foreign Language Schools in Beijing, Shanghai, and Guangzhou. Like the Translation College, the new university trained civil degree holders in Western subjects suitable for government service. The court chose William Martin, a distinguished missionary who had worked in the Beijing School, as the dean of the Western faculty in 1895.[102]

The curriculum at the Imperial University comprised eight fields: classical studies, politics, literature, medicine, science (*gezhi*), agriculture, engineering, and commerce. Six courses defined the science field: mathematics, astronomy, physics, zoology, botany, and geology. The Imperial University still referred to science courses in light of the investigation of things, although the facilities included modern laboratories equipped with the latest instruments for physics, geometry, and chemistry. This promising development was short-lived, however, because north China rebels associated with the Boxer Rebellion smashed everything in sight at the university in the summer of 1900. European armies were not any kinder during their occupation of Beijing after the Boxer siege of the foreign legations was lifted.[103]

The Qing race to establish Chinese institutions of higher learning that would stress modern science accelerated after the occupation of the capital by Western and Japanese troops in 1900. The Boxer popular rebellion in north China and the response of the Western powers and Japan to it unbalanced the power structure in the capital so much that foreigners were able to put considerable pressure on provincial and metropolitan leaders such as Li Hongzhang. Foreign support of reform and Western education thus strengthened the political fortunes of provincial reformers such as Yuan Shikai (1859–1916) and Zhang Zhidong, who had opposed the Boxers.[104]

Fryer and the Missionary Response to the Sino-Japanese War

In a May 22, 1895, letter to President Kellogg concerning the chair of Oriental Languages at Berkeley University, which he would be offered in July, John Fryer explained that his position in China had been strengthened because of China's defeat in the war. A "strong tide of demand for Western learning" was now evident among Chinese literati, who were "becoming aware of their own gross ignorance of modern arts and sciences." He added to Kellogg:

> My translations are being bought up as fast as they can be printed, and education conducted on Western principles is becoming the order of the day. It is for this tide that I have waited patiently year after year, and now that it has begun to flow it would seem almost wrong to absent myself from the country that has so long afforded me a home and for those whose enlightenment I have so long been working.

Why then entertain a teaching position at Berkeley University at this promising time? We have seen that Fryer in 1880 had rejected the possibility that English would become a universal language or that China would be ruled by foreign powers. In his 1895 letter to President Kellogg, however, Fryer explained why he now entertained accepting the Berkeley position:

> However necessary it may be for China to have the arts and sciences of the West translated into the native language and disseminated throughout the country in the first instance, it stands to reason that this will only succeed up to a point. Beyond that point no amount of translation can keep pace with the requirements of this age of progress.

The "complete education of China" had begun through translation, Fryer quipped, but that was only a first step.

The man who had tirelessly translated several score of works on science and technology into Chinese now assumed a more strident tone in his private letters. The war had proven to him and the Chinese that their efforts since 1865 had been a failure. Fryer now became a voice of doom for China's future:

> Of course this looks to the gradual decay of the Chinese language and literature, and with them the comparative uselessness of my many years of labor. Their doom seems to be inevitable, for only the fittest can survive. It may take many generations to accomplish, but sooner or later the end must come, and English be the learned language of the Empire.[105]

This intriguingly timed Spencerian perspective belied the religious message of a natural theology that informed earlier missionary translations of botany and biology for the Chinese.

On the eve of his departure for California, however, Fryer remained involved in China's affairs. He also gave the impression that his move to Berkeley might not be permanent. For instance, he publicly announced a competition for new-age novels (*xin xiaoshuo*) in Chinese that would enhance the morals of China and eviscerate the triple evils of opium, stereotypical examination essays, and footbinding. This appeal for a new literature written in "easy and clear language with meaningful implications and graceful style" attracted the interest of Liang Qichao and other reformers who called for a new culture in China, premised on the failure of traditional Chinese civilization. The bound feet of Chinese women symbolized this failure. In the 1890s, Qing radicals and revolutionaries increasingly adapted the three evils campaigns to discredit the Manchu regime.

The Boxer Rebellion of 1900 confirmed the fears of many missionaries, such as the devoted William Martin. In the 1868 preface for his *Elements of Natural Philosophy and Chemistry*, Martin had hoped to rescue "the intellect of the Chinese" from the "barren field" of belles lettres. Now he sounded more shrill: "Let this pagan empire be partitioned among Christian powers." Unlike Fryer, Martin stayed on in China and continued his educational work. In a peculiar manner, China's "defeat" in the Sino-Japanese War also represented John Fryer's and William Martin's failure "to change China."[106]

In our concluding chapter, we address the remaining steps in the Chinese construction of modern science in late imperial China. After the debacle of the Sino-Japanese War, that construction increasingly denigrated traditional natural studies and classical medicine. When we look at the changes in scientific terminology that occurred in fields such as botany and chemistry after 1900, we find that Japanese scientific terminology decisively replaced Protestant terms in China. Although the role of Chinese students who studied science in the United States and Europe should not be underestimated, a sea change in favor of Japanese science and technology took hold during the last years of the Qing dynasty. We tend to underestimate what the Japanese wave left in its wake, namely the hundreds of translations that Protestants such as Fryer, Martin, and others had labored on since 1865, which had also made their way to Yokosuka and to virtually every prefectural library in Meiji Japan.[107]

11

Displacement of Traditional Chinese Science and Medicine in the Twentieth Century

Despite the relative success of traditional Chinese natural studies and modern Western science in developing together as objects of study by a select number of literati in the late nineteenth century, they and their Protestant informants largely ignored the role of laboratories in modern science to discover and test new findings. For Catholic or Protestant missionaries and literati mathematicians, natural studies were rarely more than translating technical knowledge and memorizing and applying newly available texts. The techno-science practiced in the arsenals was exceptional, but the practical focus there on producing arms and ships precluded cutting-edge research.

Japan quickly replaced England and France as the nation the Qing dynasty should emulate in science and technology. During this period, for example, the Chinese preferred translating Japanese mathematics texts, which proved to be a convenient shortcut to modern mathematics. The conversion to Western mathematics was also aided by the many Chinese students who returned from studies abroad, particularly from Japan after 1895. Over ten thousand Chinese traveled to Japan to study from 1902 to 1907. Some 90 percent of the foreign-trained students who joined the Qing civil service after 1905, for instance, graduated from Japanese schools.[1]

Western Learning Mediated through Japan

Since 1865, literati inside and outside the bureaucracy had distinguished between Chinese learning (*Zhongxue*), which they presented as the whole of native learning, and Western studies (*Xixue*). Neither term was successful as a monolithic designation. Each was politically charged in the 1890s, when they were used by conservatives and radicals in the struggle for or against modernity. When Western learning gained momentum as a model for science and modern institutions, literati displaced mediating terms for science such as investigating things and extending knowledge (*gezhi xue*). They considered educational institutions that used such accommodations old-fashioned.

The ever-increasing numbers of overseas Chinese students in Japan, Europe, and the United States perceived that outside of China the normal language of modern science drew on universal concepts and terms that superseded traditionalist literati notions of Chinese natural studies and was disengaged from classical learning. Nevertheless, continuities remained. For example, Japanese scholars during the early Meiji period, influenced by the rise of industrial Germany, demarcated the new sciences by creating a new term for *Wissenschaft* (= *scientia*) as a broad sense of European science (*kagaku*, lit., "classified learning based on technical training"). They still presented natural studies using traditional terminology such as the exhaustive study of the principles of things (*kyūri, qiongli*).[2] The latter term, long associated with the classical stress on the investigation of things popular in early Tokugawa, was reinterpreted later based on the Dutch Learning tradition of the late eighteenth century, when Japanese scholars interested in Western science still used terms from sinology (*kangaku*) to assimilate European natural studies and medicine.[3]

After 1895, Chinese students and scholars adopted the Japanese bifurcation between technical learning and natural studies. Yan Fu, for instance, rendered the terms "science" or "sciences" using Japanese terminology in his 1900–1902 translation of John Stuart Mill's *System of Logic,* while translating "natural philosophy" as "the investigation of things" (*gewu*). Similarly, when regulations for modern schools were promulgated in 1903, the Chinese term for "science" (*gezhi*) referred collectively to the sciences in general, while the Japanese term (*kagaku*) designated the sciences as individual, technical disciplines. This two-track compromise in terminology lasted through the end of the Qing dynasty and continued after the 1911 revolution. Chinese students who returned from abroad increasingly emphasized a single, modern Japanese term, *kagaku,* for the Western sciences, abandoning the earlier accommodation between traditional Chinese natural studies and modern science.[4]

As traditional accommodational terminology receded from use, so too the multiple identity of science and technology as native and Western disappeared. "Modern science" now simply meant "Western science." When the Qing court launched its New Governance policies in 1901, however, the latter were not called a Westernizing policy. Just as the early Qing court had refused to call the Jesuit calendar Western, so now the late Qing court also called its reform policies new rather than Western.

Although hundreds of Protestant translations printed from 1865 to 1900 had delineated the unique nature of modern science, many conservatives in Qing official circles still asserted the strategic myth that all Western learning could be traced back to ancient China. Late Qing reformers, such as Chen Chi (1855–1900), Zheng Guanying (1842–1923), and Tang Zhen (fl. ca. 1896), also argued that new political institutions and conceptions from the West were rooted in classical sources. In this vein, Kang Youwei could also argue that his recovery of New Text classical learning affirmed that Confucius's political ideals were compatible with modern republics and nation-states.[5]

During the seventeenth and eighteenth centuries, traditionalistic rhetoric was plausible because it legitimated the recovery of early Chinese mathematics and ancient learning. In the late nineteenth century, Wylie's and Fryer's introduction of the calculus to China had shocked many literati mathematicians, such as Li Shanlan and Xu Shou. Until 1895, however, the success of a new development such as the calculus depended on its political packaging as ancient Chinese mathematics. After the Sino-Japanese War, which further ate away at Chinese self-confidence about their native traditions and institutions, such packaging more and more appeared as part of the problem rather than a necessary compromise.[6]

Both Liang Qichao and Zhang Zhidong, for instance, quickly changed the title of their earlier accounts of Western learning (*Xixue*) to new learning (*xinxue*). As in the late Ming, the binary opposition between native and European learning was transformed in favor of the dichotomy between old and new learning. Lacking the notion of Chinese origins, however, this move left the Qing state and its conservative literati bereft of a rationale for encouraging study of ancient learning as a complement for Western learning. Native learning now was simply ancient learning, with no ties to the new learning from the West.[7]

Science and the 1898 Reformers

Westernizing elites who played important roles in the 1898 Reform Movement, such as Kang Youwei, Tan Sitong, and Liang Qichao, enunciated a more-assertive approach to modern science. Chen Chi, a secretary in the Bureau of Revenue and a reform advocate, believed that scientific knowledge was a prerequisite for economic productivity in both industry and agriculture. To anticipate the objections of political and cultural conservatives, Chen still couched his arguments in favor of science and technology within "Chinese origins" rhetoric. Kang Youwei, like Chen Chi, believed that the military successes of Meiji Japan served as a model for China. Expanded education in the sciences and industry was required.[8]

In his 1905 essay on industrialization, for instance, Kang Youwei emphasized that China like Japan needed to master Western forms of mining, industry, and commerce. Because machines had augmented the power of European states and enhanced the welfare of the people, Kang contended that the Qing dynasty had to change its goals. He advocated educating the people in technology and not just building factories and arsenals based on foreign models. Often, Kang and the reformers misunderstood and demeaned the results achieved when the foreign affairs movement had promoted industrialization from 1865–1895.

In the post-1895 political environment, reformers could claim they were championing unprecedented policies, when in fact their calls for science and industry built on the efforts of their predecessors. Where Kang Youwei and others did break new ground, however, was in their demand that traditional, subsistence agriculture should be mechanized. A new industrial-commercial society was the goal. In this endeavor, the reformers were as impractical as the self-strengtheners had been a generation earlier. Kang's focus on an educational transformation that would increase the numbers of those trained to industrialize China through science and technology was on target, however. By 1905, the dynasty abrogated the old civil examinations, and most literati now faced a new world of career expectations that drew them away from the classical curriculum that had remained prestigious until 1904.[9]

In particular, Kang was influenced by late Qing translations of Western political economy, such as Joseph Edkins's *Policies for Enriching the Dynasty and Nourishing the People,* which was included in the *Primers for Science Studies* published in Beijing and later reprinted in Shanghai. The missionary Timothy Richard (1845–1919), who unlike Martin and Fryer had confidence in the Qing reforms underway since the Sino-Japanese War, also influenced Kang. In 1895, Richard published the influential essay "New Policies" (*Xin zhengce*) in the *Review of the Times,* which had received as much attention as had Young J. Allen's account of the Chinese defeat by Japan. Richard also prepared a series of forward-looking essays on policy matters, which the reformers published in Shanghai under the title *Tracts for the Times* (*Shishi xinlun,* lit., "New views of contemporary affairs"). Kang Youwei and Liang Qichao consulted with Richard and drew on his essays for inspiration during the 1898 Reforms.[10]

Kang was influenced by the science translations from the Jiangnan Arsenal when he visited Shanghai in 1882. As a result Kang established mathematics as part of the curriculum in his Guangzhou academy, where in the early 1890s he tried to apply its geometrical axioms to the political and philosophical views in his *Complete Book of True Principles and Public Laws* (*Shili gongfa quanshu*). Straying from mathematics, Kang Youwei contended that primal *qi* was the creator of Heaven and earth. He also believed that *qi* was a sort of pervasive ether that lent spiritual form to both electricity and lightning, an interesting trend we will discuss below.

Likewise, Tan Sitong was an amateur mathematician and advocate of science when he established the Liuyang Mathematical Academy in 1895, one of the first such academies, with the help of his teacher Ouyang Zhonggu. A typical reformer, Tan believed that mathematics was the foundation for both science and technology. Tan also read both traditional and Western mathematical works and became interested in geometry and algebra. In his major published work, *Studies of Benevolence* (*Renxue*), for example, Tan relied on mathematics as an authority more than as a tool in his writings to reveal the unity of all learning. Moreover, like Kang Youwei, he regarded ether as the "element of elements."[11]

When Tan traveled to Beijing in 1893, he had his first contact with Westerners. When he saw Fryer in Beijing in 1896, Fryer introduced him to fossils, adding machines, the X-ray, and a device that purported to measure brain waves. Tan was most impressed with Fryer's translation of a work on psychology by Henry Woods (1834–1909) titled *Ideal Suggestion Through Mental Photography* (*Zhixin mianbing fa*, lit., "Method of Avoiding Illness by Controlling the Mind"), which later informed Tan's stress on the dynamics of mental power as an example of the power of benevolence. In *Studies of Benevolence*, Tan declared, "Ether and electricity are simply means whose names are borrowed to explain mental power." At first sight, such pronouncements appear as airy, metaphysical claims out of touch with the tenor of modern science.

Tan's views of the ether were drawn from trends in physics before 1905, when electromagnetic fields were better understood. Before then, physicists explained the motion of electricity and magnetism in terms of mechanical motion within a medium of ether. Both William Thomson (Lord Kelvin, 1824–1907) and James Maxwell (1831–1879), for example, thought that ether was the key to a physical theory that explained electromagnetic phenomena. In addition, literati fiction writers appropriated the translations of modern science to present a technological utopia in late Qing science fiction novels.[12]

Tan Sitong and the Ether

Tan's use of the ether (*yitai*) began in an essay published in 1897, in which he unified the materialistic framework for the chemical elements by making ether the origin of all the elements. Thomson had theorized that ether might turn out to be the universal substratum for all physical phenomena. Drawing on the perennial Chinese notion of *qi* as an undifferentiated, primal stuff from which all things derived, Tan redefined "ether" as an unchanging essence behind all external phenomenal forms, which connected the physical, mental, and spiritual realms through the action of benevolence. Through its principal property of mutuality-interpenetration (*tong*), benevolence operated through mental power (*xinli*). The interpenetration of ether and mental power made benevolence manifest.[13]

Tan Sitong's understanding of Western science drew on notions of the ether and electricity held by eminent Western scientists at the time, which had been presented first in Carl Kreyer's and Zhao Yuanyi's 1876 translation of *Optics* (*Guangxue*) for the Jiangnan Arsenal. Moreover, Tan's use of the concept of ether in his *Studies of Benevolence* maintained the traditional Chinese notion of a continuum between the physical and spiritual realms, which had troubled the Jesuits in China.[14]

Aristotle had complemented the four elements with ether (= *aither*, *quinta essentia*) as a fifth substance to account for movement and change in the sublunar world. The Stoics had used *pneuma* to account for the coherence of matter and the interaction of all parts of the cosmos. In early modern Europe, Descartes postulated three types of corpuscles, with air or ether as the third, within which the planets swirled through vortices. Newton used the concept of ether to explain the action of gravity at a distance. By 1745, all significant British electricians postulated a special electrical medium analogous to the universal Newtonian ether (aether).[15]

In the nineteenth century, the wave theory of light, the discovery of electromagnetic fields, and the propagation of electromagnetic waves revived theories of the ether to explain theoretical problems in physics. Since 1854, William Thomson and others in Britain were searching for a consistent physical theory of ether and matter that would explain the continuum of matter through space. Earlier the notion that ether was a carrier of waves of light and radiant heat had been developed, but its electrical, magnetic, and thermodynamic characteristics were not understood.

In 1867, Lord Kelvin regarded atoms as vortexes in a plenum-filling ether. Dmitrii Mendeleev (1834–1907) saw ether as a real gas and eventually included it in a revised version of his periodic table as late as 1902. James Maxwell, whose 1860s kinetic theory of gases enabled him to explain thermodynamics, treated the magnetic field as rotating vortex tubes in the ether, which he linked to electricity in his 1873 *Treatise on Electricity and Magnetism*. John Thomson (1856–1940) postulated in 1883 that atoms consisted of interconnected vortex rings in stable structures within an ether continuum, while James Jeans (1877–1946) explained radioactivity in 1905 as a rearrangement of ether structure. The 1880s experiments of Albert Michelson (1852–1931) and Edward Morley using an interferometer had already shown that ether was undetectable scientifically. When Albert Einstein (1879–1955) formulated and completed his ether-free relativity theories between 1905 and 1915, such notions became irrelevant.[16]

Accordingly, Tan's appropriation of the ether paralleled attempts by European physicists to affirm ether as the fundamental unity of spiritual and material phenomena. Balfour Stewart, whose English primers on physics had been translated by Young J. Allen (1875) and Joseph Edkins (1886) into Chinese, had claimed as a coauthor of an 1875 volume titled *The Unseen Universe or Speculations on a Future State* (New York: Macmillan) that the human personality survived after death in a parallel universe. Similarly, Maxwell and his supporters believed that ether was not reducible to any known substance. Moreover, the attractiveness of idealist thought in nineteenth-century Europe among those who sought to redirect the mechanistic physics of atomists toward a spiritually informed world, made it possible for the concept of ether to appeal to both specialists and nonspecialists. In China, as Tan Sitong and Kang Youwei drew on this view of ether as an active medium in the universe and molded its content in light of Chinese natural philosophy.[17]

In addition, Henry Woods's work on psychology suggested that ether, unlike ordinary materials, transmitted not only magnetism, electricity, and heat, but also nervous impulses. From this plausible connection of ether to the transmission of nervous impulses through space, Tan derived his notions of the interpenetration of ether and mental power. This link was enhanced when he read Fryer's translation of Woods's *Ideal Suggestion Through Mental Photography*, which although it did not mention ether, promoted the curative powers of electricity.

Using the scientific terminology he derived from Western translations produced by Fryer at the Jiangnan Arsenal and promoted by the Society for Diffusion of Useful Knowledge, Tan presented scientific terms such as "atom" (*zhidian*) and "element" (*yuanzhi*) in his essays. Tan also used these translations to ground his notion of benevolence metaphysically in the ether. Fryer's notion of the ether came from Woods's work, which he translated in 1896 to describe the space vacuums filled by ether. Fryer associated *qi* with ether to appeal to Chinese literati.[18]

Reformist works appropriated the methods, logic, and nomenclature of science. Tan's *Studies of Benevolence,* for example, opened with a deductive system of twenty-seven definitions of benevolence and other terms, which echoed Kang Youwei's use of axioms to present his political and philosophical views. Both Tan and Kang took modern science seriously as a theoretical enterprise, but they used its concepts and methodology very loosely to reconcile Western science with conventional wisdom and Buddhist notions of production and destruction. In addition, this strategy allowed Tan to affirm the "Chinese origins" theory, while at the same time he advocated the study of modern science to enable China to catch up with the West. Neither Tan

Sitong nor Kang Youwei was well educated in the sciences, however. They were publicists, classical scholars, and sometime officials, never practitioners of science.[19]

From Traditional to Modern Mathematics

Since 1865, Li Shanlan and Hua Hengfang had presented Chinese students with a Sino-Western amalgam of traditional Chinese mathematics and modern mathematics, which in the minds of Chinese administrators at the many arsenal schools represented a hybrid of two traditions. This merging of Chinese and Western mathematics was usually overlooked by Western teachers and translators who—except for Alexander Wylie—looked down on such traditionalistic impulses. Such accommodations are usually mentioned without comment by Western historians of the arsenals and schools.[20]

Mathematicians such as Hua Hengfang quickly rejected the Chinese origins theory when they realized that Western mathematics had evolved independently of traditional methods. Ding Fubao most strongly attacked the claims of Chinese origins. His 1899 annotated catalog of works on traditional mathematics rejected all the claims that had recently appeared. At the same time, however, literati mathematicians tried to explain the convergence of Chinese and Western mathematics despite their separate origins. For this, they used a universalist argument borrowed from classical learning to the effect that "the same mind produces the same principles" (*tongxin tongli*) to explain Western developments.[21]

Other mathematicians such as Zou Zunxian still taught Chinese methods in mathematics until the 1901 civil service examination reforms forced them to use Western mathematics for teaching problem solving. After the reforms added the new fields of foreign arts and sciences, the questions in mathematics presumed knowledge of Western mathematics. Students thus needed training in the new field. Zou's 1904 textbook, *Applying Algebra to Various Types of Problems* (*Fenlei yandai*), was prepared for an accelerated course in mathematics to fill this need, but at the same time he merged Chinese and Western mathematics. Hence, his text presented Chinese solutions first so the students would be aware of traditional methods before using modern algebra to solve traditional problems.

At the same time, however, Zou refrained from insisting on the pedagogical priority of traditional mathematics. By 1905, algebra and the calculus had replaced traditional mathematics. Since the late 1890s, the annual examinations on mathematics held at the Beijing School of Foreign Languages no longer included questions on traditional equations (*kaifang*) or four-unknowns procedures. After 1868, Japanese mathematicians had been forced to teach Western mathematics, but until 1900 Chinese were able to carry out research using traditional Chinese mathematics.[22]

When the Qing government promulgated New Governance policies for an educational system composed of primary, middle, and high schools in 1902, the reformed curriculum prioritized seven fields of learning: politics, literature, science, agriculture, industry, commerce, and medicine. Science was subdivided into six areas: astronomy, geology, arithmetic, chemistry, physics, and zoology-botany. These were fields mostly in name, however. Few teachers were available until later to teach such specialty subjects.[23]

In 1905, when even the civil examinations were eliminated, further regulations put into place a new curriculum and textbooks for the schools that were established. Via the reform of the education system, China after 1905 was fully converted to Western mathematics. The Qing government rearranged curricula according to the four Western school levels for mathematics courses:

1. Junior primary school: four arithmetical operations and decimals
2. Senior primary school: fractions, ratios, areas, and volumes
3. Middle school: algebra, geometry, and trigonometry
4. High school: analytical geometry and calculus

This changeover entailed the westernization of mathematics textbooks and Western formats for numbers (1, 2, 3), known quantities (*a*, *b*, *c*), unknown quantities (*x*, *y*, *z*), and derivatives (dy/dx) and integrals ($\int y dx$).[24]

Du Yaquan, Ding Wenjiang (1887–1936), Cai Yuanpei (1868–1940) and others, quickly left behind the typical classical education of a literatus. Yan Fu, whose poor prospects in the civil examinations provoked him to enter the School of Navigation of the Fuzhou Shipyard in 1866, associated the power of the West with modern schools where students were trained in modern subjects requiring practical training in the sciences and technology. For Yan Fu and the post-1895 reformers, Western schools and Westernized Japanese education were examples China should emulate. The extension of mass schooling within a standardized classroom system stressing science courses seemed to promise a way out of the quagmire of the imperial education and civil examination regime, whose educational efficiency was suspect in the 1890s. By 1911, middle and high schools were required by the Ministry of Education to evaluate students in ten areas of instruction:

1. Philosophy
2. Chinese Literature
3. World Literature

4. Art and Music
5. History and Government
6. Mathematics and Astronomy
7. Physics and Chemistry
8. Animal and Plant Biology
9. Geography and Geology
10. Sports and Crafts[25]

Overall, until 1923 the new educational system mandated required courses in five areas of specialization:

1. Language
2. Social Sciences
3. Natural Sciences
4. Mathematics
5. Engineering

Among those affected by the educational changes, Ren Hongjun (1886–1961) was one of the founders of the Science Society of China (*Zhongguo kexue she*) in 1914. He had passed the last county civil examinations in 1904, and by 1907 he was a student in Shanghai, where he met the future iconoclast Hu Shi (1891–1962). Ren then traveled to Japan in 1908 and in 1909 entered the Higher Technical College of Tokyo as a student subsidized by the Qing government. The Qing dynasty had reached a fifteen-year agreement with the College to send forty students annually to Tokyo.[26]

While in Tokyo, Ren also joined Sun Yat-sen's early partisans in the Alliance Society (*Tongmeng hui*) and rose to an important position in the Tokyo Sichuan branch. When he returned to China after the 1911 revolution, Ren served in Sun's provisional government. He received a Qinghua University fellowship in 1912 to study chemistry at Cornell. Ren completed his B.A. in chemistry from Cornell, where he studied from 1912 to 1916, and his M.A. in chemistry from Columbia in 1917, during which time he assumed a leading role in forming a Chinese science organization that would replace the old-style literary societies.[27]

Modern Medicine in China

Those trained in modern, Western medicine derided classical Chinese medicine, which was the largest field of the Chinese sciences during the transition from the late Qing to the Republican era, 1895–1911. Traditional physicians were more successful in retaining their prestige than Chinese astronomers, geomancers, and alchemists, who were dismissed by most modern scholars for practicing superstitious forms of knowledge. Since Xu Shou had criticized the traditional concepts used in traditional Chinese medicine in an 1874 article, Chinese scholars increasingly called for a cosmopolitan synthesis of Western experimental procedures with traditional Chinese medicine.

In 1884, for instance, Tang Zonghai addressed what he considered the dismal state of Chinese medicine in his *Convergence of the Essential Meaning of Chinese and Western Medicine to Explain the Classics* (*Zhongxi huitong yijing jingyi*, Shanghai, 1884, 1892). Later, in about 1890, Yu Yue prepared the first overall attack on Chinese medicine, titled *On Abolishing Chinese Medicine* (*Fei yi lun*), which was perhaps prompted by the deaths of his wife and children due to illness. Using evidential research methods, Yu concluded that the oldest Chinese materia medica was valueless, and he contended that there were no essential differences between popular priests and physicians. Despite his support for Western medicine, however, Yu Yue still critiqued Western science as derivative.[28]

Chinese physicians of traditional medicine, however, remained very large in numbers and very influential despite the inroads of missionary physicians, Western hospitals, and the success of anatomy in mapping the internal venues for bodily illnesses. Coexisting for several decades since 1850, Western-style doctors and Chinese physicians had remarked upon the limitations in each other's theories of illness and therapeutic practices, but for the most part each was practiced in its own institutional matrix of care-giving traditions. Moreover, until anesthesiology was introduced in the early twentieth century and miracle drugs were discovered in the 1940s, the curative power of Western medicine, especially surgery, remained problematic when compared to the noninvasive pharmacopoeia traditions of Chinese physicians.[29]

One aim of the New Governance policies after 1901 was a plan—never realized—to increase Qing state involvement in policing public health. Unlike the Song dynasties, when the government created local medical bureaus to deal with epidemics, state involvement in local health issues since the late Ming markedly diminished. Local elites filled the vacuum by making charitable contributions to deal with health emergencies at a time when literati also took an increased interest in medical knowledge. Late Qing public health policies signified an intention to break with this long-term secular decline of dynastic intervention in local affairs, a devolution that abandoned medical issues to local gentry and Chinese literati-physicians.

Treaty ports such as Shanghai and Tianjin became venues that linked local elite initiatives to the increasing numbers of foreigners who favored sanitation reform and public medicine. By the end of the nineteenth century, the Qing government increasingly saw its role in statist terms. Using Germany as a model, which Meiji Japan had also emulated, the Qing government began, in the short time left to it, to use quarantine and isolation hospitals to deal with epidemics of infectious disease. During the Ming and Qing, when local physicians had faced southern epidemics they associated with "heat factor" causes, the state was minimally involved.[30]

Qing public health policies accomplished little, however, until an epidemic of plague took some sixty thousand lives from 1910 to 1911 in northeast China. The Qing state turned to Wu Liangde (1879–1960) to bring the epidemic under control. Trained in medicine at Cambridge University, Wu dramatically demonstrated to officials and the public, through substantial immunization and exacting quarantine measures, the superiority of Western medicine. As a result, the slow emergence of the modern Chinese state during the Qing-Republican transition was tied to the extension of Western medicine and the appropriation of Western models for state-run public health systems.[31]

One by-product of government involvement in public health was that Western-style physicians and classical Chinese doctors organized into separate medical associations. They drew the state into the contest for medical legitimacy between them. Hence, the modernizing state was progressively tied to Western medical theories and institutions, while Western-style doctors controlled the new Ministry of Public Health. When the Guomindang-sponsored Health Commission proposed to abolish classical Chinese medicine (*Zhongyi*) in February 1929, however, traditional Chinese doctors immediately responded by calling for a national convention in Shanghai on March 17, 1929, which was supported by a strike of pharmacies and surgeries nationwide. The protest succeeded in having the proposed abolition withdrawn, and the Institute for National Medicine (*Guoyi guan*) was subsequently established. One of its objectives, however, was to reform Chinese medicine along Western lines.[32]

The consequences of increased state involvement in Chinese medical policy after 1901 were significant for both Western and Chinese medicine. After 1929, the government established two parallel institutions, one Western and one Chinese, politically and educationally distinctive. This dichotomy survived both the Guomindang Republic and the Communist People's Republic. The bifurcation also entailed the modernization of traditional Chinese medicine, which Bridie Andrews has called the "reinvention" of classical Chinese medicine in the early decades of the century. Nathan Sivin refers to modern Chinese medicine after 1950 as "Traditional Chinese Medicine" (*Zhongyi*), which the People's Republic endowed with a distinct institutional, educational, and occupational base from Western medicine.[33]

The influence of Western medicine in early Republican China presented a substantial challenge to traditional Chinese doctors. The practice of Western medicine in China was assimilated by individual Chinese doctors in a number of different ways. Some defended traditional Chinese medicine, but they sought to update it with Western findings. Others tried to equate Chinese practices with Western knowledge and equalized their statuses as medical learning. The sinicization of Western pharmacy by Zhang Xichun (1860–1933), for example, was based on the rich tradition of pharmacopoeia in the Chinese medical tradition. Another influential group associated with the Chinese Medical Association, which stressed Western medicine, criticized traditional Chinese medical theories as erroneous because they were not scientifically based.[34]

In this cultural encounter, Chinese practitioners such as Cheng Dan'an (1899–1957) modernized techniques like acupuncture. Cheng's research enabled him to follow Japanese reforms by using Western anatomy to redefine the location of the needle entry points. His redefinitions of acupuncture thus revived what had become from his perspective a moribund field that was rarely practiced in China and, when used, also served as a procedure for bloodletting. Indeed, some have argued that acupuncture may have originally evolved from bloodletting. In this manner, acupuncture survived as an artisanal procedure among the populace.[35]

This Western reform of acupuncture, which included replacing traditional coarse needles with the filiform metal needles in use today (see Figure 11.1), ensured that the body points for inserting needles were no longer placed near major blood vessels. Instead, Cheng Dan'an associated the points with the Western mapping of the nervous system. A new scientific acupuncture sponsored by Chinese research societies emerged alongside traditional acupuncture. The former presented a better map of the human body that would enhance diagnosis of its vital and dynamic aspects.[36]

Similarly, the Chinese assimilated the discourse of nerves and the theory of germ contamination from Western medicine. These new views provided ways for Chinese physicians to discuss older illnesses such as leprosy, depletion disorder, or the wasting sickness. As a description of debilitated nerves, sexual neurasthenia now explained the illness that Chinese physicians associated with the depletion of the body's vital essences of *qi*. Multiple interpretations of germ theory enabled the Chinese to equate the attack of tuberculosis germs as a contingent, external cause, which was brought on by the susceptibility of a weakened body whose natural vitality had wasted away.[37]

Influence of Meiji Japan on Modern Science in China

In the late nineteenth century, an increasing familiarity with Western learning exposed the Chinese to the limits of traditional categories for scientific terminology. Increasingly, the claim that Western learning derived from ancient China was unacceptable. Younger literati perceived in the revival of tradi-

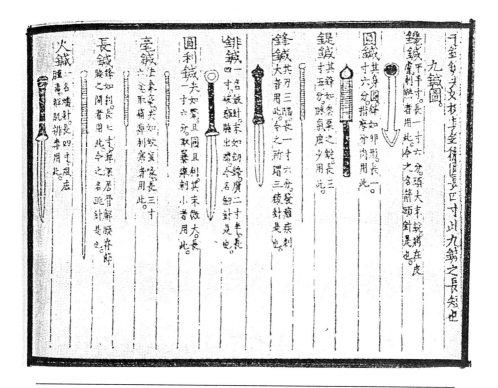

Figure 11.1. Traditional coarse needles and the filiform metal needles used in premodern (on this page) and modern (on next page) acupuncture.
Source: Zhenjiu daquan, 1601.

tional positions after the Sino-Japanese War, which represented the third stage of the Chinese origins argument, a latent conservatism that obstructed the introduction of modern science and technology rather than facilitating it. Hence, those students who studied abroad after 1895 began to question the use of investigating things and extending knowledge (*gezhi*) as a traditional trope of learning to accommodate modern science.

Instead, many turned to Japanese terminology for the modern sciences to make a complete break with the Chinese past. The Japanese neologism *kagaku* (pronounced *"kexue"* in Chinese, lit., "knowledge classified by field"), for example, was perceived as a less loaded term for science than "investigating things," which had so many semantic links to classical learning and the Cheng-Zhu orthodoxy still in place as the curriculum for the civil service examinations until 1905. By 1903, state and private schools increasingly borrowed

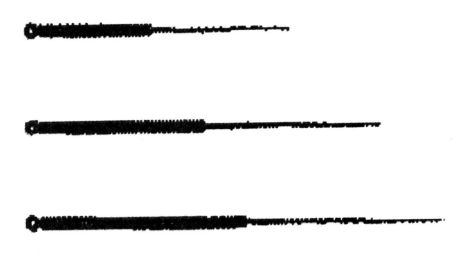

Figure 11.1. The filiform needles are widely used at present in clinics. They are made of gold, silver, alloy, etc., but most of them are made of stainless steel. A filiform needle may be divided into five parts.

from Japanese translations to enunciate the modern classifications of the social sciences (*shehui kexue*), natural sciences (*ziran kexue*), and applied sciences (*yingyong kexue*).[38]

The Impact of Science Translations in Qing China on Japan

Before 1894, Japan had imported many European books on science from Qing China, particularly after 1720 when the shogun Yoshimune relaxed the Tokugawa prohibition of all books related to Christianity. Many had been translated during the Ming and Qing after the Japanese expelled the Jesuits for their meddling in the late sixteenth-century civil wars there. Ricci's mappa mundi, Chinese translations of Euclid's geometry, and Tychonic astronomy, for example, made their way to Tokugawa Japan.[39]

Works from the late Ming collectanea *Works on Mathematical Astronomy of the Chongzhen Reign* and the Kangxi-era *Compendium of Observational and Computational Astronomy* arrived in Japan via the Ningbo-Nagasaki trade after the 1720s. The Japanese also avidly imported eighteenth-century Chinese terminology for Sino-Western mathematics when the second edition of Mei Wending's complete works were imported in 1726 and translated two years later. Physics, chemistry, and botany books, imported from Europe via the Dutch trading enclave in Nagasaki harbor in the early nineteenth century, were also translated into Japanese from Dutch.[40]

In addition, the translations on science prepared under the auspices of Protestant missionaries such as Macgowen and Hobson in the treaty ports were immediately coveted by the Meiji government. Prominent translations into Chinese of works dealing with symbolic algebra, calculus, Newtonian mechanics, and modern astronomy quickly led to Japanese editions and Japanese translations of these works. Macgowen's 1851 *Philosophical Almanac* and Hobson's 1855 *Treatise of Natural Philosophy* came out in Japan in the late 1850s and early 1860s. Four of Hobson's other medical works from 1851–1858 came out in Japan between 1858 and 1864.[41]

Issues from the 1850s *Shanghae Serial* (*Liuhe congtan*) published by Inkstone Press were also republished in Japan, along with the translations Fryer and others completed in the Jiangnan Arsenal and the publications from the Beijing School of Foreign Languages. The Wylie-Li translations of algebra (1859/1872), calculus (1859/1872), and Martin's *Natural Philosophy* (1867/1869) were all quickly available to scholars and officials in Meiji Japan. Arguably, these works had greater influence in Japan than in China, and today they can still be readily located in Japanese libraries, while they are rare in China.[42]

Many Japanese scholars still preferred Chinese scientific terms in early Meiji times over translations derived from Dutch Learning. The Chinese name for chemistry (*huaxue*), for example, replaced the term *chemie* [*semi* in Japanese] derived from Dutch. Similarly, the impact of Jiangnan Arsenal publications can be seen in the choice of Chinese terminology for metallurgy (*jinshi xue*) used in Japanese publications, which were later changed in Japan and reintroduced to China as a new term for mining (*kuangwu xue*).

Japan's Iwakura mission also visited Shanghai in September 1873 at the end of its journey to Europe and the United States and took a tour of the Jiangnan Arsenal on September 4. The report described the shipyard, foundry, school, and translation bureau there in very positive terms. The mission noted how the shipyard was operated by British managers initially. The latter were aided by Chinese who had trained abroad. The account added that "now the entire management of the yard is in the hands of Chinese" and concluded, "This one yard would be capable of carrying out any kind of work, from ship repair to ship construction."[43]

When the diplomat Yanagihara Sakimitsu (1850–1894) visited China, he purchased many of the Chinese scientific translations. On his third visit in 1872, for instance, he bought twelve titles on science and technology in thirty-one volumes from the Jiangnan Arsenal. These included works on chemistry, ship technology, geography, traditional mathematics, mining, and Chinese trigonometry (*gougu*). The Japanese government continued to buy arsenal books until 1877. In 1874, Yanagihara received twenty-one newly translated books from China. Despite the influence of Dutch learning and translations from China, and even though the Japanese began teaching modern Western science on a large scale in the 1870s, the Chinese did not borrow many scientific terms from Japan before the Sino-Japanese War.

Unlike the Chinese translations that were readily transmitted to Japan, Tokugawa authorities kept Dutch Learning translations secret. While much has been made of the contributions of Dutch Learning to Japanese science during the Tokugawa period, we have seen in Chapter 10 that the Yokosuka Dockyard was still dependant on French engineering advisors until the 1880s and British technical aid in the 1890s. There is no evidence that Dutch Learning per se enhanced the Yokosuka enterprise or determined the course of Meiji science and technology. Moreover, the impact of Dutch Learning, while important among samurai elites in the late eighteenth and early nineteenth centuries, was not sufficient to touch off in Tokugawa Japan the sort of technological revolution based on Newtonian mechanics and French analytical mathematics that we have described as the engineer's tool kit in Chapter 4.

Indeed, the concrete advantages that Dutch Learning provided in the rise of modern, industrial science during the Tokugawa-Meiji transition remain undocumented. Japan's overwhelming triumph in the Sino-Japanese War created an environment in which most accounts since 1895 have simply assumed that Dutch Learning gave Tokugawa Japan a scientific head start over the Qing dynasty.[44]

Japanese Science in China after 1895

From 1896 to 1910, the Chinese translated science books that the Japanese no longer worked with foreigners to translate. By 1905, the new Qing Ministry of Education was staunchly in favor of science education and textbooks based on the Japanese scientific system. Instead of the West represented by Protestant missionaries such as Martin and Fryer, Japan now mediated the West for Chinese literati and officials.[45]

After the Sino-Japanese War, reformers encouraged Chinese students to study in Japan. Kang Youwei promoted Meiji Japan scholarship in his *Annotated Bibliography of Japanese Books* (*Riben shumu zhi*) and in his reform memorials to the Guangxu emperor. He recommended 339 works in medicine and 380 works in the sciences (*lixue*), which now replaced the prize essays for the 1894 Shanghai Polytechnic essay competition that listed the best Western books. The Guangxu emperor's (r. 1875–1908) edict of 1898 encouraged study in Japan.[46]

As a publicist while in exile in Japan, Liang translated Japanese materials into Chinese at a fast clip. In addition to his antiquarian interests, Luo Zhenyu (1866–1940), for example, published the *Agricultural Journal* (*Nongxue bao*) from 1897 to 1906 in 315 issues. The articles were mainly drawn from Japanese sources on science and technology. Luo also compiled the *Collectanea of Agricultural Studies* (*Nongxue congshu*) in eighty-eight works, with forty-eight based on Japanese books. Du Yaquan edited journals in 1900 and 1901 that translated science materials from Japanese journals. These were the first science journals edited solely by a Chinese. The massive translation by Fan Diji in Shanghai of a Japanese encyclopedia took several years. When it appeared in 1904, the encyclopedia contained over one hundred works with twenty-eight in the sciences and nineteen in applied science.

Post-Boxer educational reforms of 1902–1904 were also crucial in the transformation of education in favor of Japanese-style science and technology. The last bastion of modern science as Chinese science (*gezhi*) remained in for the civil examinations, where the Chinese-origins approach to Western learning remained obligatory. After the examination system was abolished in 1904, Japanese science texts finally became models for Chinese education at all levels of schooling. In 1886–1901, for instance, Japan officially approved eleven different texts on physics. Eight of those, which were produced after 1897, were translated for Chinese editions. In 1902–1911, twenty-two different physics texts were approved in Japan, and seven were translated into Chinese.

Similarly, in chemistry from 1902 to 1911, seventy-one Japanese texts were translated into Chinese. Most were produced for middle schools and teacher's colleges. Twelve middle school chemistry texts were produced in Japan between 1886 and 1901. Of these, six were translated into Chinese. Eighteen Japanese middle school chemistry texts were produced between 1902 and 1911. Five were translated into Chinese. Japanese scientists were also invited to lecture in China. The Chinese also translated more technical physics and chemistry works from the Japanese. Translators completed Iimori Teizō's (1851–1916) edited volume on *Physics* (*Wulixue, Butsurigaku*) in Chinese at the Jiangnan Arsenal from 1900–1903. They were aided by the Japanese educator Fujita Toyohachi (1870–1929). Iimori's influence on Chinese physics grew out of this project.[47]

The Chinese also compiled updated Sino-Japanese dictionaries such as the 1903 *New Progress Toward Elegance* (*Xin Erya*), which modernized ancient Chinese lexicons. By 1907, when Yan Fu was in charge of the Qing Ministry of Education's committee for science textbooks, he approved the use of Japanese scientific terms. We should not underrate the historical importance of Japanese translations for the development of modern science in China. Japanese translations were much more widely available in China than those produced earlier by the Jiangnan Arsenal had been. In addition, the new Japanese science textbooks contained more up-to-date content than the 1880s arsenal and missionary translations, which were already antiquated by European standards in the 1890s. The introduction of post-1900 science via Japan, which included new developments in chemistry and physics, went well beyond what Fryer and others had provided to the emerging Chinese scientific community.[48]

Chinese presses also published in greater numbers the translations of Japanese texts, which were easier to read because only Chinese compiled them. Moreover, the quality of the translations from works by Japanese scientists improved over the earlier *Science Primers*. Chinese translators themselves could understand the Japanese originals. In addition, the Japanese texts were available to a new and wider audience of students in the new public schools and teacher's colleges that the Qing government established after 1905 as part of its education reforms. The Imperial University in Beijing also invited Japanese professors to join its faculty.[49]

Finally, to make the new translations more easily understood than standard classical translations, Chinese translators helped produce a new literary form for presentation of the sciences, which contributed to the rise of the vernacular for modern Chinese scholarly and public discourse. Among urbanites, especially in Beijing and Shanghai, the first decade of the twentieth century provided the basic education in modern science via Japanese textbooks for the generation that matured during the New Culture Movement of 1915 and the May Fourth era after 1919.[50]

The Delayed Emergence of Physics as a Technical Field in China

When we compare the development of modern physics in Meiji Japan and Qing China, we find that scholars in both countries had started to master Western studies in the early and mid-nineteenth century. The Translation Bureau at the Jiangnan Arsenal and the Dutch Translation Bureau in Tokugawa Japan had produced Chinese books on physics beginning in the 1850s in China and in Japanese from 1811 in Japan. Although the introduction of Dutch Learning in the seventeenth and eighteenth centuries enabled an earlier start in Japan, the materials on physics in the Protestant translations produced in China after 1850—quickly transmitted to Japan—made those earlier studies out of date. Moreover, the *Primer Series* produced in the 1870s and early 1880s in China remained superior overall to their Meiji counterparts until the 1890s.[51]

Despite the range of science translations in Qing China through the 1880s, physics textbooks were not available in China until they were first published in Japan. Much of this had to do with the way the Protestant missionaries such as Martin and Fryer had introduced the physical sciences to literati audiences since 1860. Rather than a unified field of physics, or natural philosophy as it was often called by Euro-American specialists until the 1860s, missionary translators first introduced the disaggregated branches of physics. Accordingly, mechanics (*lixue* or *zhongxue*), optics (*guangxue*), acoustics (*shengxue*), electricity (*dianxue*), and thermodynamics (*rexue*) were presented as independent fields in China. By presenting the subfields of physics independently, the translators made it difficult for the Chinese later to appreciate the unity of physics. Moreover, introducing the branches first made it more complicated later to reach a consensus for a more general term for physics.

Often physics was equated with investigating things (*gewu*). Others preferred calling physics investigating things and extending knowledge (*gezhi*), which frequently overlapped vaguely with the general term for science and created substantial misunderstanding. Edkins's 1886 *Science Primers* associated "investigating the materiality of things" (*gezhi zhixue*) with physics. In 1895, the school of physics in the Beijing Foreign Language School changed its name from the Hall for Investigating Things (*Gewu guan*) to the Hall for Investigating and Extending Knowledge (*Gezhi guan*).

Unlike the Japanese, who developed independent translation techniques, the Chinese remained dependent on their Protestant informants into the 1890s. This dependency placed severe limits on what the Chinese alone could translate. Overall, the Western translations prepared by Macgowen, Hobson, and Martin in China dealt with physics in very general, textbook terms and never produced useful handbooks.[52]

The Qing state also was slower in reforming its educational system. Meiji Japan's new educational system was established in 1868. Qing education reforms were not comparable until 1902. A Japanese Ministry of Education (Mombusho) followed in 1871, while its Qing counterpart was not established until 1905. Similarly, Tokyo University was founded as Japan's key modern teaching institution in 1877, but the Imperial University of Beijing did not exist until 1898. Courses in physics had already started in 1875 in Japan when the Tokyo school that evolved into the university shifted from foreign language lectures by Europeans to lectures in Japanese by students who had studied physics abroad. The first Japanese students trained in Japan graduated in physics in 1883.

Chinese science faculties were not established at the Imperial University of Beijing until 1910, but even then only classes in chemistry and geology were taught. Physics was added in 1912. Of 387 students recruited in the sciences, only 54 received diplomas in 1913. Beijing recruited Japanese science teachers to the University since 1902, but they left in 1908–1909 after their six-year contracts expired. From 1898 to 1911, only two hundred students were trained in the sciences at the Imperial University, and the initial absence of faculties of mathematics and physics remained a serious problem in training scientists. We have seen above that the science curriculum was formalized in terms of requirements at the high school level beginning in 1911. In Japan, there were few students of physics when compared to the more popular fields of law and medicine. Between 1882 and 1912, however, Tokyo University graduated 186 in physics.[53]

Japan's educational system had a head start in editing and translating physics textbooks. China by comparison lacked textbook materials to teach physics at all levels of the education system. Similar delays occurred in other technical fields, such as chemistry and geology. By 1873, the Japanese taught physics in the new Meiji schools, and Tokyo University had a physics program from 1877. By comparison, the Beijing School of Foreign Languages asked only occasional physics questions on examinations from 1868, which were based on Martin's elementary *Natural Philosophy*. The subfields of physics were taught separately as mechanics, hydraulics, acoustics, pneumatics, heat, optics, and electricity. In addition, the military and arsenal schools also taught some physics, especially its subfields.

Meiji educators produced physics textbooks in the 1870s, but none were available in China until the 1890s. Although the Japanese relied on Protestant translations from China initially, the Mombusho ordered Katayama Junkichi (1837–1887) to compile an official physics textbook when physics (*butsuri*, *wuli*) became a specialized discipline. Katayama's textbook was added to the Japanese curriculum in 1876 and republished many times. Moreover, Japan invited Western scientists to Japan. K. W. Gratama (1831–1888) served in the Chemistry Bureau from 1869. He was succeeded by H. Ritter (d. 1874). Later, Iimori Teizō completed his edition of *Physics* by consulting the works on physics published by the German J. Müller.

In the late 1890s, the Qing recognized the need to translate physics textbooks. As a result of the 1898 reforms, the government decided to copy the Meiji model for education and create a public school system for science education, rather than simply rely on schooling in the arsenals, navy yards, and factories. Full implementation of this program was not feasible until the civil examination system was scrapped, and the new school system replaced it in 1904–1905. The Sino-Japanese War had taught the Qing government that it was insufficient to rely on arsenals to modernize.[54]

Because there were few science textbooks in China and none that dealt chiefly with physics, the Chinese immediately translated Japanese texts such as Iimori's *Physics*. Direct Chinese translations of the best physics texts by the most famous Japanese physicists became the most efficient means in the early twentieth century to prepare textbooks for the new Qing school system. This policy also guaranteed that the Chinese would no longer rely on Western informants for specialized translations in important fields such as physics. But China's dependency on Japan was reconsidered after 1915 when Japan's policies toward the Republic of China became increasingly predatory.

Although high-level education in physics began at the Beijing Imperial University in 1912, the best-trained physicists studied in the United States and Japan: Li Fuji (b. 1885) studied in the United States; He Yujie (1882–1939), Xia Yuanli (1884–1944), Li Yuebang (1884–1940?), and Hu Gangfu (1892–1966) in Japan. When Beijing University was reorganized in 1912, it had formal divisions between the humanities and the sciences, with the latter including the three fields of mathematics, chemistry, and physics. An independent physics department was not created until 1917, however. The greater scope of physics texts in the school system after 1905, however, did provide for wider knowledge of the field in China than had been the case before 1900.[55]

Japan also had a lead over China in research in physics, the unification of technical terminology, and research associations by 1900. For instance, Japanese scholars started publishing in physics in 1880s. More than two hundred articles in the various subfields of physics had appeared by the end of Meiji era in 1912. Moreover, several Japanese physicists had emerged who were approaching Western levels of expertise in physics.

The terms for physics were first unified in the 1870s when translators chose the official Meiji designation for physics as (*wuli xue*) in 1872. Terminology in Japanese physics achieved a final unification with the 1888 publication of an official list of technical terms with foreign counterparts. The committee for systematizing the translation of terms for physics, which began in 1885, was led by three of the first Japanese graduates in physics from Tokyo University. Scholars unified terms for a total of 1,700 items from English, French, and German, which they then translated into Japanese and published. Chinese started using the Japanese term for physics in 1900 when a Japanese book by that name was published in China. Before then, the term had usually referred to the principles of things as part of the traditional fields of natural studies.[56]

Academics created the first mathematics society in Tokyo in 1877 with fifty-five members. In 1884, ten of its seventy-five members specialized in physics. When the Tokyo Mathematics-Physics Society was formed in 1884, it started with eighty-two members, twenty-five of whom were physicists. The latter changed its name in 1919 to the Japan Mathematics-Physics Society, which survived as an organization until it separated into two parts in 1948. Smaller specialized groups in physics were also formed in Japan in the 1880s.

China was also later than Japan in training physicists and organizing associations. The Chinese had to study physics abroad, and the research institutes for physics at the Academia Sinica, the Beijing Institute, and the Qinghua Institute were not formed until 1928–1929. Although Chinese terms for physics were unified in 1905, they were not finally settled until the 1920s. Moreover, the Chinese Science Society and its journal were not founded until 1915, and that took place abroad in the United States at Cornell University. Physicists did not form the Chinese Physics Society until 1932.[57]

The belief that Western science represented a universal application of objective methods and knowledge was increasingly articulated in the journals associated with the New Culture Movement after 1915. The journal *Science* (*Kexue*), which the newly founded Science Society of China created in 1914, assumed that an educational system based on modern science was the panacea for all of China's ills because of its universal knowledge system. Meiji Japan served as the model for that panacea until 1915, when Japanese imperialism, like its European predecessor, forced Chinese officials, warlords, and intellectuals to reconsider the benefits of copying Japan.[58]

Despite the late Qing curriculum changes described above, which had prioritized science and engineering in the new public schools since 1902 and in private universities such as Qinghua, many Chinese university and overseas students were by 1910 increasingly radical in their political and cultural views, which carried over to their convictions about science. Traditional natural studies became part of the failed history of traditional China to become modern, and this view now asserted that the Chinese had never produced any science. How premodern Chinese had demarcated the natural and the anomalous vanished, when both modernists and socialists in China accepted the West as the universal starting place of all science.[59]

After 1911, many radicals such as Ren Hongjun linked the necessity for Chinese political revolution to the claim that a scientific revolution was also mandatory. Those Chinese who thought a revolution in knowledge required Western learning not only challenged classical learning, or what they now called Confucianism (*Kongjiao*), but they also unstitched the patterns of traditional Chinese natural studies and medicine long accepted as components of imperial orthodoxy.[60]

As Chinese elites turned to Western studies and modern science, fewer remained to continue the traditions of classical learning (Han Learning) or Cheng-Zhu moral philosophy (increasingly called Neo-Confucianism in the twentieth century) that had been the basis for imperial orthodoxy and literati status before 1900. Those who still focused on traditional learning, such as Gu Jiegang (1893–1980) in Beijing and others elsewhere, often did so by reconceptualizing ancient learning in light of "doubting antiquity" and applying new, objective procedures for historiography that they derived from the sciences. Thereafter, the traditional Chinese sciences, classical studies, and Confucianism survived as vestigial native learning in the public schools established by the Ministry of Education after 1905. They have endured as contested scholarly fields taught in the vernacular in universities since 1911.[61]

The Great War from 1914 to 1919 acted as a profound intellectual boundary between those modernists who still saw in science a universal model for the future and the "New Confucian" (*Xinru*) traditionalists, such as Zhang Junmai (Carson Chang, 1886–1969), who showed renewed sympathy for distinctly Chinese moral teachings after the devastation visited on Europe. The former reformer and now scholar-publicist Liang Qichao, who was then in Europe leading an unofficial group of Chinese observers at the 1919 Paris Peace Conference, visited a number of European capitals. They witnessed the war's deadly technological impact on Europe. They also met with leading European intellectuals, such as the German philosopher Rudolf Christoph Eucken (1846–1926), Zhang Junmai's teacher, and the French philosopher Henri Bergson (1859–1941), to discuss the moral lessons of the war.[62]

In his influential *Condensed Record of Travel Impressions While in Europe* (*Ouyou xinying lu jielu*), Liang Qichao related how the Europeans they met regarded World War I as a sign of the bankruptcy of the West and the end of the "dream of the omnipotence of modern science." Liang found that Europeans now sympathized with what they considered the more spiritual and peaceful "Eastern civilization" and bemoaned the legacy in Europe of an untrammeled material and scientific social order that had fueled the world war. Liang's account of the spiritual decadence in post-war Europe indicted the materialism and the mechanistic assumptions underlying modern science and technology. A turning point had been reached, and the dark side of "Mr. Science" had been exposed. Behind it lay the colossal ruins produced by Western materialism.[63]

In the early twenty-first century, we tend to forget the degree of skepticism that Joseph Needham's remarkable collectanea, *Science and Civilization in China*, initially provoked five decades ago. The consensus when Needham's first volume appeared in 1954 drew on heroic accounts of the rise of Western science to demonstrate that premodern China had no science. Some accused him of doctrinaire Marxism. Others dismissed the embryologist's foray into the history of Chinese science as a dead end, a project they felt revealed Needham's wishful thinking about premodern China. For some, the Needham project simply reiterated the Chinese origins of Western learning approach.[64]

Many twentieth-century scholars were convinced that premodern China had no industrial revolution and had never produced capitalism. Therefore, they contended, the Chinese could never have produced modern science on their own. While Needham granted that China lacked the capitalist catalyst required for modern science, he did not stop there. Few besides Needham, his collaborators, and Nathan Sivin stopped to consider what the rich archives in Taiwan, China, and Japan might yield if someone bothered to go through them. As the evidence of a rich tradition of natural studies and medicine accrued in volume after unrelenting volume of the *Science and Civilisation in China* project from 1954 to 2000, it became harder and harder to gainsay it all as superstition, irrationality, or inductive luck.

The largest archive of premodern records for the study of nature remains in China. By better understanding the history of imperial Chinese natural studies, technology, and medicine, and the cultural mystifications that undergirded them, we can be more perceptive about ourselves and the mystifications that undergird our contemporary versions of modern science. "Chinese science" has grown in respectability among academics. The romanticized story of European science, whether capitalist or socialist in genre, has slowly unraveled under the onslaught of fifty years of nonpositivist research. Younger scholars now probe the surface of self-satisfied rhetoric about science as fundamentally Western and go beyond simplistic appeals to the Greek deductive logic by armchair philosophers of Western science.[65]

In this volume I have reconsidered the scholarly consensus about the alleged failed history of science in China and the alleged victorious history of science in Europe and Japan. Both histories are pieces of a larger, yet unwritten global narrative of science on the planet.[66] With the exception of a modernized version of traditional Chinese medicine that is now flourishing as one version of holistic medicine, the traditional fields of natural studies in imperial China did not survive the impact of modern science between 1850 and 1920. We have tried to show that earlier accounts of the historical construction of Western science, technology, and medicine in Republican China depended on specious claims that the Chinese sciences failed during the Ming and Qing dynasties.

The Chinese construction of modern science, medicine, and technology on their own terms is a remarkable achievement, even if they did not initiate the internal and external revolutions that provoked that construction. Early modern Europe, after all, borrowed much from Asia and Islam before its own scientific revolutions. To be sure, China's plans to send space expeditions to the moon and Mars in the twenty-first century are in part a response to the

Ming navy that allegedly—and anachronistically—failed to take advantage of its commercial and military opportunities in the early fifteenth century. Withdrawing from the Indian Ocean after 1450 and unintentionally leaving the Pacific to become a site for the rise of Spanish, Portuguese, Dutch, and English naval power certainly is a lesson that contemporary Chinese will not soon forget. The shock of Western and Japanese imperialism in China remains a tragedy for the Chinese, but their accruing triumphs in contemporary science, medicine, and technology should not be placed in a time frame that overlooks the era before 1900. In this century, modern science in China will hopefully work more benevolently than did Euro-American techno-science.

Appendix 1

Tang Mathematical Classics

The Ten Mathematical Classics were manuals used at the Imperial Academy to teach mathematics during the Sui and Tang dynasties.[1]

1. *The Gnomon of the Zhou Dynasty and Classic of Computations* (*Zhoubi suanjing*), compiled in first century B.C. with reworking by commentators in medieval times: commented on by Zhao Shuang (ca. third century A.D.), Zhen Luan (fl. sixth century A.D., ca. 570), who was a calendrical astronomer known for his commitment to Buddhism, and Li Chunfeng (602–670), who edited the text for inclusion in the mathematical classics in 656. Important for mathematical aspects of Chinese astronomy based on "enveloping heavens" (*gaitian*) cosmology; corroborates Pythagoras's right triangle theorem; Zhao Shuang commentary added fifteen algorithms for solving right triangles. Recovered by Dai Zhen from the *Great Compendium of the Yongle Era*.[2]

2. *Computational Methods in Nine Chapters* (*Jiuzhang suanshu*), compiled from 200 B.C. to A.D. 300, with later commentaries added: commented on by Liu Hui (late third century), Xu Yue (ca. early third century), Zhen Luan (fl. ca. 570), Zu Chongzhi (429–500) in sixth century, Li Chunfeng and Yang Hui in late thirteenth century. Its 246 problems set the model for the mathematical language of linear equations (*fangcheng*) and presented the pattern for computations; intended as means to apply right triangle methods (*gougu*) to calculate areas of fields of various shapes. Recovered by Dai Zhen from the *Great Compendium of the Yongle Era*.[3]

3. *Sea Island Computational Canon* (*Haidao suanjing*) by Liu Hui. Contains computational prescriptions for surveying inaccessible points using circle divisions with inscribed figures; applies principle of proportion to right triangles; first of nine problems determined highest point of an island in the sea from which the title was derived. Recovered by Dai Zhen from the *Great Compendium of the Yongle Era*.[4]

4. *Sunzi's Computational Canon* (*Sunzi suanjing*), ca. fifth century: commented on by Li Chunfeng. Gives details for arithmetic operations using counting rods. Recovered by Dai Zhen from the *Great Compendium of the Yongle Era*, although a Southern Song edition also survived.[5]

5. *Computational Canon of the Five Administrative Departments* (*Wucao suanjing*, i.e., agriculture, military, customs, granaries, and finance), compiled ca. fifth cen-

tury and credited to Zhen Luan: commented on by Li Chunfeng. Intended as a textbook on applied mathematics for (1) surveying land, (2) management of troops, (3) tax collection, (4) granaries, and (5) money based on medieval *Five Administrative* system. Added to mathematical classics in the Song period. Recovered by Dai Zhen from the *Great Compendium of the Yongle Era*.[6]

6. *Xiahou Yang's Computational Canon* (*Xiahou Yang suanjing*), n.d., perhaps ca. fourth to eighth centuries. Apocryphal collection of elementary problems of taxes solved through simplified computational techniques using counting rods. Added to mathematical classics in Tang or Song period. Recovered by Dai Zhen from the *Great Compendium of the Yongle Era*. Qian Bacong argues that Dai's version is different from the original book, which was lost by 1084.[7]

7. *Zhang Qiujian's Computational Canon* (*Zhang Qiujian suanjing*), compiled ca. Northern Wei, 466–485: commented on by Zhen Luan, Li Chunfeng, and Liu Xiaosun. Gives solutions for quadratic equations, arithmetic progressions, and an indeterminate problem known in history of mathematics as the "hundred fowl problem." Recovered by Dai Zhen based on the late Ming Mao family edition from the Southern Song dynasty.[8]

8. *Computational Rules of the Five Classics* (*Wujing suanshu*), compiled ca. 566: commented on by Li Chunfeng and Zhen Luan; glossed by Zhen Luan to interpret calendrical data in the Classics, the "great expansion" [method for solving simultaneous congruencies] (*dayan*) method of divination in the *Change Classic*, mythical geography, weights and measures, and musical pitch pipes. Recovered by Dai Zhen from the *Great Compendium of the Yongle Era*.[9]

9. *Computational Canon of the Continuation of Ancient Techniques* (*Jigu suanjing*), by calendricist Wang Xiaotang (ca. 650–750) in the seventh century: commented on by Li Chunfeng. Gives astronomical problems and problems for construction of astronomical tower and embankments associated with long canals where thousands of workers were mobilized. Also gives solution of right triangles leading to polynomial equations of the second or third degree. Recovered by Dai Zhen based on the late Ming Mao family edition from the Southern Song dynasty.[10]

10. *Techniques for Calculations by Combination* (*Zhuishu*) by Zu Chongzhi (429–500): commented on by Li Chunfeng; annotated by Minggatu in Qing.

11. *Notes on the Mathematical Heritage* (*Shushu jiyi*), by Xu Yue (fl. early third century A.D., ca. 220): commented on by Zhen Luan. Likely an apocryphal work containing expressions of large and very large numbers in different number systems. Related to Buddhist infinite cycles of reincarnation. SKQS version presented by the governor-general of Jiangsu and Zhejiang.[11]

12. *Arts of the Three Degrees* (*Sandeng shu*), by Dong Quan (fl. ca. sixth to seventh century A.D.): notation for large numbers; commented on by Zhen Luan. Lost by Song times.

Note: Numbers 11 and 12 were studied at the Sui-Tang Imperial Academy along with numbers 1–10, but were not specified as official parts of the ten "Tang Mathematical Classics."

Appendix 2

Some Translations of Chemistry, 1855–1873

Benjamin Hobson, trans., *Bowu xinbian* (Treatise of natural philosophy). Shanghai: Mohai shuguan, 1855. Canton, 1851.

William Martin, trans., *Gewu rumen* (Elements of natural philosophy and chemistry; lit., "Introduction to the investigation of things"). Beijing: Tongwen guan, 1868.

John Fryer and Xu Shou, trans., *Huaxue jianyuan* (Mirror of the origins of chemistry). Shanghai: Jiangnan zhizaoju, 1871. The translation was completed in 1869. Sequel in 1875; supplement in 1882.

John Kerr and He Liaoran, trans., *Huaxue chujie* (Preliminary steps in chemistry). Guangzhou: Boji yiju, 1870.

Anatole Billequin, *Huaxue zhinan* (Guide to chemistry). Beijing: Tongwen'guan, 1873.

Appendix 3

Science Outline Series, 1882–1898

1. *Shengxue xuzhi* (Acoustics, First Part—1)
2. *Tianwen xuzhi* (Astronomy, First Part—2)
3. *Huaxue xuzhi* (Chemistry, First Part—3)
4. *Dianxue xuezhi* (Electricity and Magnetism, First Part—4)
5. *Dili xuzhi* (Physical Geography, First Part—5)
6. *Dizhi xuzhi* (Political Geography, First Part—6)
7. *Dixue xuzhi* (Geology, First Part—7)
8. *Qixue xuzhi* (Pneumatics, First Part—8)
9. *Daishu xuzhi* (Algebra, Second Part—1)
10. *Suanfa xuzhi* (Arithmetic, Second Part—2)
11. *Weiji xuzhi* (Calculus, Second Part—3)
12. *Quxian xuzhi* (Conic Sections, Second Part—4)
13. *Huaqi xuzhi* (Drawing Instruments, Second Part—5)
14. *Zhongxue xuzhi* (Mechanics, Second Part—6)
15. *Liangfa xuzhi* (Mensuration, Second Part—7)
16. *Sanjiao xuzhi* (Trigonometry, Second Part—8)
17. *Guangxue xuzhi* (Optics, Third Part—1)
18. *Lixue xuzhi* (Dynamics, Third Part—2)
19. *Shuixue xuzhi* (Hydraulics, Third Part)
20. *Kuangxue xuzhi* (Mineralogy, Third Part—5)
21. *Quanti xuzhi* (Physiology and Anatomy, Third Part—7)
22. *Fuguo xuzhi* (Political Economy, Fifth Part)
23. *Xili xuzhi* (Western Etiquette—What to do, Fifth Part—7)
24. *Jieli xuzhi* (Western Etiquette—What to avoid, Fifth Part—8)

Appendix 4

Partial Chronological List of Arsenals, etc., in China, 1861–1892

For an illustration of the locations of arsenals, factories and shipyards during the Self-Strengthening period, see Map 10.2.[1]

1. Anqing Arsenal (1861), set up by Zeng Guofan.
2. Jiangnan Arsenal (1865), set up in Shanghai by Zeng and Li Hongzhang.
3. Jinling Arsenal (1865) in Nanjing used for making breech rifles and steel; gunpowder factory from 1881.
4. Fuzhou Shipyard (1866), the base for the Southern Fleet, established by Zuo Zongtang; machine manufacture from 1870.
5. Tianjin Arsenal (1867), under Li Hongzhang used to manufacture gunpowder and acid.
6. Xi'an Arsenal (1869), used to manufacture bullets and gunpowder.
7. Lanzhou Arsenal (1871–1872), used to manufacture bullets and gunpowder.
8. Guangzhou Arsenal (1874), used to manufacture bullets and repair ships; manufactured gunpowder from 1875.
9. Hunan Arsenal (1875), used to manufacture acid, gunpowder, and cannon.
10. Shandong Arsenal (1875), used for gun purchase, making acid and gunpowder.
11. Sichuan Arsenal (1877), used to manufacture acid, cannon, bullets and gunpowder.
12. Jilin Arsenal (1881), used to manufacture bullets, gunpowder, and acid.
13. Lüshun, Port Arthur Naval Station (1881–1882).
14. Weihaiwei Shipyard (1882) for the Beiyang Fleet.
15. Beijing Field Force Arsenal (1883).
16. Shanxi Machine Shop (1884), used for making foreign gunpowder.
17. Yunnan Arsenal (1884), for making bullets and gunpowder.
18. Hangzhou Arsenal (1885), used for gun purchase and making bullets, ship mines, and gunpowder.
19. Taiwan Machine Shop (1885), used for making bullets and gunpowder.
20. Taiwan Arsenal (1885).
21. Daye Iron Mine (1890), in Hubei.
22. Hanyang Ironworks, in Hubei (1890), established by Zhang Zhidong, used for iron and steel production.
23. Hanyang Arsenal (1892), used for making weaponry.
24. Zhejiang Machine Shop (1893).

Appendix 5

Table of Contents for the 1886 *Primers for Science Studies* (*Gezhi qimeng*)

1. Western learning (*Xixue lueshu*)
2. Introductory for the Science Primers (*Gezhi zongxue*), by Thomas Huxley
3. Political geography (*Dizhi*)
4. Physical geography (*Dili zhixue*)[1]
5. Geology (*Dixue*), by Archibald Geikie
6. Botany (*Zhiwuxue*), by J. D. Hooker
7. Physiology (*Shenli*), by M. Foster
8. Zoology (*Dongwuxue*)
9. Chemistry (*Huaxue*), by Henry Roscoe
10. Physics (*Gezhi zhixue*), by Balfour Stewart
11. Astronomy (*Tianwen*), by J. N. Lockyer
12. Political economy (*Fuguo yangmin ce*, lit., "Policies for enriching the dynasty and nourishing the people"), by W. Stanley Jevons
13. Greek history (*Xila zhilue*)
14. Roman history (*Luoma zhilue*)
15. Logic (*Bianxue*), by W. Stanley Jevons
16. European history (*Ouzhou shilue*)

Appendix 6

Twenty-three Fields of the Sciences in the 1886 *Primers for Science Studies*

1. Astronomy
2. Physics
3. Geology
4. Zoology
5. Mineralogy
6. Electricity and Magnetism
7. Chemistry
8. Climatology
9. Light
10. Mechanics
11. Fluid Mechanics
12. Gaseous Mechanics
13. Anatomy
14. Comparative Anatomy
15. Physiology
16. Botany
17. Medicine
18. Geometry
19. Mathematics
20. Algebra
21. Calendrics
22. Archaeology
23. Folklore

Appendix 7

Science Compendia Published in China from 1877 to 1903

- 1876–1877, 1881–1882, 1890–1892: *Gezhi huibian* (*The Chinese Scientific Magazine*; also called *The Chinese Scientific and Industrial Magazine*). In the 1890s the first four volumes were reprinted as full sets. In 1901–1902 the past issues of the journal were reorganized topically and edited under the title *Collectanea of Science* (*Gezhi congshu*) by Xu Jianyin and reprinted in Shanghai.
- 1886: *Chouren zhuan sanbian* (Biographies of mathematical astronomers, third edition); adds 128 Qing entries, compiled by Zhu Kebao (1845–1903). *Chouren zhuan sibian* (fourth edition), 1898: adds 40 Qing entries, compiled by Huang Zhongjun. Included an appendix for 284 Chinese and 157 Westerners.
- 1888: *Huangchao jingshi wen xubian* (Sequel to the collection of Qing dynasty writings on statecraft), edited by Ge Shijun. Shanghai: Tushu jicheng shu.
- 1889: *Zhongxi suanxue dacheng* (Great anthology of Chinese and Western mathematics). Compiled by Liu Sheng'an. Shanghai: Tongwen shuju.
- 1898: *Zhongxi wenjian lu xuanbian* (Selections from *The Peking Magazine*). Edited by William Martin. Four vol. Beijing, 1877. Followed by the *Wenjian lu xinbian* (New edition of *The Peking Magazine*). Beijing.
- 1889: *Xixue dacheng* (Great anthology of Western learning), compiled by Wang Tao. Shanghai: Datong shuju, 1888. Reprint, Shanghai: Cuiwen tang shufang. Includes many new editions of works originally produced by the Jiangnan Arsenal Translation Department. Followed by *Xu xixue dacheng* (Continuation), published in Shanghai: Feihong ge shuju, 1897, which also included many earlier Jiangnan Arsenal publications. A reprint, edited by Wang Xiqing and Lu Tiqing, appeared in Shanghai: Zuiliu tang, 1895.
- 1890: *Xixue jicun* (Edited depository of Western learning), compiled by Wang Tao. Shanghai: Songyin lu. Contains translations of early scientific texts by Alexander Wylie and Wang Tao.

In addition, the works below became available after 1893 and were sources for students preparing for the civil examinations and schooling based on the "New Government" (*Xinzheng*) education reforms from 1901–04.

- 1896: *Xixue qimeng* (Primers for Western learning). Shanghai: Zhuyi tang. Reprint of Edkins, GZQM (1886). Parts included in Shanghai: *Tushu jicheng*, 1898.

- 1897: *Zhong Xi xinxue daquan* (Comprehensive collection of new knowledge in China and the West). Shanghai: Hongwen shuju.
- *Gezhi guwei* (Ancient subtleties of science), by Wang Renjun (1866–1914). Wuchang: Zhixuehu (Physics society).
- 1897: *Gezhi jinghua lu* (Record of the essence of science), by Wang Renjun. Shanghai.
- 1897: *Gezhi keyi huibian* (Compendium of prize essays on science from Shanghai Polytechnic), compiled by Wang Tao. Shanghai. Based on earlier version of essays entitled, *Gezhi shuyuan keyi* (Shanghai Polytechnic prize essay competition), 1886–1893. Another edition compiled by Wang Tao: *Gezhi shuyuan keyi* (Shanghai Polytechnic prize essay competition). Arranged by topics in 15 vols., covering 1886–94. Shanghai: Fuqiangzhai shuju, 1898.
- 1897: *Xizheng congshu* (Collectanea of the Western art of government), compiled by Liang Qichao.
- 1897: *Xixue fuqiang congshu* (Collectanea of Western learning and political economy), compiled by Zhang Yinhuan (1837–1900). 64 vols. Wuchang: *Zhixue hui*. Also reprinted in Shanghai: Xiaocang shanfang, 1899. Contains more than eighty works on sciences, history, and law. Many science works reprinted from Jiangnan Arsenal Translation Department editions.
- 1898: *Qiangxue huibian* (Compendium of knowledge that strengthens [the dynasty]). Shanghai: Wenrui lou.
- 1898: *Xixue gezhi daquan* (Complete collection of Western science). Hong Kong: Xianggang shuju. Contains complete set of twenty science translations from Fryer's "Science Outline Series"; see Appendix 3.
- 1898: *Xifa cexue huiyuan* (Compendium for answering policy questions on Western methods). Shanghai: Hongbao shuju.
- 1898: *Huangchao jingshi wen sanbian* (Third collection of Qing dynasty writings on statecraft), edited by Chen Zhongyi. Baowen shuju. Includes examination papers from Shanghai Polytechnic. Reprint. Taibei: Guofeng, 1965.
- 1898: *Huangchao jingshi wen xinbian* (New collection of Qing dynasty writings on statecraft), edited by Liang Qichao. Shanghai: Datong yi shuju.
- 1901: *Huangchao jingji wenbian* (Collection of Qing dynasty writings on political economy). Shenji shuzhuang. Later version: *Huangchao jingji wen xinpien* (New collection of Qing dynasty writings on political economy). Reprint. Taibei: Wenhai chubanshe, 1987.
- 1902: *Huangchao jingshi wen xinbian xuji* (Sequel to the New collection of Qing dynasty writings on statecraft), edited by Chu Guishan. Yiji shuju.
- 1902: *Geguo zhengzhi yixue quanshu* (Complete works on international government and the arts), compiled by Ding Guanxi. 32 vols. Shanghai: Dongshan shju. Reprints many science translations from the Jiangnan Arsenal Translation Department and from Fryer's "Science Outline Series."
- 1903: *Zhongwai shiwu cewen leibian dacheng* (Great compendium of policy questions on Chinese and foreign affairs classified topically).
- 1903: *Xinxue da congshu* (Great collectanea on new studies), compiled by Tang Caichang and Liang Qichao. 32 vols. Shanghai: Jishan qiaoji shuju. Contains essays on the sciences by Chinese and Japanese writers.

- 1903: *Xinxue dacheng shu* (Great compendium of works on new studies). Shanghai: Jishan qiaoji shuju.
- After 1900: *Gewu Zhongfa* (Chinese methods for the investigation of things, i.e., science), by Liu Yueyun (1849–1917).

Appendix 8

Some Officially Selected Chinese Prize Essay Topics from the Shanghai Polytechnic

Spring, 1887, "Theme" by Xu Xingtai, Zhejiang Provincial Administration Commissioner of Hangzhou: "Compare the sciences of China and the West, showing their points of difference and similarity."

Spring, 1889, "Theme" by Gong Zhaoyuan, Zhejiang Provincial Surveillance Commissioner of Hangzhou: "What are key points in the detailed strengths and cursory weaknesses in contemporary translations of Western science?"

Spring, 1889, "Extra Theme" by Li Hongzhang: "With respect to the 'Science' referred to in the 'Great Learning,' from Ching-kang-ching downwards, there have been several tens of scholars who have written on the subject. Do any of them happen to agree with Western scientists? Western science began with Aristotle in Greece; then came Bacon in England who changed the previous system and made it more complete. In later years, Darwin's and Spencer's writings have made it still more comprehensible. Give a full sketch of the history and bearings of this whole subject."[1]

Spring, 1891, "Extra Theme" by Li Hongzhang: "Compare the similarities between the *Zhoubi suanjing* (The Gnomon of the Zhou dynasty and classic of computations) and Western techniques of trigonometry for measuring segments of a circle."

Spring, 1892: Question on importance of science for naval and industrial production; draws on precedents in the "Artificer's Record" (*Kaogong ji*) chapter in the *Rituals of Zhou* (*Zhouli*).

Summer, 1893, "Theme" by Wu Yinsun, Zhejiang Provincial Circuit Attendant for Ningbo and Shaoxing: "Try to prove in detail the following: When did Western medical techniques begin and who were they transmitted by? Are there differences in the various ways each country treats illnesses? What are the strengths and weaknesses of Chinese versus Western medical principles?"

Spring, 1894, "Theme" by Nie Jigui, Shanghai Circuit Attendant and Zhejiang Provincial Surveillance Commissioner of Hangzhou: "Itemize and demonstrate using scholia that the "Jingshang" [Classic, first part] and "Shuoshang" [Expositions, first part] chapters from the *Mozi* had already raised the Western principles of calendrical studies, optics, and mechanics."

Appendix 9

Scientific Societies Formed between 1915 and 1927

The National Medical Association: 1915 in Beijing. Became the chief voice of opposition against traditional Chinese medicine.

The Geological Society of China: 1922. Important role of Ding Wenjiang and geologists at Academia Sinica. Developed centrality after 1911 Geological Survey of China was initiated. Seventy-seven members in 1922. Grew to 195 in 1925, more than 300 by 1933. By 1939 published more than 2,300 articles, monographs, and books, 56 percent (1,168) authored by Chinese. See David Reynolds, "The Advancement of Knowledge and the Enrichment of Life: The Science Society of China and the Understanding of Science in the Early Republic, 1914–1930" (Ph.D. diss., University of Wisconsin, Madison, 1986), pp. 303–311.

The Wissen and Wissenschaft Society: founded in 1916 by Chinese students in Tokyo. Published journal called *Xueyi* (*Wissen und Wissenschaft*) from April 1919 in Tokyo, first as a quarterly and then as a monthly. Moved to Shanghai in 1920 and took on new name and organization in 1923. Had seven hundred members by 1936.

The Société Astronomique de Chine: 1922 in Beijing. Sponsored books on astronomy, standardizing astronomical terminology, observing eclipses, and sponsoring lectures. Over 400 members by 1944.

The Chinese Meteorological Society: 1924 in Qingdao. More than three hundred members in 1944. Had fifty meteorological stations by 1935. More than four hundred articles published by Chinese between 1919 and 1949.

The Chinese Chemical Industry Society: organized by thirty-seven foreign-trained Chinese chemists in 1922 under direction of Zheng Pinzen, Dean of Beijing University. Grew to more than one thousand members in 1948. Members may also have been drawn from Huanghai Research Institute of the Chemical Industry, also founded in 1922 in Danggu, Hebei, the major salt-producing area of north China.

Academia Sinica: began functioning in late 1927, although Sun Yatsen had advocated its formation in 1924.

Chinese Engineering Society and *Engineer's Society of Peiping:* (founded 1912) were amalgamated in August 1931 to form *Chinese Institute of Engineers.* CES had 254 members in 1923, 304 in 1924, and more than 700 in 1927. Figures for ESP are unavailable. CIE had 2,500 members in 1934 and 8,292 in 1944.

Foreign Societies in China before 1927

The Anatomical and Anthropological Association of China: founded in 1920 as first nonmissionary science society in China, by Canadian physician Davidson Black, then director of the Anatomical Department of Peking Union Medical College.

The China Society of Science and the Arts: developed in 1923 through Arthur Sowerby's efforts to establish a Biological Society of China. Published a journal in English. In 1923 it became the official organ to publish papers produced in the Shanghai Chemical Society, which had been formed in 1922. Represented enormous interest by Westerners in Chinese natural history, which supported expeditions there. Dissolved in 1934 because of competition with Royal Asiatic Society in Shanghai and founding of Shanghai Museum in 1933.

Peking Natural History Society: founded in 1925 under A. W. Grabau's leadership. Began with thirty-eight charter members, thirteen of them Chinese. Membership grew to 101 in 1926, 176 in 1929, 195 in 1935.

Notes

Preface

1. See Itō Shūichi, "Kindai Chūgoku ni okeru kagaku gijutsu no chii—sono shisōshi teki kōsatsu," *Tōyō gakujutsu kenkyū* 5, 5–6 (1967): 65–77. Compare Maurice Crosland, "Science and the Franco-Prussian War," *Social Studies of Science* 6 (1976): 185–214.
2. See Lo Jung-pang, "The Decline of the Early Ming Navy," *Oriens Extremus* 5, 2 (1958): 147–168. Compare SCC, vol. 4, part 3, pp. 477–553, and ECCP, pp. 108–109. See also Wang Gungwu, *Anglo-Chinese Encounters since 1800* (Cambridge: Cambridge University Press, 2003), pp. 13–42. Compare Ronald P. Toby, *State and Diplomacy in Early Modern Japan: Asia in the Development of the Tokugawa Bakufu* (Stanford, Calif.: Stanford University Press, 1991).
3. For background on the nineteenth century, see Bruce Ellemen, *Modern Chinese Warfare, 1795–1989* (London: Routledge, 2001), pp. 57–68.
4. ZGJD, pp. 576–589, presents prewar Qing views. Marianne Bastid, "Fuzhou chuanzheng ju de jishu yinjin," in *Suxian ji* (Beijing: Beijing daxue chuban she, 1993), pp. 246–247, describes the positive French views of the Fuzhou Shipyard.
5. See William Theodore de Bary and Richard Lufrano, eds., *Sources of Chinese Tradition,* 2nd ed. vol. 2 (New York: Columbia University Press, 2001), pp. 63–66.
6. Huang Zheng, "Fujian chuanzheng ju lishi diwei lunlue," *Chuanshi yanjiu* 3 (1987): 57–64, makes this last point. See also Pamela Crossley, *Orphan Warriors: Three Manchu Generations and the End of the Qing World* (Princeton, N.J.: Princeton University Press, 1990), pp. 145–146.
7. Compare Sydney Ross, "*Scientist:* The Story of a Word," *Annals of Science* 18, 2 (June 1962): 65–71, who notes that the term "scientist" was not commonly used in England until the mid-nineteenth century. See also the *Oxford English Dictionary Online,* for the early modern genealogy of science in English.
8. Nicholas Jardine, "The Forging of Modern Realism: Clavius and Kepler Against the Sceptics," *Studies in History and Philosophy of Science* 10 (1979): 147–148. See also Peter Dear, "Jesuit Mathematical Science and the Reconstitution of Experience in the Early Seventeenth Century," *Studies in History and Philosophy of Science* 18 (1987): 141–146, and Peter Dear, *Discipline and Experience: The Mathematical Way in the Scientific Revolution* (Chicago: University of Chicago Press, 1995).

9. See James Weisheipl, "Classification of the Sciences in Medieval Thought," *Medieval Studies* 27 (1965): 54–55, 58–68, 81–90, and David Lindberg, "The Transmission of Greek and Arabic Learning to the West," in David Lindberg, ed., *Science in the Middle Ages* (Chicago: University of Chicago Press, 1978), pp. 52–90. See also David King, *Islamic Mathematical Astronomy* (London: Variorum Reprints, 1986). Logic as a formal field was not introduced by Jesuits or Protestants until the late nineteenth century. See Joachim Kurtz, "Coming to Terms with Logic: The Naturalization of an Occidental Notion in China," in Michael Lackner, Iwo Amelung, and Joachim Kurtz, eds., *New Terms for New Ideas: Western Knowledge & Lexical Change in Late Imperial China* (Leiden: E. J. Brill, 2001), pp. 147–175.

10. See Nathan Sivin, "Introduction," in Nathan Sivin, ed., *Science & Technology in East Asia* (New York: Science History Publications, 1977), pp. xi–xxiv, and Jan P. Hogendijk and A. I. Sabra, "Introduction," in Jan P. Hogendijk and A. I. Sabra, eds., *The Enterprise of Science in Islam* (Cambridge, Mass.: MIT Press, 2003), pp. vii–ix.

11. See Nathan Sivin, "Ailment and Cure in Traditional China," unpublished work. See also Livia Kohn, *Daoism Handbook* (Leiden: E. J. Brill, 2000).

12. See Joseph Needham, *The Grand Titration: Science and Society in East and West* (London: Allen & Unwin, 1969), pp. 14–54. See also Toby E. Huff, *The Rise of Early Modern Science: Islam, China, and the West*, rev. ed. (Cambridge: Cambridge University Press, 2003). See also my review of the 1993 edition in *American Journal of Sociology* (November 1994): 817–819.

13. See the succinctly argued account in Nathan Sivin, "Copernicus in China," in *Science in Ancient China*, vol. 4 (Great Yarmouth: Variorum, 1995), pp. 63–122, especially pp. 85–89 on "putative Copernicans" and pp. 101–102 on literati reactions to Benoist.

14. For the religious focus of the Jesuits, see Jacques Gernet, *China and the Christian Impact: A Conflict of Cultures*, trans. Janet Lloyd (Cambridge: Cambridge University Press, 1985). For the Protestants, see Norman Girardot, *The Victorian Translation of China: James Legges' Oriental Pilgrimage* (Berkeley: University of California Press, 2002). I have benefited from James Hevia's forthcoming review of the latter and Liam Brockey's work in progress on Portuguese Jesuits in China.

15. Compare Lydia Liu, *Translingual Practice: Literature, National Culture, and Translated Modernity—China, 1900–1937* (Stanford, Calif.: Stanford University Press, 1995), pp. 3–42, and Douglas Howland, *Translating the West: Language and Political Reason in Nineteenth-Century Japan* (Honolulu: University of Hawai'i Press, 2002), pp. 61–93.

16. Liu, *Translingual Practice*, p. xv. See also James Hevia, *English Lessons: The Pedagogy of Imperialism in Nineteenth-Century China* (Durham, N.C.: Duke University Press, 2003), pp. 57–61, on translation as a form of violence.

17. Scott Montgomery, *Science in Translation: Movements of Knowledge through Cultures and Time* (Chicago: University of Chicago Press, 2000), pp. 189–226.

18. See Nathan Sivin, "Calendar Reform and Occupation Politics: The Season-Granting System of 1280" (presented at the University of Erlangen, March 25, 2003).

19. See Benjamin A. Elman, *From Philosophy to Philology: Intellectual and Social Aspects of Change in Late Imperial China* (Los Angeles: UCLA Asia Pacific Monograph Series, 2001), pp. 76–82. See also Roger Cooter and Stephen Pumfrey, "Separate Spheres and Public Places: Reflections on the History of Science Popularization and Science in Popular Culture," *History of Science* 32 (1994): 237–267.

20. Compare Zhang Baichun, "An Inquiry into the History of the Chinese Terms *Jiqi* (Machine) and *Jixie* (Machinery)," in Lackner, Amelung, and Kurtz, eds., *New Terms for New Ideas,* pp. 177–195.

21. On the "Awesome Taboo," see Nathan Sivin's review of Christoph Harbsmeier, *Science and Civilisation in China,* vol. 7, part 1, *Language and Logic in Traditional China* (Cambridge: Cambridge University Press, 1998), in *East Asian Science, Technology, and Medicine,* 17 (2000): 123–124.

22. Alain Peyrefitte's *The Immobile Empire: The First Great Collision of the East and West,* trans. Jon Rothschild (New York: Albert Knopf, 1992), paints a dark picture of Qing cultural xenophobia and conservatism. Compare Huang Yinong (Yilong), "Long yu shi duiwang de shijie—yi Majiaerni shituan fanghua hou de chubanwu wei li," *Gugong xueshu jikan* (Taiwan) 21, 2 (Winter 2003): 266–297. For correctives, see Joanna Waley-Cohen, *The Sextants of Beijing: Global Currents in Chinese History* (New York: W. W. Norton & Co., 1999), pp. 92–128; Willard Peterson, "Changing Literati Attitudes toward New Learning in Astronomy and Mathematics in Early Qing," *Monumenta Serica* 50 (2002): 375–390; and Minghui Hu, "Cosmopolitan Confucianism: China's Road to Modern Science" (Ph.D. diss., University of California–Los Angeles 2004), chapter 1.

23. For Jesuit uses of astronomy in the Islamic world and in Vietnam, see Avner Ben-Zaken, "Hebraist Motives, Pythagorean Itineraries and the Galilean Agendas of Naples: On the Margins of Text and Context" (paper presented at the UCLA History of Science Program Colloquium, Los Angeles, December 2, 2002).

24. See William Theodore de Bary, "Reflections on the Chinese Rites Controversy," in D. E. Mungello, ed., *The Chinese Rites Controversy: Its History and Meaning* (Nettetal: Steyler Verlag, 1994), p. 291. De Bary views foreign matters as marginal on p. 292.

25. W. L. Idema, "Cannon, Clocks and Clever Monkeys: Europeana, Europeans and Europe in Some Ch'ing Novels," in E. B. Vermeer, ed., *Development and Decline of Fukien Province in the 17th and 18th centuries* (Leiden: E. J. Brill, 1990), p. 468.

26. Sivin, "Copernicus in China," p. 8.

27. John Barrow, *Travels in China* (London: T. Cadell and W. Davies, 1804), p. 110. I have also benefited from Joanne Tong, "From the Macartney Mission to the Opium Wars: The British Legacy in China" (graduate paper, UCLA, 2001). See also Zeng Jingmin, "Scientific Aspects of the Macartney Embassy to China, 1792–1794" (Ph.D. diss., Newcastle University, 1998).

28. See J. L. Cranmer-Byng and Trevor H. Levere, "A Case Study in Cultural Collision: Scientific Apparatus in the Macartney Embassy to China, 1793," *Annals of Science* 38 (1981): 503–525. I have also benefited from Simon Schaffer, "The As-

tronomer's Inventory: British Instruments as Cargo in China and the Pacific" (paper presented at the Davis Center, Princeton University, December 3, 2003).

29. George Macartney, *An Embassy to China; being the journal kept by Lord Macartney during his embassy to the Emperor Ch'ien-lung, 1793–1794*, ed., J. L. Cranmer-Byng (London: Longmans, 1962), pp. 310–311. See also Cranmer-Byng and Levere, "A Case Study in Cultural Collision," p. 515.

30. Huang, "Long yu shi duiwang de shijie," pp. 291–297, contends that the Qing state in 1793 was too closed-minded to deal with imperialism. With hindsight, such views appear incontrovertible, but we will see that there were many external factors that help explain why the Newtonian revolution came to China only after 1850.

31. Susan Thorne, " 'The Conversion of Englishmen and the Conversion of the World Inseparable': Missionary Imperialism and the Language of Class in Early Industrial Britain," in Frederick Cooper and Ann Stoler, eds., *Tensions of Empire: Colonial Cultures in a Bourgeois World* (Berkeley: University of California Press, 1997), pp. 238–262.

32. Federico Masini, *The Formation of Modern Chinese Lexicon and Its Evolution toward a National Language: The Period from 1840 to 1898* (Berkeley: University of California, Journal of Chinese Linguistics Monograph Series, 1993), pp. 15–34. For Wei Yuan's views, see the translations in de Bary and Lufrano, *Sources of Chinese Tradition*, vol. 2, p. 211.

33. LHCT, vol. 1, no. 6 (1857.6.22): 3a–5a.

34. ZGJD, pp. 152–154.

35. Kume Kunitake, comp., *The Iwakura Embassy 1871–73: A True Account of the Ambassador Extraordinary & Plenipotentiary's Journey of Observation through the United States of America and Europe*, 5 vols. (Matsudo, Chiba: The Japan Documents, 2002), vol. 5, p. 352.

36. On the inadequacy of China's army and navy, see Ralph Powell, *The Rise of Chinese Military Power* (Princeton, N.J.: Princeton University Press, 1955), pp. 36–50. Compare Richard Smith, "Reflections on the Comparative Study of Modernization in China and Japan: Military Aspects," *Journal of the Hong Kong Branch of the Royal Asiatic Society* 16 (1976): 11–23.

37. Douglas Reynolds, *China, 1898–1912: The Xinzheng Revolution and Japan* (Cambridge, Mass.: Harvard University Press, 1993), pp. 131–150.

38. Yue Meng, "Hybrid Science *versus* Modernity: The Practice of the Jiangnan Arsenal, 1864–1897," *East Asian Science, Technology, and Medicine* 16 (1999): 43–45.

39. See Min-chih Maynard Chou, "Science and Value in May Fourth China: The Case of Hu Shih" (Ph.D. diss., University of Michigan, 1974), pp. 23–35.

40. Michael Adas, *Machines as the Measure of Men: Science, Technology, and Ideologies of Western Dominance* (Ithaca, N.Y.: Cornell University Press, 1989). My thanks to Christopher Cullen and R. Bin Wong for voicing their concerns about uncritically affirming Chinese scientific success.

Prologue

1. Michel Foucault, *The Order of Things: An Archaeology of the Human Sciences* [*Les Mots et les choses*] (New York: Vintage Books, 1973), pp. xv–xix, and Jorge Luis

Borges, *Other Inquisitions,* trans. Ruth L. Simms (London: Souvenir Press, 1973), p. 103. Kuhn's encyclopedia may be fictional. Zhang Longxi, *Mighty Opposites* (Stanford, Calif.: Stanford University Press, 1998), pp. 19–22, believes that Borges made it all up. My thanks to Tim Barrett for his advice on the matter. Compare Robert Wardy, "Chinese Whispers," *Proceedings of the Cambridge Philological Society* 38 (1992): 149–151.

2. Geoffrey Lloyd and Nathan Sivin, *The Way and the Word: Science and Medicine in Early China and Greece* (New Haven, Conn.: Yale University Press, 2002), p. 200, note the lack of a notion of nature in traditional China. On parallel problems in Japan, see Frederik Cryns, "The Influence of Herman Boerhaave's Mechanical Concept of the Human Body in Nineteenth-Century Japan," in W. F. Vande Walle and Kasuhiko Kasaya, eds., *Dodonæus in Japan: Translation and the Scientific Mind in the Tokugawa Period* (Leuven: Leuven University Press, 2001), pp. 348–353. Chinese terminology was later refracted through Christian notions of God and nature by first the Jesuits and then Protestants. See parts II and III. For the impact of British natural history in nineteenth-century China, see Fa-ti Fan, "Hybrid Discourse and Textual Practice: Sinology and Natural History in the Nineteenth Century," *History of Science* 38 (2000): 25–56.

3. The *locus classicus* is the Great Learning, one of the Four Books. The Four Books were classical repositories that included the *Mencius, Analects,* Doctrine of the Mean, and the Great Learning during the Song dynasty. The last two were from the *Record of Rites,* one of the Five Classics associated since the Han dynasty (206 B.C.–A.D. 220) with Confucius (551–479 B.C.). See Andrew Plaks, *Ta Hsüeh* and *Chung Yung (The Highest Order of Cultivation* and *On the Practice of the Mean)* (London: Penguin Books Ltd., 2003), pp. 9–10. Plaks translates *gewu* as "extending to all things in the world the correct conceptional grid," and for *zhizhi* he gives "maximizing one's range of comprehension." See also Tang Yung-tung, "On 'Ko-yi,' The Earliest Method by which Indian Buddhism and Chinese Thought Were Synthesized," in W. R. Inge et al., eds., *Radhakrishnan: Comparative Studies in Philosophy Presented in Honor of His Sixtieth Birthday* (London: George Allen and Unwin Ltd., 1951), pp. 276–286. My thanks to Yang Lu for this source.

4. Cheng Yi, *Erh Cheng yishu* (SKQS edition), 15.1b. See also D. C. Lau, "A Note on *Ke Wu,*" *Bulletin of the School of Oriental and African Studies* 30 (1967): 303–357. For Sino-Jesuit accommodation in natural philosophy, see Shang Zhicong, *Mingmo Qingchu (1582–1687) de gewu qiongli zhi xue* (Chengdu: Sichuan jiaoyu chubanshe, 2003).

5. On the evolution of the ancient meaning of *wu* = things, see Du Zhengsheng, "Gudai wuguai zhi yanjiu (shang)," *Dalu zazhi* 104, 1 (January 15, 2002): 1–14. Du points out that the added meaning of affairs emerged in early medieval times. See also Craig Clunas, *Superfluous Things: Material Culture and Social Status in Early Modern China* (Urbana: University of Illinois Press, 1991), pp. 75–90, on words about things, and Yangwen Zheng, *The Social Life of Opium in China, 1483–1999* (Cambridge: Cambridge University Press), forthcoming.

6. Clunas, *Superfluous Things,* pp. 75–90.

7. See Manfred Porkert, *The Theoretical Foundations of Chinese Medicine: Systems of Correspondence* (Cambridge, Mass.: MIT Press, 1974), pp. 9–54. Compare Paula

Findlen, *Possessing Nature: Museums, Collecting, and Scientific Culture in Early Modern Europe* (Berkeley: University of California Press, 1994). The term *bowu* was used for the Chinese notion of museum after 1870, namely *bowu guan*, literally, "hall of things and phenomena." See also Lorraine Daston and Katharine Park, *Wonders and the Order of Nature 1150–1750* (New York: Zone Books, 1998), pp. 173–214.

8. Shao Yong, *Huangji jingshi shu* (SKQS edition), 11.6a, 12.17b, 12.18a, which draws on the *Change Classic* (*Yijing*), one of the Five Classics. See also Anne Birdwhistell, *Transition to Neo-Confucianism: Shao Yung on Knowledge and Symbols of Reality* (Stanford, Calif.: Stanford University Press, 1989), pp. 168–169.

9. For discussion, see Benjamin A. Elman, "Rethinking 'Confucianism' and 'Neo-Confucianism' in Modern Chinese History," in Benjamin A. Elman, John Duncan, and Herman Ooms, eds., *Rethinking Confucianism: Past and Present in China, Japan, Korea, and Vietnam* (Los Angeles: Asia Pacific Monograph Series, 2002), pp. 518–554. In his study of the *Change Classic*, Zhu Xi, for instance, never mentioned Shao Yong's theory of "perceiving things" (*guanwu*). See Zhu Xi, *Zhouyi benyi* (repr., Taibei: Hualian chubanshe, 1971), 2.1b.

10. See *Zhuzi yulei* (1473; repr., Taibei: Zhongzheng Bookstore, 1973), 34.13b. See also *Shangshu tongjian* (repr., Beijing: Chaori wenxian chubanshe, 1982), 25.0124–0127, from the "Hounds of Lü," one of the forged Old Text chapters, on the conflict between virtue and things. James Legge, *The Chinese Classics, III: The Shoo King* (Taibei: Wenshizhe chubanshe, 1972), p. 348, translates the passage as "By trifling with things, he ruins his virtue; by finding his amusement in things he ruins his aims."

11. See *Zhuzi yulei*, 18.14b–15a. Compare Yung Sik Kim, *The Natural Philosophy of Chu Hsi 1130–1200* (Philadelphia: American Philosophical Society, 2000), pp. 19–27, and Yü Ying-shih, "Some Preliminary Observations on the Rise of Ch'ing Confucian Intellectualism," *Tsing Hua Journal of Chinese Studies*, New Series 11, 1 & 2 (December 1975): 105–146.

12. Daniel Gardner, *Chu Hsi and the Ta-hsueh: Neo-Confucian Reflection on the Confucian Canon* (Cambridge, Mass.: Harvard University Council on East Asian Studies, 1986), pp. 27–59.

13. Yü Ying-shih, "Some Preliminary Observations," p. 125.

14. Wang Yangming, *Wang Wencheng quanshu* (SKQS edition), 2.9a. See also Legge, *The Chinese Classics, III: The Shoo King*, p. 348.

15. Cheng, *Erh Cheng yishu*, 18.19a.

16. *Wang Yangming quanji* (Taipei: Kaozheng chubanshe, 1972), pp. 3, 93, in Wing-tsit Chan, trans., *Instructions for Practical Living and Other Neo-Confucian Writings by Wang Yang-ming* (New York: Columbia University Press, 1963), p. 249. Like other parts of Wang's enlightenment narrative, the "bamboo" story may be apocryphal. See Adam Schorr, "The Trap of Words: Political Power, Cultural Authority, and Language Debates in Ming Dynasty China" (Ph.D. diss., University of California–Los Angeles, 1994), pp. 226–244.

17. *Wang Yangming quanji*, pp. 4–5, and Wing-tsit Chan, trans., *Instructions for Practical Living*, pp. 14–15. See also Willard Peterson, "'Chinese Scientific Philosophy' and Some Chinese Attitudes towards Knowledge about the Realm of

Heaven-and-Earth," *Past and Present* 87 (May 1980): 29, and Jennifer Eichman, "Spiritual Seekers in a Fluid Landscape: A Chinese Buddhist in the Wanli-period (1573–1620)" (Ph.D. diss., Princeton University, 2005), chapter 6.

18. See Bruce Rusk, "The Rogue Classicist: Feng Fang (1493–1566) and His Forgeries" (Ph.D. diss., University of California–Los Angeles, 2004). See also Wang Fan-shen, "The 'Daring Fool' Feng Fang (1500–1570) and His Ink Rubbing of the Stone-inscribed *Great Learning*," *Ming Studies* 35 (August 1995): 74–91.

19. See Susan Cherniack, "Book Culture and Textual Transmission in Sung China," HJAS 54, 1 (June 1994): 5–125, and Ankeney Weitz, "Collecting the Records of the Sages: Archaic Bronzes as Sacred Text and Economic Commodity in Song-Dynasty China (960–1276)" (presented at the conference The Age of Antiquaries in Europe and China, Bard Graduate Center, New York, March 25, 2004). See also Jessica Rawson, "The Ancestry of Chinese Bronze Vessels," in Steven Lubar and W. David Kingery, eds., *History from Things: Essays on Material Culture* (Washington, D.C.: Smithsonian Institution Press, 1993), pp. 53–73.

20. Ankeney Weitz, "Notes on the Early Yuan Antique Art Market in Hangzhou," *Ars Orientalis* 27 (1997): 27–38; Clunas, *Superfluous Things,* pp. 12, 86–87, 109–115; and He Chuanxin, "Mingmo Qingchu huihua de fanggu feng," *Gugong wenwu yuekan* 249 (December 2003): 42–53. See also Stephen West, "Text and Ideology: Ming Editors and Northern Drama," in Paul Smith and Richard von Glahn, *The Song-Yuan-Ming Transition in Chinese History* (Cambridge, Mass.: Harvard University Asia Center Publications, 2003), pp. 329–373, and Lucille Chia, *Printing for Profit: The Commercial Publishing of Jianyang, Fujian (11th–17th centuries)* (Cambridge, Mass.: Harvard University Asia Center Publications, 2002).

21. See James T. C. Liu, *Reform in Sung China: Wang An-shih (1021–1086) and His New Policies* (Cambridge, Mass.: Harvard University East Asian Studies, 1959), pp. 1–10.

22. Mi Chu Wiens, "Changes in the Fiscal and Rural Control Systems in the Fourteenth and Fifteenth Centuries," *Ming Studies* 3 (1976): 53–69.

23. Edward Farmer, "Social Regulations of the First Ming Emperor," in Kwang-Ching Liu, ed., *Orthodoxy in Late Imperial China* (Berkeley: University of California Press, 1990), pp. 116–123. Compare Ray Huang, *Taxation and Government Finance in Sixteenth-Century China* (Cambridge: Cambridge University Press, 1974), pp. 1–6. Ming economic developments were also based in part on the Tang-Song economic revolution, and the traditional trading and tax system of the early empires. See Yü Ying-shih, *Trade and Expansion in Han China* (Berkeley: University of California Press, 1967), and Mark Elvin, *The Pattern of the Chinese Past* (Stanford, Calif.: Stanford University Press, 1973).

24. Compare Clunas, *Superfluous Things,* pp. 141–165, and Tim Brook, *The Confusions of Pleasure: Commerce and Culture in Ming China* (Stanford, Calif.: Stanford University Press, 1999), pp. 129–138, who perceptively—if unwittingly—deprioritize Zhu Xi's and Wang Yangming's influence during the late Ming. See also Cynthia Brokaw, *The Ledgers of Merit and Demerit: Social Change and Moral Order in Late Imperial China* (Princeton, N.J.: Princeton University Press, 1991), pp. 17–27.

25. Clunas, *Superfluous Things,* pp. 91–115, and Brook, *Confusions of Pleasure,* pp. 190–228.

26. See the items presented and discussed in *Through the Prism of the Past: Antiquarian Trends in Chinese Art of the 16th to 18th Century* (Taibei: National Palace Museum, 2003). See also Wu Renshu, "Wan Ming de lüyou huodong yu xiaofei wenhua—yi Jiangnan wei taolun zhongxin," *Jindai shi yanjiu suo jikan* (Academia Sinica, Taiwan) 41 (2003): 87–141.

27. Wang Yangming, *Wang Wencheng quanshu*, 2.9a.

28. Craig Clunas, "Jade Carvers and Their Customers in Ming China," *Transactions of the Oriental Ceramic Society* 50 (1985–1986): 69–85.

29. See Francesca Bray, *Technology and Society in Ming China (1368–1644)* (Washington, D.C.: American Historical Association, 2001). See also Mark Elvin, "Skills and Resources in Late Traditional China," in Mark Elvin, ed., *Another History: Essays on China from a European Perspective* (Canberra: Wild Peony, 1996), p. 73.

30. Brook, *Confusions of Pleasure*, pp. 30–57.

31. Craig Dietrich, "Cotton Culture and Manufacture in Early Modern China," in W. E. Willmont, ed., *Economic Organization in Chinese Society* (Stanford, Calif.: Stanford University Press, 1972), and Nishijima Sadao, "The Formation of the Early Chinese Cotton Industry," trans. Linda Grove, in Linda Grove and Christian Daniels, eds., *State and Society in China: Japanese Perspectives on Ming-Qing Social and Economic History* (Tokyo: Tokyo University Press, 1984), pp. 17–77. Compare Tanaka Masatoshi, "Rural Handicraft in Jiangnan in the Sixteenth and Seventeenth Centuries," trans. Grove, in Grove and Daniels, eds., *State and Society in China*, pp. 79–100. For the cultural implications, see Brook, *Confusions of Pleasure*, pp. 218–222.

32. Shigeta Atsushi, "The Origins and Structure of Gentry Rule," trans. Christian Daniels, in Grove and Daniels, eds., *State and Society in China*, pp. 335–385.

33. Mi Chu Wiens, "Cotton Textile Production and Rural Social Transformation in Early Modern China," *Journal of the Institute of Chinese Studies of the Chinese University of Hong Kong* 7, no. 2 (1974): 515–534.

34. Shih Min-hsiung, *The Silk Industry in Ch'ing China*, trans. E-tu Zen Sun (Ann Arbor: Center for Chinese Studies, University of Michigan, 1976), pp. 9–28, describes later aspects of silk production. Compare Francesca Bray, *Technology and Gender: Fabrics of Power in Later Imperial China* (Berkeley: University of California Press, 1997).

35. Brook, *Confusions of Pleasure*, pp. 219–222, and Clunas, *Superfluous Things*, pp. 153–158.

36. Angela Leung, "Organized Medicine in Ming-Ch'ing China," *Late Imperial China* 8, 1 (1987): 134–166.

37. Bray, *Technology and Society*, pp. 2–6. See also Andre Gunder Frank, *ReOrient: Global Economy in the Asian Age* (Berkeley: University of California Press, 1998), pp. 108–117.

38. John Ayers and Rose Kerr, *Blanc De Chine: Porcelain from Dehua* (Chicago: Art Media Resources, 2002); Chuimei Ho, "The Ceramic Trade in Asia, 1602–82," in A. J. H. Latham and Heita Kawakatsu, eds., *Japanese Industrialization and the Asian Economy* (London: Routledge, 1994), p. 39; and Brook, *Confusions of Pleasure*, p. 206. See also the discussion in Chapter 5.

39. See *Mingdai taoci daquan* (Taibei: Yishujia chuban she, 1983), pp. 8–71. See also *Zhongguo taoci shi* (Beijing: Wenwu chuban she, 1982), pp. 357–414, and Craig Clunas, "The Cost of Ceramics and the Cost of Collecting Ceramics in the Ming Period," *Bulletin of the Oriental Ceramic Society of Hong Kong* 8 (1986–1988): 47–53.

40. Michael Dillon, "Jingdezhen as a Ming Industrial Center," *Ming Studies* 6 (Spring 1978): 37–44, and Song Yingxing, *T'ien-kung k'ai-wu: Chinese Industry in the Seventeenth Century*, trans. E-tu Zen Shun and Shiou-chuan Sun (University Park: Pennsylvania State University, 1966).

41. Kwang Tsing Wu, "Ming Printing and Printers," HJAS 7 (1943): 203–260. See also Brook, *Confusions of Pleasure*, pp. 129–134.

42. Brokaw, "Commercial Publishing in Late Imperial China," pp. 49–92. See also Chia, *Printing for Profit*, for detailed discussion of the Jianyang printers in Ming Fujian.

43. Elman, *From Philosophy to Philology*, pp. 178–208.

44. Sakai Tadao, "Confucianism and Popular Education Works," in William Theodore de Bary et al., *Self and Society in Ming Thought* (New York: Columbia University Press, 1970), pp. 331–341. *Leishu* usually presented verbatim quotations rather than summaries of accepted knowledge. See Michael Loewe, *The Origin and Development of Chinese Encyclopedias* (London: China Society Occasional Paper 25, 1987).

45. Wang Zhenghua, "Shenghuo, zhishi yu wenhua shangpin: wan Ming Fujian ban 'riyong leishu' yu qi shuhua men," *Jindai shi yanjiu suo jikan* (Academia Sinica, Taiwan) 41 (2003): 1–83.

46. See Patrick Hanan, *The Chinese Vernacular Story* (Cambridge, Mass.: Harvard University Press, 1981), pp. 60–61, and Patrick Hanan, "The Sources of the *Chin P'ing Mei*," *Asia Major* 10 (1963): 60ff. Compare Shang Wei, "*Jin Ping Mei Cihua* and Late Ming Print Culture," in Judith Zeitlin and Lydia Liu, eds., *Writing and Materiality in China: Essays in Honor of Patrick Hanan* (Cambridge, Mass.: Harvard Asian Monograph Series, 2003), pp. 187–231, and Shang Wei, "The Making of the Everyday World: *Jin Ping Mei* and Encyclopedias for Daily Use," in David Wang and Shang Wei, eds., *Dynastic Decline and Cultural Innovation: From the Late Ming to the Late Qing* (Cambridge, Mass.: Harvard University Asia Center), forthcoming.

47. Compare Irene Bloom, "On the 'Abstraction' of Ming Thought: Some Concrete Evidence From the Philosophy of Lo Ch'in-shun," in William Theodore de Bary and Irene Bloom, eds., *Principle and Practicality: Essays in Neo-Confucianism and Practical Learning* (New York: Columbia University Press, 1979), p. 106.

48. Onozawa Seiichi et al., *Ki no shisō: Chūgoku ni okeru shizenkan to ningenkan no tenkai* (Tokyo: Tokyo University Press, 1978), pp. 473–489, and Benjamin A. Elman, "The Revaluation of 'Benevolence' *(Jen)* in Ch'ing Dynasty Evidential Research," in Richard Smith and Daniel Kwok, eds., *Cosmology Ontology, & Human Efficacy: Essays in Ch'ing Thought* (Honolulu: University of Hawai'i Press, 1993), pp. 59–80. See also Dorothy Ko, *Teachers of the Inner Chambers* (Stanford, Calif.: Stanford University Press, 1994), pp. 68–112.

49. Erik Zürcher, "The Jesuit Mission in Fujian in Late Ming Times: Levels of Response," in E. B. Vermeer, ed., *Development and Decline of Fukien Province in the 17th and 18th Centuries* (Leiden: E. J. Brill, 1990), pp. 443–449. See also Qiong

Zhang, "About God, Demons, and Miracles: The Jesuit Discourse on the Supernatural in Late Ming China," *Early Science and Medicine* 4, 1 (February 1999): 6–23, and Qiong Zhang, "Demystifying *Qi:* The Politics of Cultural Translation and Interpretation in the Early Jesuit Mission to China," in Lydia Liu, ed., *Tokens of Exchange: The Problem of Translation and Interpretation in the Early Jesuit Mission to China* (Durham, N.C.: Duke University Press, 1999), pp. 74–106.

50. Katharine Park and Lorraine Daston, "Unnatural Conceptions: The Study of Monsters in Sixteenth-Century France and England," *Past and Present* 92 (August 1981): 20–54.

51. Lorraine Daston, "Marvelous Facts and Miraculous Evidence in Early Modern Europe," *Critical Inquiry* 18 (1991): 93–124. See also Daston and Park, *Wonders and the Order of Nature,* pp. 215–253.

52. On the Jesuits, see Zhang, "About God, Demons, and Miracles," pp. 35–36. See also Benjamin A. Elman, *A Cultural History of Civil Examinations in Late Imperial China* (Berkeley: University of California Press, 2000), pp. 346–354.

53. Keith Hutchison, "Supernaturalism and the Mechanical Philosophy," *History of Science* 21 (1983): 297–333. For the Jesuits in China, see Erik Zürcher, "'In the Beginning': 17th-Century Chinese Reactions to Christian Creationism," in Chun-chieh Huang and Erik Zürcher, eds., *Time and Space in Chinese Culture* (Leiden: E. J. Brill, 1995), pp. 144–145, 161–162.

54. See Margaret Jacob, "Christianity and the Newtonian Worldview," in David Lindberg and Ronald Numbers, eds., *God and Nature: Essays on the Encounter between Christianity and Science* (Berkeley: University of California Press, 1986), pp. 238–255.

55. Hutchison, "Supernaturalism and the Mechanical Philosophy," 319–333.

56. Lorraine Daston, "The Nature of Nature in Early Modern Europe," *Configurations* 6 (1998): 149–172, and Steven Shapin, *A Social History of Truth: Civility and Science in Seventeenth-Century England* (Chicago: University of Chicago Press, 1994), pp. 122–123.

57. See Florence Hsia, "Some Observations on the *Observations:* The Decline of the French Jesuit Scientific Mission in China," in Éric Brian, ed., *Revue de synthèse. Les Jésuites dans le monde moderne: Nouvelles approches* (Paris: CNRS, 1999), pp. 305–333.

58. See *Mingdai dengke lu huibian,* 22 vols. (Taibei: Xuesheng Bookstore, 1969), vol. 18, pp. 9822–9823, 9932–9948.

1. Ming Classification on the Eve of Jesuit Contact

1. Compare Nicholas Jardine and Emma Spary, "The Natures of Cultural History," in Nicholas Jardine, J. A. Secord, and E. C. Spary, eds., *Cultures of Natural History,* pp. 3–13.

2. Yü Ying-shih, "Some Preliminary Observations," pp. 105–146.

3. Reiko Shinno, "Promoting Medicine in the Yuan Dynasty (1206–1368)" (Ph.D. diss., Stanford University, 2002), pp. 198–204. See also Chao Yuan-ling, "The Ideal Physician in Late Imperial China: The Question of *Sanshi,*" *East Asian Sci-*

ence, Technology, and Medicine 17 (2000): 66–93, particularly on medical professionalization.

4. SKQSZM, 104.3a–4a. See the translation by Yang Shou-zhong and Duan Wu-jin, *Extra Treatises Based on Investigation and Inquiry: A Translation* (Boulder, Colo.: Blue Poppy Press, 1994). Compare Wu Yiyi, "A Medical Line of Many Masters: A Prosopographical Study of Liu Wansu and His Disciples from the Jin to the Early Ming," *Chinese Science* 11 (1993–94): 36–65, and Asaf Goldschmidt, "The Transformations of Chinese Medicine During the Northern Sung Dynasty (960–1126 A.D.): The Integration of Three Past Medical Approaches into a Comprehensive Medical System Following a Wave of Epidemics" (Ph.D. diss., University of Pennsylvania, 1999), pp. 129–138, 147–153.

5. See Yu Chang, "Xu" to his *Yuyi cao* (1643 ed.), pp. 1a–3a. See also Charlotte Furth, "Producing Medical Knowledge through Cases: History, Evidence, and Action" (presented at the Colloquium on Science and Culture in Late Traditional China, School of Historical Studies, Institute for Advanced Study, Princeton, N.J., March 17–18, 2000).

6. Chu Pingyi, "Tongguan Tianxue, yixue yu ruxue: Wang Honghan yu Ming-Qing zhi ji Zhongxi yixue de jiaohui," *Lishi yuyan yanjiu suo jikan* 70, 1 (1999): 165–201.

7. See Sir David Percival, trans., *Chinese Connoisseurship, the Ko Ku Yao Lun: The Essential Criteria of Antiquity* (London: Faber, 1971). Compare Clunas, *Superfluous Things*, pp. 11–13.

8. Ma Huan, *Ying-yai Sheng-lan, "The Overall Survey of the Ocean's Shores" [1433]*, trans. and edited by J. V. G. Mills (Cambridge: Cambridge University Press, Extra Series No. XLII, 1970), pp. 69–180.

9. See SKQSZM, 78.15a–15b, 123.4a–5a. See also DMB, pp. 1026–1027. See also Clunas, *Superfluous Things*, pp. 11–12.

10. See Donald Lach, *Asia in the Making of Europe. Volume II. A Century of Wonder, Book 3* (Chicago: University of Chicago Press, 1977), pp. 446–489, 556–566.

11. Michael Ryan, "Assimilating New Worlds in the Sixteenth and Seventeenth Centuries," *Comparative Studies in Society and History* 23 (1981): 519–527. Compare Pamela Smith and Paula Findlen, eds., *Merchants and Marvels: Commerce, Science, and Art in Early Modern Europe* (London: Routledge, 2001), and Anthony Grafton, *New Worlds, Ancient Texts: The Power of Tradition and the Shock of Discovery* (Cambridge, Mass.: Belknap Press, 1992), pp. 148–157.

12. Ryan, "Assimilating New Worlds," pp. 527–538. See also D. E. Mungello, *Curious Land: Jesuit Accommodation and the Origins of Sinology* (Honolulu: University of Hawai'i Press, 1989), p. 145, and Zürcher, "'In the Beginning,'" pp. 152–153.

13. See also Mark Elvin, *The Structure of the Chinese Past*, pp. 203–234. For a critical review, see Nathan Sivin, "Imperial China: Has Its Present Past a Future," HJAS 38, 2 (December 1978): 449–480.

14. See Sakai Tadao, "Confucianism and Popular Education Works," in William Theodore de Bary et al., *Self and Society in Ming Thought* (New York: Columbia University Press, 1970), pp. 331–341. See also Elman, *A Cultural History of Civil Examinations*, pp. 239–294, and Chia, *Printing for Profit*, pp. 234–239.

15. Sakade Yoshinobu, "Kaisetsu—Mindai nichiyō ruisho ni tsuite," in Sakade Yoshinobu et al., *Chūgoku nichiyō ruisho shūsei*, 14 vols. (Tokyo: Kyūko shoin, 1999–2003), vol. 1, pp. 11–14. Compare Stephen West, "*Leishu* (Encyclopedias) & the Textualization of Quotidian Life in the 12th Century" (paper presented at the conference Print, Anthologies, & the Shape of Knowledge in Late Imperial China, sponsored by the UCLA Center for Chinese Studies in conjunction with the Southern California China Colloquium, Los Angeles, October 19, 2002). See also Chia, *Printing for Profit*, pp. 138, 144.

16. See Elman, *A Cultural History of Civil Examinations*, pp. 223, 226–227.

17. Sakai Tadao, "Mindai no nichiyō ruisho to shomin kyōiku," in Hayashi Tomoharu, ed., *Kinsei Chūgoku kyōikushi kenkyū* (Tokyo: Kokutosha, 1958), pp. 27–154. Compare Alexander Wylie, *Notes on Chinese Literature* (1867; repr., Taibei: Book Shop Limited, 1970), pp. 185–187, on Yuan-Ming encyclopedias.

18. Chia, *Printing for Profit*, pp. 234–239. See also Cynthia Brokaw, "Commercial Publishing in Late Imperial China: The Zou and Ma Family Businesses of Sibao, Fujian," *Late Imperial China* 17, 1 (June 1996): 49–92.

19. Sakade, "Kaisetsu—Mindai nichiyō ruisho ni tsuite," pp. 15–27. See also Kathryn Lowry, trans., "Personal Letters in Seventeenth-Century Epistolary Guides," in Susan Mann and Yu-yin Cheng, eds., *Under Confucian Eyes: Writings on Gender in Chinese History* (Berkeley: University of California Press, 2001), pp. 155–160.

20. See Robert F. Campany, *Strange Writing: Anomaly Accounts in Early Medieval China* (Albany: State University of New York Press, 1996), pp. 49–52. See also, Qiong Zhang, "Nature, Supernature, and Natural Studies in Sixteenth- and Seventeenth-Century China" (paper presented at the colloquium sponsored by the Center for the Cultural Studies of Science, Medicine, and Technology, UCLA History Department, Los Angeles, November 16, 1998).

21. Nathan Sivin, "Li Shih-chen," in Charles Gillispie, ed., *Dictionary of Scientific Biography* (New York: Charles Scribner's and Sons, 1973), 8, p. 390, and DMB, p. 859.

22. See Elman, *A Cultural History of Civil Examinations*, pp. 125–142.

23. Ibid., pp. 133–42. See also Ellen Widmer and Kang-i Sun Chang, eds., *Writing Women in Late Imperial China* (Stanford, Calif.: Stanford University Press, 1997).

24. Ōba Osamu, *Edo jidai ni okeru Chūgoku bunka juyō no kenkyū* (Tokyo: Dō bōsha, 1984), and Fujitsuka Chikashi, *Shinchō bunka tōden no kenkyū* (Tokyo: Kokusho kankōkai, 1975), pp. 6–48. During the late sixteenth-century Japanese invasion of Korea, however, Ming books did not travel to Korea or Japan very easily, and many Korean editions were lost.

25. DMB, p. 861, and Sivin, "Li Shih-chen," pp. 394–395. According to Yong Sik Kim, only a 1762 Chinese edition survives in Korea.

26. See Li Shizhen, "Fanli," in Li Shizhen, *Bencao gangmu*, 4 vols. (1596; repr. Beijing: Renmin weisheng chuban she, 1977–1981), p. 33. See also Sivin, "Li Shih-chen," pp. 391, 396. Compare Goldschmidt, "The Transformations of Chinese Medicine," pp. 40–41, 147–150; Angela Leung, "Organized Medicine in Ming-Ch'ing China," pp. 134–166; and Shinno, "Promoting Medicine in the Yuan Dynasty," pp. 5–10.

27. See Wu, "A Medical Line of Many Masters," p. 41n13; Chao Yuan-ling, "Medicine and Society in Late Imperial China: A Study of Physicians in Suzhou" (Ph.D. diss., University of California–Los Angeles, 1995), pp. 87–88n37; and Angela Leung, "Medical Learning from the Song to the Ming," in Paul Smith and Richard von Glahn, eds., *The Song-Yuan-Ming Transition in Chinese History* (Cambridge, Mass.: Harvard Asian Center Monograph, 2003), pp. 374–398.

28. Li Shizhen, "Fanli," *Bencao gangmu* (1603 edition), vol. 1, part A, pp. 7b–9b. This version is in the Library of Congress. See also DMB, p. 862, and Georges Métailé, "The *Bencao gangmu* of Li Shizhen: An Innovation for Natural History?" in Elisabeth Hsu, ed., *Innovation in Chinese Medicine* (Cambridge: Cambridge University Press, 2001), pp. 221–261.

29. See Sivin, "Li Shih-chen," pp. 395–397, and Métailé, "The *Bencao gangmu* of Li Shizhen," pp. 224–225. See also Métailé, "Botanical Terminology of Li Shizhen in *Bencao gangmu*," in Hakim Mohammed Said, ed., *Essays on Science* (Karachi: Hamdard Foundation Pakistan, 1990), pp. 140–153.

30. SCC, vol. 5, part 3, pp. 117–220.

31. See Li Jianmin, "*Bencao gangmu huobu* kaoshi," *Lishi yuyan yanjiu suo jikan* 73, 3 (September 2002): 395–441.

32. Sivin, "Li Shih-chen," pp. 390–398, and Métailé, "The *Bencao gangmu* of Li Shizhen," pp. 227–240. Compare Robert Visser, "Dodonæus and the Herbal Tradition," in Vande Walle and Kasaya, eds., *Dodonæus in Japan*, pp. 46–56.

33. Li, "Fanli," *Bencao gangmu*, vol. 1, part A, pp. 7b–9b. See also Métailé, "The *Bencao gangmu* of Li Shizhen," pp. 226–227, 240–242.

34. Li, "Fanli," *Bencao gangmu*, vol. 1, part A, p. 9b. On Li Shizhen's treatment of bamboo, see the entries in the new edition of the *Bencao gangmu*, 2 vols. (Beijing: Huaxia chuban she, 2002), pp. 1130, 1448. See also Willard Peterson, "Confucian Learning in Late Ming Thought," in Denis Twitchett and Frederick W. Mote, eds., *The Cambridge History of China*. Vol. 8: *The Ming Dynasty, 1368–1644*, part 2 (Cambridge: Cambridge University Press, 1998), pp. 782–784.

35. See Wang Baoping, "Mindai no kakushoka Ko Bunkan ni kan suru kōsatsu," *Kyūko* 36 (1999): 47–57. For editions in Japan, see Wang Baoping, "Riben Hu Wenhuan congshu jingyan lu," in Lu Jian and Wang Yong, eds., *Zhongguo dianji liuchuan yu yingxiang* (Hangzhou: Hangzhou daxue chuban she, 1990), pp. 322–347. On Chinese collectanea, see Elman, *From Philosophy To Philology*, pp. 190–191.

36. SKQSZM, 134.14b–15b. On the two hundred versions, see Wu Weizu, comp., *Siku caijin shumu* (Beijing: Commercial Press, 1960), p. 81. My thanks to Roger Hart for pointing out this source. See also Clunas, *Superfluous Things*, pp. 34, 37–38, 118.

37. Although Matteo Ricci passed through Hangzhou in 1598–99, not until 1611, with the arrival of three Jesuits at the invitation of a Hangzhou literatus, did Hangzhou become a Jesuit center for Christianity. See David Mungello, *The Forgotten Christians of Hangzhou* (Honolulu: University of Hawai'i Press, 1994), pp. 3, 15–18, and Matteo Ricci, *China in the Sixteenth Century: The Journals of Matteo Ricci: 1583–1610,* trans. Father Nichola Trigault (Latin) and Louis J. Gallagher, S.J. (English) (New York: Random House, 1953).

38. John Haeger, "The Significance of Confusion: The Origins of the *T'ai-p'ing yü-lan*," *Journal of the American Oriental Society* 88, 3 (1968): 401–410. See Y. W. Ma, "*Pi-chi*," in William H. Nienhauser Jr., ed., *The Indiana Companion to Traditional Chinese Literature* (Bloomington: Indiana University Press, 1986), pp. 650–652, and Peter K. Bol, "A Literati Miscellany and Sung Intellectual History: The Case of Chang Lei's *Ming-tao tsa-chih*," *Journal of Sung-Yuan Studies* 25 (1995): 121–151.

39. Daiwie Fu, "A Contextual and Taxonomic Study of the 'Divine Marvels' and 'Strange Occurrences' in the *Mengxi bitan*," *Chinese Science* 11 (1993–1994): 3–35. Compare Nathan Sivin, "Shen Kua," in Nathan Sivin, *Science in Ancient China*, vol. 3 (Great Yarmouth: Variorum, 1995), p. 10. See also Russell Kirkland, "A World in Balance: Holistic Synthesis in the *T'ai-p'ing kuang-chi*," *Journal of Sung-Yuan Studies* 23 (1993): 43–70.

40. Wang, "Mindai no kakushoka Ko Bunkan," pp. 47–57.

41. See Elman, *A Cultural History of Civil Examinations*, pp. 646, 661–665.

42. Wang, "Mindai no kakushoka Ko Bunkan," pp. 53–56.

43. Compare Daston and Park, *Wonders and the Order of Nature*, pp. 68–88, for the changing nature of collectors in early modern Europe.

44. Wang, "Mindai no kakushoka Ko Bunkan," pp. 52–55. See also Chum Shum, comp., *An Annotated Catalog of Chinese Rare Books in the Harvard-Yenching Library* (Shanghai: Cishu chuban she, 1998), vol. 1, pp. 312 (#054), 339 (#0588 and 0590), which includes individually sold volumes of the GZCS.

45. See Wang, "Mindai no kakushoka Ko Bunkan," p. 51, for a discussion of Hu Wen-huan's 1609 "Preface" to the GZCS. For discussion of the earliest edition, see Yu Weigang, "Hu Wenhuan yu Gezhi congshu," *Tushuguan zazhi* 4 (November 1982): 63–65. My thanks to Soren Edgren for his help on clarifying the various editions.

46. See Clunas, *Superfluous Things*, pp. 75–90, on words about things. Compare Foucault, *The Order of Things*, pp. 130–132.

47. W. S. Coblin, "*Erh ya*," in Michael Loewe, ed., *Early Chinese Texts: A Bibliographical Guide* (Berkeley: University of California Institute of East Asian Studies, 1993), pp. 94–104. See also Michael Carr, "A Linguistic Study of the Flora and Fauna Sections of the *Erh ya* (Ph.D. diss., University of Arizona, 1972), which delineates early Chinese botanical terms.

48. GZCS (late Ming Wanli edition preserved in the National Library of Taiwan), vol. 1. See also Bernhard Karlgren, "The Early History of the *Chou Li* and *Tso Chuan* Texts," *Bulletin of the Museum of Far Eastern Antiquities* 3 (1931): 44–54.

49. GZCS, vol. 1, and SKQSZM, 40.1b–6b. See also Coblin, "*Erh ya*," pp. 95–96.

50. See Lu Dian, "Preface" (*Xu*), pp. 1a–2b, in his *Piya*, in GZCS, vol. 5.

51. *Lunyu yinde* (repr., Taibei: Chinese Materials and Research Aids Service Center Inc., 1966), 36/17/8. Compare James Legge, trans., *The Four Books* (repr., New York: Paragon, 1966), p. 261: "The Master said, 'My children, why do you not study the *Poetry*? The *Poetry* stimulates the mind, enables one to view things, how to form groups, how to vent resentments. You learn the duty of serving one's father, and further the serving of one's lord. One becomes more knowledgeable with the names of birds, animals, herbs, and trees.'"

52. Lu, *Piya*, 20.1a–3a on thunder, and 20.3a–4a on lightning. Cf. Charles Hammond, "The Interpretation of Thunder," *Journal of Asian Studies* 53, 2 (May 1994): 487–503. A modern account for lightning and thunder is "When lightning flashes across the sky, it heats the air in its path to temperatures as high as 55,000 degrees F (30,500 degrees C)—over five times hotter than the surface of the sun. The air expands at supersonic speed, sending off shock waves that weaken into sound waves and greet us as the boom of thunder." See "Ask Us," *National Geographic* 199, 1 (January 2001), p. 1. For seventeenth- and eighteenth-century explanations of lightning as the same phenomenon as thunder, see J. L. Heilbron, *Electricity in the 17th and 18th Century: A Study of Early Modern Physics* (Berkeley: University of California Press, 1979), pp. 339–340.

53. Lu, *Piya*, 20.1a–3a.

54. See *Huang Ming ceheng*, comp. Mao Wei (Wu-hsing, 1605 ed.), 15.7a.

55. Yves Hervouet, ed., *A Sung Bibliography* (Hong Kong: Chinese University Press, 1978), p. 24. On Zhu Xi and the *Poetry*, see Jianhua Chen, "The 'Licentious Poems': Poetic Hermeneutics and Problems of Autonomy in Chu Hsi's Confucianism," *Papers on China* (Harvard) 7 (Spring 1998): 19–39.

56. See Elman, *A Cultural History of Civil Examinations*, pp. 451–459, and Elman, *From Philosophy to Philology*, pp. 251–259. See also William Bolz, "*Shuowen jiezi*," in Loewe, *Early Chinese Texts*, pp. 429–442, and Ssu-yü Teng and Knight Biggerstaff, comps., *An Annotated Bibliography of Selected Chinese Reference Works*, 2nd ed. (Cambridge, Mass.: Harvard University Press, 1971), pp. 39–41.

57. Foucault, *The Order of Things*, pp. 153–162. Compare Fa-ti Fan, *British Naturalists in Qing China: Science, Empire, and Cultural Encounter* (Cambridge, Mass.: Harvard University Press, 2004), pp. 91–121, who notes that Chinese literati did not have a scholarly tradition equivalent to Euro-American notions of natural history, botany, or zoology until the late nineteenth century.

58. SKQSZM, 68.1a–1b. See Campany, *Strange Writing*, pp. 34–36, and Riccardo Fracasso, "*Shan hai ching*," in Loewe, *Early Chinese Texts*, pp. 357–367. Compare Anne Birrell, "Introduction" to *The Classic of Mountains and Seas* (London: Penguin Classics, 1999), pp. xiii–xlvi, and Richard Strassberg, *A Chinese Bestiary: Guideways through Mountains and Streams* (Berkeley: University of California Press, 2002), pp. 43–57.

59. See Guo Pu, "*Xu*" (Preface), pp. 1a–4b, in GZCS, vol. 9. Another view is that the texts were originally captions to illustrations.

60. See *Shi Kuang qinjing*, p. 1a, and *Shoujing*, p. 1a, both in GZCS, vol. 10. See also Campany, *Strange Writing*, pp. 284–293, on Zhang Hua.

61. Campany, *Strange Writing*, pp. 43–45, 84–85.

62. See Ye Fang, "Xu fangshi shi," pp. 1a–1b, and the "Houxu," pp. 1a–2a, both in GZCS, vol. 10.

63. See Roger Greatrex, trans., *The Bowu zhi: An Annotated Translation* (Stockholm: Föreningen för Orientaliska Studier, 1987).

64. See Nicolas Standaert, S.J., "The Classification of Sciences and the Jesuit Mission in Late Ming China," in Jan de Meyer and Peter Engelfriet, eds., *Linked Faiths: Essays on Chinese Religions and Traditional Culture in Honor of Kristofer Schipper* (Leiden: E. J. Brill, 2000), pp. 287–317; Robert Wardy, *Aristotle in China: Lan-*

guage, Category and Translation (Cambridge: Cambridge University Press, 2000); and Rev. Justus Doolittle, *Vocabulary and Hand-book of the Chinese Language*, 2 vols. (Fuzhou: Rozario, Marcal and Company, 1872), p. 312.

65. See Yan Jing's "Preface" (*Xu*) to the *Shiwu jiyuan*, pp. 1a–b, in GZCS, vol. 17. See also Campany, *Strange Writing*, pp. 51–52.

66. Yan, "Preface" (*Xu*), pp. 2a–3a. See SCC, vol. 6, p. 213, and Teng and Biggerstaff, *An Annotated Bibliography*, pp. 102–103.

67. See Luo Qi, "Preface" (*Xu*), pp. 1a–2a, and Cai Qin, "Afterword" to the *Origins of Things*, pp. 1a–b, both in GZCS, vol. 15. See also SCC, p. 213n.

68. See Wang San'pin, "Preface" (*Xu*), to the *Gujin shiwu kao*, p. 1a–b, and Wang, *Gujin shiwu kao*, 1.1b–2b, 1.4b–11a, 2.3b–17b, in GZCS, vol. 19.

69. See Zhou Yong's "*Xu*" to the *Tiandi wanwu zaohua lun*, pp. 1a–3b, in GZCS, vol. 15. See also D. C. Lau, trans., *Confucius: The Analects* (New York: Penguin Books, 1979), p. 96.

70. Zhou, "Xu," pp. 1a–3b.

71. See Wang Bo, *Tiandi wanwu zaohua lun*, pp. 8a, 11b, in GZCS, vol. 15.

72. Ibid., pp. 1a–13a, especially p. 1a.

73. See *Liezi duben*, ed. by Zhuang Wanshou (Taipei: Sanmin shuju, 1996), pp. 50–51, and A. C. Graham, "The Date and Composition of Liehtzyy," *Asia Major*, n.s. 8 (1960–61): 139–161. The passage also refers to the *Change Classic* as the embodiment of these origins and changes.

74. For discussion, see Elman, *A Cultural History of Civil Examinations*, pp. 61–65.

75. *Zhuzi yulei*, 2.10a–10b, 3.1b. See also Zhang, "About God, Demons, and Miracles," pp. 7–8, and Kim, *The Natural Philosophy of Chu Hsi*, pp. 33, 98–99, 157–159.

76. Wang, *Tiandi wanwu zaohua lun*, pp. 4b–5a. Early modern European explanations will be addressed in Chapter 6.

77. Charlotte Furth, *A Flourishing Yin: Gender in China's Medical History, 960–1665* (Berkeley: University of California Press, 1999), pp. 150, 187–223.

78. See Ko Hung, *Alchemy, Medicine, Religion in the China of A.D. 320: The Nei P'ien of Ko Hung (Pao-P'u Tzu)*, trans. James Ware (Cambridge, Mass.: MIT Press, 1967). See also SCC, vol. 5, part 3, pp. 117–220. Compare Furth, *A Flourishing Yin*, pp. 190–223, and Shigehisa Kuriyama, *The Expressiveness of the Body and the Divergence of Greek and Chinese Medicine* (New York: Zone Books, 2002), pp. 217–227. See Chia, *Printing for Profit*, p. 315 (Table C.4), for medical imprints available in the late Ming.

79. Nathan Sivin, *Chinese Alchemy: Preliminary Studies* (Cambridge, Mass.: Harvard University Press, 1968), pp. 1–34, and Kuriyama, *The Expressiveness of the Body*, pp. 242–259. See also Richard Wilhelm, trans., *The Secret of the Golden Flower: A Chinese Book of Life* (London: Kegan Paul, 1931), and Michel Strickmann, *Chinese Magical Medicine*, ed. Bernard Faure (Stanford, Calif.: Stanford University Press, 2002), pp. 1–10.

80. Nathan Sivin, "Chinese Alchemy and the Manipulation of Time," in Nathan Sivin ed., *Science & Technology in East Asia*, pp. 109–111. See also Mirceau Eliade, *The Forge and the Crucible: The Origins and Structures of Alchemy*, trans. Stephen Corrin (New York: Harper Torchbooks, 1971), pp. 109–126. Yin and yang evolved from notions of lightness versus darkness to balanced polarities of yin

(cold, dark, female, negative, passive, etc.) and yang (warm, light, male, positive, active, etc.).

81. See Brokaw, *The Ledgers of Merit and Demerit,* pp. 17–27, 231–232, and Elman, *A Cultural History of Civil Examinations,* pp. 110, 306–307, 324, 346–354.

82. Furth, *A Flourishing Yin,* pp. 191–217.

83. See Wang Baoping, "Zhongguo Hu Wenhuan congshu jingyan lu," *Zhong Ri wenhua luncong* (*Hangzhou*) (1991): 15–17.

84. SKQSZM, 134.14b–15a.

85. See Zhu Zhifan, "Preface" (*Xu*), to the BJMS (1603 edition), pp. 5b–6a. The titles of the two collectanea were often conflated because of the overlap in works each contained.

86. SKQSZM, 134.14a. Numerous different collections are extant in libraries in China, Japan, Taiwan, and the United States. None is apparently extant in Korea, however. My thanks to Soren Edgren, Martin Heijdra, Chun Shum, and Yong Sik Kim for their help on this bibliographic puzzle.

87. See Ruth Rogaski, *Hygienic Modernity: Meanings of Health and Disease in Treaty-Port China* (Berkeley: University of California Press, 2004), chap. 1.

88. *Mingshi,* 237.6174–6176 (1590). See also DMB, pp. 152–153.

89. DMB, p. 521.

90. Feng Yingjing, *Huang Ming jingshi shiyong bian* (Xiuning, Anhui: Ming Wanli edition, 1603), 1.1a–2b. See also DMB, p. 1141.

91. Feng, *Huang Ming jingshi shiyong bian,* 1.2b–3b. There were limits to Feng's Christianization, however. He maintained, for example, that when *qi* congealed, ghosts and spirits were born, and that when *qi* scattered, they perished. See 1.15a–15b.

92. Feng, *Huang Ming jingshi shiyong bian,* "Table of contents," and 26.1a–2a. See also Elman, *A Cultural History of Civil Examinations,* pp. 380–420, and Kobayashi Tatsuhiko, "What Kind of Mathematics and Terminology Was Transmitted into 18th-Century Japan from China?," *Historia Scientiarum* 76 (July 2002): 1–17.

93. In his 1593 "Preface," Wu Jishou insisted on the centrality of mathematics as one of the "six arts." See Wu, "Preface" (*Xu*), pp. 1a–2b, in Cheng Dawei, *Suanfa tongzong,* new ed. (Xin'an: Saoye shanfang, 1883). See also ECCP, pp. 117–118.

94. Jean-Claude Martzloff, *A History of Chinese Mathematics,* trans. Stephen Wilson (New York: Springer-Verlag, 1997), pp. 15, 123, 159–163. See also ECCP, p. 117–118, and Peter Engelfriet, *Euclid in China: The Genesis of the First Chinese Translation of Euclid's Elements, Books I–VI* (Leiden: E. J. Brill, 1998), pp. 98–102.

95. Martzloff, *A History of Chinese Mathematics,* pp. 22, 161, 375–376. See also Li Yan and Du Shiran, *Chinese Mathematics: A Concise History,* trans. John Crossley and Anthony Lun (Oxford: Clarendon Press, 1987), pp. 196–201.

96. Feng, *Huang Ming jingshi shiyong bian,* 26.2a–10b. See also Martzloff, *A History of Chinese Mathematics,* pp. 124, 161.

97. See Feng, *Huang Ming jingshi shiyong bian,* 26.50a–53a. See also Benjamin A. Elman, *Classicism, Politics, and Kinship: The Ch'ang-chou School of New Text Confucianism in Late Imperial China* (Berkeley: University of California Press, 1990), pp. 76–85, and Engelfriet, *Euclid in China,* pp. 119, 138–142.

98. Compare William Ashworth Jr., "Emblematic Natural History of the Renaissance," in Jardine, Secord, and Spary, eds., *Cultures of Natural History,* pp. 17–37.

99. Compare Paula Findlen, "Courting Nature," in Jardine, Secord, and Spary, eds., *Cultures of Natural History,* pp. 57–74.

100. Findlen, *Possessing Nature,* pp. 1–3, 94, 396, 407.

101. Ibid., pp. 8, 15, 98, 121, 293.

102. Ibid., pp. 153–156, 198–201.

103. Ibid., pp. 208, 232, 286, and Daston, "Marvelous Facts and Miraculous Evidence," pp. 93–124.

104. Findlen, *Possessing Nature,* pp. 10, 18–27, 32, 46, 398–405, and Anthony Grafton, *Defenders of the Text: The Traditions of Scholarship in an Age of Science, 1450–1800* (Cambridge, Mass.: Harvard University Press, 1991), pp. 1–5.

105. See Elman, *From Philosophy to Philology,* pp. 89–92.

106. See Xu's "Preface," in GZHB, Vol. 1, pp. 3–4 (1, 3 [February 1876]: 1a–1b).

2. The Late Ming Calendar Crisis and Gregorian Reform

1. On precision and accuracy, see Nathan Sivin, "On the Limits of Empirical Knowledge in the Traditional Chinese Sciences," in J. T. Fraser, N. Lawrence, and F. C. Haber, eds., *Time, Science, and Society in China and the West* (Amherst: University of Massachusetts Press, 1986), p. 163. The ancient Chinese calendar depended on an astronomical system of mathatical computations similar to the medieval computus, described below, that informed the European Latin calendar.

2. See Thatcher E. Deane, "The Chinese Imperial Astronomical Bureau: Form and Function of the Ming Dynasty 'Qintianjian' from 1365 to 1627" (Ph.D. diss., University of Washington, 1989), pp. 1–15. For the Former Han (207 B.C.–A.D. 8), see Christopher Cullen, "The Birthday of the Old Man of Jiang County and Other Puzzles: Work in Progress on Liu Xin's *Canon of the Ages,*" *Asia Major,* Third Series 14, 2 (2001): 27–60.

3. Christopher Cullen, *Astronomy and Mathematics in Ancient China: The Zhoubi suanjing* (Cambridge: Cambridge University Press, 1996), pp. 4–6, and Edward Schafer, *Pacing the Void: T'ang Approaches to the Stars* (Berkeley: University of California Press, 1977), pp. 84–119.

4. Cullen, *Astronomy and Mathematics in Ancient China,* pp. 7–20. See also Nathan Sivin, "State, Cosmos, and Body in the Last Three Centuries B.C.," HJAS 55, 1 (June 1995): 7.

5. Olaf Pederson, "The Ecclesiastical Calendar and the Life of the Church," in GRC, pp. 27–28. Compare Paul Richard Blum, "The Jesuits and the Janus-faced History of Natural Sciences," in Jurgen Helm and Annette Winkelmann eds., *Religious Confessions and the Sciences in the Sixteenth Century* (Leiden: E. J. Brill, 2001), pp. 19–34.

6. Liam Brockey, "The Harvest of the Vine: The Jesuit Missionary Enterprise in China, 1579–1710" (Ph.D. diss., Brown University, 2002), pp. 22–37. See also Willard Peterson, "What to Wear? Observation and Participation by Jesuit Missionaries in Late Ming Society," in Stuart Schwartz, ed., *Implicit Understandings: Observing, Reporting, and Reflecting on the Encounters between Europeans*

and Other Peoples in the Early Modern Era (Cambridge: Cambridge University Press, 1994), pp. 416–417.

7. Nicolas Standaert, *Handbook of Christianity in China, Volume One: 635–1800* (Leiden: E. J. Brill, 2001), p. 493. See Jurgis Elisonas (George Elison), "Christianity and the Daimyo," in John Hall, ed., *The Cambridge History of Japan*. Vol. 4. *Early Modern Japan* (Cambridge: Cambridge University Press, 1991), pp. 332–333; and Derek Massarella, "The Jesuits and Japan," in Anthony Disney and Emily Booth, eds., *Vasco da Gama and the Linking of Europe and Asia* (Oxford: Oxford University Press, 2000), pp. 233–247.

8. See Shigeru Nakayama, *A History of Japanese Astronomy: Chinese Background and Western Impact* (Cambridge, Mass.: Harvard University Press, 1969), p. 69.

9. Cullen, *Astronomy and Mathematics,* pp. 35–39. See also Nakayama, *A History of Japanese Astronomy,* p. 69, and Deane, "The Chinese Imperial Astronomical Bureau," pp. 15–16, 43–48.

10. Ho Peng Yoke, "The Astronomical Bureau in Ming China," *Journal of Asian History* 3, 2 (1969): 135–157, and Elman, *A Cultural History of Civil Examinations,* pp. 157–163.

11. Morris Rossabi, "The Muslims in the Early Yuan Dynasty," in John Langlois Jr., ed., *China Under Mongol Rule* (Princeton, N.J.: Princeton University Press, 1981), pp. 258–295, and Yabuuchi Kiyoshi, "Islamic Astronomy in China during the Yuan and Ming Dynasties," trans. and ed. Benno van Dalen, *Historia Scientiarum* 7, 1 (1997): 11–43.

12. Thatcher E. Deane, "The Chinese Imperial Astronomical Bureau," pp. 203–232. See also Zvi Aziz Ben-dor, "The 'Dao of Muhammad': Scholarship, Education, and Chinese Literati Identity in Late Imperial China" (Ph.D. diss., University of California–Los Angeles, 2000).

13. SCC, III, pp. 370–372.

14. Aydin Sayili, *The Observatory in Islam and Its Place in the General History of the Observatory* (Ankara: Türk Tarih Kurumu Basimevi, 1960), pp. 197–205, 211–218, 361–364, 397.

15. Willy Lardner, "The Astronomical Instruments of Cha-ma-lu-ting, Their Identification and Their Relations to the Instruments of the Observatory of Marâgha," *Isis* 41 (1950): 184–194. See also SCC, III, pp. 372–382.

16. Compare Sayili, *The Observatory in Islam,* especially pp. 187–223 (on Marâgha), pp. 232–236, and George Saliba, "The Astronomical Tradition of Maragha: A Historical Survey and Prospects for Future Research," *Arabic Science and Philosophy* 1, 1 (1991): 67–99.

17. See also Priscilla Soucek, "Ceramic Production as Exemplar of Yuan-Ilkhanid Relations," *RES Anthropology and Aesthetics* 35 (Spring 1999): 125–141.

18. See the unsubstantiated views of Yabuuchi Kiyoshi on this question in Yabuuti, "Islamic Astronomy in China." Joseph Needham (SCC, III, p. 105) supposed that these Muslim versions were trans. into Chinese. My thanks to Nathan Sivin for his advice.

19. Lardner, "The Astronomical Instruments of Cha-ma-lu-ting," pp. 184–194. See also Karine Chemla, "Similarities between Chinese and Arabic Mathematical Writings: (I) Root Extraction," *Arabic Science and Philosophy* 4 (1994): 205–266.

20. See Lam Lay-Yong, "Chu Shih-chieh's *Suan-hsueh ch'i-meng* (Introduction to Mathematical Studies)," *Archive for History of Exact Sciences* 21, 1 (1979): 1–31.

21. Keizō Hashimoto, *Hsü Kuang-ch'i and Astronomical Reform—The Process of the Chinese Acceptance of Western Astronomy 1629–1635* (Osaka: Kansai University Press, 1988), p. 35.

22. Deane, "The Chinese Imperial Astronomical Bureau," pp. 255–262. See also Chu Pingyi, "Scientific Dispute in the Imperial Court: The 1664 Calendar Case," *Chinese Science* 14 (1997): 28–29.

23. Deane, "The Chinese Imperial Astronomical Bureau," pp. 262–269, which on the whole controverts Wolfram Eberhard, "The Political Function of Astronomy and Astronomers in Han China," in John K. Fairbank, ed., *Chinese Thought and Institutions* (Chicago: University of Chicago Press, 1957), pp. 33–70. See also Nathan Sivin, "Cosmos and Computation in Early Chinese Mathematical Astronomy," in Nathan Sivin, *Science in Ancient China: Researches and Reflections* (Brookfield, Vt.: Variorium, 1995). For the political context of the 1404 anomaly, see Benjamin A. Elman, "The Formation of 'Dao Learning' as Imperial Ideology during the Early Ming Dynasty," in Theodore Huters, R. Bin Wong, and Pauline Yu, eds., *Culture and the State in Chinese History* (Stanford, Calif.: Stanford University Press, 1997), pp. 58–82.

24. Deane, "The Chinese Imperial Astronomical Bureau," pp. 269–282.

25. Cullen, *Astronomy and Mathematics,* pp. 10–20, and Deane, "The Chinese Imperial Astronomical Bureau," pp. 118, 283–284.

26. Deane, "The Chinese Imperial Astronomical Bureau," pp. 285–304. Compare Shi Yunli and Lü Lingfeng, "Lizhi, chuanjiao yu jiaoshi ceyan: Qing Qintianjian dang'an zhong de jiaoshi jilu toushi," *Ziran bianzheng fa tongxun* 24, 6 (2002): 44–50.

27. Joseph Needham, Wang Ling, and Derek de Solla Price, *Heavenly Clockwork: The Great Astronomical Clocks of Medieval China,* 2nd ed. (Cambridge: Cambridge University Press, 1986), pp. 48–59. See also Joseph Needham, *The Grand Titration: Science and Society in East and West* (Toronto: University of Toronto Press, 1969), pp. 78–82.

28. Huang Yi-long, "Court Divination and Christianity in the K'ang-hsi Era," *Chinese Science* 10 (1991): 4. Compare Deane, "The Chinese Imperial Astronomical Bureau," pp. 305–341, 356, 360–64.

29. Deane, "The Chinese Imperial Astronomical Bureau," pp. 353–400.

30. Chinese calculation techniques to solve numerical problems of one or more unknown quantities were not abstracted from their practical applications to form a systematic science of numbers for problem solving. As a craft, Chinese counting-rod techniques conveyed the result by example rather than elucidating the process as in modern algebra and calculus. See Michael S. Mahoney, "The Royal Road: The Development of Algebraic Analysis from 1550 to 1600" (Ph.D. diss., Princeton University, 1967), chapter 3.

31. Deane, "The Chinese Imperial Astronomical Bureau," p. 400–402.

32. SCC, vol. 3, pp. 390–408. Compare Cullen, *Astronomy and Mathematics,* pp. 24–27.

33. Sivin, "Cosmos and Computation in Early Chinese Mathematical Astronomy," II, pp. 14–15, 20. See also SCC, vol. 3, pp. 406–407. Compare Cullen, *Astron-*

omy and Mathematics, pp. 24–25, and Cullen, "The Birthday of the Old Man of Jiang County," pp. 30–38.

34. Deane, "The Chinese Imperial Astronomical Bureau," pp. 405–409.

35. See Bernard Goldstein, "A Note on the Metonic Cycle," *Isis* 57 (1966): 155–156. See also Alexander Philip, *The Calendar: Its History, Structure and Improvement* (Cambridge: Cambridge University Press, 1921), pp. 7–9.

36. J. L. Heilbron, *The Sun in the Church: Cathedrals as Solar Observatories* (Cambridge, Mass.: Harvard University Press, 1999), p. 24.

37. Pederson, "The Ecclesiastical Calendar," pp. 17–18.

38. Pederson, "The Ecclesiastical Calendar," pp. 47–49, 54–55, and Philip, *The Calendar,* pp. 43–45. See also Heilbron, *The Sun in the Church,* pp. 29–36.

39. Florence Hsia, "French Jesuits and the Mission to China: Science, Religion, History" (Ph.D. diss., University of Chicago, 1999), pp. 207–215.

40. MS, 31.518–519 (167). See also Nakayama, *A History of Japanese Astronomy,* pp. 123–139.

41. See Deane, "The Chinese Imperial Astronomical Bureau," pp, 409–418, and Willard Peterson, "Calendar Reform Prior to the Arrival of Missionaries at the Ming Court," *Ming Studies* 21 (Spring 1986): 47–49.

42. Deane, "The Chinese Imperial Astronomical Bureau," pp. 417–22. The original astronomical instruments were still in Nanjing.

43. A similar question on the calendar appeared on the 1594 Guizhou provincial examination. See Elman, *A Cultural History of Civil Examinations,* pp. 468–477, 664.

44. Peterson, "Calendar Reform," pp. 46–47.

45. Elman, *A Cultural History of Civil Examinations,* pp. 469–470. See also Sivin, "Cosmos and Computation," p. 63, which speaks of "a commitment to continued observation, which Chinese calendrical astronomy was dedicated to transcend." See also Elman, *Classicism, Politics, and Kinship,* pp. 29–32, on late Ming politics.

46. Elman, *A Cultural History of Civil Examinations,* p. 470

47. Ibid., p. 471.

48. Ibid., p. 472.

49. DMB, pp. 1243–1254, and Elman, *Classicism, Politics, and Kinship,* pp. 76–83.

50. Elman, *Classicism, Politics, and Kinship,* pp. 78–79.

51. See Han Qi, "Bai Jin de Yijing yanjiu he Kangxi shidai de Xixue Zhongyuan shuo," *Hanxue yanjiu* (Taiwan) 16, 1 (June 1998): 185–201. Compare Elman, *From Philosophy to Philology,* pp. 116–120.

52. MS, 31.520 (167), and Deane, "The Chinese Imperial Astronomical Bureau," pp. 425–427. See also Hashimoto, *Hsu Kuang-ch'i,* pp. 74–77, and Roger Hart, "Quantifying Ritual: Political Cosmology, Courtly Music, and Precision Mathematics in Seventeenth-Century China" (presented at the School of Historical Studies, Institute for Advanced Study, Princeton, N.J., December 3, 1999).

53. MS, 31.527–528 (169). See also Peterson, "Calendar Reform," pp. 54–55, and Deane, "The Chinese Imperial Astronomical Bureau," pp. 421–433.

54. Nakayama, *A History of Japanese Astronomy,* p. 69.

55. Eclipses also mattered in both Latin and Orthodox Christendom, but for astrological rather than calendrical purposes. See Kurt Vogel, "Byzantine Science," in

J. M. Husey, ed., *The Cambridge Medieval History*, vol. 4, *The Byzantine Empire*, part II, new ed. (Cambridge: Cambridge University Press, 1966), pp. 264–305.

56. Pederson, "The Ecclesiastical Calendar," pp. 29–30, 55–62.

57. Jerzy Dobrzycki, "Astronomical Aspects of the Calendar Reform," pp. 117–120, in GRC. See also Heilbron, *The Sun in the Church*, pp. 37–40.

58. J. D. North, "The Western Calendar—'Intolerabilis, Horribilis, et Derisibilis'; Four Centuries of Discontent," pp. 95–100, in GRC. See also Dobrzycki, "Astronomical Aspects," pp. 118–23; Noel Swerdlow, "The Origin of the Gregorian Civil Calendar," *Journal for the History of Astronomy* 5 (1974): 48–49; and Edoardo Proverbio, "Copernicus and the Determination of the Length of the Tropical Year," pp. 129–134, in GRC. Compare Robert Westman, "The Copernicans and the Churches," in Lindberg and Numbers, eds., *God and Nature*, p. 77.

59. North, "The Western Calendar," pp. 101–102, and Philip, *The Calendar*, p. 21.

60. Pederson, "The Ecclesiastical Calendar," pp. 65, 69. See also August Ziggelaar, S.J., "The Papal Bull of 1582 Promulgating a Reform of the Calendar," pp. 201–209, and Gordon Moyer, "Alosius Lilius and the 'Compendium Novae Rationis Restitueni Kalendarium,'" p. 182–183, both in GRC.

61. Heilbron, *The Sun in the Church*, p. 41.

62. See Grafton, *Defenders of the Text*, pp. 104–144, and R. Po-chia Hsia, *The World of Catholic Renewal 1540–1770* (Cambridge: Cambridge University Press, 1998), pp. 1–9. See also Ziggelaar, "The Papal Bull of 1582," pp. 226–231, and Michael Hoskin, "The Reception of the Calendar by Other Churches," in GRC, pp. 259–262.

63. H. M. Nobis, "The Reaction of the Astronomers to the Gregorian Calendar," pp. 243–250, in GRC. See also Hoskin, "The Reception of the Calendar," and Owen Gingerich, "The Civil Reception of the Gregorian Calendar," in GRC, pp. 265–279.

64. Nobis, "The Reaction of the Astronomers," pp. 250–251, and Hoskin, "The Reception of the Calendar," pp. 256–259. The disruptions of World War I brought the Gregorian calendar to Eastern Europe and Greece, but it was not brought to Russia until the Bolshevik government in 1918. See Gingerich, "The Civil Reception," p. 276.

65. Ugo Baldini, "Christoph Clavius and the Scientific Scene in Rome," pp. 145–251, 161, 165, GRG and Engelfriet, *Euclid in China*, pp. 49–50. See Richard Westfall, "The Rise of Science and the Decline of Orthodox Christianity: A Study of Kepler, Descartes, and Newton," in Lindberg and Numbers, eds., *God and Nature*, pp. 218–237.

66. See Ugo Baldini, "The Academy of Mathematics of the Collegio Romano from 1553 to 1612," in Mordechai Feingold, ed., *Jesuit Science and the Republic of Letters* (Cambridge, Mass.: MIT Press, 2002), and Mario Biagioli, "The Social Status of Italian Mathematicians, 1450–1600," *History of Science* 27 (1989): 41–95. In 1597–1598, Ricci translated the first book of Euclid's *Elements* with the help of Qu Rukui, but this version was later lost. Later Ricci adapted Clavius's Latin commentary on the *Elements* for a new Chinese translation in 1607 with the help of Xu Guangqi.

67. Baldini, "Christoph Clavius," pp. 143–145, 157, and North, "The Western Calendar," p. 81. Copernicus was perceived as a tolerable reformer by Clavius and

others until Galileo and the 1616 proscription. See Westman, "The Copernicans," pp. 76–113, and James Lattis, *Between Copernicus and Galileo: Christopher Clavius and the Collapse of the Ptolemaic Cosmology* (Chicago: University of Chicago Press, 1994), pp. 30–60.

68. See Sivin, "Cosmos and Computation," pp. 1–73.

69. See Nathan Sivin, "Copernicus in China," in *Colloquia Copernica II. Etudes sur l'audience de la theorie heliocentrique* (Warsaw: Union Internationale d' Historie et Philosophie des Sciences, 1973), pp. 63–122.

70. Nicolas Standaert, "Jesuit Corporate Culture as Shaped by the Chinese," in John W. O'Malley et al., eds., *The Jesuits: Cultures, Sciences, and the Arts 1540–1773* (Toronto: University of Toronto Press, 1999), pp. 352–362. In recent work, Robert Westman contends that the "Copernican revolution" was a revolution in both astronomy and astrology. See Robert Westman, *The Copernican Question: Prognostication, Scepticism and Celestial Order, 1470–1610* (Chicago: University of Chicago Press, forthcoming), chap. 1.

71. Peterson, "What to Wear?," pp. 403–421.

72. Philip, *The Calendar*, p. 51. See also Pederson, "The Ecclesiastical Calendar," pp. 50–51.

73. Chu, "Scientific Dispute," pp. 12, 25. The largely Ming calendar materials presented in 1646 to the Shunzhi emperor were titled "Works on calendrical studies according to new Western methods" (*Xiyang xinfa lishu*). See also Hashimoto, *Hsü Kuang-ch'i*, pp. 33–34. On Japan, see Elisonas, "Christianity and the Damyo," pp. 326–356.

74. Gabor Lukacs, "The Early Penetration of French Science into Japan during the Tokugawa Period," in Vande Walle and Kasaya, eds., *Dodonæus in Japan*, pp. 314–317.

75. Marcus Heller, "Jesuit Physics in Eighteenth-Century Germany: Some Important Continuities," in O'Malley et al., eds., *The Jesuits*, pp. 538–554. See also Hsia, *The World of Catholic Renewal*, pp. 32–33, and Engelfriet, *Euclid in China*, pp. 11–13. Compare Jesús López-Gay, "Las primeras relaciones del Colegio Romano con el extremo oriente," *Zeitschrift für Missionswissenschaft und Religionswissenschaft* 86, 4 (2002): 277–285.

76. Peter Engelfriet, "The Chinese Euclid and Its European Context," in Catherine Jami and Hubert Delahaye, eds., *L'Europe en Chine: Interactions scientifique, religieuses et culturelles aux xviie et xviiie siecles* (Paris: Institute des Hautes Études Chinoises, College de France, 1993), p. 114. See Arnold Sprenger, "Johann Adam Schall's Educational Foundation and the Intellectual Climate of His Time," in Roman Malek, ed., *Western Learning and Christianity in China*, 2 vols. (Sankt Augustin: China Zentrum, 1998), pp. 42–43, and Catherine Jami, "From Clavius to Pardies: The Geometry Transmitted to China by Jesuits (1607–1723)," in Masini, ed., *Western Humanistic Culture*, pp. 182–183. For Verbiest's early education and training, see Wang Bing, *Qinmin zhi shi: Nan Huairen* (Beijing: Kexue chuban she, 2000), pp. 14–22. On Ricci, see Engelfriet, *Euclid in China*, pp. 17–23, 50–53.

77. See Brockey, "The Harvest," pp. 217–248.

78. J. S. Cummins, "Two Missionary Methods in China: Mendicants and Jesuits," *Archivo Ibero-Americano* 37, 149–152 (1978): 33–108.

79. Engelfriet, *Euclid in China*, pp. 14–17, 332–333, 434–435. See also Claudia von Collani, "Charles Maigrot's Role in the Chinese Rites Controversy," in D. E. Mungello, ed., *The Chinese Rites Controversy: Its History and Meaning* (Nettetal: Steyler Verlag, 1994), pp. 150–151.

80. Steven Harris, "Transposing the Merton Thesis: Apostolic Spirituality and the Establishment of the Jesuit Scientific Tradition," *Science in Context* 3, 1 (1989): 39–52. Compare Rivka Feldhay, "The Cultural Field of Jesuit Science," in O'Malley et al., eds., *The Jesuits*, pp. 107–130, which describes the compromises Jesuit mathematicians made with the entrenched religious interests of the Society of Jesus. See also Feingold, ed., *Jesuit Science and the Republic of Letters*, which stresses the scientific contributions of the Jesuit and the constraints under which they worked.

81. Hashimoto, *Hsü Kuang-ch'i*, pp. 14–15. See also Nicolas Standaert, "The Jesuit Presence in China (1580–1773): A Statistical Approach," *Sino-Western Cultural Relations Journal* 13 (1991): 4–17, which notes that the number of Jesuits in China peaked in 1700.

82. MS, 31.528 (169).

83. See *Xu Guangqi nianpu* (Shanghai: Guji chuban she, 1981), p. 95, citing the MS, 31.528 (169). See also Deane, "The Chinese Imperial Astronomical Bureau," pp. 401–434; Hashimoto, *Hsü Kuang-ch'i*, pp. 13–14; and Peterson, "Calendar Reform," p. 55.

84. Brockey, "The Harvest," p. 232n28. See also Hsia, "French Jesuits," pp. 10–20.

85. *Xu Guangqi nianpu*, pp. 98–99, and Xu Guangqi, *Nongzheng quanshu* (repr., Shuhai lou, 1843). Compare Engelfriet, *Euclid in China*, pp. 326–329. See also Peterson, "Learning from Heaven," p. 806, and Baldini, "Christoph Clavius," p. 142.

86. *Xu Guangqi nianpu*, p. 104.

87. Pasquale M. d'Elia, *Galileo in China: Relations through the Roman College between Galileo and the Jesuit Scientist-Missionaries (1610–1640)*, trans. R. Suter and M. Sciascia (Cambridge, Mass.: Harvard University Press, 1960), pp. 21–23; Deane, "The Chinese Imperial Astronomical Bureau," pp. 435–438; and Hashimoto, *Hsü Kuang-ch'i*, pp. 14–15.

88. *Xu Guangqi nianpu*, p. 104, and Hashimoto, *Hsü Kuang-ch'i*, pp. 16–19. See also Elman, *A Cultural History of Civil Examinations*, pp. 97–105.

89. Adrian Dudink, "The Inventories of the Jesuit House at Nanking Made up during the Persecution of 1616–1617 (Shen Que, *Nangong shudu*, 1620)," in Federico Masini, ed., *Western Humanistic Culture Presented to China by Jesuit Missionaries (XVIII–XVIII Centuries)* (Rome: Institutum Historicum S.I., 1996), pp. 121–123. Compare Brockey, "The Harvest," pp. 58–89. On the Jianwen reign, see Elman, *A Cultural History of Civil Examinations*, pp. 97–105, 119–121.

90. *Xu Guangqi nianpu*, pp. 112–113, Engelfriet, *Euclid in China*, p. 330, and DMB, pp. 1177–1178, 1332–1333. Galileo had observed sunspots, the four moons circling Jupiter, and the rough surface of the moon.

91. *Xu Guangqi nianpu*, pp. 135–142, 158–160, ECCP, p. 453, and Hashimoto, *Hsü Kuang-ch'i*, pp. 19–21. See also Jonathan Spence, *To Change China: Western Advisers in China 1620–1960* (Harmondsworth: Penguin Books, 1980), p. 9; Mungello, *The Forgotten Christians of Hangzhou*, pp. 3, 15–16; Engelfriet, *Euclid*

in China, p. 335; and Paul Rule, *K'ung-tzu or Confucius? The Jesuit Interpretation of Confucianism* (London: Allen & Unwin, 1986), pp. 66–67. On late Ming politics, see Benjamin A. Elman, "Imperial Politics and Confucian Societies in Late Imperial China: The Hanlin and Donglin Academies," *Modern China* 15, 4 (1989): 379–418. On the Jesuit production of cannon for the Ming and Qing, see Standaert, *Handbook of Christianity in China*, pp. 771–777.

92. See *Xu Guangqi ji* (Taibei: Mingwen shuju, 1986), pp. 319–323, 332–338, and *Xu Guangqi nianpu*, pp. 163–167. See also Hasihimoto, *Hsü Kuang-ch'i*, pp. 41–44, and Engelfriet, *Euclid in China*, p. 344.

93. *Xu Guangqi nianpu*, pp. 104, 173, ECCP, p. 317; Deane, "The Chinese Imperial Astronomical Bureau," pp. 440–441; and Hashimoto, *Hsü Kuang-ch'i*, pp. 20–25, 41.

94. See *Xu Guangqi nianpu*, p. 148, and Standaert, "The Classification of Sciences and the Jesuit Mission in Late Ming China," pp. 287–317, and Bernhard Hung-Kay Luk, "Aleni Introduces the Western Academic Tradition to Seventeenth-Century China: A Study of the *Xixue fan*," in T. Lippiello and R. Malek, eds., *"Scholar from the West": Giulio Aleni S.J. (1582–1649) and the Dialogue between Christianity and China* (Nettetal: Steyler Verlag, Monumenta Serica Monograph Series XLII, 1997), pp. 479–518.

95. Hashimoto, *Hsü Kuang-ch'i*, pp. 23–27.

96. See Sivin, "Copernicus in China," in Sivin, *Science in Ancient China*, vol. 4, pp. 1–52.

97. Hashimoto, *Hsü Kuang-ch'i*, pp. 24, 28–31, 168–178. The switch to the Tychonic world system lasted until the middle of the eighteenth century, when it was replaced by the Keplerian system.

98. Sivin, "Cosmos and Computation," pp. 11–19, and Cullen, *Astronomy and Mathematics*, pp. 50–54. See also Nakayama, *A History of Japanese Astronomy*, pp. 24–43, for a discussion of early Chinese cosmology.

99. *Xu Guangqi ji*, pp. 374–375, 377–378, and Hashimoto, *Hsü Kuang-ch'i*, pp. 46–51, 227. Compare Han Qi, "Astronomy, Chinese and Western: The Influence of Xu Guangqi's Views in the Early and Mid-Qing," in Catherine Jami, Peter Engelfriet, and Gregory Blue, eds., *Statecraft and Intellectual Renewal in Late Ming China: The Cross-Cultural Synthesis of Xu Guangqi (1562–1633)* (Leiden: E. J. Brill, 2001), pp. 360–373.

100. *Xu Guangqi ji*, pp. 356–358, *Xu Guangqi nianpu*, pp. 169–178, and Hashimoto, *Hsü Kuang-ch'i*, pp. 35–45, 50.

101. *Xu Guangqi ji*, pp. 385–386, 402–404, 424–426, and Hashimoto, *Hsü Kuang-ch'i*, pp. 62–72.

102. Hashimoto, *Hsü Kuang-ch'i*, pp. 70–71.

103. Ibid., pp. 63–72. See also Chu Pingyi, "Scientific Dispute," p. 13.

104. Martzloff, *A History of Chinese Mathematics*, pp. 204, 211, 386. Hashimoto, *Hsü Kuang-ch'i*, pp. 66–68.

105. Compare Hu Mingie, "Merging Chinese and Western Mathematics: The Introduction of Algebra in China (Ph.D. diss., Princeton University, 1998), pp. 35–36, 42, Catherine Jami, "From Clavius to Pardies," pp. 175–199, and Engelfriet, "The Chinese Euclid," pp. 111–135.

106. Sivin, "Copernicus in China," in Sivin, *Science in Ancient China,* vol. 4, pp. 14–16, and Hashimoto, *Hsü Kuang-ch'i,* pp. 1–4, 75–81. See also Baldini, "Christoph Clavius," pp. 50–51.

107. Hashimoto, *Hsü Kuang-ch'i,* pp. 75–102, 125–127, 170–171, 228.

108. Ibid., pp. 84–95, 100–108, and Sivin, "Copernicus in China," in *Colloquia Copernica II,* p. 80.

109. *Catalogue de la Bibliothèque du Pè-T'ang* (Beijing: Imprimarie des Lazaristes, 1949), p. 560, indicates the tables were used to predict solar and lunar eclipses in late 1646. See also Chu Pingyi, "Trust, Instruments, and Cross-Cultural Scientific Exchanges: Chinese Debate over the Shape of the Earth, 1600–1800," *Science in Context* 12, 3 (1999): 395.

110. See Chu Pingyi, "Sanjiao hanshu biao yu Mingmo de Zhongxi lifa zhi zheng—shang xia," *Dalu zazhi* 99, 5 (1999): 233–240, and 99, 6 (1999): 249–258; Han Qi, "Sino-British Relations through Jesuits in the seventeenth and eighteenth centuries," in Michel Cartier, ed., *La Chine entre amour et haine* (Institut Ricci, Desclèe de Brouwer, 1998), p. 57; and Hashimoto, *Hsü Kuang-ch'i,* pp. 140–141. See also Tsuen-hsuin Tsien, "Western Impact on China through Translation," *Far Eastern Quarterly* 13 (1954): 310, and Li Yan and Du Shiran, *Chinese Mathematics,* pp. 208–209.

111. SCC, III, pp. 378–380.

112. Hashimoto, *Hsü Kuang-ch'i,* pp. 142–146, 151–162. See also Joseph Needham, "The Peking Observatory in A.D. 1280 and the Development of the Equatorial Mounting," *Vistas in Astronomy* 1 (1955): 67–83.

113. Liu Lu, ed., *Qinggong xiyang yiqi* (Hong Kong: Commercial Press, 1998), p. 54.

114. Some *Astronomiae* documents are described in *The Library of Philip Robinson,* part 2: *The Chinese Collection* (London: Sotheby's Auction Catalogue, 1988), pp. 135–139. See also Zhang Baicun, "The Introduction of European Astronomical Instruments and the Related Technology into China during the Seventeenth Century," *East Asian Science, Technology, and Medicine* 20 (2003): 99–131; Allan Chapman, "Tycho Brahe in China: the Jesuit Mission to Peking and the Iconography of European Instrument-Making Processes," *Annals of Science* 41 (1984): 417–443; and Liu, ed., *Qinggong xiyang yiqi,* p. 8. Compare Isaia Iannaccone, "Syncretism between European and Chinese Culture in the Astronomical Instruments of Ferdinand Verbiest in the Old Beijing Observatory," in John Witek, ed., *Ferdinand Verbiest (1623–1688): Jesuit Missionary, Scientist, Engineer and Diplomat* (Nettetal: Steyler Verlag, 1994), pp. 93–121, and Noël Golvers, *The Astrononia Europaea of Ferdinand Verbiest, S.J. (Dillingen, 1687): Text, Translation, Notes and Commentaries* (Nettetal: Steyler Verlag, 1993).

115. Hashimoto, *Hsü Kuang-ch'i,* pp. 186–189, 212–216, and SCC, III, pp. 444–445.

116. Catherine Jami, "Western Devices for Measuring Time and Space: Clocks and Euclidian Geometry in Late Ming and Ch'ing China," in Chun-chieh Huang and Erik Zürcher, eds., *Time and Space in Chinese Culture* (Leiden: E. J. Brill, 1995), pp. 186–187. See also Hashimoto, *Hsü Kuang-ch'i,* pp. 216–217, and Chapman, "Tycho Brahe in China," p. 421.

117. Standaert, *Handbook of Christianity in China*, p. 493, notes that Ricci brought clocks, a clavichord, and other curiosa to Beijing in the hope of presenting them to the Wanli emperor.

118. Joseph Needham, *Clerks and Craftsmen in China and the West* (Cambridge: Cambridge University Press, 1970), p. 205, and Jami, "Western Devices," pp. 169–200. See also Jonathan Spence, *Emperor of China: Self-Portrait of K'ang-hsi* (New York: Vintage Books, 1975), p. 68.

119. Chapman, "Tycho Brahe in China," p. 420. See also Iannaccone, "Syncretism between European and Chinese Culture," p. 100.

120. Chapman, "Tycho Brahe in China," pp. 424–426.

121. Ibid., pp. 439–440. See also Allan Chapman, "The Accuracy of Angular Measuring Instruments Used in Astronomy between 1500 and 1850," *Journal for the History of Astronomy* 14 (1983): 133–137.

122. Dava Sobel and William Andrews, *The Illustrated Longitude* (London: Fourth Estate, 1999), pp. 15–41.

123. Engelfriet, *Euclid in China*, pp. 53–55. For the Jesuits, only an Aristotelian natural philosophy could explain movement.

124. Florence Hsia, "Some Observations on the *Observations*: The Decline of the French Jesuit Scientific Mission in China," *Revue de synthèse* 2–3 (1999): 305–333.

125. Chapman, "Tycho Brahe in China," pp. 441–442. See also Zhang, "The Introduction of European Astronomical Instruments," pp. 106–107. Compare Hsia, "French Jesuits and the Mission to China," pp. 5–60.

126. Engelfriet, *Euclid in China*, pp. 56, 86. Ricci still needed the forbearance of Chinese elites and the court to protect the Jesuit missionaries who were successful in converting the common folk. See Brockey, "The Harvest."

3. Sino-Jesuit Accommodations During the Seventeenth Century

1. Latin-Chinese glossaries compiled by the Jesuits and their collaborators were forerunners of modern dictionaries of the Chinese language. See Federico Masini, "Using the Works of the Jesuit Missionaries in China to Study the Chinese Language: A Research Project" (presented at the international conference Translating Western Knowledge into Late Imperial China, Göttingen University, December 6–9, 1999), and Masini, "Some Preliminary Remarks on the Study of Chinese Lexicographic Material Prepared by the Jesuit Missionaries in China in the Seventeenth Century," in Masini, ed., *Western Humanistic Culture*, pp. 235–245. On Ricci, see Martzloff, *A History of Chinese Mathematics*, pp. 20–21.

2. See Sprenger, "Johann Adam Schall's Educational Foundation," pp. 42–43, and Jami, "From Clavius to Pardies," pp. 182–183.

3. SCC, vol. 3, p. 110. Compare Karine Chemla, "Theoretical Aspects of the Chinese Algorithmic Tradition (first to third century)," *Historia Scientiarum*, 42 (1991): 75–98.

4. Lay-Yong Lam, *A Critical Study of the Yang Hui Suan Fa, a Thirteenth-Century Mathematical Treatise* (Singapore: Singapore University Press, 1977), pp.

240–241. See also Ulrich Libbrecht, *Chinese Mathematics in the Thirteenth Century: The Shu-shu Chiu-chang of Ch'in Chiu-shao* (Cambridge, Mass.: MIT Press, 1973), p. 11. Compare Mei Rongzhao, "The Decimal Place-Value Numeration and the Rod and Bead Arithmetics," in *Ancient China's Technology and Science*, pp. 57–65.

5. Li and Du, *Chinese Mathematics*, pp. 194, 196–201, and Martzloff, *A History of Chinese Mathematics*, pp. 22, 375–376, 383. See also Limin Bai, "Children and the Survival of China: Liang Qichao on Education Before the 1898 Reform," *Late Imperial China* 22, 2 (December 2001): 139.

6. See Peterson, "Learning From Heaven," pp. 806–807; Engelfriet, *Euclid in China*, pp. 17–35, 44; Jonathan Spence, *The Memory Palace of Matteo Ricci* (New York: Viking Penguin, Inc., 1985), pp. 143–144; and Standaert, "The Classification of Sciences and the Jesuit Mission in Late Ming China," pp. 287–317. See also Baldini, "Christoph Clavius," p. 145; Jardine, "The Forging of Modern Realism," p. 168; and William B. Ashworth, Jr., "Catholicism and Early Modern Science," in Lindberg and Numbers, eds., *God and Nature*, pp. 136–166.

7. Dear, "Jesuit Mathematical Science," p. 134. On the historical context, see Lukacs and Consentino, *Church, Culture & Curriculum*, pp. 17–46. See also Daston, "Marvelous Facts and Miraculous Evidence," and R. Po-chia Hsia, *The World of Catholic Renewal 1540–1770*, pp. 106–121.

8. Standaert, "The Classification of Sciences," p. 306. See also David Mungello, "The First Complete Translation of the Confucian Four Books in the West," in Lo Kuang, ed., *International Symposium on Chinese-Western Cultural Exchange in Commemoration of the 400th Anniversary of the Arrival of Matteo Ricci, S.J. in China* (Taibei: Furen University Press, 1983), p. 519, and Andrew Plaks, "The mean, nature, and self-realization: European translations of the *Zhongyong*," in Viviane Alleton and Michael Lackner, eds., *De 'un au multiple: Traductions du chinois vers les langues européennes* (Paris: Maison des sciences de l'homee, 1999), pp. 313–331.

9. Wardy, "Chinese Whispers," p. 159, and David Mungello, "The Seventeenth-Century Jesuit Translation Project of the Confucian Four Books," in Charles Ronan, S.J., and Bonnie Oh, eds., *East Meets West: the Jesuits in China*, pp. 252–272. See also Jerome Heyndrick, ed., *Philippe Couplet, S.J. (1623–1693): The Man Who Brought China to Europe* (Nettetal: Steyler Verlag, 1990).

10. Standaert, "The Classification of Sciences," pp. 5–7, and "The Investigation of Things and the Fathoming of Principles," p. 407. Compare Eugenio Menegon, *Un solo Cielo: Giulio Aleni S.J. (1582–1649)* (Rome: Grafo, 1994), pp. 157–162.

11. Paul Rule, "Goa—Macao—Beijing: The Jesuits and Portugal's China Connection," in Anthony Disney and Emily Booth, eds., *Vasco da Gama and the Linking of Europe and Asia* (Oxford: Oxford University Press, 2000), pp. 248–260. For discussion, see Nicolas Standaert, "Methodology in View of Contact between Cultures: The China Case in the 17th Century," *CSRCS Occasional Paper* (Chinese University of Hong Kong, Centre for the Study of Religion and Chinese Society) 11 (December 2002): 1–64.

12. See Alfonso Vagnoni's *Kongji gezhi*, in *Tianzhu jiao dongchuan wenxian sanpian* (Taibei: Wenjin chuban she, 1979), A.1a. See also Standaert, "The Classification of Sciences," pp. 3–5, and Wardy, *Aristotle in China*, pp. 76, 136, and ECCP, pp. 807–809.

13. See Tsuen-hsuin Tsien, "Western Impact on China through Translation," *Far Eastern Quarterly* 13 (1954): 307–310, and Harris, "Transposing the Merton Thesis," pp. 29–65. See also Benjamin A. Elman, "Geographical Research in the Ming-Ch'ing Period," *Monumenta Serica* 35 (1981–1983): 1–18.

14. Willard Peterson, "Fang I-chih: Western Learning and the 'Investigation of Things,'" in William Theodore de Bary et al., eds., *The Unfolding of Neo-Confucianism* (New York: Columbia University Press, 1975), pp. 369–411.

15. Peterson, "Learning from Heaven," pp. 789–839. Xu Haisong, *Qingchu shiren yu xixue* (Beijing: Dongfang chubanshe, 2000), pp. 72–82, discusses the dissemination of European learning in the seventeenth century as literary gifts, book sales, and inclusion in Chinese collectanea.

16. James Cummins, "Palafox, China and the Chinese Rites Controversy," *Revista de Historia de America* (1961): 395–419. See also Mungello, ed., *The Rites Controversy*, and Michael Sievernich, "Von der Akkommodation zur Inkulturation. Missionarische Leitideen der Gesellschaft Jesu," *Zeitschrift für Missionswissenschaft und Religionswissenschaft* 86, 4 (2002): 260–276.

17. See Standaert, "The Investigation of Things," pp. 395–420.

18. SKQSZM, 124.10a-11b. For the Jesuit glossing of *xuewen* = "*scientia*," see the *Dictionarium latino-sinico-mantchou*, circa 1758, in the Oriental Manuscript Department of the Bibliotheque Nationale, Paris. My thanks to Han Qi for providing me with this source.

19. See Pan Jixing, "The Spread of Georgius Agricola's *De Re Metallica* in Late Ming China," *T'oung Pao*, 57 (1991): 108–118, and Pan Jixing, Hans Ulrich Vogel, and Elisabeth Theisen-Vogel, "Übersetzung und Verbreitung von Georg Agricolas 'De re metallica' im China der späten Ming-Zeit (1366–1644)," *Journal of the Economic and Social History of the Orient* 32 (1989): 153–202. See also Roger Hart, "Local Knowledges, Local Contexts: Mathematics in Yuan and Ming China" (paper presented at the Song-Yuan-Ming Transitions Conference, Lake Arrowhead, Calif., June 5–11, 1997), and Pamela Long, "Of Mining, Smelting, and Printing," *Technology and Culture* 44, 1 (January 2003): 97–101.

20. See Zürcher, "The Jesuit Mission in Fujian," pp. 417–457, and Elman, *From Philosophy to Philology*, 2nd ed., pp. 96–100. See also Brockey, "The Harvest," pp. 571–585.

21. See Louis Gallagher, trans. *China in the Sixteenth Century: The Journals of Matthew Ricci, 1583–1610* (New York: Random House, 1953), p. 95. For discussion, see Gerald Dunne, *Generation of Giants: The Story of the Jesuits in China in the Last Decades of the Ming Dynasty* (South Bend, Ind.: University of Notre Dame Press, 1962), pp. 27–43, and Knud Lundbaek, "The Image of Neo-Confucianism in *Confucius Sinarum Philosophus*," *Journal of the History of Ideas* 44, 1 (January–March 1983): 19–30.

22. Howard Goodman and Anthony Grafton, "Ricci, the Chinese, and the Toolkits of Textualists," *Asia Major*, Third Series, 3, 2 (1990): 99–106.

23. Ibid., pp. 118–120, and Ryan, "Assimilating New Worlds in the Sixteenth and Seventeenth Centuries," pp. 519–538. See also Han Qi, "Bai Jin de Yijing yanjiu," pp. 185–201.

24. Engelfriet, "The Chinese Euclid," pp. 114–118. Compare Karine Chemla, "What Is at Stake in Mathematical Proofs from 3rd-Century China?," *Science in Context* 10, 2 (1997): 227–251.

25. Standaert, "The Classification of Sciences," p. 16. See also Peterson, "Learning from Heaven," p. 799.

26. Mungello, *The Forgotten Christians of Hangzhou*, pp. 70–98, and Mungello, *Curious Land*, pp. 60–62, 333–334.

27. Michael Lackner, "Jesuit Figurism," in Thomas Lee, ed., *China and Europe: Images and Influences in Sixteenth to Eighteenth Centuries* (Hong Kong: The Chinese University Press, 1991), pp. 129–150. See also Paul Rule, *K'ung-tzu*, pp. 150–182. Compare Florence Hsia, "Anthanasius Kircher's *China Illustrata* (1667): an *apologia pro sua vita*," in Paula Findlen, ed., *Baroque Imaginary* (New York: Routledge), forthcoming.

28. Mungello, "The First Complete Translation of the Confucian Four Books," pp. 523–524, and Mungello, "The Seventeenth-Century Jesuit Translation Project," pp. 264–268.

29. See Peterson, "Fang I-chih: Western Learning," pp. 369–411, and SCC, III, pp. 438–439.

30. Qiong Zhang, "Demystifying *Qi*," pp. 74–106. See also Qiong Zhang, "Translation as Cultural Reform: Jesuit Scholastic Psychology in the Transformation of the Confucian Discourse on Human Nature," in O'Malley et al., eds., *The Jesuits*, pp. 364–379.

31. Ping-kuen Yu, comp., *Chinese Collections in the Library of Congress*. 3 vols. (Washington, D.C.: Center for Chinese Research Materials, 1974), pp. 592–594. See also Xu Guangtai, "Mingmo Qingchu xifang gezhixue de chongji yu fanying: yi Xiong Mingyu *Gezhi cao* wei li," in *Shibien chunti yu geren* (Taibei: Taiwan National University, 1996), pp. 236–258, and Engelfriet, *Euclid in China*, p. 357. Compare Shang, *Mingmo Qingchu de gewu qiongli zhi xue*, pp. 239–263.

32. See Xiong Mingyu, "Zixu" to the *Gezhi cao* (1648 edition in the U.S. Library of Congress Asian Library), for efforts to inscribe Chinese views of natural change with the teleologies of the Jesuits. See also Feng Jinrong (Fung Kam-wing), "Mingmo Xiong Mingyu Gezhi cao neirong tanzhe," *Ziran kexue shi yanjiu* 16, 4 (1997): 304–328.

33. Despite the common Western view, the *wuxing* were not elements. See Lloyd and Sivin, *The Way and the Word*, pp. 253–266.

34. Nathan Sivin, "The Myth of the Naturalists," in Sivin, *Medicine, Philosophy, and Religion in Ancient China* (Brookfield, Vt.: Variorum, 1995) IV, 1–29, notes that "by the first century B.C., a coherent theory of yin-yang and the Five Phases as aspects of ch'i [*qi*] emerged." Aihe Wang, *Cosmology and Political Culture in Early China* (Cambridge: Cambridge University Press, 2000), pp. 81–128, relates how a correlative cosmos was formulated in light of the five phases.

35. Compare Kim, *The Natural Philosophy of Chu Hsi*, pp. 91–103.

36. See the critique of Vagnoni by eighteenth-century Chinese literati in SKQSZM, 125.34a–b. Compare the account in Zhang, "Demystifying *Qi*," pp. 74–106.

37. Algonso Vagnoni, *Kongji gezhi*, A.1b–2b. See also James Bogen, "Fire in the Belly: Aristotelian Elements, Organisms, and Chemical Compounds," *Pacific*

Philosophical Quarterly 76 (1995): 370–404. My thanks to Minghui Hu for drawing my attention to this article.

38. Vagnoni, *Kongji gezhi*, A.4a–6a.

39. Vagnoni, *Kongji gezhi*, A.16a–20a.

40. Zhang, "Demystifying *Qi*," pp. 91–98, presents an overstated modernist critique. Compare R. Dunch, "Beyond Cultural Imperialism: Cultural Theory, Christian Missions, and Global Modernity," *History and Theory* 41, 3 (October 2002): 301–325.

41. See Vagnoni's *Kongji gezhi*, A.1a–13b. See also Willard Peterson, "Western Natural Philosophy Published in Late Ming China," *Proceedings of the American Philosophical Society* 117.4 (August 1973): 295–322. Compare David Wright, "The Great Desideratum: Chinese Chemical Nomenclature and the Transmission of Western Chemical Concepts," *Chinese Science* 14(1997): 39–40.

42. Fang Yizhi, *Wuli xiaozhi* (Preliminary record of phenomena and their patterns of occurrence) (SKQS edition), vol. 867, p. 759.

43. La Vallee Poussin, "Cosmogony and Cosmology (Buddhist)," in James Hastings, ed., *Encyclopedia of Religion and Ethics*, 13 vols. (New York: Charles Scribner's Sons, 1928), vol. 4, pp. 131–32. See also Ben-dor, "The 'Dao of Muhammad.'"

44. Christopher Cullen, "The Science/Technology Interface in Seventeenth-Century China: Song Yingxing on *Qi* and the *Wuxing*," *Bulletin of the School of Oriental and African Studies* 53, 2 (1990): 295–318. Song may have had some knowledge of Aristotelian cosmology, because the five phases for Song resembled the theory of elements introduced by the Jesuits. In his published writings, Song discussed the Jesuits' view of a round earth suspended in space, which he thought unreasonable.

45. See Yang Lien-sheng's review of Yabuuchi Kiyoshi, ed., *Tenkō kaibutsu no kenkyū* [Studies on The T'ien-kung k'ai-wu], HJAS 17, 1 (1954): 307–316, and Song Yingxing, *T'ien-kung k'ai-wu, Chinese Technology in the Seventeenth Century*, trans. E-tu Zen Sun and Shiou-Chuan Sun (College Park: Pennsylvania State University Press, 1966), pp. viii–ix. On the Jesuits and botany, see Standaert, *Handbook of Christianity in China*, vol. 1, pp. 803–808.

46. Chu Pingyi, "Kua wenhua zhishi chuanbo de ge an yanjiu—Mingmo Qingchu guanyu diyuan shuo de zhengyi, 1600–1800," *Zhongyang yanjiu yuan lishi yuyan yanjiu suo jikan* (Taiwan) 69, 3 (1998): 589–670.

47. For recent discoveries, see Chen Meidong, *Zhongguo gu xing tu* (Shenyang: Liaoning jiaoyu chubanshe, 1996).

48. SCC, III, pp. 497–561, and Cordell Yee, "Traditional Chinese Cartography and the Myth of Westernization," in J. B. Harley, and David Woodward, eds., *The History of Cartography*, vol. 2, book 2: *Cartography in the Traditional East and Southeast Asian Societies* (Chicago: University of Chicago Press, 1994), pp. 170–202. See also Standaert, *Handbook of Christianity in China*, vol. 1, pp. 752–770, for sources, and Standaert, "Methodology in View of Contact between Cultures," pp. 12–18, 40–43, for a critique of Yee's views, pp. 62–68. See also Wylie, *Notes on Chinese Literature*, pp. 43–67, for a bibliography of premodern works on geography.

49. Richard Smith, "Mapping China's World: Cultural Cartography in Late Imperial Times," in Yeh Wen-hsin, ed., *Landscape, Culture and Power in Chinese Society* (Berkeley: University of California, Center for East Asian Studies, 1998), pp. 67–68.

50. SCC, vol. 3, 533–551, vol. 4, part 3, pp. 500–502, and DMB, p. 982. Compare Gavin Menzies's well-orchestrated science fiction titled *1421: The Year China Discovered America* (New York: HarperCollins, 2002).

51. Originally completed in 1562, the *Maritime Survey* was reedited by Hu Zongxian's descendants and reissued in 1624. The editors of the later editions dropped Zheng Ruozeng's name as the author and replaced it with Hu Zongxian, their revered ancestor. See DMB, pp. 204–208, 980–984. Compare the account in SKQSZM, 69.31a–32b.

52. DMB, pp. 206–207, and SCC, III, pp. 514–517.

53. See J. V. G. Mills's "Introduction," in Ma Huan, *Ying-yai Sheng-lan*, pp. 1–66. Compare Jan Julius Lodewijk Duyvendak, *Ma Huan Re-examined* (Amsterdam: Noord-Hollandsche uitgeversmaatschappij, 1933), pp. 8–9.

54. See *Hsing-ch'a-sheng-lan: The Overall Survey of the Star Raft by Fei Hsin*, trans. J. V. G. Mills, and rev., ann., and ed. Roderich Ptak (Amsterdam: Noord-Hollandsche uitgeversmaatschappij, 1933). Compare DMB, pp. 440–441.

55. SKQSZM, 78.14b–19a.

56. SKQSZM, 71.3b–4b, 78.14b–15a; DMB, pp. 194–204, 546–547, 1296–1297, 1405–1406; SCC, III, pp. 511, 557–558; and ECCP, pp. 314–316. See also Ting Wen-chiang, "On Hsu Hsia-k'o (1586–1641). Explorer and Geographer," *New China Review* 3, 5 (October 1921): 325–337. Detailed discussion of Hsu Hsia-k'o's travels can be found in Li Chi, *The Travel Diaries of Hsu Hsia-k'o* (Hong Kong: Chinese University, 1974).

57. Ricci's world map is reproduced in SCC, III, pp. 582–583 (Plates XC and XCI). A 1608 color version is included in *Symbols of Power: Masterpieces from the Nanjing Museum* (Anaheim, Calif.: Bowers Museum, 2002).

58. Helen Wallis, "The Influence of Father Ricci on Far Eastern Cartography," *Imago Mundi* 19 (1965): 35–45, and Kenneth Ch'en, "Matteo Ricci's Contribution to, and Influence on, Geographical Knowledge in China," *Journal of the American Oriental Society* 59, 3 (September 1939): 325–359. Compare Theodore Foss, "A Western Interpretation of China Jesuit Cartography," in Ronan and Oh, eds., *East Meets West*, pp. 209–251.

59. Zhang Huang, *Tushu bian* (repr., Taibei: Chengwen chuban she, 1971), 19.33a–40a. See also Ch'en, "Matteo Ricci's Contribution," pp. 343–347, and DMB, pp. 83–84. For the initial reaction of Chinese literati, see Pasquale M. d'Elia, "Recent Discoveries and New Studies (1938–60) on the World Map in Chinese of Father Matteo Ricci S.J.," *Monumenta Serica* 20 (1961): 82–164.

60. *The Library of Philip Robinson*, part 2 (London: Sotheby's Auction Catalogue, 1988), pp. 77–95, especially pp. 76–77, 78–79. The 1593 world map was engraved by Liang Zhou, an education official in Wuxi, and printed in Nanjing. See also John Day, "The Search for the Origins of the Chinese Manuscript Copies of Matteo Ricci's Maps," *Imago Mundi* 47 (1995).

61. *The Library of Philip Robinson*, part 2, pp. 79–80. See also Lionel Giles, "Translations from the World Map of Father Ricci," *Geographical Journal* 52, 6 (1918): 367–385, and 53, 1 (1919): 19–30.

62. Wylie, *Notes on Chinese Literature*, p. 58.

63. See Bernard Hung-kay Luk, "A Study of Giulio Aleni's *Chih-fang wai-chi*," *Bulletin of Oriental and African Studies* 40 (1977): 58–84, and Menegon, *Un solo Cielo*, pp. 141–146. See also Richard Smith, *Chinese Maps: Images of All Under Heaven* (Oxford: Oxford University Press, 1996), pp. 49–54.

64. *The Library of Philip Robinson*, part 2, pp. 83, 85–86, 89, 95. Martini was in China from 1643 to 1646. See also Boleslaw Szczesniak, "The Seventeenth Century Maps of China: An Inquiry into the Compilations of European Cartographers," *Imago Mundi* 13 (1956): 116–136.

65. Yee, "Traditional Chinese Cartography," pp. 170–202. Smith, "Mapping China's World," pp. 71–83, usefully introduces the "Complete Maps" genre but overstresses the limits of Chinese interest in Jesuit cartography. For further discussion, see Chapter 5 below.

66. See Luk, "A Study of Giulio Aleni's *Chih-fang wai-chi*," pp. 76–77. See also Chen Minsun, "Ferdinand Verbiest and the Geographical Works by Jesuits in Chinese 1584–1674," pp. 123–164, and Lin Tongyang, "Ferdinand Verbiest's Contribution to Chinese Geography and Cartography," pp. 135–164, both in Witek, ed., *Ferdinand Verbiest (1623–1688)*.

67. Walter Fuchs, "Der Jesuiten-Atlas der Kanghsi-Zeit," *Monumenta Serica*, Monograph Series 4 (1943): 60–75. See also James A. Millward, *Beyond the Pass: Economy, Ethnicity, and Empire in Qing Central Asia, 1759–1864* (Stanford, Calif.: Stanford University Press, 1998), pp. 70–72.

68. Smith, "Mapping China's World," pp. 77–81.

69. SCC, vol. 3, pp. 551–586. Maps based on latitudinal and longitudinal degrees were essential for ocean-going vessels, which the Chinese did not prioritize for their coastal fleet.

70. George H. Wong, "The Anti-Christian Movement in China: Late Ming and Early Ch'ing," *Tsing Hua Journal of Chinese Studies* (Taiwan), new series 3, 1 (1956): 214–215n22.

71. Spence, *Emperor of China*, pp. xvii–xix, 15–16, 74–75, and Chu Pingyi, "Scientific Dispute in the Imperial Court," 7–34.

72. See Elman, *From Philosophy to Philology*, 2nd ed., pp. 85–92. On the Cheng-Zhu revival see Kai-wing Chow, *The Rise of Confucian Ritualism in Late Imperial China* (Stanford, Calif.: Stanford University Press, 1994), pp. 164–165. Compare Lundbaek, "The Image of Neo-Confucianism," pp. 26–28. For the Jesuit role contra Galileo in the early seventeenth-century papal court culture in Rome, see Mario Biagioli, *Galileo Courtier: The Practice of Science in the Culture of Absolutism* (Chicago: University of Chicago Press, 1993), pp. 267–311.

73. See Westman, *The Copernican Question: Prognostication*, on the role of astrology and astronomy in the Copernican revolution in Europe.

74. ECCP, pp. 889–890, and John D. Young, "An Early Confucian Attack on Christianity: Yang Kuang-hsien and His *Pu-te-i*," *Xianggang Zhongwen daxue xuebao* 3 (1975): 162.

75. Yung Sik Kim, "Chu Hsi (1130–1200) on Calendar Specialists and Their Knowledge: A Scholar's Attitude toward Technical Scientific Knowledge in Traditional China," *T'oung Pao* 78 (1992): 94–115. See also Grete Moortgat, "Substance

Versus Function (*Ti* vs. *Yong*): The Humanistic Relevance of Yang Guangxian's Objection to Western Astronomy," in Masini, ed., *Western Humanistic Culture*, pp. 260–275, and Eugenio Menegon, "Yang Guangxian's Opposition to Johann Adam Schall: Christianity and Western Science in His Work *Budeyi*," in Malek, ed., *Western Learning and Christianity in China*, pp. 311–337.

76. Huang Yilong (Yinong), "Qingchu tianzhujiao yu huijiao tianwenjia de cheng-dou," *Jiuzhou xuekan* 5, 3 (1993): 47–69. See also Huang Yilong, "Court Divination and Christianity in the K'ang-hsi Era," *Chinese Science* 10 (1991): 1–20. See also Chu Pingyi, "Scientific Dispute," pp. 10–14; SCC, III, p. 404; and Hu, "Cosmopolitan Confucianism," pp. 180–185.

77. Huang Yilong, "Tang Ruowang yu Qingchu xili zhi zhengtong hua," in Wu Jiali and Ye Hongsa, eds., *Xinbian Zhongguo keji shi* (Taibei: Yinmu wenhua shiye gongsi, 1990), pp. 465–490. See also Huang Yilong and Chang Chih-ch'eng, "The Evolution and Decline of the Ancient Practice of Watching for the Ethers," *Chinese Science* 13 (1996): 99. Although some modern scholars have assumed that the Jesuits' method for such computations, like their eclipse techniques, were European, they simply applied the intercalation technique that the bureau had used since the computational system of 1280. After the Gregorian reforms, Europeans no longer worried about intercalating months. My thanks to Nathan Sivin for his advice on this matter.

78. Huang, "Qingchu tianzhujiao yu huijiao tianwenjia de chengdou," pp. 47–69. See also Chu, "Scientific Dispute," p. 15.

79. The "Great Plan" is a chapter in the *Documents Classic*.

80. Huang, "Court Divination and Christianity," pp. 4–12.

81. ECCP, pp. 890–891. Fang Chao-ying speculates that the Manchus were susceptible to such claims because of their beliefs in shamanism and the power of evil spells.

82. Isaia Iannaccone, "From N. Longobardo's Explanation of Earthquakes as Divine Punishment to F. Verbiest's Systematic Instrumental Observations: The Evolution of European Science in China in the Seventeenth Century," in Masini, ed., *Western Humanistic Culture*, pp. 167–169.

83. Derk Bodde, "The Chinese Cosmic Magic known as Watching for the Ethers," in Sören Egerod and Elise Glahn, eds., *Studia Serica Bernhard Karlgren Dedicata, Sinological Studies Dedicated to Bernhard Karlgren on His Seventieth Birthday October Fifth 1959* (Copenhagen: Ejnar Munksgaard, 1959), pp. 14–35, notes that the practice was in disrepute in the sixteenth century.

84. DMB, pp. 368–369, and Huang and Chang, "The Evolution and Decline," pp. 87–88. Although not an alternative for *houqi*, Zhu elaborated in mathematical terms the equal temperament of the twelve semitones represented by the traditional pitch pipe series.

85. Huang, "Court Divination," pp. 6–7, and Huang and Chang, "The Evolution and Decline," pp. 82–85, 90–96, 104. See also SCC, III, p. 405, and Tiziana Lippiello, "Astronomy and Astrology: Johann Adam Schall von Bell," in Malek, ed., *Western Learning and Christianity in China*, pp. 403–430.

86. Huang Yilong, "Court Divination," pp. 6–10, and Huang and Chang, "The Evolution and Decline," pp. 91–94. See also Joanna Waley-Cohen, "Religion, War, and

Empire-Building in Eighteenth-Century China," *International History Review* 20, 2 (June 1998): 340–341.

87. Huang, "Court Divination," pp. 9–11.

88. Ibid., pp. 12–14.

89. See Wong, "The Anti-Christian Movement," 207–210, and Young, "An Early Confucian Attack," pp. 173–189. See also ECCP, p. 890.

90. Edwin J. Van Kley, "Europe's 'Discovery' of China and the Writing of World History," *American Historical Review* 76, 2 (April 1971): 358–385, and Mungello, *Curious Land*, pp. 91, 102–103, 125–126.

91. Ryan, "Assimilating New Worlds," pp. 532–534. See also Mungello, *Curious Land*, pp. 127, 137–145, and David Porter, *Ideographia: The Chinese Cipher in Early Modern Europe* (Stanford, Calif.: Stanford University Press, 2001), pp. 43–49. Compare Bruce Rusk, "Old Scripts, New Actors: Ming Philology before the Jesuits" (paper presented at the History of Science Workshop, Global Science and Comparative History: Jesuits, Science, and Philology in China and Europe, 1550–1850, Princeton University, December 6, 2002).

92. Rule, *K'ung-tzu*, pp. 98–100.

93. Ibid., pp. 98–100, and Mungello, *Curious Land*, pp. 93–94.

94. Chu, "Scientific Dispute," pp. 16–17. See also ECCP, p. 891, and Paul Cohen, *China and Christianity: The Missionary Movement and the Growth of Chinese Antiforeignism, 1860–1870* (Cambridge, Mass.: Harvard University Press, 1963), pp. 24–27. The pope beatified the 1664 Chinese martyrs in 2000 despite protests from the PRC government.

95. Huang and Chang, "The Evolution and Decline," pp. 95–97. Again, my thanks to Nathan Sivin for his clarifications.

96. Ibid., pp. 95–97, and Chu, "Scientific Dispute," pp. 19–20.

97. Chu, "Scientific Dispute," pp. 20–22.

98. Ibid., pp. 28–31. See also Hashimoto, *Hsü Kuang-ch'i*, p. 35.

99. Compare Verbiest's calendar to the Ten Thousand Years Calendar (*Wannian li*) prepared by Muslims in the Yuan Astro-calendric Bureau. On Verbiest's impact, see Witek, ed. *Ferdinand Verbiest*. See also Standaert, *Handbook of Christianity in China*, vol. 1, p. 719.

100. Chu, "Scientific Dispute," pp. 22–27; Standaert, *Handbook of Christianity in China*, vol. 1, p. 496; and Wylie, *Notes on Chinese Literature*, p. 129. The published collection of missionary writings that survives is *Xiyang xinfa lishu*; there is much confusion about editions and titles. See Sivin, "Copernicus in China," *Colloqula Copernica II*, pp. 89n56, and 90n58.

101. Standaert, "The Investigation of Things," pp. 407–408. See also Noël Golvers, "Verbiest's Introduction of *Aristoteles Latinus* (Coimbra) in China: New Western Evidence," in Noël Golvers, ed., *The Christian Mission in China in the Verbiest Era: Some Aspects of the Missionary Approach* (Leuven: Leuven University Press, 1999), pp. 33–53.

102. Shang Zhicong, "Nan Huairen Qiongli xue de zhuti neirong yu jiben jiegou," *Qingshi yanjiu* 3 (2003): 73–84. See also Hugh Shapiro, "How Different Are Western and Chinese Medicine? The Case of Nerves," in Helaine Selin, ed.,

Medicine across Cultures: The History of Non-Western Medicine (Boston: Kluwer Academic Publishers, 2003), pp. 8–9.

103. See Elman, *A Cultural History of Civil Examinations,* pp. 105–123, 125–142, and Ad Dudink and Nicolas Standaert, "Ferdinand Verbiest's *Qiongli xue* (1683)," in Golvers, ed., *The Christian Mission in China,* pp. 11–31. See also Wong, "The Anti-Christian Movement," pp. 190–191.

104. See Golvers, "Verbiest's Introduction of *Aristoteles Latinus* (Coimbra) in China," pp. 33–53. My thanks to Joachim Kurtz for his clarifications of Verbiest's petition.

105. The emperor's comments are cited in Dudink and Standaert, "Ferdinand Verbiest's *Qiongli xue,*" pp. 15–17. See also Standaert, "The Investigation of Things," pp. 417–418. See Kuriyama, *The Expressiveness of the Body,* for an account of the different visions of the body informing classical medicine in early modern Europe and late imperial China.

106. *Mengzi yinde* (repr., Taibei: Chengwen Bookstore, 1966), 46/6A/15, and D. C. Lau, trans., *Mencius* (Harmondsworth: Penguin, 1970), p. 168. See also Li Shizhen, *Bencao gangmu* (SKQS edition). Compare Yong-xin Zhu, "Historical Contributions of Chinese Scholars to the Study of the Human Brain," *Brain and Cognition* 11 (1989): 133–138, and Wolfgang Behr, "Some Traditional Descriptions of Brain Organization and Functioning" (presented at eighth ICHCS Conference, Berlin, August 24–27, 1998), pp. 1–10.

107. See Standaert, "The Investigation of Things," pp. 407–408, and Dudink and Standaert, "Ferdinand Verbiest's *Qiongli xue,*" pp. 12, 22–29.

108. Zhu, "Historical Contributions of Chinese Scholars," pp. 134–36. On the Han Kitab, see Ben-dor, "The 'Dao of Muhammad.'" See also Friedrich Zimmerman, "Al-Farabi und die philosophische Kritik an Galen von Alexander zu Averroes," in A. Dietrich, ed., *Akten des VIII Kongresses für Arabistik und Islamwissenschaft* (Abhandlungen des Akademie des Wissenschaften in Göttingen, Phil.-His., Dritte Folge Nr. 98, 1976), pp. 401–414, and Heinrich von Staden, *Herophilus: The Art of Medicine in Early Alexandria* (Cambridge: Cambridge University Press, 1989), pp. 155–161, 247–250. The latter discusses the Greek notion of the brain as a controlling faculty or command center within the body that informed the Galenic position, which Jesuits in China drew on for their Coimbra commentaries.

109. Xu Haisong, "The Reaction of Scholars to the Work of Ferdinand Verbiest, S.J., during the Kangxi-Qianlong Reign," in Golvers, ed., *The Christian Mission in China,* pp. 73–83. See also Noël Golvers, *Ferdinand Verbiest, S.J. (1623–1688) and the Diffusion of the Chinese Heaven* (Leuven: Leuven University Press, 2003), chap. 9. Compare Biagioli, *Galileo Courtier,* pp. 211–244.

110. Huang and Chang, "The Evolution and Decline," p. 99. See also Chu Pingyi, "The Last Crusade of Chinese Astronomical Methods: Zhang Yongjing's *Dinglu yuheng,*" forthcoming.

111. Dunne, *Generation of Giants,* p. 360.

112. See Douglas Lancashire, "Anti-Christian Polemics in Seventeenth Century China," *Church History* 38, 2 (1969): 218–241; Adrian Dudink, "The Inventories of the Jesuit House at Nanking Made up During the Persecution of

1616–1617 (Shen Que, *Nangong Shudu*, (1620)," in Masini, ed., *Western Humanistic Culture*, pp. 119–157, notes that Xu omitted materials and texts were not always accurately copied.

113. Young, "An Early Confucian Attack," pp. 159–160, 173. See also Wong, "The Anti-Christian Movement," pp. 187–222, and Sivin, "Copernicus in China," in *Science in Ancient China*, IV, p. 50n80.

114. See Han Qi, "Gewu qiongli yuan yu Mengyang zhai: shiqi shiba shiji zhi Zhong-Fa kexue jiaoliu," *Faguo hanxue* 4 (1999): 302–324, and Catherine Jami, "Learning Mathematical Sciences during the Early and Mid-Ch'ing," in Benjamin H. Elman and Alexander Woodside, eds., *Education and Society in Late Imperial China, 1600–1900* (Berkeley: University of California Press, 1994), pp. 231–233, 238–241.

115. Xu, *Qingchu shiren yu xixue*, pp. 319–339, presents the earlier stages of "Chinese origins" rhetoric. Compare Catherine Jami, "History of Mathematics in Mei Wending's (1633–1721) Work," *Historia Scientiarum* 4, 2 (1994): 159–174, and John Henderson, "The Assimilation of the Exact Sciences into the Ch'ing Confucian Tradition," *Journal of Asian Affairs* 5, 1 (Spring 1980): 15–31.

116. See Limin Bai, "Mathematical Study and Intellectual Transition in the Early and Mid-Qing," *Late Imperial China* 16, 2 (December 1995): 23–61, and Minghui, "Cosmopolitan Confucianism," chap. 4.

4. The Limits of Western Learning in the Early Eighteenth Century

1. Compare Sivin, "Copernicus In China," in *Science in China*, IV, pp. 35–50. See also R. Étiemble, *Les Jésuites en Chine (1552–1773), la querelle des Rites* (Paris: Julliard, 1966), and John W. O'Malley, S.J., "The Historiography of the Society of Jesus: Where Does It Stand Today?" in O'Malley et al., eds., *The Jesuits*, pp. 11–14. In England and America, where papal power was opposed, the Jesuit order remained intact throughout the eighteenth century. On the cultural and political implications of the "absence of knowledge," see Robert Proctor, *Cancer Wars: How Politics Shapes What We Know and Don't Know About Cancer* (New York: Basic Books, 1995), chap. 5.

2. Bai, "Mathematical Study and Intellectual Transition in the Early and Mid-Qing," pp. 23–61. Compare Biagioli, *Galileo Courtier*, pp. 54–59.

3. See Mark Elliot, *The Manchu Way: The Eight Banners and Ethnic Identity in Late Imperial China* (Stanford, Calif.: Stanford University Press, 2001).

4. Han Qi, "'Zili' jingshen yu lisuan huodong: Kang-Qian zhi ji wenren dui Xixue taidu zhi gaibian ji qi beijing," *Ziran kexue shi yanjiu* 21, 3 (2002): 210–221.

5. Wang Ping, "Qingchu lisuan yanjiu yu jiaoyu," *Jindai shi yanjiu suo jikan* 3 (1972): 365–375. See also Li Yan, "Qingdai shuxue jiaoyu zhidu," *Xueyi* 13, 4 (1934): 38–41.

6. Vladimir Miasnikov, "Ferdinand Verbiest and His Role in the Formation of Sino-Russian Diplomatic Relations," pp. 273–279, and Ku Weiying, "Between the Court and the Church: Ferdinand Verbiest in the Catholic History of China," pp. 329–347, both in Witek, ed., *Ferdinand Verbiest (1623–1688)*.

7. Engelfriet, *Euclid in China*, pp. 434–436. See also Wann-sheng Horng, "Chinese Mathematics at the Turn of the 19th Century: Jiao Xun, Wang Lai, and Li Rui," in Cheng-hung Lin and Daiwie Fu, eds., *Philosophy and Conceptual History of Science in Taiwan* (Netherlands: Kluwer Academic Publishers, 1993), pp. 176–177.

8. Catherine Jami, "Western Influence and Chinese Tradition in an Eighteenth-Century Chinese Mathematical Work," *Historia Mathematica* 15 (1988): 311–331. Compare Li Yan and Du Shiran, *Chinese Mathematics*, pp. 216–218.

9. Martzloff, *A History of Chinese Mathematics*, pp. 218–219. See also ECCP, pp. 473–475.

10. Jean-Claude Martzloff, "Space and Time in Chinese Texts of Astronomy and of Mathematical Astronomy in the Seventeenth and Eighteenth Centuries," *Chinese Science* 11 (1993–1994): 71.

11. Chu Pingyi, "Technical Knowledge, Cultural Practices and Social Boundaries: Wan-nan Scholars and the Recasting of Jesuit Astronomy, 1600–1800" (Ph.D. diss., University of California–Los Angeles, 1994). See also Wylie, *Notes on Chinese Literature*, p. 121; Engelfriet, *Euclid in China*, pp. 383–431; Hashimoto Keizō, "Bai Buntei no rekisan gaku," *Tōhō gakuhō* (Kyoto) 41 (March 1970): 495–496.

12. Han, "'Zili' jingshen yu lisuan huodong," pp. 217–218.

13. Engelfriet, *Euclid in China*, pp. 406–428, and Hashimoto, "Bai Buntei," pp. 492–493. See also Horng, "Chinese Mathematics," p. 173, and Chu, "Technical Knowledge," chap. 4, for an account of Mei's rise to celebrity status.

14. Hashimoto, "Bai Buntei," pp. 495–496, and ECCP, p. 570. See also Chu, "Technical Knowledge," pp. 153, 158, and Han, "'Zili' jingshen yu lisuan huodong," p. 217.

15. Chu, "Technical Knowledge," chap. 5, especially pp. 200–206.

16. Mei Wending, *Jixue tang shiwen chao* (Hefei: Huangshan shushe, 1995), p. 54. I have modified the translation in Engelfriet, *Euclid in China*, p. 431. For discussion of Mei's argument in light of recent critiques of Xu Guangqi for his shallow knowledge of the Chinese mathematical tradition, see Roger Hart, "Proof, Propaganda, and Patronage: A Cultural History of the Dissemination of Western Studies in Seventeenth-Century China" (Ph.D. diss., University of California–Los Angeles, 1997), chap. 2.

17. Mei Wending, *Lixue yiwen bu shang*, SKQS, vol. 794, pp. 56–57. Chu, "Technical Knowledge," pp. 200–209, dates the supplement after the 1705 meeting between Mei and the emperor.

18. Mei, *Lixue yiwen bu shang*, pp. 56–57. See also Han Qi, "Joachim Bouvet's Study of the *Yijing* and the Theory of the Chinese Origin of Western Learning in the Kangxi Period" (presented at the conference Europe in China III, between Ming and Qing, Berlin, April 22–26, 1998), pp. 1–16.

19. See Wang, "Qingchu lisuan yanjiu yu jiaoyu," p. 368, and Li and Du, *Chinese Mathematics*, pp. 214–216. See also Martzloff, *A History of Chinese Mathematics*, pp. 25, 80, and ECCP, pp. 570–571.

20. Qian Baocong, ed., *Qian Baocong kexue shi lunwen xuanji* (Beijing: Kexue chuban she, 1983), pp. 10–16, presents the evolution of Chinese mathematical equations. Roger Hart is preparing a history of linear algebra that describes the origins of determinants in ancient Chinese mathematics.

21. See Martzloff, *A History of Chinese Mathematics*, p. 119, and Standaert, *Handbook of Christianity in China*, vol. 1, p. 745. Compare Catherine Jami and Han Qi, "The Reconstruction of Imperial Mathematics in China during the Kangxi Reign (1662–1722)," *Early Science and Medicine* 8, 2 (2003): 96–104, which describes the mathematics Antoine Thomas introduced to the court after 1688. See also Michael Mahoney, "The Beginnings of Algebraic Thought in the Seventeenth Century," in S. Gaukroger, ed., *Descartes: Philosophy, Mathematics and Physics* (Totowa, N.J.: Barnes and Noble Books, 1980), chap. 5.

22. Jean-Claude Martzloff, *Recherches sur L'oeuvre mathématique de Mei Wending (1633–1721)* (Paris: Institut des hautes études chinoises, 1981). See also ECCP, pp. 570–571; Chu, "Technical Knowledge," pp. 163–167; and Engelfriet, *Euclid in China*, pp. 428–431.

23. Hashimoto, "Bai Buntei," pp. 497–499, especially p. 517n15. See also Martzloff, *A History of Chinese Mathematics*, p. 25, and Li and Du, *Chinese Mathematics*, pp. 215–216.

24. Sivin, "Wang Hsi-shan," in Sivin, *Science in Ancient China*, V, pp. 7–12, and Hashimoto, *Xu Guangqi*, pp. 53–54. Sivin contends that by 1700 there had been a fundamental redefinition of what the problems and solutions were in Chinese mathematical astronomy, which he calls an intellectual scientific revolution, comparable to the Copernican revolution.

25. Mei Wending, "Lun lixue gushu jinmi," in *Lixue yiwen*, in *Meishi congshu jiyao* (Qing Qianlong edition, 1761), 46.8a–8b.

26. Mei Wending, "Lun Zhoubi zhong ji you diyuan zhi li," in SKQS, vol. 794, pp. 57–58.

27. Chu, "Technical Knowledge," pp. 191–199.

28. Mei Wending, "Lun Zhong-Xi erfa zhi tong," in *Lixue yiwen*, in *Meishi congshu jiyao*, 46.9a–b. See Lü Lingfeng, "Mingmo ji Qingdai jiaoshi jilu yanjiu" (Ph.D. diss., University of Science and Technology of China, 2002), pp. 62–65. See also Hashimoto, "Bai Buntei," p. 499, and Hashimoto, *Xu Guangqi*, pp. 53–62. Compare Han Qi, "*Shuli gezhi* de faxian—jian lun 18 shiji Niudun xiangguan zhuzuo zai Zhongguo de chuanbo," *Zhongguo keji shiliao* 19, 2 (1998): 79–80.

29. Mei Wending, *Jixue tang shiwen chao*, pp. 34–35. See also Engelfriet, *Euclid in China*, p. 430, and Wylie, *Notes on Chinese Literature*, pp. 110–111.

30. Hashimoto, "Bai Buntei," pp. 500–501.

31. Ibid., pp. 502–505.

32. See ibid., pp. 507–514; Engelfriet, *Euclid in China*, pp. 407–428; Hu "Merging Chinese and Western Mathematics," p. 42, and Li and Du, *Chinese Mathematics*, pp. 213–215. Compare Catherine Jami, "Aleni's Contribution to Geometry in China: A Study of the Jiheyaofa," in Tanaka Awashi, ed., *Chūgoku gijutsu shi no Kenkyū* (Kyoto: Kyoto Institute for Humanistic Studies, 1998), pp. 775–796, and Jami, "History of Mathematics," pp. 166–167.

33. Wylie, *Notes on Chinese Literature*, pp. 111–113, summarizes the contents of the *Lisuan quanshu*. See also Andrew C. West, *Catalogue of the Morrison Collection of Chinese Books* (London: School of Oriental and African Studies, 1998), pp. 108, 124–126.

34. Li and Du, *Chinese Mathematics*, pp. 214–216, and Martzloff, *Recherches sur L'oeuvre mathématique*. Although some Mathematical Classics were in late Ming

catalogs, these were in imperial archives or other inaccessible collections. See also Margaret Baron, *The Origins of the Infinitesimal Calculus* (New York: Dover Publications, 1969), pp. 1–9.

35. Hashimoto, "Bai Buntei," pp. 491, 516; Martzloff, *A History of Chinese Mathematics*, p. 119; Engelfriet, *Euclid in China*, p. 432; and Horng, "Chinese Mathematics," p. 175.

36. Hsia, "Some Observations on the *Observations*," p. 312.

37. Chu, "Technical Knowledge," pp. 224–229, and D. E. Mungello, "An Introduction to the Chinese Rites Controversy," in Mungello, ed., *The Chinese Rites Controversy*, pp. 3–12.

38. See Menegon, "Yang Guangxian's Opposition," p. 335, and John Henderson, "Ch'ing Scholars' Views of Western Astronomy," HJAS 46, 1 (1986): 121–148. See also Catherine Jami, "From Louis XIV's Court to Kangxi's Court: An Institutional Analysis of the French Jesuit Mission to China (1688–1722)," in Hashimoto Keizō, Catherine Jami, and Lowell Sklar, eds., *East Asian Science: Tradition and Beyond* (Osaka: Kansai University Press, 1995), pp. 493–499.

39. Cummins, "Two Missionary Methods in China," pp. 33–108, and John D. Young, "Chinese Views of Rites and the Rites Controversy, 18th–20th Centuries," in Mungello, ed., *The Chinese Rites Controversy*, p. 86.

40. See *The Library of Philip Robinson*, part 2, p. 42, and James S. Cummins, *A Question of Rites: Friar Domingo Navarrete and the Jesuits in China* (Aldershot, Hants: Scholar Press, 1993), pp. 215–264. Compare A. S. Rosso, *Apostolic Legations to China of the Eighteenth Century* (South Pasadena, Calif.: Perkins, 1948), for the Franciscan viewpoint.

41. F. Villarroel, "The Chinese Rites Controversy: Dominican Viewpoint," *Philippiniana Sacra* 28 (82) (1993): 5–61. Rule, *K'ung-tzu or Confucius?*, is critical of anti-Jesuit figures such as Navarrete, Maigrot, and their modern defender A. S. Rosso.

42. Documents concerning the Propaganda Fide in Asia are discussed in *The Library of Philip Robinson*, part 2, pp. 31–32.

43. Collani, "Charles Maigrot's Role in the Chinese Rites Controversy," pp. 149–183.

44. Chu, "Technical Knowledge," pp. 225–226.

45. See Hsia, *The World of Catholic Renewal*, pp. 207–208, and Porter, *Ideographia*, pp. 81–83, 108–121. On probabilism, Jesuits argued that it was permissible to initiate a doubtful action if there was some probable opinion in its favor.

46. For a description of the relevant documents concerning the de Tournon mission, see *The Library of Philip Robinson*, part 2, pp. 23, 25.

47. Collani, "Charles Maigrot's Role," pp. 149, 161–181, and Chu, "Technical Knowledge," pp. 227–228.

48. See *The Library of Philip Robinson*, part 2, p. 25.

49. Collani, "Charles Maigrot's Role," p. 182, and Edward Malatesta, S.J., "A Fatal Clash of Wills: The Condemnation of the Chinese Rites by the Papal Legate Carlo Tommaso Maillard de Tourrnon," in Mungello, ed., *The Chinese Rites Controversy*, pp. 211–245. For the text of the *Regula*, see pp. 221–224.

50. In 1939 the papacy reversed itself by allowing East Asian converts to observe Shinto according to the pronouncements of the Japanese and Manchuguo governments, which claimed that Shinto rites were civil rather than religious.

51. Paul Rule, "Towards a History of the Chinese Rites Controversy," in Mungello, ed., *The Chinese Rites Controversy,* pp. 249–266, and Collani, "Charles Maigrot's Role," pp. 182–183. See also Mungello, "An Introduction to the Chinese Rites Controversy," pp. 7–8; *The Library of Philip Robinson,* part 2, pp. 31, 98; and Hsia, *The World of Catholic Renewal,* pp. 92–105.

52. *The Library of Philip Robinson,* part 2, p. 115. See also Jonathan Spence, *Treason by the Book* (New York: Viking Press, 2001).

53. Standaert, *Handbook of Christianity in China,* vol. 1, pp. 362–363, 498–499, 683–684, and Knud Lundbaek, "Joseph de Prémare and the Name of God in China," in Mungello, ed., *The Chinese Rites Controversy,* pp. 129–145. The obedience oaths are described in *The Library of Philip Robinson,* part 2, pp. 23–25.

54. See Hsia, *The World of Catholic Renewal,* p. 32. See also *The Library of Philip Robinson,* part 2, pp. 7, 14. The Robinson Library, before its sale in 1988, included important portions of the Clermont collection. On the Jesuit mission's collapse, compare Dunne, *Generation of Giants,* p. 287, and M. Hay, *Failure in the Far East* (London: Neville Spearman, 1956).

55. On the calculus as the mathematics of the state professional in late eighteenth-century France, see Albin Brian, *La Mesure de l'Etat: Administrateurs et géomètres au XVIIIe siècle* (Paris: Albin Michel, 1994), pp. 112–144, 230–255. See also L. W. B. Brookliss, *French Higher Education in the Seventeenth and Eighteenth Centuries* (Oxford: Clarendon Press, 1987), p. 366.

56. Han Qi, "Gewu qiongli yuan," pp. 302–324. See also Chu, "Scientific Dispute," p. 25; Jami, "Western Devices," p. 190; and Foss, "A Western Interpretation of China," pp. 222–240. Materials from the Jesuit survey were published in Jean Baptiste du Halde's (1674–1743) 1735 encyclopedic survey of China in four volumes. See *The Library of Philip Robinson,* part 2, p. 53.

57. Elman, *A Cultural History of Civil Examinations,* pp. 468–477, 719–720.

58. See Hsia, "Some Observations on the *Observations,*" pp. 305–333.

59. For discussion of court compilations in mathematical astronomy, see Elman, *From Philosophy to Philology,* pp. 79–80. Compare Chu, "Technical Knowledge," pp. 235–236, who perceptively indicates that the emperor wished to encourage independent study of these fields. See also Deane, "The Chinese Imperial Astronomical Bureau," p. 259.

60. See Elman, *A Cultural History of Civil Examinations,* pp. 482–483, and Chu, "Technical Knowledge," p. 231.

61. Han Qi, "*Shuli gezhi de faxian,*" pp. 78–79. See also Hu, "Cosmopolitan Confucianism," pp. 150–159.

62. Li and Du, *Chinese Mathematics,* p. 223, note that literati absorbed the achievements of Western science before the Rites Controversy but went no further.

63. Lo-shu Fu, comp., *A Documentary Chronicle of Sino-Western Relations (1644–1820)* (Tucson: University of Arizona Press, 1966), pp. 93, 105, 122.

64. Hsia, "French Jesuits and the Mission to China," pp. 121–131, and Foss, "A Western Interpretation of China," p. 233. See also Standaert, "Jesuit Corporate Culture," p. 362n6.

65. Hsia, "French Jesuits and the Mission to China," pp. 131–138, and Hsia, "Some Observations on the *Observations,*" pp. 305–333.

66. Bruce Rusk's "Old Scripts, New Actors: Ming Philology Before the Jesuits," *East Asian Science, Technology, and Medicine,* forthcoming, shows how Jesuit understanding of the Chinese written language was mediated by Chinese primitivist perceptions of their own language.

67. Mungello, *Curious Land,* p. 263. See also Claudia von Collani, *P. Joachim Bouvet, S.J.: Sein Leben und sein Werk* (Nettetal: Steyler Verlag, 1985), and Porter, *Ideographia,* pp. 55–58.

68. Han Qi, "Bai Jin de yijing yanjiu," pp. 188–195.

69. D. E. Mungello, *The Great Encounter of China and the West, 1500–1800* (Lanham, Md.: Rowman & Littlefield Publishers, 1999), pp. 68–72. Compare Rusk, "Old Scripts, New Actors."

70. Goodman and Grafton, "Ricci, the Chinese, and the Toolkit of Textualists," 139–140.

71. Rule, *K'ung-tzu or Confucius?* pp. 150–82, and Mungello, *Curious Land,* pp. 19, 29–31, 310–316. See also Hsia, "French Jesuits and the Mission to China," pp. 201–205.

72. On Gaubil's letters and manuscripts, see *The Library of Philip Robinson,* part 2, pp. 45–50.

73. Hsia, "French Jesuits and the Mission to China," pp. 206–215.

74. See J. W. Witek, *Controversial Ideas in China and Europe: A Biography of Jean-Francois Foucquet (1665–1741)* (Rome: Institutum Historicum S.I., 1982), and Collani, *P. Joachim Bouvet.* See Minghui Hu "Cosmopolitan Confucianism," chap. 3.

75. *Mingshi,* 31.544–45 (173–174). See also Mei Wending, "Lun Zhongtu lifa de chuanru Xiguo zhi you," in *Lixue yiwen bu shang,* pp. 56–57, which influenced the compilers of the *Ming History.* Mei cited Sima Qian's (ca. 100 B.C.) *Shiji,* pp. 1258–1259 (*juan* 26) for the account.

76. *Xu Guangqi ji,* pp. 74–78, trans. in Engelfriet, *Euclid in China,* pp. 291–93. Xu was unaware of Song-Yuan algebra in part because he did not have access to past mathematical texts, which were not fully reconstituted until the eighteenth century. See chap. 7. Compare Xu Haisong, *Qingchu shiren yu xixue,* pp. 320–338.

77. Engelfriet, *Euclid in China,* pp. 353–356.

78. See Han, "Bai Jin de yijing yanjiu," pp. 199–200, and Cullen, *Astronomy and Mathematics,* pp. 74–92. Compare Xu, *Qingchu shiren yu xixue,* pp. 339–372, and David Mungello, *Leibniz and Confucianism: The Search for Accord* (Honolulu: University of Hawai'i Press, 1977).

79. Han, "Bai Jin de yijing yanjiu," pp. 186–191.

80. Engelfriet, "The Chinese Euclid," pp. 133–134, notes the differences among the Greek original, the Latin version used by Ricci, and the French version used by Gerbillon and Bouvet. Compare Liu Dun, "*Shuli jingyun* zhong *Jihe yuanben* de diben wenti," *Zhongguo keji shiliao* 12, 3 (1991): 88–96.

81. See Elman, *Classicism, Politics, and Kinship,* pp. 128–130. Compare Lam, *A Critical Study of the Yang Hui Suan Fa,* pp. 293–311, for magic squares of three to ten orders.

82. See *Yuzhi Shuli jingyun,* in ZKGD, vol. 3, pp. 12–21, and *Xu Guangqi ji,* pp. 74–75.

83. Mungello, "The Seventeenth-Century Jesuit," pp. 263–265; Han, "Bai Jin de yi-jing yanjiu," pp. 185–201; and Hsia, "French Jesuits and the Mission to China," pp. 107–116.

84. Wann-sheng Horng, "Li Shan-lan: The Impact of Western Mathematics in China during the Late 19th Century" (Ph.D. diss., City University of New York, 1991), pp. 16–17.

85. See Jami, "From Louis XIV's Court to Kangxi's Court," pp. 493–499, and Han, "'Zili' jingshen yu lisuan huodong," pp. 211–214. See also Catherine Jami, "Imperial Control and Western Learning: The Kangxi Emperor's Performance," *Late Imperial China* 23, 1 (June 2002): 28–49.

86. See Li and Du, *Chinese Mathematics,* p. 218; Martzloff, *A History of Chinese Mathematics,* pp. 218–219; and Jami, "Learning Mathematical Sciences during the Early and Mid-Ch'ing," pp. 231, 238–240. See also Han, "Gewu qiongli yuan," pp. 302–324.

87. Martzloff, *A History of Chinese Mathematics,* pp. 163–166, and Li and Du, *Chinese Mathematics,* p. 219.

88. SCC, vol. 3, pp. 141–145. See also Han, "Gewu qiongli yuan," pp. 302–324, and Han Qi, "Chuanjiaoshi Weilie Yali zai Hua de kexue huodong," *Ziran bianzheng fa tongxun* 20, 114 (1998.2): 61. Compare Jami, "Aleni's Contribution to Geometry in China," pp. 778–781; Jami, "Western Influence and Chinese Tradition," pp. 311–331; and Engelfriet, *Euclid in China,* p. 438. Wylie successfully reintroduced algebraic symbols in 1859.

89. Sivin, "Copernicus in China, *Colloquia Copernica II,*" pp. 63–75, 89–92.

90. Hashimoto Keizō, "Rekisho Kōsei no seiritsu: Shindai shoki no tembun rekisan gaku," in Yabuuchi Kiyoshi and Yoshida Mitsukuni, eds., *Min Shin jidai no kagaku gijutsu shi* (Kyoto: Research Institute for Humanistic Studies, 1970), pp. 49–92.

91. Wang, "Qingchu lisuan yanjiu yu jiaoyu," p. 369, and Han, "'Zili' jingshen yu lisuan huodong," pp. 215–219.

92. Foss, "A Western Interpretation of China," p. 220, and Hsia, "Some Observations on the *Observations,*" p. 314.

93. Hashimoto, "Rekisho Kōsei no seiritsu," pp. 49–92.

94. Wang, "Qingchu lisuan yanjiu yu jiaoyu," pp. 370–371. See also Hu, "Cosmopolitan Confucianism," chap. 3.

95. Roger Hahn, *The Anatomy of a Scientific Institution: The Paris Academy of Sciences, 1666–1803* (Berkeley: University of California Press, 1971), pp. 275–285.

96. See Charles Gillispie, *Science and Polity in France at the End of the Old Regime* (Princeton, N.J.: Princeton University Press, 1980). Montgomery, *Science in Translation,* pp. 227–250, discusses the introduction of Newtonian mechanics to Japan via Dutch learning.

97. Michael Mahoney, "Infinitesimals and Transcendent Relations: The Mathematics of Motion in the Late Seventeenth Century," in David Lindberg and Robert Westman, eds., *Reappraisals of the Scientific Revolution* (Cambridge, Mass.: Harvard University Press, 1990), chap. 12, notes that a problem is analytic when it is solved algebraically via a theory of equations.

98. Keith Baker, *Condorcet: From Natural Philosophy to Social Mathematics* (Chicago: University of Chicago Press, 1975), pp. viii–ix, 4–16, 114, and A. Rupert Hall,

Philosophers at War: The Quarrel between Newton and Leibniz (Cambridge: Cambridge University Press, 1980), pp. 13–31, 254–259.

99. Roger Hahn, "Laplace and the Mechanistic Universe," in Lindberg and Numbers, eds., *God & Nature*, pp. 256–276.

100. See Michael S. Mahoney, "Diagrams and Dynamics: Mathematical Perspectives on Edgerton's Thesis," in John W. Shirley and F. David Hoeniger, eds., *Science and the Arts in the Renaissance* (Washington, D.C.: Folger Shakespeare Library, 1985), chap. 10. Compare Margaret Jacob, *Scientific Culture and the Making of the Industrial West* (Oxford: Oxford University Press, 1997), pp. 94, 126.

101. J. A. Bennett, "The Mechanics' Philosophy and the Mechanical Philosophy," *History of Science* 24 (1986): 1–28. See also Larry Stewart, *The Rise of Public Science: Rhetoric, Technology, and Natural Philosophy in Newtonian Britain* (Cambridge: Cambridge University Press, 1992).

102. See Hu, "Merging Chinese and Western Mathematics," chaps. 3 and 4.

103. Jacob, *Scientific Culture*, pp. 110–111.

104. Ibid., pp. 4–9.

105. Maurice Daumas et al., *A History of Technology & Invention. Progress through the Ages*, vol. 3, *The Expansion of Mechanization*, trans. Eileen Hennessy (New York: Crown Publishers, 1979), pp. 39–80. Jack Goldstone describes these developments in terms of a linkage between "engine science" and scientific culture, in "Efflorescences and Economic Growth in World History: Rethinking the 'Rise of the West' and the Industrial Revolution," *Journal of World History* 13, 2 (Fall 2002): 362–377.

106. Jacob, *Scientific Culture*, pp. 99–115.

107. Gillispie, *Science and Polity in France*, pp. 81–89.

108. Jacob, *Scientific Culture*, pp. 78–93, 136.

109. Ibid., p. 139. See also R. Rappaport, "Government Patronage of Science in 18th Century France," *History of Science* 8 (1969): 119–136, and Jean G. Dhombres, "French Textbooks in the Sciences 1750–1850," *History of Education* 13 (1984): 153–161. Compare Jean-Laurent Rosenthal, "The Political Economy of Absolutism Reconsidered," in Robert Bates et al., *Analytic Narratives* (Princeton, N.J.: Princeton University Press, 1998), pp. 64–108, on why France did not evolve similarly to the British institutional model.

110. Ervan Garrison, *A History of Engineering and Technology: Artful Methods* (Boca Raton, Fla.: CRC Press, 1991), pp. 130–139. Compare Daumas et al., *A History of Technology*, vol. 3, pp. 235–270.

111. Gillispie, *Science and Polity in France*, pp. 479–548. See also Karl Alder, "French Engineers Become Professionals; or, How Meritocracy Made Knowledge Objective," in William Clark, Jan Golinski, and Simon Schaffer, eds., *The Sciences in Enlightened Europe* (Chicago: University of Chicago Press, 1999), pp. 94–125.

112. Jacob, *Scientific Culture*, pp. 138–141, 178–86. See also Jean Dhombres, "L'enseignement des mathématiques par al 'methode révolutionaire.' Les leçons de Laplace à l'Ecole normale de l'an III," *Revue d'histoire des sciences* 33 (1980): 315–348, and Dhombres, *Naissance d'un pouvoir: sciences et savants en France (1793–1824)* (Paris: Payot, 1989). Norton Wise is engaged in a study that will describe the rise of German science and technology in the nineteenth

century and its roots in the Napoleonic models for education and institutional-ization that the Prussian state emulated.

113. Macartney, *An Embassy to China*, p. 266. Compare Jingmin Zeng, "Scientific Aspects of the Macartney Embassy to China 1792–1794," pp. 37–104. See also Tong, "From the Macartney Mission to the Opium Wars"; Han, "Chuanjiaoshi Weilie Yali," 62–63; and James Hevia, *Cherishing Men from Afar: Qing Guest Ritual and the Macartney Embassy of 1793* (Durham, N.C.: Duke University Press, 1995).

114. See John Fryer's June 1880 discussion in GZHB, vol. 1, pp. 349–350, which includes mention of Xu Shou's reactions after reading Hobson's textbook.

5. The Jesuit Role as Experts in High Qing Cartography and Technology

1. Li Bin, "Xishi wuqi dui Qingchu zuozhan fangfa de yingxiang," *Ziran bianzheng fa tongxun* 24, 4 (2002): 45–53, and Fu, *A Documentary Chronicle*, p. 48. See also Nicola Di Cosmo, "Did Guns Matter? Firearms and the Qing Formation," in Lynn Struve, ed., *The Qing Formation in World-Historical Time* (Cambridge, Mass.: Harvard University Asia Center, 2004), pp. 121–154, and Don Wagner, "Chinese Monumental Iron Casting," *Journal of East Asian Archaeology* 2, 3–4 (2000): 213–218. Compare Shu Liguang, "Ferdinand Verbiest and the Casting of Cannons in the Qing Dynasty," in Witek, ed., *Ferdinand Verbiest*, pp. 227–244, and Idema, "Cannon, Clocks and Clever Monkeys," pp. 464–466.

2. See Sivin, "On the Limits of Empirical Knowledge," p. 163. Compare Hsia, "French Jesuits and the Mission to China," pp. 66–72.

3. Cary Liu, "Architects and Builders in the Qing Dynasty Yuanming Yuan Impe-rial Garden-Palace," *Hong Kong University Museum Journal* 1 (2002): 38–59.

4. Heilbron, *Electricity in the 17th and 18th Centuries*, pp. 121, 352, 405. See also Wu Yiyi, "Chinese Technology in Eighteenth-Century American Eyes," *The Gest Library Journal* 6, 2 (Winter 1993): 31–54, and SCC, vol. 3, pp. 450–451. Compare Stephen Shapin and Simon Schaffer, *Leviathan and the Air-Pump: Hobbes, Boyle, and the Experimental Life* (Princeton, N.J.: Princeton University Press, 1989).

5. For the locus classicus, See *Hanshu*, 5.2410. See also Elman, *From Philosophy to Philology*, pp. 72–122.

6. See Frank Swetz, trans., *The Sea Island Mathematical Manual: Surveying and Mathematics in Ancient China* (University Park: Pennsylvania State University Press, 1992), pp. 7–16.

7. Sun Laichen, "Qi Jiguang and the Japanese Invasion of Korea (1592–98)" (pre-sented at the panel Perspectives on the First Greater East Asian War and Its Af-termath, annual meeting of the Association for Asian Studies, San Diego, March 4, 2004).

8. Kenneth Swope, "Crouching Tigers, Secret Weapons: Military Technology Employed in the Japanese Invasion of Korea, 1592–1598," *Journal of Military History*, forthcoming. See also DMB, pp. 168–172.

9. John Shepherd, *Statecraft and Political Economy on the Taiwan Frontier* (Stanford, Calif.: Stanford University Press, 1993), pp. 47–90, and Cordell Yee, "A Cartography of Introspection: Chinese Maps as Other than European," *Asian Art* 5 (1992): 29–45. See also Tonio Andrade, "Commerce, Culture and Conflict: Taiwan Under European Rule, 1623–1662" (Ph.D. diss., Yale University, 2001).

10. Cheng K'o-ch'eng, "Cheng Ch'eng-kung's Maritime Expansion and Early Ch'ing Coastal Prohibition," in Vermeer, ed., *Development and Decline of Fukien*, pp. 228–244. See also Shepherd, *Statecraft and Political Economy*, pp. 91–103.

11. John Wills Jr., "Maritime China from Wang Chih to Shih Lang," in Jonathan Spence and John Wills, eds., *From Ming to Ch'ing: Conquest, Region, and Continuity in Seventeenth-Century China* (New Haven, Conn.: Yale University Press, 1979), pp. 228–234. See also Shepherd, *Statecraft and Political Economy*, pp. 103–108.

12. Wills, "Maritime China," p. 233, and Peter Perdue, "Boundaries, Maps, and Movement: Chinese, Russian, and Mongolian Empires in Early Modern Central Eurasia," *International History Review* 20, 2 (June 1998): 263–286.

13. Ch'en, "Matteo Ricci's Contribution," pp. 347–359, and ECCP, pp. 762–767. See also the discussion in Chapter 8.

14. ECCP, pp. 522, 909.

15. *Huangchao wenxian tongkao*, in *Shitong* (Shanghai: Commercial Press, 1935–1937), vol. 2, 298.7469–7470. See also Chen Minsun, "Ferdinand Verbiest," pp. 129–131, on the idealization of Europe in Jesuit writings.

16. *Huangchao wenxian tongkao*, vol. 2, 298.7469–7474, 298.7481–7489.

17. SKQSZM, 69.22a–23a. See also Libbrecht, *Chinese Mathematics*, pp. 122–149, and SCC, vol. 3, pp. 569–579.

18. SKQSZM, 70.4b. Compare Sivin, "On the Limits of Empirical Knowledge," p. 163. See also DMB, pp. 1286–1287, and Minghui Hu, "Measuring the Cosmos: The Rise of Precision Narratives in Qing China" (paper presented at the colloquium sponsored by the Center for the Cultural Studies of Science, Medicine, and Technology, UCLA History Department, May 6, 2002).

19. On the *fenye* system, see Schafer, *Pacing the Void*, pp. 75–84. Unofficial life and popular culture paid no attention to literati efforts to distance such correspondences. See Richard Smith, *Fortune-Tellers & Philosophers: Divination in Traditional Chinese Society* (Boulder, Colo.: Westview Press, 1991), and Elman, *A Cultural History*, pp. 295–370

20. See Shigeru Nakayama, *A History of Japanese Astronomy*, pp. 47–48; Schafer, *Pacing the Void*, pp. 75–84; and John Major, "A Note on the Translation of Two Technical Terms in Chinese Science: *Wu-hsing* and *Hsiu*," *Early China* 2 (Fall 1976): 1–3.

21. See Yan Ruoju, *Qianqiu zhaji* (Taibei: Commercial Press, 1973), 3.2a. See also Yamada Keiji, *Shushi no shizengaku* (Tokyo: Iwanami Shoten, 1978), pp. 241–242, and SCC, vol. 3, pp. 231–252.

22. John Henderson, *The Development and Decline of Chinese Cosmology* (New York: Columbia University Press, 1984), pp. 82–87, 219–225.

23. On magic squares, see Schuyler Cammann, "The Evolution of Magic Squares in China," *Journal of the American Oriental Society* 80 (1960): 116–124.

24. Michael Saso, "What Is the *Ho-t'u?*" *History of Religions* 17, 3–4 (February–May 1978): 399–416. See also SCC, vol. 3, pp. 55–59.

25. Dai Zhen, "Liyan," in *Fenzhou fuzhi* (1771 ed.), pp. 5b–6a. See also Hilde De-Weerdt, "Regional Descriptions: Administrative and Scholarly Traditions," in Patrick Hanan, ed., *Treasures of the Yenching* (Cambridge, Mass.: Harvard-Yenching Library, 2003), pp. 139–142.

26. See Rong Zhaozu, "Yan Ruoju de kaozhengxue," *Lingnan xuebao* 1, 4 (1930): 90–91.

27. Du Weiyun, *Qing Qianjia shidai zhi shixue yu shijia* (Taibei: Wenshi zongkan, 1962), pp. 11–12, gives a comprehensive list of such supplements. See also David Nivison, *The Life and Thought of Chang Hsueh-ch'eng* (Stanford, Calif.: Stanford University Press, 1966), pp. 195, 216.

28. ECCP, p. 421. See also Henderson, *The Development and Decline of Chinese Cosmology,* pp. 231–253, and Sivin, "Wang Hsi-shan," pp. 159–162.

29. Smith, "Mapping China's World," pp. 71–77, usefully documents the same trends but focuses instead on Chinese disdain of "barbarians." For problems with this perspective, see Lydia Liu, "Legislating the Universal: The Circulation of International Law in the Nineteenth Century," in Liu, ed., *Tokens of Exchange,* pp. 127–164. Though frequently antiforeign, Qing rhetoric about "barbarians" is a later European construction.

30. See Mark Elliot, "The Limits of Tartary: Manchuria in Imperial and National Geographies," JAS 59, 3 (August 2000): 603–646.

31. See *The Library of Philip Robinson,* part 2, p. 95, and Smith, *Chinese Maps,* pp. 54–59. Compare Philippe Forêt, *Mapping Chengde: The Qing Landscape Enterprise* (Honolulu: University of Hawai'i Press, 2000), chap. 6.

32. Perdue, "Boundaries, Maps, and Movement," pp. 267–268, and Laura Hostetler, *Qing Colonial Enterprise: Ethnography and Cartography in Early Modern China* (Chicago: University of Chicago Press, 2001), pp. 66–71. See also Hahn, *The Anatomy of a Scientific Institution,* pp. 90, 96–97. Compare Elliot, "The Limits of Tartary," pp. 619–620.

33. Miasnikov, "Ferdinand Verbiest," pp. 274–275.

34. See Sun Jiji, *Kang-Yong-Qian shiqi yutu cehui yu jiangyu xingcheng yanjiu* (Beijing: Zhongguo renmin daxue chuban she, 2003). Compare Deepak Kumar, "Problems in Science Administration: A Study of the Scientific Surveys in British India," in Patrick Petitjean, Catherine Jami, and Anne Marie Moulin, eds., *Science and Empires: Historical Studies about Scientific Development and European Expansion* (Boston: Kluwer Academic Publishers, 1992), pp. 269–280.

35. See Walter Fuchs, "Materialen zur Kartographie der Mandju-Zeit, part 1," *Monumenta Serica* 1 (1935–1936): 395–396, and Walter Fuchs, "Materialen zur Kartographie der Mandju-Zeit, part 2," *Monumenta Serica* 3 (1938). Compare Elliot, "The Limits of Tartary," pp. 621–632.

36. Perdue, "Boundaries, Maps, and Movement," pp. 263–286, and Peter Perdue, "Military Mobilization in Seventeenth and Eighteenth-Century China, Russia and Mongolia," *Modern Asian Studies* 30, 4 (1996): 782–790. Compare Joanna Waley-Cohen, "Commemorating War in Eighteenth-Century China," *Modern Asian Studies* 30, 4 (1996): 869–899.

37. Peter the Great later expelled the Jesuits from Russia because he thought they had betrayed Russian interests in the Nerchinsk treaty. See Miasnikov, "Ferdinand Verbiest," pp. 273–279, and Hao Zhenhua, "Ferdinand Verbiest and Sino-Russian Relations," in Witek, ed., *Ferdinand Verbiest (1623–1688)*, pp. 282–296. Compare Mark Mancall, *Russia and China: Their Diplomatic Relations to 1728* (Cambridge, Mass.: Harvard University Press, 1971), pp. 98–109.

38. See Perdue, "Boundaries, Maps, and Movement," pp. 263–286, and Laura Hostetler, "Qing Connections to the Early Modern World: Ethnography and Cartography in Eighteenth-Century China," *Modern Asian Studies* 34, 3 (2000): 623–662. See also Elliot, *The Manchu Way*.

39. Perdue, "Boundaries, Maps, and Movement," pp. 274–275, and Hostetler, *Qing Colonial Enterprise,* pp. 17–18, 76. Compare David Turnbull, "Cartography and Science in Early Modern Europe: Mapping and the Construction of Knowledge Spaces," *Imago Mundi* 48 (1996): 5–24.

40. See Hahn, *The Anatomy of a Scientific Institution*, pp. 66–67, 90, 96–97.

41. Hostetler, *Qing Colonial Enterprise,* pp. 71–75. See also Perdue, "Military Mobilization," pp. 757–793, and James Millward, "'Coming onto the Map': 'Western Regions' Geography and Cartographic Nomenclature in the Making of the Chinese Empire in Xinjiang," *Late Imperial China* 20, 2 (December 1999): 61–98.

42. Hostetler, *Qing Colonial Enterprise,* pp. 74–79, and Elliot, "The Limits of Tartary," p. 626.

43. Thomas Barfield, *The Perilous Frontier: Nomadic Empires and China, 221 B.C. to A.D. 1757* (Cambridge, Mass.: Barfield Publishers, 1989), pp. 277–294.

44. Mancall, *Russia and China*, pp. 149–159, 209–210. The translation of the treaty is on pp. 280–283. See also Eric Widmer, *The Russian Ecclesiastical Mission in Peking during the Eighteenth Century* (Cambridge, Mass.: Harvard University East Asian Research Center, 1976), pp. 45–58, 174–178.

45. Catherine Pagani, *"Eastern Magnificence & European Ingenuity": Clocks of Late Imperial China* (Ann Arbor: University of Michigan Press, 2001), pp. 39–57, 70–74. On earlier clockwork in China, see SSC, vol. 4, pp. 220–266.

46. Jami, "Western Devices for Measuring Time and Space," pp. 169–200. See also Standaert, *Handbook of Christianity in China*, vol. 1, pp. 840–850.

47. Pagani, *"Eastern Magnificence,"* pp. 26–57, 181–184, and Craig Clunas, "Ming and Qing Ivories: Useful and Ornamental Pieces," in William Watson, ed., *Chinese Ivories from the Shang to the Qing* (London: Oriental Ceramic Society, British Museum Publications, 1984), p. 122.

48. Pagani, *"Eastern Magnificence,"* pp. 58–98.

49. Ibid., pp. 76–78, 91–93. See also Idema, "Cannon, Clocks and Clever Monkeys," pp. 459–488.

50. Sobel and Andrews, *The Illustrated Longitude*, pp. 121–195. See also Michael Mahoney, "Charting the Globe and Tracking the Heavens: Navigation and the Sciences in the Early Modern Era," in Brett Steele and Tamera Dorland, eds., *The Heirs of Archimedes: Technology, Science and the Art of Warfare through the Age of Enlightenment* (Cambridge, Mass.: MIT Press), forthcoming.

51. Emily Curtis, "Plan of the Emperor's Glassworks," *Ars Asiatiques* (Paris) 56 (2001): 81–90.

52. Peter Lam, "The Glasswork of the Qing Imperial Household Department," in *Elegance and Radiance: Grandeur in Qing Glass, The Arthur K. F. Lee Collection* (Hong Kong: Art Museum, Chinese University of Hong Kong, 2000), pp. 46–47, and Yang Boda, "Qingdai boli gaishu," *Gugong bowu yuan yuankan* (1983): 13–16. See also Yang Boda, "An Account of Qing Dynasty Glassmaking," in Robert H. Brill and John H. Martin, eds., *Scientific Research in Early Chinese Glass* (Corning, N.Y.: The Corning Museum of Glass, 1991), p. 144.

53. Zhang Rong, "Imperial Glass of the Yongzheng Reign," in *Elegance and Radiance*, p. 64. See also Rosemary Scott, "Eighteenth Century Overglaze Enamels: The Influence of Technological Development on Painting Style," in Rosemary Scott and Graham Hutt, eds., *Colloquies on Art & Archaeology in Asia* (London: Percival David Foundation of Chinese Art, 1987), pp. 156–158.

54. Zhang, "Imperial Glass of the Yongzheng Reign," p. 63. See also Peter Lam, *Elegant Vessels for the Lofty Pavilion: The Zande Lou Gift of Porcelain with Studio Marks* (Hong Kong: Art Gallery, Chinese University of Hong Kong, 1993), pp. 33–36. Compare Emily Curtis, "Glass for the Qing Court: The Jesuit Workshop" (paper presented at the colloquium, "Art Brokering for China: The Missionary Connection," sponsored by the UCLA Center for Chinese Studies in conjunction with the Southern California China Colloquium, May 4, 2002).

55. See the special publication of *Yuanming yuan* 1 (November 1981). See also Liu, "Architects and Builders," pp. 38–59, which stresses the role of Chinese architects from the Lei family for the both the Chinese- and European-style pavilions.

56. Richard Rudolph, "Early China and the West: Fertilization and Fetalization," in Richard Rudolph and Schuyler Cammann, *China and the West: Culture and Commerce* (Los Angeles: Clark Library, 1977), pp. 4–5. See also Gauvin Alexander Bailey, "Art et architecture des Jésuites en Extrême-Orient, 1542–1773," in Giovanni Sale, ed., *L'Art des Jésuites* (Paris: Mengès, 2003), pp. 277–296.

57. See Wong Young-tsu, *A Paradise Lost: The Imperial Garden Yuanming Yuan* (Honolulu: University of Hawai'i Press, 2001), pp. 59–65, and the drawings included in Régine Thiriez, "The Qianlong Emperor's European Palaces," in *The Delights of Harmony: The European Palaces of the Yuanmingyuan & the Jesuits at the 18th Century Court of Beijing* (Worcester, Mass.: Cantor Art Gallery, College of Holy Cross, 1994).

58. See Victoria Siu, "Castiglione and the Yuanming Yuan Collections," *Orientations* (Hong Kong) 19 (November 1988): 72–79.

59. Wong, *A Paradise Lost*, p. 59.

60. See George Loehr, "The Sinicization of Missionary Artists and Their Works at the Manchu Court during the Eighteenth Century," *Cahiers D'histoire Mondiale* 8 (1963): 795–803. Compare Forêt, *Mapping Chengde*, pp. 15–23. See also Pamela Crossley, *A Translucent Mirror: History and Identity in Qing Imperial Ideology* (Berkeley: University of California Press, 1999), pp. 282–283.

61. See Lydia Liu, "Robinson Crusoe's Earthenware Pot," *Critical Inquiry* 25 (Summer 1999): 749–750, and *The Gardiner Museum of Ceramic Art: A Collection of Collections* (Toronto: The Gardiner Museum, 2002), pp. 30–37. Compare Karl Berling, ed., *Meissen China: An Illustrated History* (repr., New York: Dover Books, 1972), pp. 1–6, and Han Qi, *Zhongguo kexue jishu de xichuan ji qi yingxiang* (Beijing: Hebei chubanshe, 1999), pp. 152–161, on French-made porcelain. See also ECCP, p. 691.

62. See Ayers and Kerr, *Blanc De Chine*. See also Lydia Liu, "The Quest for True Porcelain" (paper prepared for the conference, "Discourses and Practices of Everyday Life in Imperial China," Columbia University, New York, October 25–27, 2002), p. 4, and J. H. Plumb, *In the Light of History* (Boston: Houghton Mifflin Co., 1973), pp. 58–59.

63. See *Zhongguo taoci shi*, pp. 415–454. See also Stephen Little, *Chinese Ceramics of the Transitional Period: 1620–1683* (New York: China Institute in America, 1983), pp. 1–28, and Patricia Ferguson, *Cobalt Treasures: The Bell Collection of Chinese Blue and White Porcelain* (Toronto: Gardiner Museum of Ceramic Art, 2003).

64. Michel Beurdeley and Guy Raindre, *Qing Porcelain* (New York: Rizzoli, 1986), pp. 8–10. Compare Michael Butler et al., *Treasures from an Unknown Reign: Shunzhi Porcelain 1644–1661* (Alexandria, Va.: Art Services International, 2002).

65. Beurdeley and Raindre, *Qing Porcelain*, pp. 10, 24.

66. Ibid., pp. 30, 119–124, 127. See also Helmut Brinker and Albert Lutz, *Chinese Cloisonné: The Pierre Uldry Collection*, trans. Susanna Swoboda (New York: The Asia Society Galleries, 1989), p. 23, and Liu, "The Quest for True Porcelain," p. 13.

67. See Elisabetta Corsi, "Nian Xiyao's (1671–1738) Rendering of Western Perspective in the Prologues to 'Science of Vision,'" in Antonio Forte and Federico Masini, eds., *A Life Journey to the East: Sinological Studies in Memory of Guliano Bertuccioli* (Kyoto: Scuola Italiana di Studi sull'Asia Orientale, 2002), pp. 201–233. Compare Agustin Munoz Vidal, "Pintores jesuitas en la corte de los emperadores Qing durante el siglo XVIII," *Goya* 269 (March–April 1999): 93–102.

68. Beurdeley and Raindre, *Qing Porcelain*, pp. 86, 92–95, 112–115, 124–132, 136–137, 141–142. See also S. W. Bushell, "A Description of Ching-te-chen," in Robert Tichane, *Ching-te-chen: Views of a Porcelain City* (New York: NYS Institute for Glaze Research, 1983), pp. 175–185, and ECCP, pp. 441–442, 590.

69. Beurdeley and Raindre, *Qing Porcelain*, pp. 32–34. See also Tang Ying, "The Twenty Illustrations of the Manufacture of Porcelain," trans. S. W. Bushell with "Comments," in Tichane, *Ching-te-chen*, pp. 129–130, 131–132.

70. Beurdeley and Raindre, *Qing Porcelain*, pp. 35–37, 143–144, 192–210.

71. See Dorothy Bernstein, "Hunts, Processions, and Telescopes: A Painting of an Imperial Hunt by Lang Shining (Giuseppe Castiglione)," *RES* 35 (Spring 1999): 171–184. See also Joanna Waley-Cohen, "China and Western Technology in the Late Eighteenth Century," *American Historical Review* 98 (December 1993): 1527.

72. See Jonathan Hay, "Culture, Ethnicity, and Empire in the Work of Two Eighteenth-Century 'Eccentric' Artists," *RES* 35 (Spring 1999): 214–215.

73. Ibid., pp. 201–223. See also Nicolas Standaert, "A Chinese Translation of Ambroise Paré's *Anatomy*," *Sino-Western Cultural Relations Journal* 21 (1999): 9–33. Standaert adds that the drawings may have been drawn from Vesalius's original because the Rho manuscript was never printed. How Luo got access to the drawings is unclear.

74. Hay, "Culture, Ethnicity, and Empire," pp. 219–222. See also Chu Pingyi, "Shenti, linghun yu tianzhu: Mingmo Qingchu Xixue zhong de renti shengli zhishi," *Xin shixue* 7, 2 (1996): 47–98.

75. Macartney, *An Embassy to China*, p. 299. See also James Hevia, "Looting Beijing: 1860, 1900," in Liu, ed., *Tokens of Exchange*, pp. 192–199.

6. Evidential Research and the Restoration of Ancient Learning

1. See Noel M. Swerdlow, "The Recovery of the Exact Science of Antiquity," in Anthony Grafton, ed., *Rome Reborn: The Vatican Library and Renaissance Culture* (Washington, D.C.: Library of Congress, 1993), pp. 125–168, and Noel M. Swerdlow, "Science and Humanism in the Renaissance: Regiomontanus's Oration on the Dignity and Utility of the Mathematical Sciences," in Paul Horwich, ed., *World Changes: Thomas Kuhn and the Nature of Science* (Cambridge, Mass.: MIT Press, 1993), pp. 131–168.

2. Julia Ching, ed., *The Record of Ming Scholars by Huang Tsung-hsi* (Honolulu: University of Hawai'i Press, 1987), pp. 59–60. For a discussion of Huang's interest in European learning, see Xu Haisong, *Qingchu shiren yu xixue*, pp. 278–306.

3. See Willard Peterson, "The Life of Ku Yen-wu (1613–1682)," HJAS 28 (1968): 114–156, and 29 (1969): 201–247, and Willard Peterson, "Changing Literati Attitudes toward New Learning in Astronomy and Mathematics in Early Qing," in Zhou Zhiping and Willard Peterson, eds., *Guoshi fuhai kaixin lu: Yu Yingshi jiaoshou rongtui lunwen ji* (Taibei: Lianjing, 2002), pp. 379–403. Compare Xu Haisong, *Qingchu shiren yu xixue*, pp. 314–318, which presents Gu's critical views of European learning. See also Sivin, "On the Limits of Empirical Knowledge," in Fraser et al., eds., *Time, Science, and Society*, p. 163.

4. SKQSZM, 98.1a–2a, 98.8b–9b, 98.11a–b. A Qing defender of Cheng-Zhu learning, Fang Dongshu (1772–1851), in his *Hanxue shangdui* (Taibei: Commercial Press, 1968), pp. 49–50, noted that evidential scholars foolishly accepted Zang Lin's critique of Zhu Xi's emendation of the Great Learning.

5. See Goldschmidt, "The Transformations of Chinese Medicine," pp. 72–118. See also Chu Pingyi, "Tongguan Tianxue, yixue yu ruxue," pp. 165–201, for a unique example of the impact of European medicine in the seventeenth century. Compare Rudolph, "Early China and the West: Fertilization and Fetalization," pp. 17–24.

6. On the circulation tracts, see Porkert, *The Theoretical Foundations of Chinese Medicine*, pp. 197–216, and Asaf Goldschmidt, "Changing Standards: Tracing Changes in Acu-moxa Therapy during the Transition from the Tang to the Song Dynasties," *East Asian Science, Technology, and Medicine* 18 (2001): 91–92. See also Nathan Sivin, *Traditional Medicine in Contemporary China* (Ann Arbor: Center for Chinese Studies, University of Michigan, 1987), pp. 249–272. Compare Kuriyama, *The Expressiveness of the Body*, pp. 31–60, and von Staden, *Herophilus*.

7. On the literati-physician, see Robert Hymes, "Not Quite Gentlemen? Doctors in Sung and Yuan," *Chinese Science* 7 (1986): 11–85. For the "push and pull" aspects that made medicine intellectually attractive in the Song-Yuan period, see

Shinno, "Promoting Medicine in the Yuan," pp. 157–158. See also Nathan Sivin, "Text and Experience in Classical Chinese Medicine," in Don Bates, ed., *Knowledge and the Scholarly Medical Traditions* (Cambridge: Cambridge University Press, 1995), pp. 177–204, especially 190–195, which stresses the perennial effort in traditional Chinese medicine to recapture the genuine meaning of the medical classics.

8. Nathan Sivin, *"Huang ti nei ching,"* in Loewe, ed., *Early Chinese Texts,* pp. 198–199, and SCC, vol. 6, part 6, pp. 74–76. See also Yasuo Otsuka, "Chinese Traditional Medicine in Japan," in Charles Leslie, ed., *Asian Medical Systems: A Comparative Study* (Berkeley: University of California Press, 1976), pp. 323. A variant of the *Inner Canon* called the *Great Basis (Taisu)* was discovered among the Mawangdui manuscripts, circa 168 B.C., in 1973. See Donald Harper, *Early Chinese Medical Literature: The Mawangdui Medical Manuscripts* (London: Kegan Paul, 1998).

9. Goldschmidt, "The Transformations of Chinese Medicine," pp. 21–35, 41–56. Compare Kuriyama, *The Expressiveness of the Body,* pp. 237–242, 251–259, on the etiology of "wind" as the "chief of the hundred diseases" in Chinese medicine.

10. See Goldschmidt, "The Transformations of Chinese Medicine," pp. 36–45, 56–67; Porkert, *The Theoretical Foundations of Chinese Medicine,* pp. 42–43; and Sivin, *Traditional Medicine in Contemporary China,* pp. 460–461. Compare Marta Hanson, "Robust Northerners and Delicate Southerners: The Nineteenth-Century Invention of a Southern Medical Tradition," *positions* 6, 3 (Winter 1998): 544.

11. Sivin, *Traditional Medicine in Contemporary China,* pp. 89–90. See also Chao, "Medicine and Society in Late Imperial China," pp. 73–74, and Goldschmidt, "The Transformations of Chinese Medicine," pp. 28–35. On the *Inner Canon* as heterogeneous short texts originally, see David Keegan, *"Huang-ti nei-ching.* The Structure of the Compilation, the Significance of the Structure" (Ph.D. diss., University of California–Berkeley, 1988).

12. Goldschmidt, "The Transformations of Chinese Medicine," pp. 35–71. See also Yasuo Otsuka, "Chinese Traditional Medicine," p. 325, and Porkert, *The Theoretical Foundations,* p. 43.

13. Wu, "A Medical Line," pp. 40–45. See also SKQSZM, 104.3a–4a. Compare Paul Unschuld, *Medicine in China: A History of Ideas* (Berkeley: University of California Press, 1985), pp. 168–169, 172–174.

14. Wu, "A Medical Line," pp. 56–57, and Hymes, "Not Quite Gentlemen?" pp. 11–85.

15. See T. J. Hinrichs, "The Medical Transforming of Governance and Southern Customs in Song Dynasty China (960–1279 C.E.)" (Ph.D. diss., Harvard University, 2003).

16. See Robert Hartwell, "Demographic, Political, and Social Transformations of China, 750–1550," *HJAS* 42, 2 (1982): 365–426, and Goldschmidt, "The Transformations of Chinese Medicine," pp. 13–14. William McNeill, *Plagues and Peoples* (New York: Anchor Books, 1976), pp. 132–175, discusses the consequences of the Mongol conquests on shifting disease balances in Asia and Europe. See also Janet L. Abu-Lughod, *Before European Hegemony* (Oxford: Oxford University Press, 1989), pp. 94–96, 171–174, 339–343, on the epidemiology of the plague

in the fourteenth century. Compare Alfred Crosby, *Ecological Imperialism: The Biological Expansionism of Europe, 900–1800* (Cambridge: Cambridge University Press, 1986).

17. See Furth, *A Flourishing Yin*, pp. 135–137.

18. Ibid., pp. 137, 146. See also Hugh Shapiro, "The Puzzle of Spermatorrhea in Republican China," *positions* 6, 3 (Winter 1998): 557–560. The Chinese did not have a concept of fever until very late, although the notion of a heat disorder referred to a patient's inner sensations, which according to biomedicine were mainly fevers. My thanks to Nathan Sivin for clarifying this point.

19. See Charlotte Furth, "Health, Embodiment and Morality in the Medical Thought of Zhu Zhenheng (1281–1358)" (presented at the "Symposium on Medicine and Society in Traditional China," sponsored by the Institute of History and Philology, Academia Sinica, Taiwan, June 26–28, 1998). See also Angela Leung, "Medical Instruction and Popularization in Ming-Qing China" (paper presented at the International Colloquium on "Education et instruction en Chine," organized by the Institut National des Langues et Civilisations Orientales, Paris, June 29–July 2, 1999). Pulse-reading in China differed from the Greek and Roman tradition. See Kuriyama, *The Expressiveness of the Body*, pp. 19–60.

20. Statistics by Japanese scholars are cited in Helen Dunstan, "The Late Ming Epidemics: A Preliminary Survey," *Ch'ing-shih wen-t'i* 3, 3 (November 1975): 1–65.

21. Wu, "A Medical Line," p. 47.

22. Chao, "Medicine and Society," pp. 72–80, 113–123, and Sivin, *Traditional Medicine in Contemporary China*, p. 151. Compare Dunstan, "The Late Ming Epidemics," pp. 45–53.

23. Hanson, "Robust Northerners," p. 548n43.

24. Joanna Grant, "Medical Practice in the Ming Dynasty—A Practitioner's View: Evidence from Wang Ji's *Shishan yi'an*," *Chinese Science* 15 (1998): 37–80, and Joanna Grant, *A Chinese Physician. Wang Ji and the 'Stone Mountain Medical Case Histories'* (Needham Research Institute Series, 2. London: RoutledgeCurzon, 2003). See also Christopher Cullen, "Yi'an (case statements): The Origins of a Genre of Chinese Medical Literature," in Hsu, ed., *Innovation in Chinese Medicine*, pp. 297–323.

25. Chao, "Medicine and Society," pp. 63–68. See also Unschuld, *Medicine in China*, pp. 208–210. Compare Elman, *From Philosophy to Philology*, pp. 35–38, 82–84.

26. Chao, "Medicine and Society," pp. 51–57.

27. Catherine Despeux, "The System of the Five Circulatory Phases and the Six Seasonal Influences (*wuyun liuqi*), a Source of Innovation in Medicine under the Song (960–1279)," trans. Janet Lloyd, in Hsu, ed., *Innovation in Chinese Medicine*, pp. 121–165. On the diversity of classical Chinese medicine, see Volker Scheid, *Chinese Medicine in Contemporary China: Plurality and Synthesis* (Durham, N.C.: Duke University Press, 2002), pp. 27–61.

28. SKQSZM, 104.45a–48b, and 104.43b–45a. See also Sivin, *Traditional Medicine*, pp. 460–461.

29. ECCP, pp. 322–324. See also *Qingshi gao*, vol. 46, p. 13,878.

30. SKQSZM 104.51a–55b, and *Qingshi gao*, vol. 46, p. 13,879. See also Chao, "Medicine and Society," pp. 90–91. Compare Paul Unschuld, trans., *Forgotten*

Traditions of Ancient Chinese Medicine: The I-hsueh Yüan Liu Lun of 1757 by Hsü Ta ch'un (Brookline, Mass.: Paradigm Publication Co., 1990), which fails to grasp the full scope of evidential learning in Xu's time.

31. Sheng Zengxiu et al., *Wenbing xuepai si dajia yanjiu* (Beijing: Zhongguo Zhongyi yao chuban she, 2000), pp. 12–22, 113–126. See also ECCP, pp. 323, 902–903.

32. SKQSZM, 104.45a–48b. Compare Marta Hanson, "The *Golden Mirror* in the Imperial Court of the Qianlong Emperor, 1739–1743," *Early Science and Medicine* 8, 2 (2003): 112–147.

33. SKQSZM 103.1a. See also SCC, vol. 6, part 6, p. 27.

34. Chao, "Medicine and Society," pp. 66–70, 97, and SCC, vol. 6, part 6, p. 26. See also Hanson, "Robust Northerners," pp. 541–542, and her forthcoming *Chinese Medical Geographies: Regional Cultures, Therapeutic Practices, and Imagined Bodies in Early Modern China.*

35. See Lu Shiyi, "Sibianlu lunxue," in Wei Yuan, ed., *Huangchao jingshi wenbian* (1827 and 1873 ed.; repr., Taibei: World Bookstore, 1964), 3.7–3.9.

36. See *Kangxi jixia gewu bian* (Tongxuezhai ed., n.d.), which was translated by the French Jesuits in 1779. See *Memoires* 4 (1776–1814): 452–483.

37. SKQSZM, 136.17a–25a.

38. Elman, *A Cultural History,* pp. 106–109, 122–123, and ECCP, pp. 96–97.

39. Chen Yuanlong, "Preface" to his GZJY (1735 ed.), pp. 1a–2b, and SKQSZM, 136.26a.

40. ECCP, pp. 93–94, 922. See also Han Qi, "Emperor, Prince, and Literati: Role of the Princes in the Organization of Scientific Activities in the Early Qing Period," in Yung Sik Kim and Francesca Bray, eds., *Current Perspectives in the History of Science in East Asia* (Seoul: Seoul National University Press, 1999), pp. 209–216.

41. ECCP, pp. 747–749, and Spence, *Treason by the Book,* pp. 94–95, 181–182.

42. See Han Qi, "Cong Mingshi lizhi de zuanxiu kan Xixue zai Zhongguo de chuanbo," in Liu Dun ed., *Keshi xinzhuan* (Shenyang: Liaoning jiaoyu chuban she, 1997), pp. 61–70. See also Han, "'Zili' jingshen," pp. 214–219.

43. ECCP, pp. 54–55, 240, 803, 826.

44. See Fujitsuka Chikashi, *Shinchō bunka tōden no kenkyū,* p. 11. See also SKQSZM, 134.14b–15b, 136.25a–28a.

45. See Chen, "Preface" to the GZJY (1735 edition), pp. 1a.

46. See Chen Yuanlong, "Fanli," in GZJY (1735 ed.), pp. 1a–1b. See also SKQSZM, 136.25a–26a.

47. GZJY (SKQS ed.), vol. 1031: 13 (1.20b), 7–11 (1.8b–15a).

48. Ibid., 13–14 (1.20b–21b).

49. See TPYL, 988.3a–3b, and the *Zhouli jishuo* (SKQS ed.), 9A.29a. Compare Lin Wenzhao, "Magnetism and the Compass," in *Ancient China's Technology and Science* (Beijing: Foreign Language Press, 1983), pp. 152–165.

50. See Gerald Holten, *Thematic Origins of Scientific Thought: Kepler to Einstein* (Cambridge, Mass.: Harvard University Press, 1973), pp. 72–76. See also Heilbron, *Electricity,* pp. 24–31, 61–72, 87, 107.

51. GZJY (SKQS ed.), 33.16b, and Li Shizhen, *Bencao gangmu* (SKQS ed.), 34.46b, and 37.8a–8b.

52. Heilbron, *Electricity,* pp. 169–192.

53. Compare Qiong Zhang, "About God, Demons, and Miracles," p. 36.

54. GZJY (SKQS), vol. 1031: 36–40; 3.11a–18b). Compare Hammond, "The Interpretation of Thunder," pp. 487–503, and Heilbron, *Electricity*, p. 72.

55. Heilbron, *Electricity*, pp. 121, 352, 405. See also Heller, "Jesuit Physics in Eighteenth-Century Germany," pp. 538–554.

56. See Betty Dobbs and Margaret Jacob, *Newton and the Culture of Newtonianism* (Atlantic Highlands, N.J.: Humanities Press, 1995), pp. 46–56.

57. Heilbron, *Electricity*, pp. 3, 24, 31, 61, 65, 68–69, 72, 87, 107, 180–192, 349–350.

58. See Qian Baocong, ed., *Zhongguo shuxue shi* (Beijing: Kexue chubanshe, 1964), pp. 151–174.

59. See Ruan Yuan, "Chouren zhuan fanli," in CRZ, pp. 1–5. See also Hashimoto, "Bai Buntei no rekisan gaku," pp. 514–516, and Pingyi, "Technical Knowledge," pp. 217–219. Hashimoto anachronistically reads evidential research into Mei's intellectual synthesis. For a corrective see Zhang Yongtang, *Mingmo Qingchu lixue yu kexue guanxi zailun* (Taibei: Xuesheng shuju, 1994), pp. 139–143. See also Hu, "Cosmopolitan Confucianism," which focuses on the rise of comprehensive literati in the late Qianlong period.

60. ECCP, p. 569, SCC, vol. 3, p. 53, and Martzloff, *A History of Chinese Mathematics*, pp. 159–163.

61. On the interpretive difficulties in describing *tianyuan* techniques as algebra, see Ken'ichi Sato, "Reevaluaton of *Tengenjutsu* or *Tianyuanshu*: In the Context of Comparison between China and Japan," *Historia Scientiarum* 5, 1 (1995): 57–67. Using counting rods rather than written equations for calculations, Chinese elided the exact steps used in reaching the correct numerical answer for an algebraic problem that Europeans described more analytically.

62. ECCP, p. 569, and Hu Mingjie, "Merging Chinese and Western Mathematics," p. 288. See also Martzloff, *A History of Chinese Mathematics*, p. 20, and SCC, vol. 3, pp. 38–50. Compare Roger Hart, "Notes on the Early History of Linear Algebra: Reconstructing Solutions to Simultaneous Linear Equations Using Determinants in the *Nine Chapters of the Mathematical Arts (Jiu zhang suan shu)*," mss.

63. Roger Hart notes that most Ming bibliographies that listed mathematical texts included mention of Li Ye's *Sea Mirror* and Qin's works on polynomial algebra. See Hart, "Proof, Propaganda, and Patronage."

64. Li and Du, *Chinese Mathematics*, pp. 223–224.

65. SKQSZM, chaps. (*juan*) 106–107. See Martzloff, *A History of Chinese Mathematics*, pp. 32–33, and ECCP, p. 637.

66. Sato Ken'ichi, "Reevaluation of *Tengenjutsu*," pp. 63–65. See also Shigeru Jochi, "The Influence of Chinese Mathematical Texts on Seki Kowa" (Ph.D. diss., London University, 1993), pp. 203–204. Compare Annick Horiuchi, *Les Mathématiques Japonaises à l'époque d'Edo* (Paris: Vrin, 1994), pp. 179–182, and Andrea Bréard, "On Mathematical Terminology: Culture Crossing in Nineteenth-Century China," in Lackner, Amelung, and Kurtz, eds., *New Terms for New Ideas*, pp. 318–320.

67. See Ruan Yuan, "Preface," pp. 1a–4a, in Zhu Shijie, *Suanxue qimeng* (repr., Shanghai: Jiangnan Arsenal, 1871). Compare Lam, "Chu Shih-chieh's *Suanhsueh ch'i-meng*," 1–31.

68. See Fujitsuka Chikashi (Rin), *Nichi Sen Shin no bunka kō ryū* (Tokyo: Chūbunkan shoten, 1947), p. 77, and Fujitsuka, *Shinchō bunka tōden no kenkyū*, pp. 111–113, 400.

69. Li and Du, *Chinese Mathematics*, pp. 225–26, and Lionel Giles, comp., *An Alphabetical Index to the Chinese Encyclopedia, Ch'in Ting ku Chin T'u Shu Chi Ch'eng* (repr., Taibei: Chengwen Publishing Co., 1969), pp. vi–viii. See the *Ripple Pavilion Collectanea (Weibo xie congshu)*, the *Collectanea from the Can't Know Enough Pavilion (Zhibuzu zhai congshu)*, and the *Collectanea of the Yijia Hall (Yijia tang congshu)*.

70. See Roger Hart, "Xu Guangqi, Memorialist" (presented at the colloquium sponsored by the Center for the Cultural Studies of Science, Medicine, and Technology, UCLA History Department, April 15, 2002).

71. Du Shiran, "Mathematical Classics," in *Ancient China's Technology and Science*, pp. 50–54. On the Tang, see also Siu Man-Keung and Alexeï Volkov, "Official Curriculum in Traditional Chinese Mathematics: How Did Candidates Pass the Examinations?" *Historia Scientiarum* 9, 1 (1999): 85–99.

72. Li and Du, *Chinese Mathematics*, pp. 228–29, and Martzloff, *A History of Chinese Mathematics*, pp. 81–82, 123–124, 136–41. Roger Hart has found mention of many of the mathematical classics in Ming bibliographies, which indicates they were not all lost. See Hart, "Xu Guangqi, Memorialist." Hart thinks that Li Chunfeng was primarily a court diviner. See also Hart, "Reconstructing Solutions to Simultaneous Linear Equations Using Determinants in Chinese Mathematical Treatises, part 1, The *Nine Chapters of Mathematical Methods (Jiu zhang suan shu)*," *Archive for the History of the Exact Sciences*, forthcoming.

73. See Qian Baocong, comp., "Xu" (Preface) to *Suanjing shishu*, 2 vols. (Beijing: Zhonghua shuju, 1963), pp. 4–5.

74. Martzloff, *A History of Chinese Mathematics*, p. 125. My thanks to Roger Hart for providing me with an updated version of his forthcoming "Table of the Ming Bibliographic Records of Mathematical Treatises."

75. Li and Du, *Chinese Mathematics*, pp. 226–227. The Needham Research Institute Library includes reissues of the Palace Editions that Dai prepared. Hart, "Reconstructing Solutions to Simultaneous Linear Equations," notes that many scholars today believe that Dai introduced many errors, elided text he did not understand, and generally misrepresented the editions he recovered from the *Great Compendium*.

76. Li and Du, *Chinese Mathematics*, pp. 227–230. Compare Swetz, trans., *The Sea Island Mathematical Manual*, pp. 27–55. See also Elman, *From Philosophy to Philology*, pp. 242–244.

77. Qian Baocong, *Qian Baocong kexue shi lunwen xuanji*, pp. 22–36, and Du, "Mathematical Classics," pp. 54–56. See also Wylie, *Notes on Chinese Literature*, p. 116, and Martzloff, *A History of Chinese Mathematics*, pp. 149–152. Compare Lam, *A Critical Study of the Yang Hui Suan Fa*, pp. 36–39.

78. Libbrecht, *Chinese Mathematics*, and Martzloff, *A History of Chinese Mathematics*, pp. 2–12, 231–247.

79. Hu Mingjie, "Merging Chinese and Western Mathematics," pp. 252–253. Compare Wylie, *Notes on Chinese Literature*, pp. 116–117.

80. Li and Du, *Chinese Mathematics*, pp. 110–114, 231; Martzloff, *A History of Chinese Mathematics*, pp. 143–149; and Hart, "Local Knowledges, Local Contexts."

Li Ye's works represent the first mention of single-unknown methods, although Li did not claim priority. See Libbrecht, *Chinese Mathematics*, p. 18.

81. Lam, *A Critical Study of the Yang Hui Suan Fa*, pp. xv–xvii. See also Li and Du, *Chinese Mathematics*, pp. 230–231; Martzloff, *A History of Chinese Mathematics*, pp. 149–152, 157–159; and Kobayashi, "What Kind of Mathematics and Terminology," pp. 2–3.

82. Sato, "Reevaluaton of *Tengenjutsu* or *Tianyuanshu*," pp. 59–60. See also Mei, "The Decimal Place-Value Numeration," pp. 63–65.

83. See Jock Hoe, "Zhu Shijie and His *Jade Mirror of the Four Unknowns*," in *First Australian Conference on the History of Mathematics: Proceedings of a Conference at Monash University* (Clayton, Australia), 6 & 7 (November 1980): 105.

84. Luo Shilin, *Siyuan yujian xicao* (1836; repr., Shanghai: Shangwu yinshuguan, 1937), *juan* 1. See also Li and Du, *Chinese Mathematics*, pp. 115–117, 231–232, and Martzloff, *A History of Chinese Mathematics*, pp. 153–157.

85. See Wylie, *Notes on Chinese Literature*, pp. 115–116, and Li and Du, *Chinese Mathematics*, pp. 242, 251.

86. Macartney, *An Embassy to China*, p. 50. See also Ssu-yü Teng and John Fairbank, eds., *China's Response to the West: A Documentary Survey, 1839–1923* (Cambridge, Mass.: Harvard University Press, 1979), pp. 19–20. Compare Hevia, *Cherishing Men From Afar*, pp. 225–248.

7. Seeking the Truth and High Qing Mathematics

1. Chu Pingyi, "Ch'eng-Chu Orthodoxy, Evidential Studies and Correlative Cosmology: Chiang Yung and Western Astronomy," *Philosophy and the History of Science: A Taiwanese Journal* 4, 2 (October 1995): 71–108. See also Wylie, *Notes on Chinese Literature*, p. 122.

2. See Elman, *From Philosophy to Philology*, pp. 93–122.

3. See Hui Shiqi, *Daxue shuo* (Lanhai shushi, Jiajing ed., 1795–1820), pp. 1a–3b. See also ECCP, pp. 356–357.

4. Liu Shipei, "Gewu jie," in *Zuo'an waiji*, vol. 3, pp. 4b–5a (p. 1592), in *Liu Shenshu xiansheng yishu* (Taibei: Daxin shuju, 1965), and CRZ, 41.512–513. See also Elman, *From Philosophy to Philology*, pp. 8–14, and Elman, *Classicism, Politics, and Kinship*, pp. 6–8.

5. Qian Daxin, *Qianyan tang wenji* (Shanghai: Guji chuban she, 1989), pp. 22–25, 284–87. See also Elman, *From Philosophy to Philology*, pp. 118–120.

6. ECCP, p. 153, and Elman, *From Philosophy to Philology*, pp. 159–160.

7. See the translation of Zhu Xi's commentary in Legge, trans., *The Four Books*, p. 360, which I have modified. For discussion, see Gardner, *Chu Hsi and the Ta-hsueh*, pp. 27–59.

8. *Dai Zhen wenji* (Hong Kong: Zhonghua shuju, 1974), pp. 251–260.

9. SKQSZM, 98.11a–b. On the Kangxi revival, see Chow, *The Rise of Confucian Ritualism*, pp. 44–53.

10. *Dai Zhen quanji* (Beijing: Qinghua daxue chuban she, 1991–1994), vol. 1, p. 75, vol. 3, pp. 1577–1581.

11. Hamaguchi Fujiō, *Shindai kokyogagu no shisōshiteki kenkyū* (Tokyo: Kokusho kankōkai, 1994), pp. 378–497.

12. Wang Zhong, "Buyi," in Wang Zhong, *Shuxue* (repr., Taibei: Guangwen shuju, 1970), 35.14a–b. Compare Elman, *From Philosophy to Philology*, pp. 160–161.

13. See also Ling Tingkan, *Jiaoli tang wenji* (Beijing: Zhonghua shuju, 1998), 16.7a–9b (pp. 607–608), and Cheng Yaotian, "Lunxue xiaoji" in *Tongyi lu* (late Qianlong ed.), pp. 24a–29b, and Jiao Xun, *Diaogu ji* (Daoguang ed., 1821–1850), 9.6a–8a, 9.15b–18a.

14. Elman, *From Philosophy to Philology*, pp. 218–221.

15. Ruan Yuan, "Daxue gewu shuo," in Ruan, *Yanjing shi ji*, 3 vols. (Taibei: Shijie shuju, 1964), vol. 1, pp. 47–48 (*juan* 2). See also Yü, "Some Preliminary Observations," pp. 105–146.

16. See Ruan, "Daxue gewu shuo," p. 48. See also Elman, *From Philosophy to Philology*, pp. 245–247.

17. SKQSZM, 136.25a–26a. See also R. Kent Guy, *The Emperor's Four Treasuries: Scholars and the State in the Late Ch'ien-lung Era* (Cambridge, Mass.: Harvard Council on East Asian Studies, 1987).

18. SKQSZM, 122.29a–29b.

19. Some accounts present thirty-seven such works.

20. SKQSZM, 106.28a–36a. See also Ji Wende, *Cong Siku quanshu tanjiu Ming Qing jian shuru zhi Xixue* (Taibei: Han Mei tushu youxian gongsi, 1991), pp. 404–436. Wu Boya also provided me with her unpublished essay on the European works evaluated for the Imperial Library project.

21. See SKQSZM, 106.1a–51a, 107.23a–24a, 125.27b–35b, 134.10a–11b, and Standaert, "The Investigation of Things," pp. 412–417. Compare W. F. Vande Walle, "Linguistics and Translation in Pre-Modern Japan and China: A Comparison," in Vande Walle and Kasaya, eds., *Dodonæus in Japan*, p. 140.

22. SKQSZM, 125.31b–34a, 125.34b–35a. See also Standaert, "The Classification of Sciences," pp. 287–317, and the translation in Luk, "Aleni Introduces the Western Academic Tradition," p. 517.

23. Ji Wende, *Cong Siku quanshu*, pp. 410–426.

24. SKQSZM, *juan* 106–107. Wylie's 1867 *Notes on Chinese Literature*, pp. 106–130, presents a catalog of Chinese works in the category of *tianwen suanfa*. See also Elman, *From Philosophy to Philology*, pp. 202–204.

25. See the editorial introduction to the "Astronomy and Mathematics" section in the SKQSZM, 106.1a–2a. See also Nakamura Shōhachi, "Chūgoku shisōshi jō ni okeru jutsusu," *Chūgoku kankei ronsetsu shiryō* 39,1 (1997): 11.

26. SKQSZM, 106.1b.

27. See Li and Du, *Chinese Mathematics*, pp. 232–233, *Qian Baocong kexueshi*, pp. 308–309, and Paul Cohen, *Between Tradition and Modernity: Wang T'ao and Reform in Late Ch'ing China* (Cambridge, Mass.: Council on East Asian Studies, Harvard University, 1987), pp. 176–177.

28. Wang Ping, "Ruan Yuan yu *Chouren zhuan*," *Jindaishi yanjiu suo jikan* 4 (1973): 601–611.

29. Sivin, "Copernicus in China," *Science in Ancient China*, IV, pp. 45–50. See also Martzloff, *A History of Chinese Mathematics*, pp. 166–172.

30. Ruan Yuan, "Chouren zhuan fanli," pp. 1–5, and Tan Tai, "Chouren jie," in CRZ, pp. 1–4. Compare Yue Meng, "Hybrid Science," pp. 26–27n38, and Pingyi

Chu, "Remembering Our Grand Tradition: The Historical Memory of the Scientific Exchanges between China and Europe," *History of Science* 41 (2003): 194–199.

31. Limin Bai, "Mathematical Study and Intellectual Transition in the Early and Mid-Qing," *Late Imperial China* 16, 2 (December 1995): 23–61.

32. See Pingyi Chu, "Western Astronomy and Evidential Study: Tai Chen on Astronomy," in Kim and Bray, eds., *Current Perspectives in the History of Science in East Asia,* p. 144.

33. Elman, *From Philosophy to Philology,* pp. 118–119.

34. In the United States, an attempt to legislate mistaken numbers for *pi,* among them π = "3.2," was ultimately unsuccessful, although it came close to passing in 1897 when a representative introduced a bill in the Indiana House of Representatives. My thanks to Bruce Rusk for pointing me to this information.

35. CRZ, 42.545. See also Horng, "Chinese Mathematics," pp. 183–190, especially p. 186.

36. Tian Miao, "*Jiegenfang, Tianyuan,* and Algebra in Qing China," *Historia Scientiarum* 9, 1 (1999): 101–119, argues that evidential scholar-mathematicians initially preferred Jesuit algebra because it was demonstrably superior.

37. See Wylie, *Notes on Chinese Literature,* pp. 122–123, Li and Du, *Chinese Mathematics,* pp. 234–240, 254, and Martzloff, *A History of Chinese Mathematics,* p. 358. Horng, "Chinese Mathematics," pp. 177–178, contends that Minggatu developed an equivalent method to the infinitesimal calculus.

38. CRZ, vol. 22, 48.623.

39. Horng, "Chinese Mathematics," pp. 167, 188–190, and Wylie, *Notes on Chinese Literature,* p. 123. See also Li and Du, *Chinese Mathematics,* pp. 240–243.

40. ECCP, pp. 144–145, and Horng, "Chinese Mathematics," pp. 167–168.

41. Horng, "Chinese Mathematics," pp. 185–186, 190. See also Wylie, *Notes on Chinese Literature,* p. 123, and Li and Du, *Chinese Mathematics,* pp. 234, 242, 251–252.

42. See CRZ, 30.366, for a critique of Gu Yingxiang and Ming mathematics. See also Wylie, *Notes on Chinese Literature,* pp. 124–125, and ECCP, p. 144.

43. Macartney, *An Embassy to China,* pp. 50, 264. Recall how James Dinwiddie described the limits of Macartney's mission: "What information could we derive respecting the arts and sciences in a country where we could not converse with the inhabitants?" The French Jesuit Gaubil on Chinese astronomy was the exception.

44. Fujitsuka, *Nichi Sen Shin no bunka kōryū,* p. 108.

45. See Elman, *From Philosophy to Philology,* pp. 126–133.

46. For a list of writings on mathematics and astronomy by Ming and Qing scholars, see Sivin, *Science in Ancient China,* V, pp. 17–20.

47. See Ruan Yuan, ed., *Huang Qing jingjie,* 20 vols. (1892; repr., Taibei: Fuxing, 1961), chaps. (*juan*) 1059–1068, especially 1067.1a–1b (vol. 15, p. 11324) on Newton.

48. See the table of contents to Ruan, ed., *Huang Qing jingjie,* vol. 1, pp. 9–32. See also Elman, *From Philosophy to Philology,* p. 243.

49. See Ruan, ed., *Huang Qing jingjie,* chaps. 485–490 (vol. 7, pp. 5305–5394), especially 488.2b–4b (pp. 5348–5349).

50. See Ruan, ed., *Huang Qing jingjie,* 1328.1a–2a (vol. 19, p. 24,459), especially 1328.22a–33a (pp. 14,471–14,472). See also CRZ, vol. 2, 48.634–637.

51. Juan Yuan et al., eds., *Gujing jingshe wenji* (Taipei, Commercial Press, 1966), pp. 29–56. See also Elman, *From Philosophy to Philology*, pp. 162–163.

52. See Yu Yue's "Zixu" to the *Huang Qing jingjie jianmu* (1886 ed.), pp. 1a–2a. My thanks to Minghui Hu for providing me with a copy of Yu's index. See also ECCP, pp. 944–945, and Elman, *From Philosophy to Philology*, p. 162.

53. Li Ruzhen, *Jinghua yuan* (Taibei: Xuehai chuban she, 1985), pp. 415–418, 484–492, 527–534. See also *Flowers in the Mirror, by Li Ju-chen*, abridged translation by Lin Tai-yi (Berkeley: University of California Press, 1965), pp. 133–141, 229–235, 242–244. Compare Idema, "Cannon, Clocks and Clever Monkeys," p. 487.

54. Li, *Jinghua yuan*, pp. 229–235. See also Maram Epstein, "Engendering Order: Structure, Gender, and Meaning in the Qing Novel Jinghua yuan," *Chinese Literature: Essays, Articles, and Reviews* 18 (December 1996): 105–131.

55. See ECCP, p. 473, and Elman, *A Cultural History of Civil Examinations*, pp. 7–12, 361–365.

56. ECCP, p. 473, and Yu Wang Luen, "Knowledge of Mathematics and Science in Ching-Hua-Yuan," *Oriens Extremus* 21, 2 (1974): 217–236. Compare Lu Xun, *A Brief History of Chinese Fiction*, trans. Yang Xianyi and Gladys Yang (Beijing: Foreign Language Press, 1964), p. 334. See also Elman, *From Philosophy to Philology*, p. 100.

57. Luen, "Knowledge of Mathematics," pp. 235–236. Compare Sato, "Reevaluaton of *Tengenjutsu* or *Tianyuanshu*," pp. 60–63.

58. Li, *Jinghua yuan*, pp. 511–512. The answers are 17 persons and 120 pieces. See also Luen, "Knowledge of Mathematics," p. 220.

59. Li, *Jinghua yuan*, pp. 532–533. See also Luen, "Knowledge of Mathematics," pp. 226–229.

60. Li, *Jinghua yuan*, pp. 645–646. See also Luen, "Knowledge of Mathematics," pp. 230–231.

61. Li, *Jinghua yuan*, pp. 531–532. See also He Shaogeng, "Method for Determining Segment Areas and Evaluation of π," in *Ancient China's Technology and Science*, pp. 90–98, and Luen, "Knowledge of Mathematics," pp. 221–224. Compare Karine Chemla, "Relations between Procedure and Demonstration: Measuring the Circle in the *Nine Chapters on Mathematical Procedures* and Their Commentary by Liu Hui (3rd Century)," in Hans Niels Jahnke et al., eds., *History of Mathematics and Education: Ideas and Experiences* (Göttingen: Vandenhoeck and Ruprecht, 1996).

62. Li, *Jinghua yuan*, pp. 533–534. See also Luen, "Knowledge of Mathematics," pp. 233–234.

63. See Elman, *A Cultural History of Civil Examinations*, p. 578.

64. Luen, "Knowledge of Mathematics," pp. 234–236. Compare Feng-ying Ming, "In Search of a Position: The Paradox of Genre Typology in the Late Qing Polygeneric Novel" (Ph.D. diss., University of California–Los Angeles, 1999), chap. 6.

65. For discussion, see Nathan Sivin, "Why the Scientific Revolution Did Not Take Place in China—or Didn't It?" reprinted in Sivin, *Science in Ancient China*, VII, pp. 45–66.

66. Tian Miao, "*Jiegenfang, Tianyuan,* and Algebra," p. 114.

67. Compare Guo Shuchun, "The Numerical Solution of Higher Equations and the *Tianyuan* Method," in *Ancient China's Technology and Science*, pp. 111–123.

68. Tian Miao, "*Jiegenfang, Tianyuan,* and Algebra," pp. 114–115.

69. Horng, "Chinese Mathematics," pp. 187, 199. See also Hu, "Merging Chinese and Western Mathematics," pp. 214–223; ECCP, pp. 538–539; and Wylie, *Notes on Chinese Literature,* p. 125.

8. Protestants, Education, and Modern Science to 1880

1. James Chandler, *England in 1819: The Politics of Literary Culture and the Case of Romantic Historicism* (Chicago: University of Chicago Press, 1998).

2. See Frederic Wakeman Jr., "The Canton Trade and the Opium War," in Denis Twitchett and John Fairbank, eds., *Cambridge History of China,* vol. 10, *Late Ch'ing, 1800–1911,* part 1 (Cambridge: Cambridge University Press, 1978), p. 173. See also David Porter, "A Peculiar but Uninteresting Nation: China and the Discourse of Commerce in Eighteenth-Century England," *Eighteenth-Century Studies* 33, 2 (1999–2000): 181–199.

3. Thorne, "The Conversion of Englishmen," pp. 240–253. See also Susan Thorne, *Congregational Missions and the Making of an Imperial Culture in Nineteenth-Century England* (Stanford, Calif.: Stanford University Press, 1999), pp. 13–14, 23–52. On the Protestant missionary movement in China, see Paul Cohen, "Christian Missions and Their Impact to 1900," in Twitchett and Fairbank, eds., *Cambridge History of China,* vol. 10, part 1, pp. 547–550.

4. See Wang Lixin, *Meiguo chuanjiao shi yu wan Qing Zhongguo xiandai hua* (Tianjin: Renmin chuban she, 1997), pp. 10–26. See also Ellsworth Carlson, *The Foochow Missionaries, 1847–1880* (Cambridge, Mass.: Harvard East Asian Research Center, 1974), and John Fairbank, ed., *The Missionary Enterprise in China and America* (Cambridge, Mass.: Harvard University Press, 1974). Compare James R. Moore, *The Post-Darwinian Controversies: A Study of the Protestant Struggle to Come to Terms with Darwin in Great Britain and America, 1870–1900* (Cambridge: Cambridge University Press, 1981), pp. 6–7; Spence, *To Change China;* and Ryan Dunch, *Fuzhou Protestants and the Making of a Modern China, 1857–1927* (New Haven, Conn.: Yale University Press, 2001).

5. Britton Rosewell, *The Chinese Periodical Press: 1800–1912* (Shanghai: Kelley & Walsh Ltd., 1933), pp. 16–22.

6. Tsien, "Western Impact on China through Translation," p. 313. See also West, *Catalogue of the Morrison Collection of Chinese Books,* pp. 108, 124–126, and Harriet Zurndorfer, *Chinese Bibliography: A Research Guide to Reference Works about China Past and Present* (Leiden: E. J. Brill, 1995), p. 104.

7. Wang Yangzong, "Liuhe congtan zhong de jindai kexue zhishi ji qi zai Qingmo de yingxiang," *Zhongguo keji shiliao* 20 (1999.3): 212–213.

8. Ting-yee Kuo and Kwang-Ching Liu, "Self-Strengthening: The Pursuit of Western Technology," in Twitchett and Fairbank, eds., *Cambridge History of China,* vol. 10, part 1, pp. 544–549. See also Rosewell, *The Chinese Periodical Press,* pp. 22–29.

9. ECCP, p. 504. See also Xiong Yuezhi, "Difficulties in Comprehension and Differences in Expression: Interpreting American Democracy in the Late Qing," *Late Imperial China* 23, 1 (June 2002): 4–5.

10. Shen Guowei, "The Creation of Technical Terms in English-Chinese Dictionaries from the Nineteenth Century," in Lackner, Amelung, and Kurtz, eds., *New Terms*, pp. 288–289.

11. Rosewell, *The Chinese Periodical Press,* pp. 30–34, and Suzanne Wilson Barnett, "The Language of National Power, Prosperity and Politics in the Publications of the Early Protestant Press in China," in Lin Zhiping, ed., *Jidujiao yu Zhongguo xiandai hua* (Taibei: Caituan faren Jidujiao yuzhou guang chuanbo zhongxin chuban she, 1994), pp. 251–315. Compare Suzanne Barnett, "Wei Yuan and Westerners: Notes on the Sources of the *Hai-kuo t'u-chih,*" *Ch'ing-shih wen-t'i* 2, 4 (November 1970): 1–20. See also ECCP, pp. 503–504, 514, 851, and Fred W. Drake, *China Charts the World: Hsü Chi-yü and His Geography of 1848* (Cambridge, Mass.: Harvard East Asia Research Center, 1975). Wei Yuan's *Treatise* was popular in Japan in the mid-nineteenth century.

12. Kuo and Liu, "Self-Strengthening," pp. 552–553.

13. Han Qi, "Chuanjiaoshi Weilie Yali," pp. 59, 68n13. See also David Wright, *Translating Science: The Transmission of Western Chemistry into Late Imperial China, 1840–1900* (Leiden: E. J. Brill, 2000), pp. 86–93, and Masini, *The Formation of Modern Chinese Lexicon,* p. 49. Compare Bridie Andrews, "The Making of Modern Chinese Medicine, 1895–1937" (Ph.D. diss., Cambridge University, 1996), pp. 31–32.

14. See Patrick Hanan, "Chinese Christian Literature: The Writing Process," in Hanan, ed., *Treasures of the Yenching,* pp. 261–283.

15. Benjamin Hobson, *Bowu xinbian* (Shanghai: Inkstone Press, 1855), 1.3a, 1.18b. See also Xiong Yuezhi, *Xixue dongjian yu wan Qing Shehui* (Shanghai: Renmin chuban she, 1994), pp. 10–11; Wang, "Liuhe congtan," p. 212; Knight Biggerstaff, *The Earliest Modern Government Schools in China* (Ithaca, N.Y.: Cornell University Press, 1961), pp. 175–176; and David Wright, "The Translation of Modern Western Science in Nineteenth-Century China, 1840– 1895," *Isis* 89 (1998): 660. Compare Wang Bing, "Jindai zaoqi Zhongguo he Riben zhi jian de wulixue jiaoliu," *Ziran kexue shi yanjiu* 15, 3 (1996): 227–233.

16. Shapiro, "How Different Are Western and Chinese Medicine?," p. 10.

17. Compare William Lockhart, *The Medicical Missionary in China: A Narrative of Twenty Years' Experience* (London: Hurst and Blackett, 1861), pp. 111–172, and Andrews, "The Making of Modern Chinese Medicine," pp. 32–34. See also Wylie, *Notes on Chinese Literature,* p. 129, and Tsien, "Western Impact on China," pp. 312–313. On Chinese translations in Tokugawa and Meiji Japan, see Nakamura Kyūshirō, "Kinsei Shina no Nihon bunka ni oyoboshitaru seiryoku eikyō," *Shigaku zasshi* 25, 12 (December 1914): 1597–1605, and 26, 2 (February 1915): 1–13. Compare Montgomery, *Science in Translation,* pp. 202–212.

18. See Sheng Zengxiu et al., *Wenbing xuepai,* pp. 282–326, and Hanson, "Robust Northerners," pp. 532–542.

19. For Tianjin in the north, see Rogaski, *Hygienic Modernity,* chap. 2. On the Taipings, see Elman, *From Philosophy to Philology,* pp. 287–292.

20. Wang Shixiong, "Preface" (of 1852), p. 2a, in Wang Shixiong, *Wenre jingwei* (Shanghai: Qianqing tang, 1863). See also Chao, "Medicine and Society," p. 65, and Yi-li Wu, "God's Uterus: Benjamin Hobson and Missionary 'Midwifery' in Nineteenth-Century China" (presented at the Disunity of Chinese Science conference, University of Chicago, May 10–11, 2002), pp. 32–34.

21. Yamada Keiji, "Anatometrics in Ancient China," *Chinese Science* 10 (1991): 39–52, describes ancient dissections, some for punitive reasons. For later developments, see Saburō Miyashita, "A Link in the Westward Transmission of Chinese Anatomy in the Later Middle Ages," in Sivin, ed., *Science & Technology,* pp. 200–204. See also Bridie Andrews, "Tailoring Tradition: The Impact of Modern Medicine on Traditional Chinese Medicine, 1887–1937," in Viviane Alleton and Alexeï Volkov, eds., *Notions et Perceptions du Changement en Chine* (Paris: Collège de France, Institut des Hautes Études Chinoises, 1994), pp. 149–166, and Goldschmidt, "The Transformations of Chinese Medicine," pp. 249–250. Compare Larissa Heinrich, "The Pathological Body: Science, Race, and Literary Realism in China, 1770–1930" (Ph.D. diss., University of California–Berkeley, 2002), pp. 143–156.

22. Hanson, "Robust Northerners," pp. 519, 532–533.

23. Warwick Anderson, "Immunities of Empire: Race, Disease, and the New Tropical Medicine, 1900–1920," *Bulletin of the History of Medicine* 70, 1 (1996): 94–118. See also Bridie Andrews, "Tuberculosis and the Assimilation of Germ Theory in China, 1895–1937," *Journal of the History of Medicine and Allied Sciences* 52, 1 (1997): 142–143. Compare the articles in Roy Macleod and Milton Lewis, eds., *Disease, Medicine, and Empire: Perspectives on Western Medicine and the Experience of European Expansion* (London: Routledge, 1988).

24. Shang-jen Li, "British Imperial Medicine in Late Nineteenth-Century China and the Early Career of Patrick Manson" (Ph.D. diss., London University, 1999), pp. 32–100, 161–162. Compare Andrews, "The Making of Modern Chinese Medicine," pp. 226–230. See Patrick Manson, *Tropical Diseases: A Manual of the Diseases of Warm Climates* (London: Cassell & Co., 1898). See also Douglas Haynes, *Imperial Medicine: Patrick Manson and the Conquest of Tropical Disease* (Philadelphia: University of Pennsylvania Press, 2001), pp. 18–27, 36–41.

25. Compare Michael Worboys, "Germs, Malaria and the Invention of Mansonian Tropical Medicine," in David Arnold, ed., *Warm Climates and Western Medicine: The Emergence of Tropical Medicine, 1500–1900* (Amsterdam: Rodopi, 1996), pp. 181–207.

26. See also Quan Hansheng, "Qingmo Xiyang yixue chuanru shi guoren suo chi de taidu," *Shihuo yuekan* 3, 12 (December 1936): 43–53.

27. Zhao Hongjun, "Chinese Versus Western Medicine: A History of Their Relations in the Twentieth Century," *Chinese Science* 10 (1991): 32. See also Shapiro, "The Puzzle of Spermatorrhea," pp. 565–571, and Andrews, "The Making of Modern Chinese Medicine," pp. 223–256.

28. See SCC, vol. 6, part 6, and Chia-feng Chang, "Aspects of Smallpox and Its Significance in Chinese History" (Ph.D. diss., London University, 1996), pp. 124–168.

29. Zhao, "Chinese Versus Western Medicine," pp. 22–23. For Western invention of China as the "cradle of smallpox," see Heinrich, "The Pathological Body," chap. 1.

30. GZHB, vol. 1, pp. 69–70 (1, 3 [April 1876]: 8a–9a). See also David Wright, "Careers in Western Science in Nineteenth-Century China: Xu Shou and Xu Jianyin," *Journal of the Royal Asiatic Society,* third series 5 (1995): 60.

31. See Andrews, "Tailoring Tradition," pp. 155–159. See also Elizabeth Hsu, "The Reception of Western Medicine in China: Examples from Yunnan," in Petitjean, Jami, and Moulin, eds., *Science and Empires,* pp. 91–94, for more xenophobic forms of resistance.

32. Benjamin Hobson and Guan Maocai, *Xiyi luelun* (Shanghai: Renji yiguan, 1857), A.1a–2b. See also Wu, "God's Uterus," pp. 1–43, and Andrews, "Tailoring Tradition," pp. 150–151. Compare Yi-li Wu, "Transmitted Secrets: The Doctors of the Lower Yangzi Region and Popular Gynecology in Late Imperial China" (Ph.D. diss., Yale University, 1998), pp. 203–239.

33. Pan Shicheng, "Conventions" ("Liyan"), for the *Treatise on Physiology (Quanti xinlun),* in Pan Shicheng, ed., *Haishan xian'guan congshu* (1852; repr., Taipei: Yiwen yinshu guan, 1967), box 16, vol. 129, 1.2a. The skeletons are illustrated on pp. 1.13a–13b. Compare Rudolph, "Early China and the West," pp. 21–24, who did not see Hobson's 1851 originals, and Wu, "God's Uterus," pp. 27–31. See also ECCP, pp. 605–606.

34. See Pan Shicheng's "Preface" ("Xu") for the *Treatise on Physiology,* in Pan, ed., *Haishan xian'guan congshu,* vol. 119. See also Heinrich, "The Pathological Body," pp. 127–142.

35. Wang Qingren, *Yilin gaicuo* (Jinreng: Shuye deji, 1847), A.7a, 20a, 21a. On Wang's indirect link to Jesuit anatomy, see Xiong, *Xixue dongjian,* pp. 74–75. See also Yong-xin Zhu, "Historical Contributions of Chinese Scholars to the Study of the Human Brain," pp. 133–138, and Andrews, "Tailoring Tradition," p. 151. Compare Heinrich, "The Pathological Body," p. 116n17.

36. See Bridie Andrews, "Wang Qingren and the History of Chinese Anatomy," *Journal of Chinese Medicine* 35 (January 1991): 30–36. For discussion, see Yi-li Wu, "Introducing the Uterus to Chinese Gynecology: Benjamin Hobson and His *Treatise on Midwifery and Diseases of Children* (Fuying xinshuo), 1858" (paper presented at the panel on "Representing 'Western Medicine' in Qing and Republican China," Association for Asian Studies Annual Meeting, Chicago, March 24, 2001).

37. See Wang, "Liuhe congtan," p. 214, and Wright, *Translating Science,* pp. 36–38.

38. Wright, *Translating Science,* pp. 43–46.

39. John Fryer, "An Account of the Department for the Translation of Foreign Books at the Kiangnan Arsenal, Shanghai," *North-China Herald,* January 29, 1880, pp. 77–81. See also Wang, "Liuhe congtan" pp. 212–214, and Wright, "Careers in Western Science," pp. 55–56.

40. Shen Guowei, "Xu," pp. 6, 40, in Shen Guowei (Shin Kokui), ed., *Rokugo sōtan no gakusai teki kenkyū* (Tokyo: Hakuteisha, 1999). Wang came to the press in 1849 and left for Hong Kong in 1862 because of his alleged links to the Taipings. See Zhang Hailin, *Wang Tao pingzhuan* (Nanjing: Nanjing daxue chuban she, 1993), pp. 24–35, 96–105, and Cohen, *Between Tradition and Modernity,* pp. 9–13, 45–55.

41. Masini, *The Formation of Modern Chinese Lexicon,* pp. 57–60. The term *liuhe* referred to heaven, earth, and the four cardinal directions, and first appeared in the *Zhuangzi.* See *Zhuangzi yinde* (Taibei: Chengwen Bookstore, 1974), 5/2/56.

The term was also used in the Western Han. See *Huainan honglie jijie*, compiled under the auspices of Liu An (d. 122 B.C.), the prince of Huainan (Taibei: Commercial Press, 1974), vol. 1, 4.1a.

42. LHCT, vol. 1, no. 2 (1857.2.24): 10a; vol. 1, no. 11 (1857.11.16): 10b–12a; and vol. 2, no. 2 (1858.6.11): 6b.

43. Masini, *The Formation of Modern Chinese Lexicon*, pp. 57–62.

44. Zhang, *Wang Tao pingzhuan*, pp. 29–35, and Han, "Chuanjiaoshi Weilie Yali," p. 58. See also Zurndorfer, *Chinese Bibliography*, p. 105.

45. See Han, "*Shuli gezhi* de faxian," pp. 78–85. The manuscript was rediscovered in Hangzhou in 1937 and then lost again in 1939. Another version, perhaps taken by Wylie when he returned to England, was discovered by Han Qi in London in 1995.

46. See the prefaces in Alexander Wylie and Li Shanlan, *Tantian* (Shanghai: Mohai shuguan, 1859). See also Adrian Bennett, *John Fryer: The Introduction of Western Science and Technology into Nineteenth-Century China* (Cambridge, Mass.: Harvard University Research Center, 1967), p. 140, and Han, "Chuanjiaoshi Weilie Yali," pp. 62–63. Earlier Newton was mentioned in China by the Jesuits I. Koegler (1680–1746) and N. Grammatici (1684–1736) in the *Supplement to the Compendium of Observational and Computational Astronomy (Lixiang kaocheng houbian)*, without presenting his theories. For the situation at Cambridge, see Crosbie Smith and M. Norton Wise, *Energy and Empire: A Biographical Study of Lord Kelvin* (Cambridge: Cambridge University Press, 1989), pp. 153–155.

47. Zeng Guofan, "Preface," pp. 1a–2a, in Wylie's and Li Shanlan's enlarged edition of the *Jihe yuanben* (Nanjing: Nanjing Arsenal, 1865).

48. Wylie, "Preface" (in English), pp. i–ii, in Alexander Wylie and Li Shanlan, *Shuxue qimeng* (Shanghai: Mohai shuguan, 1853), and Alexander Wylie, "Preface" (in Chinese), pp. 1a–3b, in Alexander Wylie and Li Shanlan, *Daishu xue* (Shanghai: Mohai shuguan, 1859). Compare Alexander Wylie, *Memorials of Protestant Missionaries* (Shanghai: American Presbyterian Mission Press, 1867), p. 138. See also Han, "Chuanjiaoshi Weilie Yali," p. 61.

49. Wylie, "Preface" (in Chinese), p. 2a, in Wylie and Li, *Shuxue qimeng*. See also Wang, "Liuhe congtan," pp. 211–212; Han, "Chuanjiaoshi Weilie Yali," p. 57; and Hu, "Merging Chinese and Western Mathematics," pp. 119–123, 136–143, 153, 352. Compare DSB, vol., 14, pp. 35–37, and Wylie, *Notes on Chinese Literature*, p. 125.

50. Yatsumimi Toshifumi, "Shizen shingaku to shizen kagaku no aide de," in Shen, ed., *Rokugo sōtan no gakusai teki kenkyū*, p. 117. See also Wang, "Liuhe congtan," p. 224–225, and Masini, *The Formation of Modern Chinese Lexicon*, pp. 84–88.

51. LHCT, vol. 1, no. 1 (1857.1.26): 1a–1b. See also Wang, "Liuhe congtan," p. 213.

52. LHCT, vol. 1, no. 1 (1857.1.26): 2a, vol. 1, no. 2 (1857.2.24): 3b–6b. The Protestants continued to use an early Jesuit translation for God, namely *Shangdi*, during the nineteenth century.

53. For one of the earliest apparent references to Darwin in China, see *Shen bao* 2 (August 21, 1873), p. 2. See also Yatsumimi, "Shizen shingaku," pp. 123, 135n27, and David Wright, "John Fryer and the Shanghai Polytechnic: Making Space for Science in Nineteenth-Century China," *British Journal of History of Science* 29 (1996): 12–13. Compare the less useful account in James R. Pusey, *China and*

Charles Darwin (Cambridge, Mass.: Harvard University Press, 1983), pp. 4–5. For discussion of Paley and the formation of "Christian Darwinism," see Moore, *The Post-Darwinian Controversies,* pp. 15–16, 329–330, 342–343.

54. LHCT, vol. 1, no. 1 (1857.1.26): 3a–8b.

55. LHCT, vol. 2, no. 2 (1858.6.11): 13a.

56. LHCT, vol. 1, no. 5 (1857.5.24): 6a–7a; vol. 1, no. 9 (1857.9.18): 5b–8a; vol. 1, no. 10 (1857.10.18): 6a–9a, vol. 1, no. 11 (1857.11.16): 4b–7b; vol. 1, no. 12 (1857.12.16): 5a–5b; vol. 1, no. 13 (1858.1.15): 10b–12a; vol. 2, no. 1 (1858.2.14): 8b–9b; vol. 2, no. 2 (1858.6.11): 6a–13a. See also Wang Yangzong, "Mingmo Qingchu Xixue Zhongyuan lun xinkao," in Liu Dun, ed., *Kexue xinzhuan: qingzhu Du Shiran xiansheng congshi kexue shi yanjiu sishi zhounian xueshu lunwen ji* (Shenyang: Liaoning Education Press, 1997), pp. 71–83.

57. Wylie, *Notes on Chinese Literature,* p. 126, and ECCP, p. 242.

58. See Wang, "Liuhe congtan," pp. 211, 224–225, and Shen Guowei's discussion in his "Preface" to Shen, ed., *Rokugo sōtan no gakusai teki kenkyū,* pp. 31–34.

59. Sakade Yoshinobu, "*Rokugo sōtan* ni mieru kagaku kiji," *Kagakushi kenkyū* 93 (1970): 38–39.

60. Masini, *The Formation of Modern Chinese Lexicon,* p. 59, reads the calculus anachronistically into the Song dynasty terms used by Wylie and Li to translate Loomis's work.

61. Alexander Wylie, "Preface" (in English), p. i, in Alexander Wylie and Li Shanlan, *Dai weiji shiji* (Shanghai: Mohai shuguan, 1859).

62. Wylie, "Preface" (in Chinese), pp. 2b–3a, in Wylie and Li, *Dai weiji shiji,* noted that Newtonian mathematical notation for the calculus was not as convenient as Leibniz's for advancing computations. See also Wylie's 1867 preface to Joseph Edkins and Li Shanlan, trans., *Zhongxue* (Nanjing: Arsenal, 1867), p. i.

63. See Elaine Koppelman, "The Calculus of Operations and the Rise of Abstract Algebra," *Archive for History of Exact Sciences* 7 (1971): 155–242, and Niccolò Guicciardini, *The Development of Newtonian Calculus in Britain 1700–1800* (Cambridge: Cambridge University Press, 1989), pp. vii, 1–6, who argues that British failures were magnified by the Cambridge reformers. Compare Bréard, "On Mathematical Terminology," pp. 305–307, 322–323.

64. See Wylie's "Preface" (in Chinese), p. 2b, in Wylie and Li, *Dai weiji shiji.* See also Richard Yeo, "William Whewell: A Cambridge Historian and Philosopher of Science," in Peter Harman and Simon Mitton, eds., *Cambridge Scientific Minds* (Cambridge: Cambridge University Press, 2002), pp. 51–62.

65. Guicciardini, *The Development of Newtonian Calculus,* pp. 139–142.

66. See Li Shanlan, "Preface" (in Chinese), p. 1a, and Wylie, "Preface" (in English), pp. i–vi, both in Wylie and Li, *Dai weiji shiji.* Calvin Mateer (1836–1908) was the only other missionary who grasped the sophistication of "Chinese algebra." Mateer translated Loomis's 1868 *A Treatise on Algebra* in 1890. Most modern accounts of Western mathematics in nineteenth-century China have failed to appreciate the important role of traditional Chinese mathematics during the Self-Strengthening period after 1865. See, for example, Biggerstaff, *The Earliest Modern Government Schools,* pp. 170–171.

67. Guicciardini, *The Development of Newtonian Calculus,* pp. 104, 120.

68. Hu, "Merging Chinese and Western Mathematics," p. 195.

69. The account here and below follows Hu, "Merging Chinese and Western Mathematics," pp. 181–215. See also Martzloff, *A History of Chinese Mathematics,* pp. 379–382, and Bréard, "On Mathematical Terminology," pp. 308–309.

70. Hu, "Merging Chinese and Western Mathematics," pp. 229–235, and Tian Miao, "*Jiegenfang, Tianyuan,* and *Daishu:* Algebra in the Qing Dynasty." See also Elman, *From Philosophy to Philology,* pp. 287–290, on the Taiping Rebellion.

71. Xiong, *Xixue dongjian,* pp. 350–351. In the *Shenbao,* the new academy was called the "academy of extensive writing" (*Hongwen shuyuan*), a reference to the special "broad learning and extensive words" (*boxue hongci*) examinations held to select distinguished scholars to serve the dynasty. Compare Jeffrey A. Auerbach, *The Great Exhibition of 1851* (New Haven, Conn.: Yale University Press, 1999).

72. See A. Hunter Dupree, "Christianity and the Scientific Community in the Age of Darwin," pp. 358–359, and James Moore, "Geologists and Interpreters of Genesis in the Nineteenth Century," pp. 328–329, both in Lindberg and Numbers, eds., *God and Nature.* See also Charles Gillispie, *Genesis and Geology: A Study of the Relations of Scientific Thought, Natural Theology, and Social Opinion in Great Britain, 1790–1850* (Cambridge, Mass.: Harvard University Press, 1951); Moore, *The Post-Darwinian Controversies,* pp. 105–106, 205–213; and Roy Porter, "Charles Lyell and the Principles of the History of Geology," *British Journal for the History of Science* 9 (1976): 91–103. See also Martin Rudwick, "The Shape and Meaning of Earth History," in Lindberg and Numbers, eds., *God and Nature,* pp. 312–317. Compare Eikoh Shimao, "Darwinism in Japan, 1877–1927," *Annals of Science* 38 (1981): 93–102.

73. W. H. Medhurst, "A Reading-Room for the Chinese," *The North-China Herald,* March 12, 1874," pp. 225–226. See Knight Biggerstaff, "Shanghai Polytechnic Institution and Reading Room: An Attempt to Introduce Western Science and Technology to the Chinese," *Pacific Historical Review* 25 (May 1956): 131–134, and Wang Ermin, *Shanghai Gezhi Shuyuan zhilue* (Hong Kong: Zhongwen daxue chuban she, 1980), pp. 5–13. Wang, pp. 18–22, presents a list of the donors to the Polytechnic, as does Xiong, *Xixue dongjian,* pp. 356–359.

74. Xiong, *Xixue dongjian,* p. 353, and Biggerstaff, "Shanghai Polytechnic Institution," pp. 133–138. See also Zhang, *Wang Tao pingzhuan,* pp. 354–376; Cohen, *Between Tradition and Modernity,* p. 182; Wright, "Careers in Western Science," pp. 71–73; and Wright, "John Fryer," p. 11n55.

75. Ferdinand Dagenais, *John Fryer's Calendar: Correspondence, Publications, and Miscellaneous Papers with Excerpts and Commentary* (Berkeley: Center for Chinese Studies, University of California, 1999), version 3, 1895:13. The curriculum is described in Wang, *Shanghai Gezhi Shuyuan zhilue,* pp. 41–48.

76. Biggerstaff, "Shanghai Polytechnic Institution," pp. 139–141. See also Wright, "John Fryer," p. 14.

77. Tian Tao, "Zhongxi wenjian lu, Gezhi huibian yingyinben xu," in *Zhongxi wenjian lu,* 4 vols. (repr., Nanjing: Guji shudian, 1992), vol. 1, pp. 2–3. See also Biggerstaff, "Shanghai Polytechnic Institution," p. 144, and San-pao Li, "Letters to the Editor in John Fryer's *Chinese Scientific Magazine,* 1876–1892: An Analysis," *Jindaishi yanjiu suo jikan* (Academia Sinica, Taiwan) 4 (1974): 737–738. Later,

Martin published *Selections from* The Peking Magazine *(Zhongxi wenjian lu xuanbian)* in four volumes in 1877. During the 1898 reform era, Martin also prepared the *New Edition of* The Peking Magazine *(Wenjian lu xinbian)*. See Appendix 7.

78. On the curriculum, see Wang Xianming, *Jindai xinxue: Zhongguo chuantong xueshu wenhua de shanbian yu chonggou* (Beijing: Shangwu yinshu guan, 2000), pp. 117–118.

79. ZGJD, pp. 71–79.

80. For mathematics questions and science essays, see *Zhongxi wenjian lu*, vol. 1, pp. 403–404 (1872.2: 10a–10b), 411–418 (1872.2: 14a–17b), 481–487 (1873.3: 16a–19b). On the earth versus the sun, see vol. 4, pp. 189–193 (1875.3: 6a–8a), 351–354 (1875.6: 7a–8b). See also ZGJD, pp. 79–105, 109–125.

81. See Zhang Jian, "Zhongxi wenjian lu shulue," *Fudan xuebao (shehui kexue ban)* 4 (1995): 57–62. Zhang disputes Xiong Yuezhi's claim that *The Peking Magazine* was mainly concerned with current news and events and literature, although Zhang acknowledges that *The Chinese Scientific and Industrial Magazine* was the more specialized journal focusing on science. Compare Xiong Yuezhi, "Gezhi huibian yu Xixue chuanbo," *Shanghai yanjiu luncong* 1 (1989): 65–66.

82. Major and his brother established the *Shen bao* in 1871 and the *Dianshi zhai Pictorial* in 1876. See Ye Xiaoqing, *The Dianshizhai Pictorial: Shanghai Suburban Life, 1884–1898* (Ann Arbor: Center for Chinese Studies, University of Michigan, 2003), pp. 4–5.

83. Wang Yangzong, "Gezhi huibian zhi Zhongguo bianjizhe kao," *Wenxian* 63 (January 1995): 237–243. See also Wright, *Translating Science*, pp. 319–325.

84. Wang Ermin, *Shanghai Gezhi Shuyuan zhilue*, pp. 103–111. On Ernest Major and the commercial success of the *Shenbao*, see Rosewell, *The Chinese Periodical Press*, pp. 63–70; Rudolph Wagner, "The Early Chinese Newspapers and the Chinese Public Sphere," *European Journal of East Asian Studies* 1, 1 (2001): 3–13, 25–26; and Joan Judge, *Print and Politics: 'Shibao' and the Culture of Reform in Late Qing China* (Stanford, Calif.: Stanford University Press, 1996), p. 20.

85. The venues where to buy issues since 1877 were listed in the journal: Beijing, Yantai, Wurong, Shanghai, Hangzhou, Shantou (Swatow), Kobe, Tianjin, Dengzhou, Jiuzhang, Suzhou, Fuzhou, Hong Kong, Yokohama, Niuzhuang, Hankou, Nanjing, Xiamen (Amoy), Guangzhou, Ji'nan, Wuchang, Zhenjiang, Ningbo, Danshui (Taiwan), and Singapore. Compare Li, "Letters to the Editor," p. 743. See also Dagenais, *John Fryer*, 1891:1; Bennett, *John Fryer*, pp. 50–55; Wright, "John Fryer," pp. 1–16; and Wright, "Careers in Western Science," pp. 49–90.

86. *Guangxu chao donghua lu* (Shanghai: Zhonghua shuju, 1909), 82.11a. See also Biggerstaff, "Shanghai Polytechnic Institution," pp. 148–149.

87. Li, "Letters to the Editor," pp. 730–731, 762.

88. See Xu Jianyin, "Fanli," in Xu Jianyin, ed., *Gezhi congshu* (Shanghai: Zhuji shuzhuang, 1900–1901), pp. 1a–2b. See also Wang, "Gezhi huibian," pp. 238, 241. Wright, "Careers in Western Science," notes that Xu Jianyin oversaw the collection of reprints on science while he was at the Fuzhou Navy Yard in 1901.

89. Biggerstaff, "Shanghai Polytechnic Institution," pp. 144–149. The school survives today in central Shanghai.

90. See "A Chinese Scientific Journal," in *Scientific American* 34, 17 (April 29, 1876): 279.

91. See Xu Shou, *"Gezhi huibian xu,"* in GZHB, vol. 1, pp. 3–4. Fryer probably dictated the English translation that was prepared by the Pacific Translation Service. See Dagenais, *John Fryer,* 1876:2–3.

92. GZHB, vol. 1, "Zong mulu" (general table of contents), pp. 6–24, vol. 2, pp. 223–230, 251–258. See Rogaski, *Hygienic Modernity,* chap. 4. The table of contents for the Nanjing reprint of the GZHB mistakenly gives 1878 for the date of the 1880 volume.

93. GZHB, vol. 1, pp. 5–14, 31–38, 55–57, 79–82, 105–109, 129–135, 153–162, 177–185, 201–208, 225–237, 251–258, 275–286. See also Biggerstaff, "Shanghai Polytechnic Institution," p. 145.

94. GZHB, vol. 1, pp. 31–38, 55–57, 79–82, 105–109, 129–135.

95. GZHB, vol. 1, pp. 153–162, 177–185, 201–208, 225–237, 251–258, 275–286.

96. GZHB, vol. 1, pp. 69–70. See Wu, "Introducing the Uterus to Chinese Gynecology," pp. 12–13, for Chinese criticisms of Hobson and Western anatomically based medicine for failing to provide any useful medical therapies to complement its diagnostic techniques.

97. GZHB, vol. 1, pp. 168–169, 195–196, 433–434. See also Kim, *The Natural Philosophy of Chu Hsi,* pp. 19–27.

98. GZHB, vol. 1, pp. 368–371, 399–400, vol. 2, pp. 27–29, 49–55, 89–91.

99. David Reynolds, "Re-Drawing China's Intellectual Map: 19th Century Chinese Images of Science," *Late Imperial China* 12, 1 (June 1991): 27–61.

9. The Construction of Modern Science in Late Qing China

1. See Kume Kunitake, comp., *The Iwakura Embassy, 1871–73,* vol. 5, "Continental Europe, 3: and the Voyage Home."

2. W. A. P. Martin, "Preface," p. i, in *Gewu rumen* (Beijing: Tongwen guan, 1868). See also Hu, "Merging Chinese and Western Mathematics," p. 82, and Bennett, *John Fryer,* p. 123. Compare Wright, *Translating Science,* pp. 266–272.

3. Xiong, *Xixue dongjian,* pp. 397–400. See also Liu Guangding, *"Gewu tanyuan* yu Wei Lianchen de Zhongwen zhuzuo," in Yang Cuihua and Huang Yinong [Yilong], eds., *Jindai Zhongguo keji shi lunji* (Taibei: Academia Sinica, Institute of Modern History, and Qinghua University, Institute of History, 1991), pp. 195–213. The *Gewu tanyuan* was published in Japan in 1878.

4. See GZQM (1879–80). A later version of this series was published by the Imperial Customs Office in 1886 in Beijing, and will be discussed below. On the role of leave in the imperial bureaucracies established by the British in China, see Haynes, *Imperial Medicine,* p. 30.

5. Ayano Hiroyuki, "H. E. Roscoe to Owens College, Manchester no seido tenkan," *Kagakushi kenkyū* 8, 212 (1999): 214–222, and DSB, vol. 11, pp. 536–538. The Royal Society was a preindustrial revolution haven for early modern gentlemen interested in natural studies, not a technical institution for training in engineering and the industrial sciences. In the nineteenth century, Cambridge University, par-

ticularly Trinity College, pioneered research linking mathematics and physics. See Harman and Mitton, eds., *Cambridge Scientific Minds.*

6. Compare Cooter and Pumfrey, "Separate Spheres and Public Places," pp. 237–267.
7. See Balfour Stewart, *Lessons in Elementary Physics,* new ed. (London: Macmillan and Co., 1878), p. iii, and Ayano Hiroyuki, "Jūkyū seiki koban no England ni okeru kagaku no kaikaku undō," *Kagakushi kenkyū* 36, 204 (1997): 209–217. See also John Fryer, "Science in China," *Nature* 601, 24 (May–October 1881): 9–11, 54–57. Compare David Reynolds, "The Advancement of Knowledge and the Enrichment of Life: The Science Society of China and the Understanding of Science in the Early Republic, 1914–1930" (Ph.D. diss., University of Wisconsin–Madison, 1986), pp. 109–112.
8. Biggerstaff, "Shanghai Polytechnic Institution," pp. 127–149. See also Wright, "John Fryer," pp. 1–16, and Wright, "Careers in Western Science," pp. 49–90. Compare Roy MacLeod, *The Creed of Science in Victorian England* (Aldershot: Variorum, 2000).
9. Dagenais, *John Fryer,* 1887:7, 1889:1.
10. Wang Yangzong, "Hexuli Kexue daolun de liangge Zhong yiben," *Zhongguo keji shiliao* 21, 3 (2000): 207–221. Fryer noted in his 1891 review that Allen's volumes on geography, physics, chemistry, and astronomy were basically the same works that Allen had translated for the Jiangnan Arsenal version, albeit more detailed. See GZHB, vol. 5, pp. 200–203.
11. GZHB, vol. 1, "Zong mulu," pp. 21–22, and vol. 5, pp. 200–203, and Li Hongzhang, "Xu," in GZQM (1886), vol. 1, pp. 1a–4b.
12. Zeng Jize, "Ji," in GZQM (1886), vol. 1, pp. 1a–3b. Compare Yan Jing's 1448 "Xu" to the *Shiwu jiyuan,* pp. 1a–b, in GZCS, cited in chap. 1.
13. See Wylie's preface to Edkins and Li Shanlan, trans., *Zhongxue,* p. i. See also Dagenais, *John Fryer,* 1894:2.
14. Joseph Edkins, "Xu," in GZQM (1886), vol. 1, pp. 1a–2a, and Li Hongzhang, "Xu," vol. 1, pp. 2a–4a. See the titles in the GZQM (1886), vol. 1, table of contents, and vol. 1, *juan* 7 for the twenty-three fields. Compare Wang Ping, *Xifang lisuan xue zhi shuru* (Taibei: Institute of Modern History, Academia Sinica, 1966), pp. 75–124, and Wright, "The Great Desideratum," p. 47.
15. GZQM (1886), vol. 2, A.4b, A.15a. Edkins used the "the supreme maker" (*zaohua zhu*) as a general translation for the deistic notion of "nature." See Wang, "Hexuli Kexue daolun," p. 218.
16. GZQM (1886), vol. 2, A16a–17b. Compare Elman, *A Cultural History of Civil Examinations,* pp. 68–69, 111–113, for the "methods of the mind" in Cheng-Zhu learning.
17. Compare T. H. Huxley, *Introductory* (for the *Science Primers*) (London: Macmillan, 1880).
18. Zhang Yuanfang, "Xixue shiliu zhong xu," *Xixue qimeng* (Shanghai: Zhuyitang, 1896), pp. 1a–b.
19. Wang, "Hexuli Kexue daolun," pp. 208–210, 220. Allen had used *gewu xue* for physics in his translation, while Edkins preferred *gezhi zhixue.* Gravity in Allen was translated as *jieli* (GZQM, 1879, vol. 2, 2.4a); for Edkins it was *sheli* (GZQM,

1886, vol. 10, p. 7a). Edkins gave sixty-seven chemical elements (1886, vol. 9, p. 53b); Allen had presented sixty-three (1879, vol. 1, p. 46a).

20. See Zou Zhenhuan, *Wan Qing Xifang dilixue zai Zhongguo* (Shanghai: Guji chuban she, 2000), pp. 89–134, and Guo Shuanglin, *Xichao jiaoyang xia de wan Qing dilixue* (Beijing: Beijing University Press, 2000), pp. 30–37, 258–267, 270–271.

21. GZQM (1886), vol. 3, 1.1a–33a, 2.34a–53a.

22. Ibid., 3.54a–71a, 4.72a–111a.

23. GZQM (1886), vol. 4, 1.1a–6a, and vol. 5, 1.1a–5a, 5.53a–61a.

24. GZQM (1886), vol. 5, 8.119a–120b.

25. See J. D. Hooker, *Botany* (New York: Appleton and Co., 1877). Compare Joe Burchfield, *Lord Kelvin and the Age of the Earth* (New York: Science History Publications, 1975), p. 72.

26. See Georges Métailié, "La création lexicale dans le premier traité de botanique occidentale publié en chinois (1858), *Documents pour l'histoire du vocabulaire scientifique* 2 (1981): 65–73, and Georges Métailié, "The Formation of Botanical Terminology: A Model or a Case Study," in Lackner, Amelung, and Kurtz, eds., *New Terms for New Ideas*, pp. 328–329. Compare Fa-ti Fan, "Hybrid Discourse and Textual Practice," pp. 26–56. Métailié has recently shown, however, that evidential scholars such as Cheng Yaotian were developing an incipient botany based on textual and empirical research. See his "The Botany of Cheng Yaotien (1725–1814): Multiple Perspectives on Plants" (presented at the conference The Age of Antiquaries in Europe and China, Bard Graduate Center, New York, March 25, 2004).

27. GZQM (1886), vol. 6, pp. 1a–b, 110b–112b.

28. Hooker, *Botany*, pp. 100–102.

29. GZQM (1886), vol. 6, pp. 1a–b, 110b–112b. See also Frederick Gregory, "The Impact of Darwinian Evolution on Protestant Theology in the Nineteenth Century," in Lindberg and Numbers, eds., *God and Nature*, pp. 220–241.

30. Compare Hooker, *Botany*, p. 101.

31. GZQM (1886), vol. 7, pp. 3a–5b, 6a–7a, 39b–79a, 131b. Compare Michael Foster, *A Course of Elementary Practical Physiology*, assisted by J. N. Langley, 3rd ed. (London: Macmillan, 1878).

32. GZQM (1886), vol. 9, pp. 51b–55a, 55a–58b, 60a–68a. Compare Huxley, *Introductory*. See also Wang, "Hexuli Kexue daolun," p. 220, and Moore, *The Post-Darwinian Controversies*, pp. 12–16, 142–152, which stresses the role of the Lamarckian theory of biological transmutation in the formation of Christian Darwinism.

33. GZQM (1886), vol. 12. See also Mori Tokihiko, "Liang Qichao de jingji sixiang," in Hazama Naoki, ed., *Liang Qichao, Mingzhi Riben, Xifang* (Beijing: Shehui kexue wenxian chuban she, 2001), pp. 219–220.

34. See Nakamura, "Kinsei Shina no Nihon bunka," pp. 4–5, for the impact of Martin's translation in Japan. For discussion, see Benjamin Schwartz, *In Search of Wealth and Power: Yen Fu and the West* (New York: Harper Torchbooks, 1969), pp. 113–129.

35. GZQM (1886), vol. 12, pp. 1a–4b.

36. Dorothy Ross, "The Social Science Disciplines," in Theodore Porter and Dorothy Ross, eds., *The Cambridge History of Science,* vol. 7: *The Modern Social Sciences* (Cambridge: Cambridge University Press, 2003).

37. GZQM (1886), vol. 12, pp. 6a–13a, 14a–b.

38. GZQM (1886), vols. 13–16. See Edkins's "Xu" to the *Bianxue qimeng,* p. 1a, in GZQM (1886), vol. 14. For discussion, see Kurtz, "Coming to Terms with Logic," pp. 158–159.

39. GZQM (1886), vol. 16, pp. 187a–188a.

40. For the 1887 "Report," see *North-China Herald,* December 28, 1887, pp. 702–703.

41. Bennett, *John Fryer,* pp. 64–66.

42. An 1887 imperial edict allowed mathematics questions in civil examinations. See W. T. A. Barber, "A Chinese Examination Paper," *The North-China Herald,* July 7, 1888, p. 15.

43. See Fryer's "Report" in the *North-China Herald,* December 28, 1889, pp. 702–703. See also Elman, *A Cultural History of Civil Examinations,* pp. 584, 728, on the 23 percent increase of gentry with regular, including purchased, degrees to 920,000 after 1850.

44. See John Fryer, "Chinese Prize Essays Report for 1886 and 1887," pp. 100–101, and Rudolph Wagner, "The Formation of Late Qing Print Culture: Ernest Major and the *Shenbao guan*" (East Asian Studies Lecture, Princeton University, April 16, 2003). Compare Biggerstaff, "Shanghai Polytechnic Institution," pp. 141–143. On prize-essay questions in eighteenth-century Europe, see Hahn, "Laplace and the Mechanistic Universe," p. 266, and the biography for Joseph Louis Lagrange, in DSB, vol. 7, pp. 559–573. See also Keith Hoskin, "Examinations and the Schooling of Science," in Roy MacLeod, ed., *Days of Judgement: Science, Examinations, and the Organization of Knowledge in Late Victorian England* (Driffield, N. Humberside: Studies in Education, 1982).

45. Fryer, "Chinese Prize Essays Report," p. 100.

46. On the eight-legged essay and examination essay collections, see Elman, *A Cultural History of Civil Examinations,* pp. 380–420.

47. John Fryer, "Second Report of the Chinese Prize Essay Scheme in Connection with the Chinese Polytechnic Institution and Reading Rooms, Shanghai, from July 1887 to July 1889," *North-China Herald,* July 20, 1889, pp. 85–86.

48. The essays were later collected by Wang Tao and published in Shanghai in 1897 under the title *Gezhi keyi huibian.* Wang Ermin, *Shanghai Gezhi Shuyuan zhilue,* pp. 54–55, presents a list of officials involved.

49. The lead article, "The Progress of Foreign Studies," in the *North-China Herald,* April 14, 1893, pp. 513–514, described some sources for answering the questions. Compare Ming, "In Search of a Position," chap. 6. See also Shang Zhicong, "1886–1894 nianjian jindai kexue zai wan Qing zhishi fenzi zhong de yingxiang—Shanghai Gezhi shuyuan Gezhi lei keyi fenxi," *Qingshi yanjiu* 3 (August 2001): 73, 82n6, and Wright, *Translating Science,* pp. 163–173.

50. Fryer, "Second Report," p. 86. Complemented by figures from GZSYKY, vols. 1–2, 1889–1893. See also Biggerstaff, "Shanghai Polytechnic Institution," p. 149.

51. Xiong, *Xixue dongjian,* pp. 385–386. Compare Shang, "1886–1894 nianjian," p. 80.

52. Xiong, *Xixue dongjian*, pp. 387–391. Compare Shang, "1886–1894 nianjian," p. 81, and Wang, *Shanghai Gezhi Shuyuan zhilue*, pp. 69–72.

53. Fryer, "Second Report," p. 85.

54. Xiong Yuezhi, "Gezhi huibian," pp. 65–66, and Wang, *Shanghai Gezhi Shuyuan zhilue*, pp. 69–72.

55. *Zhongwai shiwu cewen leibian dacheng* (1903 ed.), 16.1a–b (essay prepared by Zhong Tianwei), 18.2b–3a (civil examination question taken from Xu Xingtai's spring 1887 polytechnic essay competition topic). See also Elman, *A Cultural History of Civil Examinations*, pp. 594–602, 605–608, Wright, *Translating Science*, pp. 178–179, and Theodore Huters, "*Dongfang zazhi* and the New Republic of Letters" (presented at the conference New Light on Publishing in Early Twentieth-Century Shanghai, sponsored by the UCLA Center for Chinese Studies in conjunction with the Southern California China Colloquium, May 18, 2002).

56. See GZSYKY, vol. 1, 1887–90, vol. 2, 1891–1893. Compare Xiong, *Xixue dongjian*, p. 374n3, and Shang, "1886–1894 nianjian," pp. 79–80, with Wang, *Shanghai Gezhi Shuyuan zhilue*, pp. 56–68. Shang presents all twenty-four essay topics on science from 1886 through 1894. Fryer's collection at the University of California–Berkeley Library ends in 1893.

57. See *Zhongwai shiwu cewen*, "Mulu," pp. 13a–b.

58. See *Zhongwai shiwu cewen*, "Mulu," pp. 13a–b. For discussion of the civil examination reforms of 1901–1902, see Elman, *A Cultural History of Civil Examinations*, pp. 594–602. See also Xiong, *Xixue dongjian*, pp. 362–63.

59. See GZSYKY, vol. 1, 1889, p. 20b. The essay is cited from another edition in Xiong, *Xixue dongjian*, pp. 371–372, and Li Shuangbi, "Cong 'gezhi' dao 'kexue': Zhongguo jindai keji guan de yanbian guiji," *Guizhou shehui kexue* 137 (1995.5): 105. See also Elman, *A Cultural History of Civil Examinations*, pp. 389–399, for stylistic examples of the eight-legged essay.

60. See GZSYKY, vol. 1, 1889, p. 6b, which is also cited from another edition in Xiong, *Xixue dongjian*, p. 371.

61. See Su Jing, *Qingji Tongwen guan ji qi shisheng* (Taibei: Shanghai yinshua chang, 1985) pp. 252–255. See also Kuo and Liu, "Self-Strengthening," pp. 531–532, and Andrews, "Tailoring Tradition," p. 152.

62. See Xiaoqun Xu, "'National Essence' vs 'Science': Chinese Native Physicians' Fight for Legitimacy, 1912–37," *Modern Asian Studies* 31, 4 (1997): 848, and Cohen, "Christian Missions," pp. 574–575.

63. Bridie Andrews, "Medical Lives and the Odyssey of Western Medicine in Early Twentieth-Century China" (paper presented at the History of Science Society Annual Meeting, San Diego, Calif., November 8, 1997). See also Lu Xun, *Selected Works of Lu Hsun*, trans. Yang Hsien-yi and Gladys Yang (Beijing: Foreign Languages Press, 1956), p. 2, and Howard Boorman and Richard Howard, eds., *Biographical Dictionary of Republican China* (New York: Columbia University Press, 1967), pp. 170–189.

64. Andrews, "Tailoring Tradition," pp. 152–153.

65. GZSHKY, vol. 2, 1891, p. 1b (topic), pp. 1a–17a (prize essays).

66. GZSHKY, vol. 2, 1892, pp. 3a–b (topic), pp. 4b–6b, 33b–34b (prize essays). See also Shang, "1886–1894 nianjian," p. 81.

67. GZSHKY, vol. 2, 1893, p. 2a (topic), pp. 4a–9b (prize essays).

68. *Zhongwai shiwu cewen*, "Mulu," pp. 14a–b, 19.15a–22b, especially 18b–20b.

69. GZSHKY, vol. 1, 1889, p. 1a.

70. Ibid., p. 8b.

71. Ibid., p. 12b.

72. *Zhongwai shiwu cewen*, "Mulu," p. 12a. My thanks to Joachim Kurtz for pointing out Zhong's verbatim citations from the *Primers*.

73. Li Changli, "Wan Qing dui Xixue de liangzhong wudu—lun Zhong Tianwei de Xixue guan," *Jiangsu shehui kexue* (Nanjing) 6 (1999): 110–114.

74. For "Hundun shuo," see GZHB, vol. 2, pp. 13–15 (September 1877). See also *Gezhi keyi huibian* (Shanghai: Polytechnic, 1897), pp. 4.1a, 4.6a, 4.9b, 4.16a, 5.42b; *Dixue qianshi* (Shanghai: Jiangnan Arsenal, 1873), 13.16a; and *Shen bao* 2 (August 21, 1873), p. 2. Compare Wright, "John Fryer," p. 13. Pusey, *China and Charles Darwin*, pp. 4–5, has stressed 1895 as the time when Darwin's views were first presented.

75. See GZSYKY, vol. 1, 1889, pp. 21a–21b. The essay is also cited from another edition in Xiong, *Xixue dongjian*, pp. 365–366.

76. Yan Fu's translation of Herbert Spencer's *Study of Sociology* appeared after 1898. Pusey, *China and Charles Darwin*, p. 59, traces the first introduction of Darwin to Yan Fu's 1895 essay "On the Origins of Strength" (*Yuan qiang*). See also David Wright, "Yan Fu and the Tasks of the Translator," in Lackner, Amelung, and Kurtz, eds., *New Terms for New Ideas*, pp. 235–255, especially p. 253.

77. GZSYKY, vol. 1, 1889, p. 21a.

78. Shang, "1886–1894 nianjian," p. 81, and Li, "Wan Qing dui Xixue," pp. 112–113.

79. See *North-China Herald*, January 29, 1902, p. 180; September 22, 1905, pp. 697–698; January 25, 1907, p. 202; February 21, 1908; pp. 418–419. See also Xiong, *Xixue dongjian*, p. 373, and Chen Qiwei, "Zaitan Wang Tao he Gezhi shuyuan dui Xifang zhexue de jieshao," *Dongyue luncong* 22, 5 (2001.9): 54–57. Compare Natascha Vittinghoff, "Unity Vs. Uniformity: Liang Qichao and the Invention of a 'New Journalism' for China," *Late Imperial China* 23, 1 (June 2002): 91–143, and Rudolph Wagner, "The *Shenbao* in Crisis: The International Environment and the Conflict between Guo Songtao and the *Shenbao*," *Late Imperial China* 20, 1 (June 1999): 107–138. Both correct the account in Rosewell, *The Chinese Periodical Press*, pp. 86–101.

10. Government Arsenals, Science, and Technology in China after 1860

1. Wright, "The Great Desideratum," pp. 35–70.

2. See Hans J. van de Ven, "War in the Making of Modern China," *Modern Asian Studies* 30, 4 (1996): 737–756, which draws on Peter Paret's discussion of the new military history in Paret's *Understanding War* (Princeton, N.J.: Princeton University Press, 1992), especially pp. 209–226. For recent Chinese historiography, see Benjamin Elman, "Rethinking the Foreign Affairs Movement: Naval Warfare and the Refraction of China's Self-Strengthening Reforms into Scientific and Technological Failure, 1860–1895," *Modern Asian Studies* 38, 2 (2003): 283–291.

3. Biggerstaff, *The Earliest Modern Government Schools,* pp. 174–175, and Dagenais, *John Fryer,* 1868: 8, 1871:2. See also Wright, *Translating Science,* pp. 238–240.

4. See Jonathan Spence, *To Change China,* pp. 34–56.

5. See the title page for Martin's *Gewu rumen.* See also Han Qi, "Chuanjiaoshi Weilie Yali," p. 58, and Biggerstaff, *The Earliest Modern Government Schools,* p. 174. See also Bennett, *John Fryer,* pp. 20–21, 149n7.

6. See ECCP, p. 862. Compare Wang Xianming, *Jindai xinxue,* pp. 88–90.

7. See Spence, *To Change China,* pp. 74–75, and Hevia, "Looting Beijing," pp. 193–199.

8. Biggerstaff, *The Earliest Modern Government Schools,* pp. 6–14. See also ECCP, pp. 380–383, and Ssu-yü Teng and John Fairbank, eds., *China's Response to the West* (Cambridge, Mass.: Harvard University Press, 1954), p. 35.

9. See Su *Qingji Tongwen guan ji qi shisheng.* See also Nancy Evans, "The Canton T'ung-wen Kuan: A Study of the Role of Bannermen in One Area of Self-strengthening," *Papers on China* (Harvard University), 22A (1969): 89–103, and Masini, *The Formation of Modern Chinese Lexicon,* pp. 46–53. Compare Crossley, *Orphan Warriors,* pp. 141–150.

10. Biggerstaff, *The Earliest Modern Government Schools,* pp. 15–18, and ECCP, pp. 402–404, 721–721, 754. See also Teng and Fairbank, eds., *China's Response to the West,* pp. 61–63.

11. Kuo and Liu, "Self-Strengthening," pp. 519–521, and Spence, *To Change China,* pp. 108–111. In 1875, three sixty-eight-pound guns made by Macartney exploded while being tested.

12. David Pong, *Shen Pao-chen and China's Modernization in the Nineteenth Century* (Cambridge: Cambridge University Press, 1994), pp. 134–160, 178. See also Teng and Fairbank, eds., *China's Response to the West,* pp. 81–83.

13. Teng and Fairbank, eds., *China's Response to the West,* pp. 76–77.

14. See Hong Wansheng, "Tongwen guan suanxue jiaoxi Li Shanlan," in Yang Cuihua and Huang Yinong, eds., *Jindai Zhongguo keji shi lunji* (Taibei: Academia Sinica, Taiwan, Institute of Modern History, and National Qinghua University, Institute of History, 1991), pp. 215–259. See also Biggerstaff, *The Earliest Modern Government Schools,* pp. 19–34, and ECCP, p. 480.

15. Teng and Fairbank, eds., *China's Response to the West,* pp. 64–65.

16. Wei Xiangong, comp., *Jiangnan zhizao ju ji* (1904; repr., Taibei: Wenhai chuban she). See also Quan Hansheng, "Qingji de Jiangnan zhizao ju," *Bulletin of the Institute of History and Philology, Academia Sinica* 23 (1951): 145–159, and Mary Clabaugh Wright, *The Last Stand of Chinese Conservatism* (New York: Atheneum, 1996), pp. 211–212.

17. Meng Yue, "Hybrid Science," pp. 13–52.

18. See Thomas Kennedy, "The Establishment and Development of the Kiangnan Arsenal, 1860–95" (Ph.D. diss., Columbia University, 1968), chap. 2, and ECCP, pp. 721–722. See also Biggerstaff, *The Earliest Modern Government Schools,* pp. 165–166, and Kuo and Liu, "Self-Strengthening," pp. 519–521.

19. Masini, *The Formation of Modern Chinese Lexicon,* pp. 62–71.

20. Wright, *Translating Science,* pp. 108–109.

21. Fryer, "An Account of the Department for the Translation of Foreign Books," pp. 78–79. See also Biggerstaff, *The Earliest Modern Government Schools,* p. 175n33, and Dagenais, *John Fryer's,* 1868: 3–8.

22. See also Bennett, *John Fryer,* p. 50, and Dagenais, *John Fryer,* 1867:1, 1872:1. On books dealing with chemistry, see James Reardon-Anderson, *The Study of Change: Chemistry in China, 1840–1949* (Cambridge: Cambridge University Press, 1991), pp. 29–52.

23. Dagenais, *John Fryer,* 1886: 1–2. We will see below that Fryer's confidence was later replaced by a darker vision of China in the aftermath of the 1894–1895 Sino-Japanese War.

24. Fryer, "An Account of the Department for the Translation of Foreign Books," pp. 79–81. Compare Wright, *Translating Science,* pp. 229–240.

25. Dagenais, *John Fryer,* 1880:5, 1883:1, and Bennett, *John Fryer,* pp. 46, 60–61. See also Adrian Bennett, *Missionary Journalist in China: Young J. Allen and His Magazines, 1860–1883* (Athens: University of Georgia Press, 1983), pp. 294–296, and Wang Yangzong, "Qingmo Yizhi shuhui tongyi keji shuyu gongzuo shuping," *Zhongguo keji shiliao* 12 (1991.2): 9–19.

26. Doolittle, *Vocabulary and Hand-book of the Chinese Language,* vol. 1, pp. 175–178, 308–318. Compare Shen Guowei, "The Creation of Technical Terms," pp. 301–303.

27. Masini, *The Formation of Modern Chinese Lexicon,* pp. 50–53. See also Anders Lundgren and Bernadette Bensaude-Vincent, eds., *Communicating Chemistry: Textbooks and Their Audiences, 1789–1939* (Canton, Mass.: Science History Publications, 2000).

28. Reardon-Anderson, *The Study of Change,* pp. 377–380. See also Wright, "The Great Desideratum," pp. 35–70; Wang Yangzong, "A New Inquiry into the Translations of Chemical Terms by John Fryer and Xu Shou," in Lackner, Amelung, and Kurtz, eds., *New Terms for New Ideas,* pp. 271–283; and Shen, "The Creation of Technical Terms," pp. 294–300. For more detail, see Wright, *Translating Science,* pp. 337–365.

29. Dagenais, *John Fryer,* 1895:6, and Wright, "The Great Desideratum," pp. 51–53.

30. John Fryer, "Scientific Terminology: Present Discrepancies and Means of Securing Uniformity," in *Records of the General Conference of the Protestant Missionaries of China Held at Shanghai, May 7–20, 1890* (Shanghai: American Presbyterian Mission Press, 1890), pp. 531–549. See also Wang, "Qingmo Yizhi shuhui," 9–19; Bennett, *John Fryer,* pp. 30–33; and Wright, "The Translation of Modern Western Science," pp. 653–673.

31. See Calvin W. Mateer and W. M. Hayes, "Letter to the Committee of Education Association on Terms," and "The Committee on Terminology's Report," both in *John Fryer Papers,* Bancroft Library, University of California–Berkeley. Compare Bennett, *John Fryer,* pp. 32–33. See also Wang, "A New Inquiry into the Translations of Chemical Terms," pp. 274–281.

32. See Fryer's review of Edkins's science primer in GZHB, vol. 5, pp. 200–203.

33. Joseph Edkins and Li Shanlan, *Zhongxue* (Shanghai: Mohai shuguan, 1859). A new edition was issued in 1867 by Meihua Shuguan. Schreck's translation was aided by the Chinese convert Wang Zheng (1571–1644). See LHCT, vol. 1, no. 1

(1857.1.26): 1a–1b; vol. 2, no. 1 (1858.2.14): 10a–15a; and vol. 2, no. 2 (1858.6.11): 14a–22a; and Doolittle, *Vocabulary and Hand-book of the Chinese Language,* vol. 2, part 3, p. 177. See also Dudink and Standaert, "Ferdinand Verbiest's *Qiongli xue* (1683)," pp. 13, 25, and Iwo Amelung, "Weights and Forces: The Reception of Western Mechanics in Late Imperial China," in Lackner, Amelung, and Kurtz, eds., *New Terms for New Ideas,* pp. 198–199.

34. See Martin, *Gewu rumen,* 5.1a–1b, and Liang Qichao, "Gezhi xue yange kaolue," in Liang, *Yinbing shi wenji,* 8 vols. (Taibei: Zhonghua Bookstore, 1970), vol. 2, p. 14. See also Amelung, "Weights and Forces," pp. 204–207, 206n39, who notes that the shift to force as the standard translation for mechanics in China occurred circa 1910.

35. "The Committee on Terminology's Report." See also Calvin W. Mateer, "The Revised List of the Chemical Elements," *Chinese Recorder* 29 (February 1898): 87–94, and Wright, "The Great Desideratum," pp. 53–55.

36. Fryer, "An Account of the Department for the Translation," pp. 78–79. See also Bennett, *John Fryer,* pp. 14–17, 20–26, 69, and Biggerstaff, *The Earliest Modern Government Schools,* pp. 174–175. Compare Spence, *To Change China,* pp. 156–157.

37. See Wylie, *Notes on Chinese Literature,* pp. 118–119, 120.

38. See "Yangwu yundong dashi ji," in Xu Tailai, ed., *Yangwu yundong xinlun* (Changsha: Hunan People's Press, 1986), pp. 349–448, and ECCP, pp. 240–243, 331–333. Compare Zhang Baichun, "An Inquiry into the History of the Chinese Terms," pp. 177–195.

39. Wang, "Qingmo Yizhi shuhui," pp. 11–12, and Wang, "A New Inquiry into the Translations of Chemical Terms," pp. 282–283.

40. ZGJD, pp. 468–474. See also Fryer, "An Account of the Department for the Translation of Foreign Books," pp. 78–79. Bennett, *John Fryer,* pp. 34–35, notes that after 1880, Fryer concentrated on natural science rather than technology.

41. ZGJD, pp. 214–255. See also Biggerstaff, *The Earliest Modern Government Schools,* pp. 166–167, 173, and Bennett, *John Fryer,* pp. 18–25. The school, however, avoided the texts that Fryer and the Translation Bureau produced. See Fryer, "An Account of the Department for the Translation of Foreign Books," p. 81.

42. Meng Yue, "Hybrid Science," pp. 32–33. See also Biggerstaff, *The Earliest Modern Government Schools,* p. 79. Compare this to the classical curriculum before 1900 described in Elman, *A Cultural History of Civil Examinations,* pp. 239–294. On the histories, see pp. 485–503.

43. Biggerstaff, *The Earliest Modern Government Schools,* pp. 166–171. See also Wann-sheng Horng, "Chinese Mathematics at the Turn of the 19th Century," pp. 191–192.

44. *Jiangnan zhizao ju ji,* pp. 151–168. See Bennett, *John Fryer,* p. 18, and Meng, "Hybrid Science," pp. 29–30. See also Biggerstaff, *The Earliest Modern Government Schools,* p. 172.

45. Meng, "Hybrid Science," pp. 16–17. For recent views, see the special issue of *Chuanshi yanjiu* 8 (1995), devoted to the 130th anniversary of the Jiangnan Shipyard. Van de Wen, "War in the Making of Modern China," p. 740, notes that most are "in agreement that China's navy was superior to the Japanese," although the final verdict remains open. Others would argue that such claims are valid only

if one uses aggregate measures such as tonnage and weight of shell. The Japanese navy by 1894 was newer, faster, and equipped, as we will see later in this chapter, with faster firing guns.

46. See Takehiko Hashimoto, "Introducing a French Technological System," *East Asian Science, Technology, and Medicine* 16 (1999): 53–65. Yokosuka had a single dock by 1867.

47. *Jiangnan zhizao ju ji*, pp. 319–328, and Meng, "Hybrid Science," pp. 16–24, especially tables 1 and 2, which draw on Hansgeorg Jentschura, Dieter Jung, and Peter Michel, *Warships of the Imperial Japanese Navy: 1869–1945*, trans. Antony Preston and J. D. Brown (Annapolis, Md.: U.S. Naval Institute 1977), p. 115.

48. Meng, "Hybrid Science," p. 17. See also Pong, *Shen Pao-chen*, p. 224, and Biggerstaff, *The Earliest Modern Government Schools*, pp. 246–247.

49. Meng, "Hybrid Science," pp. 17–19. Kenneth Pomeranz, *The Great Divergence: Europe, China, and the Making of the Modern World Economy* (Princeton, N.J.: Princeton University Press, 2001), pp. 62–68, notes that China's fossil fuels were relatively inaccessible when compared to Europe's.

50. Meng, "Hybrid Science," pp. 21–23. See also *Jiangnan zhizao ju ji*, pp. 374–388, 442, and Wright, "Careers in Western Science," p. 80. In 1905, the Arsenal was separated into a shipyard and machine shop. See Liu Dajun, "Jiangnan zhizao zongju cong chengli dao 'juchang fenli' shi de yanbian," *Chuanshi yanjiu* 3 (1987): 51–55.

51. ZGJD, pp. 329–379. See also the special issue of *Chuanshi yanjiu* 10 (1996), devoted to the 130th anniversary of the Fuzhou Navy Yard.

52. Biggerstaff, *The Earliest Modern Government Schools*, pp. 200–208, and Pong, *Shen Pao-chen*, pp. 208–209. German military technology would become more conspicuous in China and Japan after the decisive German victory in the Franco-Prussian War in 1870.

53. Zhang Yufa, "Fuzhou chuanchang zhi kaichuang ji qi chuqi fazhan," *Jindai shi yanjiu suo jikan* (Academia Sinica, Taiwan) 2 (June 1971): 177–225.

54. Pong, *Shen Pao-chen*, pp. 241–243, 261.

55. Steven Leibo, *Transferring Technology to China: Prosper Giquel and the Self-Strengthening Movement* (Berkeley: University of California Press, 1985); Pong, *Shen Pao-chen*, pp. 214–225; and Biggerstaff, *The Earliest Modern Government Schools*, pp. 203–210. See also Bastid, "Fuzhou chuanzheng ju," pp. 238–239.

56. Bastid, "Fuzhou chuanzheng ju," p. 246.

57. Biggerstaff, *The Earliest Modern Government Schools*, pp. 203–211. See also Elman, *From Philosophy to Philology*, p. 256, and Elman, *A Cultural History of Civil Examinations*, pp. 135, 221–222.

58. Bastid, "Fuzhou chuanzheng ju," pp. 246–251, and Biggerstaff, *The Earliest Modern Government Schools*, pp. 214–219.

59. See Wang Xinzhong, "Fuzhou chuanchang zhi yange," *Qinghua xuebao* 8 (December 1932): 27–30, and Biggerstaff, *The Earliest Modern Government Schools*, pp. 223–241. See also Kuo and Liu, "Self-Strengthening," pp. 524–525.

60. Biggerstaff, *The Earliest Modern Government Schools*, pp. 53, 220, 239, 271, and Pong, *Shen Pao-chen*, pp. 208–209, 266–270. See also Bastid, "Fuzhou chuanzheng ju," pp. 246–255, and Kuo and Liu, "Self-Strengthening," p. 534.

61. A convenient source of Chinese naval documents is Zhang Xia et al., *Qingmo hai-jun shiliao* (Beijing: Haijun chuban she, 2001), pp. 2–9, for the Taiwan incident. See also Thomas Kennedy, *The Arms of Kiangnan: Modernization in the Chinese Ordnance Industry, 1860–1895* (Boulder, Colo.: Westview Press, 1978), pp. 150–160, and Pong, *Shen Pao-chen*, pp. 292–293, 335. Compare Kitayama Yasuo, "Chūgoku ni okeru kan'ei gunji kōgyō no ichi kōsatsu—Kōnan seizōkyoku o chūshin to shite," *Hisutoria* 9 (1954): 1–8, which contends that Zeng, Li, and Zuo built up the armaments industry mainly for their power bases and to maintain domestic security.

62. John Rawlinson, *China's Struggle for Naval Development, 1839–1895* (Cambridge, Mass.: Harvard University Press, 1967), pp. 60–61, and Wright, "Careers in Western Science," p. 81.

63. See Zhang et al., *Qingmo haijun shiliao,* pp. 35–41, 333–336, and Liao Zonglin, *Zhong Fa zhanzheng* (Tianjin: Guji chubanshe, 2002), pp. 638–649. See also Bruce Elleman, *Modern Chinese Warfare, 1795–1989* (New York: Routledge, 2001), pp. 82–93.

64. Liao, *Zhong Fa zhanzheng,* pp. 443–464.

65. Rawlinson, *China's Struggle,* pp. 109–128, and Liao, *Zhong Fa zhanzheng,* pp. 465–590. Compare Allen Fung, "Testing the Self-Strengthening: The Chinese Army in the Sino-Japanese War of 1894–95," *Modern Asian Studies* 30, 4 (1996): 1010–1015, which stresses China's lack of preparedness for the Sino-Japanese War when the Japanese navy, like the French, attacked without warning.

66. Bastid, "Fuzhou chuanzheng ju," pp. 248–249.

67. Rawlinson, *China's Struggle,* pp. 129–139, and Biggerstaff, *The Earliest Modern Government Schools,* pp. 221–222. See also Kwang-Ching Liu and Richard Smith, "The Military Challenge: The North-West and the Coast," in Fairbank and Liu, eds., *The Cambridge History of China,* vol. 11, *Late Qing, 1800–1911,* part 2, pp. 254–256.

68. In 1871 the Qing and Meiji states had signed a treaty guaranteeing full reciprocal rights, which included mutual extraterritoriality.

69. Noriko Kamachi, "The Chinese in Meiji Japan: Their Interaction with the Japanese before the Sino-Japanese War," in Akira Iriye, ed., *The Chinese and the Japanese: Essays in Political and Cultural Interactions* (Princeton, N.J.: Princeton University Press, 1980), pp. 69–72. See also Donald Keene, "The Sino-Japanese War of 1894–95 and Its Cultural Effects in Japan," in Donald Shively, ed., *Tradition and Modernization in Japanese Culture* (Princeton, N.J.: Princeton University Press, 1971), pp. 122–123.

70. See *North-China Herald,* July 27, 1894, pp. 131, 152, and *Japan Weekly Mail,* August 25, 1894, p. 227. Foreign media accounts, which generally blamed Japan for the hostilities, are presented in S. C. M. Paine, *The Sino-Japanese War of 1894–1895: Perceptions, Power and Primacy* (Cambridge: Cambridge University Press, 2003), pp. 107–134, 138–140, 154–160. Paine, pp. 125, 141–144, summarizes the size of Chinese and Japanese sea and land forces in 1894. See also Marius Jansen, Samuel Chu, Shumpei Okamoto, and Bonnie Oh, "The Historiography of the Sino-Japanese War," *International History Review* 1, 2 (April 1979): 191–227, and Elleman, *Modern Chinese Warfare,* pp. 94–115.

71. Zhang et al., *Qingmo haijun shiliao,* pp. 301–327, and Rawlinson, *China's Struggle,* pp. 163–169. See also Keene, "The Sino-Japanese War," p. 132.

72. *Japan Weekly Mail,* August 4, 1894, pp. 130–131. See also Keene, "The Sino-Japanese War," pp. 127, 132. Compare Shumpei Okamoto, "Background of the Sino-Japanese War, 1894–95," in Shumpei Okamoto, *Impressions of the Front: Woodcuts of the Sino-Japanese War, 1894–95* (Philadelphia: Philadelphia Museum of Art, 1983), p. 13. On the Japanese media before the onset of war see also Paine, *The Sino-Japanese War,* pp. 134–140.

73. *Japan Weekly Mail,* August 4, 1894, p. 132, explains the importance of the naval war. See also Paine, *The Sino-Japanese War,* pp. 132–135, 158–163.

74. Some accounts claim the Chinese Northern Fleet had fourteen warships led by two ironclad ships. Others claim the Chinese had only ten vessels. *Japan Weekly Mail* claimed eleven Japanese ships had faced twenty Chinese ships. See Okamoto, "Background of the Sino-Japanese War," p. 13. Compare Paine, *The Sino-Japanese War,* pp. 178–192, 197–198.

75. Rawlinson, *China's Struggle,* pp. 169–174, 201. See also Biggerstaff, *The Earliest Modern Government Schools,* p. 248.

76. Rawlinson, *China's Struggle,* pp. 174–197, and Paine, *The Sino-Japanese War,* pp. 194–195, 206–222, 228–238. See also Yoda Yoshiie, "Jūkyūseiki kōhan ni okeru Nisshin ryōgoku no shokusan seisaku ni tsuite," *Shakai kagaku tōkyū* 12, 3 (March 1967): 1–38. See also Louise Virgin, "Japan at the Dawn of the Modern Age," in *Japan at the Dawn of the Modern Age: Woodblock Prints from the Meiji Era, 1868–1912* (Boston: Museum of Fine Arts, 2001), pp. 66–72, 86.

77. Bastid, "Fuzhou chuanzheng ju," p. 253, and Paine, *The Sino-Japanese War,* pp. 265–266, 326–327. See also Crossley, *Orphan Warriors,* p. 145.

78. See Okamoto, "Background of the Sino-Japanese War," p. 16, and Virgin, "Japan at the Dawn of the Modern Age," p. 112. See also Paine, *The Sino-Japanese War,* pp. 192–195, 247–293, 306–309. On Russian military developments, see J. N. Westwood, *Russia against Japan, 1904–05: A New Look at the Russo-Japanese War* (Houndmills, Basingstoke, Hampshire: Macmillan, 1986).

79. See Keene, "The Sino-Japanese War," pp. 121–175. See also James Huffman, "Commercialization and Changing World of the Mid-Meiji Press," pp. 574–579, and Giles Richter, "Entrepreneurship and Culture: The Hakubunkai Publishing Empire in Meiji Japan," p. 591, both in Helen Hardacre and Adam Kern, eds., *New Directions in the Study of Meiji Japan* (Leiden: E. J. Brill, 1992).

80. ZGJD, pp. 589–604. See Wright, "John Fryer," p. 15, and Kuo and Liu, "Self-Strengthening," p. 587. See also Zhao Hongjun, "Chinese Versus Western Medicine," pp. 36–37.

81. Wang Yangzong, "1850 niandai zhi 1910 nian Zhongguo yu Riben zhi jian kexue shuji de jiaoliu shulue," *Tōzai gakujutsu kenkyūjo kiyō* (Kansai University) 33 (March 2000): 147–150.

82. Wang, "1850 niandai zhi 1910 nian," pp. 139–144. See also Tsien, "Western Impact on China through Translation," pp. 323–325, and Masini, *The Formation of Modern Chinese Lexicon,* pp. 105–108.

83. See Zhu Kaijia, "Gezhi xinbao yuanqi," *Gezhi xinbao* 1 (1898.3.13): 1a–2b, included in *Jindai Zhongguo shiliao congkan* (Taibei: Wenhai chuban she, 1987),

vol. 231. Compare Meng, "Hybrid Science," p. 43. See also *Gezhi yiwen huibao,* 4 vols. (Shanghai: Catholic Mission, 1898–1899), vol. 1, p. 105. Compare Xiong, *Xixue dongjian,* pp. 457–458.

84. See van de Ven, "War in the Making of Modern China," pp. 737–756. Paine, *The Sino-Japanese War,* pp. 64–66, 333–366, explains China's failure to industrialize in light of cultural issues. Elleman, *Modern Chinese Warfare,* p. 102, continues to stress China's scientific and technological inability to support a modern navy.

85. Fung, "Testing the Self-Strengthening," pp. 1007–1031. Compare Richard Smith, "Foreign Training and China's Self-Strengthening: The Case of Feng-huang-shan," *Modern Asian Studies* 10, 2 (1976): 195–223, which also stresses the late Qing failure to train a modern officer corps.

86. For a review of the literature in Chinese, English, and Japanese, see Elman, "Naval Warfare and the Refraction of China's Self-Strengthening Reforms," pp. 283–326. On the inadequacy of China's army and navy, see Powell, *The Rise of Chinese Military Power,* pp. 36–50. Compare Smith, "Reflections on the Comparative Study of Modernization in China and Japan," pp. 11–23.

87. John Fryer, "An Account of the Department for the Translation of Foreign Books," pp. 77–81. See also Bennett, *John Fryer,* p. 42.

88. Bennett, *John Fryer,* pp. 42–44, and Masini, *The Formation of Modern Chinese Lexicon,* p. 75.

89. See Kuo and Liu, "Self-Strengthening," pp. 519–537, and Onoue Etsuzō, "Chūgoku ni okeru kōgyō no ishoku—yōmu undō no ichi sokumen," *Rokkōdai ronshū* 5, 3 (October 1958): 67–86. On German influence in Qingdao, Shandong, see George Steinmetz, "Precoloniality and Colonial Subjectivity: Ethnographic Discourse and Native Policy in German Overseas Imperialism," *Political Power and Social Theory* 15 (2002): 181–191.

90. Thomas Kennedy, "Chang Chih-tung and the Struggle for Strategic Industrialization: The Establishment of the Hanyang Arsenal, 1884–1895," HJAS 33 (1973): 154–182.

91. Natascha Vittinghoff, "Social Actors in the Field of New Learning in Nineteenth Century China," in Michael Lackner and Natascha Vittinghoff, eds., *Mapping Meanings: The Field of New Learning in Late Qing China* (Leiden: E. J. Brill, 2004), pp. 75–118.

92. Wright, "Careers in Western Science," pp. 54, 60–62, 69–71. See also GZHB, vol. 3, pp. 30–31 [3. 7 (1880): 14b–15a], and Fryer's letter to *Nature* dated November 25, 1880, and published there on March 10, 1881: pp. 448–449. The letter is reproduced and discussed in Wright, *Translating Science,* pp. 50–53.

93. Wright, "Careers in Western Science," pp. 51–80, 88.

94. Meng, "Hybrid Science," pp. 26–27. See also Vittinghoff, "Social Actors in the Field of New Learning," pp. 75–118. On "techno-science," see Bruno Latour, *Science in Action: How to Follow Scientists and Engineers through Society* (Cambridge, Mass.: Harvard University Press, 1987), pp. 145–176.

95. Evans, "The Canton T'ung-wen Kuan," pp. 89–103.

96. Wright, "The Great Desideratum," pp. 35–70. On Lu Xun, see Boorman and Howard, eds., *Biographical Dictionary of Republican China,* p. 417, and Mary Buck, "Justice for All: The Application of Law by Analogy in the Case of Zhou

Fuqing," *Journal of Chinese Law* 7, 2 (Fall 1993): 118–127. See also Benjamin Elman, "Wang Kuo-wei and Lu Hsun: The Early Years," *Monumenta Serica, 34* (1979–80): 389–401.

97. Biggerstaff, *The Earliest Modern Government Schools,* pp. 53, 251. See also Kuo and Liu, "Self-Strengthening," p. 534.

98. Kwang-ching Liu, "Nineteenth-Century China," in Ping-ti Ho and Tang Tsou, eds., *China in Crisis,* 2 vols. (Chicago: University of Chicago Press, 1968), vol. 1, pp. 93–178, had earlier described the *Yangwu yundong* as an early stage of China's institutional reform and not a failure.

99. Huo Rusu, "Zaochuan gongye zai Zhongguo jindai shi shang de lishi diwei," *Chuanshi yanjiu* 3 (1987): 1–7.

100. See Xiong, *Xixue dongjian,* pp. 620–623. Onogawa Hidemi, *Shimmatsu seiji shisō kenkyū* (Tokyo: Misuzu shobō, 1969), pp. 52–85. Compare Vittinghoff, "Unity vs. Uniformity," p. 112.

101. The essay is abridged in Qian Zhongshu and Zhu Weizheng, eds., *Wan'guo gongbao wenxuan* (Beijing: Sanlian shudian, 1998), pp. 179–201.

102. Chao Erxun et al., comps., *Qingshi gao,* 40 vols. (Beijing: Zhonghua shuju, 1977), *juan* 107 (*xuanju er*). See also Zhuang Jifa, *Jingshi daxue tang* (Taibei: National Taiwan University, 1970), and Li Shuangbi, "Cong 'gezhi' dao 'kexue,'" p. 104.

103. See Renville Lund, "The Imperial University of Beijing" (Ph.D. diss., University of Washington, 1956), pp. 118–122, and Reardon-Anderson, *The Study of Change,* p. 109. See also Hevia, *English Lessons,* pp. 195–240.

104. Elman, *A Cultural History of Civil Examinations,* pp. 608–618.

105. Dagenais, *John Fryer,* 1895: 4–6. This gainsays Fryer's earlier view in his "An Account of the Department for the Translation of Foreign Books," pp. 77–81.

106. William Martin, *The Awakening of China* (London: Hodder and Stoughton, 1907), p. 177. Compare with Martin, "Preface," p. i, in *Gewu rumen.* See also Dagenais, *John Fryer,* 1895:7–8, 11–12, 1896:4, and Patrick Hanan, "The Missionary Novels of Nineteenth-Century China," HJAS 60, 2 (December 2000): 440–441, and Anonymous, "The New Novel before the New Novel," in Zeitlin and Liu, eds., *Writing and Materiality in China,* pp. 317–340. For the Boxer impact, see Spence, *To Change China,* pp. 158–160. Wright, "The Translation of Modern Western Science," p. 672, contends, "The transmission of science via translations between 1840 and 1895 was a failure."

107. See Métailié, "The Formation of Botanical Terminology," pp. 333–338.

11. Displacement of Traditional Chinese Science and Medicine in the Twentieth Century

1. Hu Mingjie, "Merging Chinese and Western Mathematics," pp. 367–376. See also Paula Harrell, *Sowing the Seeds of Change: Chinese Students, Japanese Teachers, 1895–1905* (Stanford, Calif.: Stanford University Press, 1992), p. 214, and Barry Keenan, "Beyond the Rising Sun: The Shift in the Chinese Movement to Study Abroad," in Laurence Thompson, ed., *Studia Asiatica* (San Francisco: Chinese Materials Center, 1975), pp. 157–169.

2. Liu, *Translingual Practice*, pp. 33, 336, presents *kexue / kagaku* as a second-hand borrowing from classical Chinese used to translate science into Japanese. Her source is the Song dynasty literatus Chen Liang (1143–1194), who used *kexue* as a shorthand reference to mean civil examination studies (*keju zhi xue = kexue*). This twelfth-century usage is unique to the Song dynasty.

3. Albert Craig, "Science and Confucianism in Tokugawa Japan," in Marius Jansen, ed., *Changing Japanese Attitudes toward Modernization* (Princeton, N.J.: Princeton University Press, 1965), pp. 139–42. See also Numata Jirō, *Western Learning: A Short History of the Study of Western Science in Early Modern Japan*, trans. R. C. J. Bachofner (Tokyo: The Japan-Netherlands Institute, 1992), pp. 60–95.

4. Reardon-Anderson, *The Study of Change*, pp. 82–87.

5. See Elman, *Classicism, Politics, and Kinship*, p. 302.

6. Hu, "Merging Chinese and Western Mathematics," pp. 214–215. See also Min Tu-ki, *National Polity and Local Power: The Transformation of Late Imperial China*, trans. Tim Brook and Philip Kuhn (Cambridge, Mass.: Harvard University Press, 1989), p. 84.

7. Meng Yue, "Hybrid Science," pp. 43–45.

8. Masini, *The Formation of Modern Chinese Lexicon*, pp. 71–83. See also Kung-chuan Hsiao, *A Modern China and a New World: Kang Yu-wei, Reformer and Utopian, 1858–1927* (Seattle: University of Washington Press, 1975), pp. 328–346, and Reynolds, *China, 1898–1912*, pp. 43–44.

9. Elman, *A Cultural History of Civil Examinations*, pp. 585–594, 608–618.

10. Hsiao, *A Modern China*, pp. 306–307, 331–346. See also ECCP, pp. 703–704.

11. Tan Sitong, "Xing suanxue yi," in Tan Sitong, *Tan Sitong quanji*, 2 vols. (Beijing: Zhonghua shuju, 1981), pp. 153–194. See also Hu, "Merging Chinese and Western Mathematics," pp. 271–279, 282–283.

12. See Chan Sin-wai, trans., *An Exposition of Benevolence: The Jen-hsueh of T'an Ssu-t'ung* (Hong Kong: The Chinese University Press, 1984), pp. 10–20, and Richard Shek, "Some Western Influences on T'an Ssu-t'ung's Thought," in Paul Cohen and John Schrecker, eds., *Reform in Nineteenth-Century China* (Cambridge, Mass.: Harvard University Research Center, 1976), pp. 200–201, 237. Compare Feng-ying Ming, "In Search of a Position," chap. 6.

13. See Tan Sitong, "Yitai shuo," in *Tan Sitong quanji*, pp. 432–444. See also Ingo Schäfer, "Natural Philosophy, Physics and Metaphysics in the Discourse of Tan Sitong: The Concepts of *Qi* and *Yitai*," in Lackner, Amelung, and Kurtz, eds., *New Terms for New Ideas*, pp. 257–269, and Luke S. K. Kwong, *T'an Ssu-t'ung, 1865–1898: Life and Thought of a Reformer* (Leiden: E. J. Brill, 1996), p. 71. Compare Daniel Siegel, "Thomson, Maxwell, and the Universal Ether," in G. N. Cantor and M. J. S. Hodge, eds., *Conceptions of Ether: Studies in the History of Ether Theories, 1740–1900* (Cambridge: Cambridge University Press, 1981), pp. 245–246.

14. David Wright, "Tan Sitong and the Ether Reconsidered," *Bulletin of the School of Oriental and African Studies* 57 (1994): 551–557. Compare Stewart, *Lessons in Elementary Physics*, p. 71.

15. Wright, "Tan Sitong and the Ether Reconsidered," pp. 558–563. Compare Clarence J. Glacken, *Traces on the Rhodian Shore: Nature and Culture in Western Thought from Ancient Times to the End of the Eighteenth Century* (Berkeley: University of California Press, 1990), pp. 7–24, and Heilbron, *Electricity*, pp. 50–53.

16. Smith and Wise, *Energy and Empire*, pp. 396–444. See also David Wilson, "Stokes and Kelvin, Cambridge and Glasgow, Light and Heat," in Harman and Mitton, eds., *Cambridge Scientific Minds*, pp. 107–121. Compare Wright, "Tan Sitong and the Ether Reconsidered," pp. 561–563.

17. Balfour Stewart followed up with a sequel to his *Unseen Universe*, called *Paradoxical Philosophy* (London: Macmillan and Co., 1878). See also Bruce Hunt, *The Maxwellians* (Ithaca, N.Y.: Cornell University Press, 1991), p. 243.

18. Sakamoto Hiroko, "Tan Sitong de *Renxue* he Wute Hengli de *Zhixin mianbing fa*," *Zhongguo zhexue* 13 (1985): 264–275. See also Schäfer, "Natural Philosophy," pp. 261–262.

19. Wright, "Tan Sitong and the Ether Reconsidered," pp. 566–575, and Schäfer, "Natural Philosophy," pp. 263–264.

20. Hu, "Merging Chinese and Western Mathematics," pp. 232–285. See also Biggerstaff, *The Earliest Modern Government Schools*, p. 171.

21. Hu, "Merging Chinese and Western Mathematics," pp. 235–245.

22. Tian Miao, "The Westernization of Chinese Mathematics: A Case Study of the *duoji* Method and Its Development," *East Asian Science, Technology, and Medicine* 20 (2003): 63–70.

23. Elman, *A Cultural History of Civil Examinations*, pp. 605–608. See also Paul Bailey, *Reform the People: Changing Attitudes towards Popular Education in Early Twentieth-Century China* (Edinburgh: Edinburgh University Press, 1990), pp. 140–141.

24. Hu, "Merging Chinese and Western Mathematics," pp. 349–364. See also Marianne Bastid, *Educational Reform in Early 20th-Century China*, trans. Paul J. Bailey (Ann Arbor: University of Michigan China Center, 1988), pp. 38–39.

25. Chen Yuanhui et al., comps., *Zhongguo jindai jiaoyu shi ziliao huibian: xuezhi yanbian* (Shanghai: Jiaoyu chubanshe, 1991), pp. 806–827. See also Su Yunfeng, *Cong Qinghua xuetang dao Qinghua daxue* (Beijing: Sanlian shuju, 2001), pp. 161–172.

26. Bastid, *Educational Reform in Early 20th-Century China*, pp. 12–13, and Y. C. Wang, *Chinese Intellectuals and the West, 1872–1949* (Chapel Hill: University of North Carolina Press, 1966), pp. 52–59. See also Elman, *A Cultural History of Civil Examinations*, pp. 585–594. Compare Reynolds, *China, 1898–1912*, pp. 41–64.

27. Ferdinand Dagenais, "Organizing Science in Republican China (1914–1950)," presented at the Ideology and Science Symposium, Center for Chinese Studies, University of California, Berkeley, October 19, 2001.

28. Yu Yue, "Xu," in Wang Renjun, *Gezhi guwei* (Wuchang: Zhixuehui, 1897), pp. 1a–2b. See Paul Unschuld, "Epistemological Issues and Changing Legitimation: Traditional Chinese Medicine in the Twentieth Century," in Charles Leslie and Allan Young, eds., *Paths to Asian Medical Knowledge* (Berkeley: University of California Press, 1992), pp. 47–49, and Zhao Hongjun, "Chinese Versus Western Medicine," pp. 24, 26. Compare Eugenia Lean, "The Modern Elixir: Medicine as

a Consumer Item in the Early Twentieth-Century Press" (M.A. paper, University of California–Los Angeles, November 1996).

29. See Yi-li Wu, "Transmitted Secrets."

30. See Leung, "Organized Medicine in Ming-Ch'ing China," pp. 134–166. Compare Kerrie MacPherson, *A Wilderness of Marshes: The Origins of Public Health in Shanghai, 1843–1893* (Oxford: Lexington Books, 2002), pp. 49–82.

31. Carl F. Nathan, "The Acceptance of Western Medicine in Early 20th Century China: The Story of the North Manchurian Plague Prevention Service," in John Bowers and Elizabeth Purcell, eds., *Medicine and Society in China* (New York: Josiah Macy Jr., Foundation, 1974), pp. 55–75, and Leung, "Medical Learning," pp. 374–398. See also Carol Benedict, "Policing the Sick: Plague and the Origins of State Medicine in Late Imperial China," *Late Imperial China* 14, 2 (December 1993): 60–77, and Ruth Rogaski, "From Protecting Life to Defending the Nation: Emergence of Public Health in Tianjin" (Ph.D. diss., Yale University, 1996). Compare Francesca Bray, "The Chinese Experience," in Roger Cooter and John Pickstone, eds., *Companion to Medicine in the Twentieth Century* (New York: Routledge, 2003), p. 724.

32. See Hsiang-lin Lei, "When Chinese Medicine Encountered the State: 1910–1949" (Ph.D. diss., University of Chicago, 1999), pp. 1–24. See also Xu, "'National Essence' vs 'Science,'" pp. 847–877, and Zhao, "Chinese Versus Western Medicine," pp. 27–28.

33. See Sivin, *Traditional Medicine*, pp. 3–4n1. Ralph Crozier, *Traditional Medicine in Modern China: Science, Nationalism, and the Tensions of Cultural Change* (Cambridge, Mass.: Harvard University Press, 1968), pp. 151–209. See aso Bridie Andrews, "Traditional Chinese Medicine as Invented Tradition," *Bulletin of the British Association for Chinese Studies* 6 (1995): 6–15, and Xu, "'National Essence' vs 'Science,'" p. 848.

34. Zhao, "Chinese versus Western Medicine," pp. 33–34. See also Andrews, "Medical Lives and the Odyssey of Western Medicine."

35. D. C. Epler, "Bloodletting in Early Chinese Medicine and Its Relation to the Origin of Acupuncture," *Bulletin of the History of Medicine* 54 (1980): 337–367. Kuriyama, *The Expressiveness of the Body*, pp. 204–206, 229–230, describes blood and *qi* as "complementary facets" of vitality.

36. Andrews, "Tailoring Tradition," pp. 154–155, 158–159. For other fields, see Laurence Schneider, "Genetics in Republican China," in J. Bowers, J. Hess, and N. Sivin, eds., *Science and Medicine in Twentieth-Century China: Research and Education* (Ann Arbor: Center for Chinese Studies, University of Michigan, 1988), pp. 3–29, and Yang Tsui-hua, "The Development of Geology in Republican China, 1912–1937," in Lin and Fu, eds., *Philosophy and Conceptual History of Science in Taiwan*, pp. 221–244.

37. Andrews, "Tuberculosis," pp. 114–157, and Shapiro, "The Puzzle of Spermatorrhea," pp. 551–596. Compare Francesca Bray, "A Deadly Disorder: Understanding Women's Health in Late Imperial China," in Bates, ed., *Knowledge and the Scholarly Medical Traditions*, p. 235–250, on amenorrhoea as blocked circulation of the vital fluids in the menstrual cycle.

38. Li Shuangbi, "Cong 'gezhi' dao 'kexue': Zhongguo jindai keji guan de yanbian guiji," pp. 105–107. Compare Masini, *The Formation of Modern Chinese Lexicon*, pp. 84–88.

39. Shigeru Jochi, "The Influence of Chinese Mathematical Texts," and Horiuchi, *Les Mathématiques Japonaises*, pp. 119–155. See also Jiang Jing, "Li Madou shijie ditu zai Riben," *Zhejiang daxue xuebao* (Renwen shehui kexue ban), 33, 5 (September 2003): 47–53.

40. See Nakamura, "Kinsei Shina no Nihon bunka," pp. 1597–1605. Compare Jurgis Elisonas (George Elison), "Christianity and the Daimyo," pp. 301–372. See also Kobayashi, "What Kind of Mathematics and Terminology," pp. 1–17.

41. See Chen Liwei, "Meiji shoki ni okeru Kan yakusho no shūyō," *Tōhōgaku* 99 (2000): 61–74, and Wang Yangzong, "1850 niandai zhi 1910 nian," pp. 139–152.

42. Wang Bing, "Jindai zaoqi Zhongguo he Riben zhi jian de wulixue jiaoliu," pp. 227–233. See also Wang, "1850 niandai zhi 1910 nian," p. 140.

43. Kume Kunitake, comp., *The Iwakura Embassy 1871–73*, vol. 5, p. 352. I have also benefited from Martin Collcutt's talk on the Iwakura mission, titled "Industrial Blueprint for Meiji Japan" (presented at the Princeton East Asian Studies Department Lunch Colloquium, February 18, 2004).

44. Nakamura, "Kinsei Shina no Nihon bunka," p. 1597, and Wang, "1850 niandai zhi 1910 nian," p. 142. Compare Wright, "The Translation of Modern Western Science," p. 671, and Masini, *The Formation of Modern Chinese Lexicon*, pp. 91–92.

45. Chen et al., comps., *Zhongguo jindai jiaoyu shi ziliao huibian*, pp. 197–232. See Wang, "1850 niandai zhi 1910 nian," pp. 139–144; Tsien, "Western Impact on China," pp. 323–325; and Masini, *The Formation of Modern Chinese Lexicon*, pp. 104–108.

46. See Zhao Yuanyi, ed., *Gezhi shuyuan jiawu keyi* (Shanghai, 1897). See also Wang, "1850 niandai zhi 1910 nian," pp. 144–145, and ECCP, p. 30. Compare Reynolds, *China, 1898–1912*, pp. 48, 58–61, and Keenan, "Beyond the Rising Sun," p. 157.

47. Chen et al., comps., *Zhongguo jindai jiaoyu shi ziliao huibian*, pp. 233–287, and Wang, "1850 niandai zhi 1910 nian," p. 146–147. See also Xiaoping Cong, "Localizing the Global, Nationalizing the Local: The Role of Teacher's Schools in Making China Modern" (Ph.D. diss., University of California–Los Angeles, 2001).

48. Masini, *The Formation of Modern Chinese Lexicon*, pp. 145–151.

49. Timothy Weston, *The Position of Power: Beijing University, Intellectuals, and Chinese Culture, 1898–1929* (Berkeley: University of California Press, 2004), pp. 50–52.

50. Wang, "1850 niandai zhi 1910 nian," pp. 147–150.

51. Wang Bing, "19 shiji zhongqi zhi 20 shiji chuqi Zhongguo he Riben de wulixue," *Ziran kexue shi yanjiu* 13, 4 (1994): 326–329.

52. GZQM, 1886, vol. 10, pp. 1a–1b. See Crosbie Smith, "The Transmission of Physics from France to Britain, 1800–1840," *Historical Studies in the Physical Sciences* 9 (1978): 1–62. Compare Iwo Amelung, "Naming Physics: The Strife to Delineate a Field of Modern Science in Late Imperial China," in Lackner and Vittinghoff, eds., *Mapping Meanings*, pp. 381–422.

53. Marianne Bastid-Bruguière, "L'enseignement des sciences à l'Université métropolitaine de Pékin, 1898–1911" (presented at Education et Instruction en China conference, Paris, June 29–July 2, 1999).

54. Wang, "19 shiji zhongqi zhi 20 shiji," pp. 329–332. Compare Reynolds, *China, 1898–1912,* pp. 65–110, 131–150.

55. Wang, "19 shiji zhongqi zhi 20 shiji," pp. 331–332.

56. Wang Bing, "On the Physics Terminology in Chinese and Japanese during Early Modern Times," in Kim and Bray, eds., *Current Perspectives in the History of Science,* pp. 517–521.

57. Wang, "19 shiji zhongqi zhi 20 shiji chuqi," pp. 332–334. See also Zhang Rongmei, "Nihon ni okeru 'butsuri gaku' to iu yakugo no keisei," *Kanazawa daigaku daigakuin shakai kankyō kagaku kenkyū* 2 (1997): 83–95.

58. Dagenais, "Organizing Science in Republican China," pp. 1–31. See also Jia Sheng, "The Origins of the Science Society of China, 1914–1937" (Ph.D. diss., Cornell University, 1995), pp. 11–12.

59. Chen et al., comps., *Zhongguo jindai jiaoyu shi ziliao huibian,* pp. 608–650. Compare Hildred Geertz," An Anthropology of Religion and Magic, I," *Journal of Interdisciplinary History* 6, 1 (Summer 1975): 71–89. She notes, "It is not the 'decline' of the practice of magic that cries for explanation, but the emergence and rise of the label 'magic.'"

60. See Elman, "The Formation of 'Dao Learning' as Imperial Ideology," pp. 58–82.

61. Elman, *A Cultural History of Civil Examinations,* pp. 608–625. See also Luo Zhitian, "Zouxiang guoxue yu shixue de 'Sai xiansheng,'" *Jindai shi yanjiu* (2000.3): 59–94, and Weston, *The Position of Power,* p. 83.

62. See Ding Wenjiang, *Liang Rengong xiansheng nianpu changbian chugao,* 2 vols. (Taibei, Shijie shuju, 1972), pp. 551–574. For African and Indian critiques, see Michael Adas, "Contested Hegemony: The Great War and the Afro-Asian Assault on the Civilizing Mission Ideology," *Journal of World History* 15, 1 (March 2004): 31–63.

63. For the *Ouyou xinying lu jielu,* see Liang Qichao, *Yinbing shi zhuanji,* 10 vols. (Taibei: Zhonghua shuju, 1972), vol. 7, pp. 10–12. For discussion, see Chow Tsetsung, *The May 4th Movement: Intellectual Revolution in Modern China* (Cambridge, Mass.: Harvard University Press, 1960), pp. 327–329, and Jerome Grieder, *Hu Shih and the Chinese Renaissance: Liberalism in the Chinese Revolution, 1917–1937* (Cambridge, Mass.: Harvard University Press, 1970), pp. 129–135.

64. See the "Review Symposia" on Needham's work in *Isis* 75, 1 (1984): 171–189.

65. See Lloyd and Sivin, *The Way and the Word.*

66. Drawing on Joseph Needham, SCC, vol. 3, p. 437, James McClellan and Harold Dorn, *Science and Technology in World History: An Introduction* (Baltimore: Johns Hopkins University Press, 1999), pp. 117–140, conclude, "With the arrival of [Matteo] Ricci in China the subsequent history of Chinese science largely becomes its integration into ecumenical, world science." Compare Arnold Pacey, *Technology in World Civilization: A Thousand-Year History* (Cambridge, Mass.: MIT Press, 1991), pp. 94–97, on "blocked technical systems," and Lewis Pyenson and Susan Sheets-Pyenson, *Servants of Nature: A History of Scientific Institutions, Enterprises, and Sensibilities* (New York: W. W. Norton & Co., 2000), pp.

382–386. See also Michel Paty, "Comparative History of Modern Science and the Context of Dependency," *Science, Technology, & Society* 4, 2 (1999): 171–204, and Roy MacLeod, ed., *Nature and Empire*, special issue of *Osiris* 15 (2001). Paty draws on Petitjean, Jami, and Moulin, eds., *Science and Empires*, but concludes that alienated Chinese elites failed to respond to the modern natural sciences transmitted by the Jesuits. The Jesuits transmitted Renaissance mathematics and Aristotelian natural studies—not modern science—to China.

Appendix 1

1. See Man-Keung and Volkov, "Official Curriculum in Traditional Chinese Mathematics: How Did Candidates Pass the Examinations?" pp. 85–99.
2. See Cullen, trans., *Astronomy and Mathematics in Ancient China*, pp. 171–205, and Li and Du, *Chinese Mathematics*, pp. 25–32.
3. See Shen Kangshen, John N. Crossley, Anthony Lun, trans., *The Nine Chapters on the Mathematical Art: Companion and Commentary* (Oxford: Oxford University Press; Beijing, People's Republic of China: Science Press, 1999), and Li and Du, *Chinese Mathematics*, pp. 33–56. See also Donald B. Wagner, "An Early Derivation of the Volume of a Pyramid: Liu Hui, Third Century A.D.," *Historia Mathematica* 6 (1979): 164–188.
4. Swetz, trans., *The Sea Island*. See Wylie, *Notes on Chinese Literature*, p. 114, and Li and Du, *Chinese Mathematics*, pp. 75–80.
5. Wylie, *Notes on Chinese Literature*, p. 114, and Martzloff, *A History of Chinese Mathematics*, pp. 136–138. See also Li and Du, *Chinese Mathematics*, pp. 92–95.
6. Li and Du, *Chinese Mathematics*, pp. 95–97; Wylie, *Notes on Chinese Literature*, p. 114; and Martzloff, *A History of Chinese Mathematics*, p. 124.
7. Li and Du, *Chinese Mathematics*, pp. 97–98; Wylie, *Notes on Chinese Literature*, p. 115; and Martzloff, *A History of Chinese Mathematics*, p. 141. See also Qian Baocong, comp., "Xu" (Preface) to *Suanjing shishu*, 2 vols. (Beijing: Zhonghua shuju, 1963), pp. 4–5.
8. Li and Du, *Chinese Mathematics*, pp. 98–100; Wylie, *Notes on Chinese Literature*, p. 115; and Martzloff, *A History of Chinese Mathematics*, pp. 124, 138–139.
9. Wylie, *Notes on Chinese Literature*, p. 115, and Martzloff, *A History of Chinese Mathematics*, pp. 124, 140.
10. Li and Du, *Chinese Mathematics*, pp. 100–104; Wylie, *Notes on Chinese Literature*, p. 115–116; and Martzloff, *A History of Chinese Mathematics*, pp. 125, 140–141.
11. Wylie, *Notes on Chinese Literature*, p. 114, and Martzloff, *A History of Chinese Mathematics*, pp. 125, 141.

Appendix 4

1. See Liu Dajun, "Jiangnan zhizao zongju," p. 46. Compare K. H. Kim, *Japanese Perspectives on China's Early Modernization: A Bibliographical Survey* (Ann Arbor: University of Michigan, Center for Chinese Studies, 1974), pp. 3–12. The arsenals, shipyards, and machine shops were also listed in the journal *Review of the Times* (*Wan'guo gongbao*), in the April–May 1891 issues.

Appendix 5

1. On the geography texts, see the discussion in Zhenhuan, *Wan Qing Xifang dilixue zai Zhongguo,* pp. 126–133.

Appendix 8

1. The translation for this question is taken from Fryer, "Chinese Prize Essays," p. 100. Fryer's "Ching-kang-ching" is the Later Han classicist, Zheng Kangcheng, i.e., Zheng Xuan (127–200), whose classical importance Fryer did not seem to recognize.

Bibliography of Chinese and Japanese Sources

Ayano Hiroyuki, "H. E. Roscoe to Owens College, Manchester no seido tenkan" (Roscoe and the institutional change at Owens College, Manchester), *Kagakushi kenkyū* 8, 212 (1999).

———, "Jūkyū seiki koban no England ni okeru kagaku no kaikaku undō" (Roscoe and the science reform movement in England in the second half of the nineteenth century), *Kagakushi kenkyū* 36, 204 (1997).

Bastid, Marianne, "Fuzhou chuanzheng ju de jishu yinjin" (Introduction of technology in the Fuzhou Shipyard), in *Suxian ji: jinian Shao Xunzheng xiansheng xueshu lunwen ji* (Collection of Jasmine: Scholarly essays honoring Mr. Shai Xunzheng). Beijing: Beijing daxue chubanshe, 1994.

Chen Li, *Dongshu dushu ji* (Chen Li's reading notes). Taibei: Shangwu yinshuguan, 1970.

Chen Liwei, "Meiji shoki ni okeru Kan yakusho no shūyō" (The acceptance of Chinese translations in the early Meiji period), *Tōhōgaku* 99 (2000).

Chen Meidong, *Zhongguo gu xing tu* (Star charts in ancient China). Shenyang: Liaoning jiaoyu chubanshe, 1996.

Chen Qiwei, "Zaitan Wang Tao he Gezhi shuyuan dui Xifang zhexue de jieshao" (Review of Wang Tao and the Shanghai Polytechnic concerning the introduction of Western philosophy), *Dongyue luncong* 22, 5 (2001.9).

Chen Yuanhui et al., comps., *Zhongguo jindai jiaoyu shi ziliao huibian: xuezhi yanbian* (Compendium of sources on the history of Chinese modern education: Changes in educational institutions). Shanghai: Jiaoyu chubanshe, 1991.

Cheng Dawei, *Suanfa tongzong* (Systematic treatise on computational methods). New ed. Xin'an: Saoye shanfang, 1883.

Cheng Yaotian, "Lunxue xiaoji" (Small notes on learning) in *Tongyi lu* (Record of comprehensive arts). Late Qianlong edition.

Cheng Yi, *Erh Cheng yishu* (Bequeathed writings of Cheng Yi and Chang Hao). SKQS edition.

Chu Pingyi, "Shenti, linghun yu tianzhu: Mingmo Qingchu Xixue zhong de renti shengli zhishi" (Body, soul and the Heavenly ruler: late Ming and early Qing knowledge of anatomy and physiology), *Xin shixue* 7, 2 (1996).

———, "Kua wenhua zhishi chuanbo de ge an yanjiu—Mingmo Qingchu guanyu diyuan shuo de zhengyi, 1600–1800" (The formation of factual knowledge in transcultural scientific transactions: the debate over the sphericity of the earth in

China, 1600–1800), *Zhongyang yanjiu yuan lishi yuyan yanjiu suo jikan* (Taiwan) 69, 3 (1998).

———, "Tongguan Tianxue, yixue yu ruxue: Wang Honghan yu Ming-Qing zhi ji Zhongxi yixue de jiaohui" (Medicine East and West: Wang Honghan's synthesis of medicine, Christianity, and Confucianism), *Lishi yuyan yanjiu suo jikan* (Academia Sinica, Taiwan) 70, 1 (1999).

———, "Sanjiao hanshu biao yu Mingmo de Zhongxi lifa zhi zheng—shang xia" (Trigonometric tables and late Ming debates concerning Western versus Chinese mathematical astronomy, parts 1 and 2), *Dalu zazhi* 99, 5 (1999), and 99, 6 (1999).

Chuanshi yanjiu (Marine history research) 8 (1995), and 10 (1996).

Dai Zhen quanji (Dai Zhen's complete collection). Beijing: Qinghua daxue chuban she, 1994.

Dai Zhen, "Liyan" (Guidelines), in *Fenzhou fuzhi* (Gazetteer of Fenzhou prefecture). 1771 edition.

Dai Zhen wenji (Dai Zhen's collected essays). Hong Kong: Zhonghua shuju, 1974.

Ding Wenjiang, *Liang Rengong xiansheng nianpu changbian chugao* (First draft of a chronological biography of Liang Qichao). 2 vols. Taibei, Shijie shuju, 1972.

Du Weiyun, *Qing Qianjia shidai zhi shixue yu shijia* (Historians and historical studies in the Qianlong and Jiaqing eras). Taibei: Wenshi zongkan, 1962.

Du Zhengsheng, "Gudai wuguai zhi yanjiu (shang) (First part of a study of ancient things that were strange), *Dalu zazhi* 104, 1 (January 15, 2002).

Duan Zhiwen, *Zhongguo xiandai kexue wenhua de xingqi* (The rise of Chinese contemporary scientific culture). Shanghai: Shanghai renmin chuban she, 2001.

Edkins, Joseph, and Li Shanlan, trans., *Zhongxue* (An elementary treatise on mechanics). First printing of 1859 destroyed. Reissued by the Nanjing Arsenal in 1867.

———, "Xu" (Preface), in GZQM (1886), vol. 1.

———, "Xu" (Preface) to the *Bianxue qimeng* (Primer on logic), p. 1a, in GZQM (1886), vol. 14.

Fang Dongshu, *Hanxue shangdui* (An assessment of Han Learning). Taibei: Commercial Press, 1968.

Fang Yizhi, *Wuli xiaozhi* (Preliminary record of phenomena and their patterns of occurrence). SKQS edition.

Feng Jinrong (Fung Kam-wing), "Mingmo Xiong Mingyu Gezhi cao neirong tanzhe" (Critical study of Xiong Mingyu's Draft for investigating things and extending knowledge), *Ziran kexue shi yanjiu* 16, 4 (1997).

Feng Yingjing, *Huang Ming jingshi shiyong bian* (Compendium of practical statecraft in the august Ming dynasty). Xiuning, Anhui: Ming Wanli edition, 1603.

Fujitsuka Chikashi (Rin), *Nichi Sen Shin no bunka kōryū* (Cultural exchange between Japan, Korea, and Qing China). Tokyo: Chūbunkan shoten, 1947.

———, *Shinchō bunka tōden no kenkyū* (Research on the eastward transmission of Qing dynasty culture). Tokyo: Kokusho kankōkai, 1975.

Gezhi keyi huibian (Compendium of prize essays on science), collected and edited by Wang Tao. Shanghai: Shanghai Polytechnic, 1897.

Gezhi yiwen huibao (*I-Wen-Lou et Revue Scientifique*). Nos. 1–50 in 4 vols. Shanghai: Catholic Mission, 1898–1899.

Gujing jingshe wenji (Prose collection of the Gujing jingshe), edited by Juan Yuan et al. Taipei, Commercial Press, 1966.

Guangxu chao donghua lu (Records from within the Eastern Gate during the Guangxu reign). Shanghai: Zhonghua shuju, 1909.

Guo Shuanglin, *Xichao jiaodang xia de wan Qing dilixue* (Western currents in late Qing geography). Beijing: Beijing University Press, 2000.

Hamaguchi Fujiō, *Shindai kōkyogagu no shisō shi teki kenkyū* (Research on the intellectual history of Qing dynasty evidential studies). Tokyo: Kokusho kankōkai, 1994.

Han Qi, "Cong Mingshi lizhi de zuanxiu kan Xixue zai Zhongguo de chuanbo" (A glance at the transmission of Western learning in China from the compilation of the calendrical treatise of the Ming History), in Liu Dun, ed., *Keshi xinchuan*. Shenyang: Liaoning jiaoyu chuban she, 1997.

———, "*Shuli gezhi* de faxian—jian lun 18 shiji Niudun xiangguan zhuzuo zai Zhongguo de chuanbo" (The Discovery of the Mathematical Principles of Natural Philosophy—A discussion of the transmission of related works in China on Newton in the 18th century), *Zhongguo keji shiliao* 19, 2 (1998).

———, "Bai Jin de Yijing yanjiu he Kangxi shidai de Xixue Zhongyuan shuo" (Bouvet's research on the Change Classic and the theory that Western learning originated in China), *Hanxue yanjiu* (Taiwan) 16, 1 (June 1998).

———, Chuanjiaoshi Weilie Yali zai Hua de kexue huodong" (The missionary Alexander Wylie's scientific activities in China), *Ziran bianzheng fa tongxun* 20, 114 (1998.2).

———, *Zhongguo kexue jishu de xichuan ji qi yingxiang* (The transmission of Chinese science and technology to the West and its influence). Beijing: Hebei chubanshe, 1999.

———, "Gewu qiongli yuan yu Mengyang zhai: shiqi shiba shiji zhi Zhong-Fa kexue jiaoliu" (The Academy of Science and Studio for the Cultivation of Youth: Sino-French scientific exchange in the seventeenth and eighteenth centuries), *Faguo hanxue* (Beijing) 4 (2000).

———, "'Zili' jingshen yu lisuan huodong: Kang-Qian zhi ji wenren dui Xixue taidu zhi gaibian ji qi beijing" (A spirit of self-reliance and computational astronomy activities: the background for changes in literati attitudes toward Western learning during the Kangxi and Qianlong eras), *Ziran kexue shi yanjiu* 21, 3 (2002).

Hanshu (History of the Former Han dynasty). Compiled by Ban Gu. 7 vols. Beijing, Zhonghua Bookstore, 1962, and Taibei: Shixue chuban she, 1974.

Hashimoto Keizō, "Bai Buntei no rekisan gaku" (Mei Wending's computational astronomy), *Tōhō gakuhō* (Kyoto) 41 (March 1970).

———, "Rekisho Kōsei no seiritsu" (The formation of the Compendium of Observational and Computational Astronomy), in Yabuuchi Kiyoshi and Yoshida Mitsukuni, eds., *Min Shin jidai no kagaku gijutsu shi* (History of science and technology in the Ming and Qing periods). Kyoto: Research Institute for Humanistic Studies, 1970.

Hatano Yoshihiro, "Chūgoku kindaishi ni kansuru mitsu no mondai—Chūgoku no kindaika wa naze okureta ka" (Three questions concerning modern Chinese history—Why was China's modernization late?), *Nagoya daigaku bungakubu kenkyū ronshū* 7 (1958).

————, *Chūgoku kindai kōgyō shi no kenkyū* (Research on the history of modern Chinese industry). Kyoto: Tōyōshi kenkyūkai, 1960.

He Chuanxin, "Mingmo Qingchu huihua de fanggu feng" (The craze of emulating the ancients in late Ming and early Qing painting), *Gugong wenwu yuekan* 249 (December 2003).

Hobson, Benjamin (He Xin), *Bowu xinbian* (Treatise of natural philosophy). Guangzhou: Hum-le-fow Hospital, 1871, and Shanghai: Mohai shuguan, 1855.

————, and Guan Maocai, *Xiyi lue lun* (Summary of Western medicine). Shanghai: Renji yiguan, 1857.

Hong Wansheng, "Tongwen guan suanxue jiaoxi Li Shanlan" (The mathematics professor Li Shanlan at the School of Translated Learning), in Yang Cuihua and Huang Yinong, eds., *Jindai Zhongguo keji shi lunji* (Collected essays on modern Chinese history of science and technology) (Taibei: Academia Sinica, Taiwan, Institute of Modern History, and National Qinghua University, Institute of History, 1991).

Hong Shaobin, ed., *Ding Wenjiang*. Beijing: Zhongguo qingnian chuban she, 1999.

Hu Shi, "Qingdai xuezhe de zhixue fangfa" (Scholarly methods of Qing dynasty scholars), in *Hushi wencun* (Preserved writings of Hu Shi) (4 vols. Taibei: Yuandong dashu gongsi, 1968), vol. 1.

Hu Wangjin, "Ba" (Afterword), in Wang Renjun, *Gezhi guwei*.

Huainan honglie jijie (Collected notes to the illustrious writings of the master of Huainan), compiled ca. 135 B.C. under the auspices of Liu An, the Prince of Huainan. Taibei: Commercial Press, 1974.

Huang Ming ceheng (Weighing civil policy examination essays during the Ming dynasty), compiled by Mao Wei. Wu-hsing, 1605 edition.

Huang Yilong (Yinong), "Tang Ruowang yu Qingchu xili zhi zhengtong hua" (Adam Schall and the legitimation of the Western calendar in the early Qing), in Wu Jiali and Ye Hongsa, eds., *Xinbian Zhongguo keji shi*. Taibei: Yinmu wenhua shiye gongsi, 1990.

————, "Qingchu tianzhujiao yu huijiao tianwenjia de chengdou" (The struggle between Catholic and Muslim astronomers in the early Qing), *Jiuzhou xuekan* 5, 3 (1993).

————, "Long yu shi duiwang de shijie—yi Majiaerni shituan fanghua hou de chubanwu wei li" (The world when the dragon encountered the lion: an example of publications after the Macartney mission to China), *The National Palace Museum Research Quarterly* (*Gugong xueshu jikan*, Taiwan) 21, 2 (Winter 2003).

Huang Zheng, "Fujian chuanzheng ju lishi diwei lunlue" (Synopsis of the historical status of the Fujian Shipyard), *Chuanshi yanjiu* 3 (1987).

Huangchao wenxian tongkao (Comprehensive analysis of civil and military institutions during the Qing dynasty), in *Shitong* (The ten comprehensive encyclopedias of civil and military governance). Shanghai: Commercial Press, 1935–1937.

Hui Shiqi, *Daxue shuo* (Explanations for the Great Learning). Langai shushi, Jiaqing edition, 1796–1820.

Huo Rusu, "Zaochuan gongye zai Zhongguo jindai shi shang de lishi diwei" (The historical status of the shipbuilding industry in modern Chinese history), *Chuanshi yanjiu* 3 (1987).

Itō Shūichi, "Kindai Chūgoku ni okeru kagaku gijutsu no chii—sono shisōshi teki kōsatsu" (The place of science and technology in modern China—an examination of its intellectual history), *Tōyō gakujutsu kenkyū* 5, 5–6 (1967).

Ji Wende, *Cong Siku quanshu tanjiu Ming Qing jian shuru zhi Xixue* (Inquiring into the importation of Western learning during the Ming-Qing transition from the point of view of the Complete Collection of the Four Treasuries). Taibei: Han Mei tushu youxian gongsi, 1991.

Jiang Jing, "Li Madou shijie ditu zai Riben" (Matteo Ricci's World Map in Japan), *Zhejiang daxue xuebao* (Renwen shehui kexue ban), 33, 5 (September 2003).

Jiangnan zhizaoju ji (Records of the Jiangnan Arsenal), compiled by Wei Xiangong. 1904. Reprint, Taibei: Wenhai chuban she.

Jiao Xun, *Diaogu ji* (Collected writing from [the Studio of] Engraved Bamboo). Daoguang edition, 1821–1850.

Kangxi jixia gewu bian (Compilation on investigating things by the Kangxi emperor in leisure). Tongxuezhai edition, n.d.

Kitayama Yasuo, "Chūgoku ni okeru kan'ei gunji kōgyō No ichi kōsatsu—Kōnan seizōkyoku o chūshin to shite" (Study of the government-operated armaments industry—with focus on the Jiangnan Arsenal), *Historia* 9 (1954).

Li Bin, "Xishi wuqi dui Qingchu zuozhan fangfa de yingxiang" (The impact of Western-style weapons on combat techniques in the early Qing), *Ziran bianzheng fa tongxun* 24, 4 (2002).

Li Changli, "Wan Qing dui Xixue de liangzhong wudu—lun Zhong Tianwei de Xixue guan" (Two kinds of misreading of Western learning in the late Qing—On Zhong Tianwei's view of Western Learning), *Jiangsu shehui kexue* (Nanjing) 6 (1999).

Li Hongzhang, "Xu" (Preface), in GZQM, vol. 1.

Li Jianmin, "*Bencao gangmu huobu* kaoshi" (Explanation of the fire section in the *Systematic Materia Medica*), *Lishi yuyan yanjiu suo jikan* (Academia Sinica, Taiwan) 73, 3 (September 2002).

Li Ruchen, *Jinghua yuan* (*Flowers in the mirror*). Taibei: Xuehai chuban she, 1985.

Li, San-pao, "Letters to the Editor in John Fryer's *Chinese Scientific Magazine, 1876–1892*: An Analysis," *Jindai shi yanjiu suo jikan* (Academia Sinica, Taiwan) 4 (1974).

Li Shizhen, "Fanli" ("Conventions"), in Li, *Bencao gangmu* (Systematic materia medica). Reprint of 1596 first edition. 4 vols. Beijing: Renmin weisheng chuban she, 1977–1981. References to the 1603 and SKQS edition are also made. See also the following convenient, new edition: *Bencao gangmu*. 2 vols. Beijing: Huaxia chuban she, 2002.

Li Shuangbi, "Cong 'gexhi' dao 'kexue': Zhongguo jindai keji guan de yanbian guiji" (From 'Investigating things and extending knowledge' to Science: Tracking the conceptual transformation in Chinese modern science and technology), *Guizhou shehui kexue* 137 (1995.5).

Li Yan, "Qingdai shuxue jiaoyu zhidu" (The educational system for mathematics in the Qing period), *Xueyi* 13, 4 (1934).

Liang Qichao, "Gezhi xue yan'ge kaolue" (Synopsis of the vicissitudes in the history of science), in Liang, *Yinbing shi wenji* (Collected writings from the Ice-Drinker's Studio). 8 vols. Taibei: Zhonghua Bookstore, 1970, vol. 2.

————, *Yinbing shi zhuanji* (Collected works from the Ice-Drinker's Studio). 10 vols. Taibei: Zhonghua shuju, 1972.

Liao Zonglin, *Zhong Fa zhanzheng* (The Sino-French War). Tianjin: Guji chubanshe, 2002.

Liezi duben (Liezi reader), edited by Zhuang Wanshou. Taipei: Sanmin shuju, 1996.

Lin Yishan, "Xu" (Preface), in Wang Renjun, *Gezhi guwei*.

Lin Zhiping, ed., *Jidujiao yu Zhongguo xiandai hua* (Protestantism and Chinese modernization). Taibei: Caituan faren Jidujiao yuzhou guang chuanbo zhongxin chuban she, 1994.

Ling Tingkan, *Jiaoli tang wenji* (Collected essays from the Hall of Collating Rituals). Beijing: Zhonghua shuju, 1998.

Liu Dajun, "Jiangnan zhizao zongju cong chengli dao 'juchang fenli' shi de yanbian" (Changes in the Jiangnan Arsenal from its time of establishment to the separation of its dockyard), *Chuanshi yanjiu* 3 (1987).

Liu Dun, "*Shuli jingyun* zhong *Jihe yuanben* de diben wenti" (The problem of the edition of Euclid's *Elements* used in the *Shuli jingyun*), *Zhongguo keji shiliao* 12, 3 (1991).

Liu Guangding, "*Gewu tanyuan* yu Wei Lianchen de Zhongwen zhuzuo" (*Natural Theology* and Williamson's Chinese works), in Yang Cuihua and Huang Yinong [Yilong], eds., *Jindai Zhongguo keji shi lunji*. Taibei: Academia Sinica, Institute of Modern History, and Qinghua University, Institute of History, 1991.

Liu Lu, ed., *Qinggong xiyang yiqi* (Scientific and Technical Instruments of the Qing dynasty). Hong Kong: Commercial Press, 1998.

Liu Shipei, "Gewu jie" (Explanation of the investigation of things), in *Zuo'an waiji* (Outer collection from the Left Cloister), in *Liu Shenshu xiansheng yishu* (Bequeathed works of Mr. Liu Shipei). 4 vols. Taibei: Daxin shuju, 1965, vol. 3.

————, "Gewu jie" (Explanation of the investigation of things), in *Zuoyi ji* (Collection from the Left Cloister), in *Liu Shenshu xiansheng yishu*, vol. 3.

Lu Shiyi, "Sibianlu lunxue" (Discussions of learning from the record of thought and debate), in Wei Yuan (1794–1856), ed., *Huangchao jingshi wenbian* (Collected writings on statecraft from the Qing dynasty). Reprint of 1827 and 1873 edition. Taibei: World Bookstore, 1964.

Lü Lingfeng, "Mingmo ji Qingdai jiaoshi jilu yanjiu" (Research on eclipse records at the end of the Ming and during the Qing dynasty). Hefei, Anhui: Ph.D. diss. in History of Science and Technology from the University of Science and Technology of China, 2002.

Lunyu yinde (Concordance to the Analects). Reprint. Taibei: Chinese Materials and Research Aids Service Center, Inc., 1966.

Luo Zhitian, "Zouxiang guoxue yu shixue de 'Sai xiansheng'" (Heading toward the "Mr. Science" of national studies and history), *Jindai shi yanjiu* (2000.3).

Ma Boying et al., *Zhongwai yixue wenhua jiaoliu shi* (History of Chinese-foreign medical exchanges). Shanghai: Wenhui chubanshe, 1993.

Martin, W. A. P., *Gewu rumen* (Elements of natural philosophy and chemistry). Beijing: Tongwen guan, 1868.

Mei Wending, "Lun lixue gushu jinmi" (On mathematical astronomy as imprecise in antiquity and precise today), in *Lixue yiwen* (Queries on mathematical astronomy), in *Meishi congshu jiyao* (Essential collectanea of Master Mei), compiled by Mei Juecheng. Qing Qianlong edition, 1761.

———, "Lun Zhongtu lifa de chuanru Xiguo zhi you" (The reason why Chinese native mathematical astronomy was transmitted to the Western countries), in *Lixue yiwen bu shang* (Supplement to Queries on mathematical astronomy, part 1), in SKQS.

———, "Lun Zhoubi zhong ji you diyuan zhi li" (The theory of the sphericity of the earth was contained in the *Zhou Gnomon*), in SKQS, vol. 794.

———, *Jixue tang shiwen chao* (Essays from the Hall of accumulating learning). Hefei: Huangshan shushe, 1994.

———, "Lun Zhong-Xi erfa zhi tong" (On the similarities between the two methods of China and the West), in *Lixue yiwen* (Queries on mathematical astronomy), in *Meishi congshu jiyao* (Essential collectanea of Master Mei), compiled by Mei Juecheng. Qing Qianlong xinssu, 1761.

Mengzi yinde (Concordance to the Mencius). Reprint. Taibei: Chengwen Bookstore, 1966.

Mingdai dengke lu huibian (Compendium of Ming dynasty civil and military examination records). 22 vols. Taibei: Xuesheng Bookstore, 1969.

Mingdai taoci daquan (Great collection of Ming porcelain). Taibei: Yishujia chuban she, 1983.

Mingshi (Ming History), compiled by Zhang Tingyu et al. Beijing: Zhonghua shuju, 1974.

Miyazaki Ichisada, "Shōshō kyoku no ryakushi—Chūgoku no dokusenteki kisen kaisha" (A Short history of the China Merchants' Company—China's Monopolistic Steamship Company), *Tōyōshi kenkyū* 11, 2 (March 1951).

Mori Tokihiko, "Liang Qichao de jingji sixiang" (Liang Qichao's economic thought), in Hazama Naoki, ed., *Liang Qichao, Mingzhi Riben, Xifang*. Beijing: Shehui kexue wenxian chuban she, 2001.

Nakamura Kyūshirō, "Kinsei Shina no Nihon bunka ni oyoboshitaru seiryoku eikyō" (The impact of early modern China on Japanese culture), *Shigaku zasshi* 25, 12 (December 1914), and 26, 2 (February 1915).

Nakamura Shōhachi, "Chūgoku shisōshi jō ni okeru jutsusu" (Numerological skills in the history of Chinese thought), *Chūgoku kankei ronsetsu shiryō* 39, I (1997).

Ōba Osamu, *Edo jidai ni okeru Chūgoku bunka juyō no kenkyū* (Research on the acceptance of Chinese culture in the Edo period). Tokyo: Dōbōsha, 1984.

Onogawa Hidemi, *Shimmatsu seiji shisō kenkyū* (Research on late Qing political thought). Tokyo: Misuzu shobō, 1969.

Onoue Etsuzō, "Chūgoku ni okeru kōgyō no ishoku—yōmu undō no ichi sokumen" (Transplanting modern industry to China—An aspect of the western affairs movement), *Rokkōdai ronshū* 5, 3 (October 1958).

Onozawa Seiichi et al., eds., *Ki no shisō: Chūgoku ni okeru shizenkan to ningenkan no tenkai* (The idea of *Qi:* The development of a view of nature and man). Tokyo, Tokyo University Press, 1978.

Pan Shicheng, ed., *Haishan xian'guan congshu* (Collectanea of the Wizard Pavilion in Haishan). Shanghai, n.p., 1852.

Qian Baocong, comp., "Xu" (Preface) to *Suanjing shishu* (Ten Mathematical Classics). 2 vols. Beijing: Zhonghua shuju, 1963.

———, ed., *Zhongguo shuxue shi* (History of Chinese mathematics). Beijing: Kexue chubanshe, 1964.

———, *Qian Baocong kexue shi lunwen xuanji* (Selected essays on history of science by Qian Baozong). Beijing: Kexue chuban she, 1983.

Qian Daxin, *Qianyan tang wenji* (Collected essays of the Hall of Subtle Research). Shanghai: Guji chuban she, 1989.

Qingshi gao (Draft history of the Qing dynasty), compiled by Chao Erh-hsun et al. 48 vols. Beijing: Zhonghua shuju, 1977.

Quan Hansheng, "Qingmo de 'Xixue yuanchu Zhongguo' shuo" (The late Qing theory that 'Western learning originated in China'), *Lingnan xuebao* 4, 2 (June 1935).

———, "Qingmo Xiyang yixue chuanru shi guoren suo chi de taidu" (Late Qing Chinese attitudes toward the transmission of Western medicine), *Shihuo yuekan* 3, 12 (December 1936).

———, "Qingji de Jiangnan zhizaoju" (The Qing period Jiangnan Arsenal), *Bulletin of the Institute of History and Philology, Academia Sinica* 23 (1951).

Rong Zhaozu, "Yan Ruoju de kaozhengxue" (Yan Ruoju's evidential research scholarship), *Lingnan xuebao* 1, 4 (1930).

Rong Xinjiang, "Yige renshi Tangchao de Bosi jingjiao jiazu" (A Tang dynasty elite Persian Nestorian family), in Ye Yiliang, ed., *Yilangxue zai Zhongguo lunwen ji*, vol. 2. Beijing: Beijing daxue chuban she, 1998.

Ruan Yuan, "Chouren zhuan fanli" (Conventions of the Biographies of mathematical astronomers), in CRZ.

———, "Xu" (Preface), in Zhu Shijie, *Suanxue qimeng* (Primer of mathematical calculations). Reprint. Shanghai: Jiangnan Arsenal, 1871.

———, ed., *Huang Qing jingjie* (Qing exegesis of the classics). 20 vols. Taibei, Fuxing Reprint of the 1892 edition, 1961.

———, "Daxue gewu shuo" (The theory of the investigation of things in the Great Learning), in Ruan, *Yanjing shi ji* (Collection from the Studio for Investigating Classics). 3 vols. Taibei: Shijie shuju, 1964.

Saitō Tetsurō, "Chi no ryōan—Chūgoku no 'kagaku to jinseikan' ronsō o chūshin ni" (Two sides of knowledge: focusing on the Chinese debate over 'science and outlooks on life'), *Chūgoku—shakai to bunka* 8 (1993).

Sakade Yoshinobu, "Kaisetsu—Mindai nichiyō ruisho ni tsuite" (Explanations—Concerning Ming dynasty daily use compendia), in Sakade Yoshinobu et al., *Chūgoku nichiyō ruisho shūsei* (Collection of Chinese daily use compendia) (14 vols. Tokyo: Kyūko shoin, 1999–2003), vol. 1.

———, Rokugo sōtan ni mieru kagaku kiji" (Records of chemistry in the LHCT), *Kagakushi kenkyū* 93 (1970).

Sakai Tadao, "Mindai no nichiyō ruisho to shomin kyōiku" (Ming dynasty daily use compendia and commoner education), in Hayashi Tomoharu, ed., *Kinsei Chūgoku kyōikushi kenkyū* (Research on education in early modern China). Tokyo: Kokutosha, 1958.

Sakamoto Hiroko, "Tan Sitong de *Renxue* he Wute Hengli de *Zhixin mianbing fa*" (Tan Sitong's Study of Benevolence and Henry Wood's Method of Avoiding Illness by Controlling the Mind), *Zhongguo zhexue* 13 (1985).

Sano Seiko, "Gosei qi to qikai sho—i o meguru shiten no sō i" (Treatises on the five agents and records of anomalies: different perspectives on strange phenomena), *Tōhō gaku* 104 (July 2002).

Shang Zhicong, "1886–1894 nianjian jindai kexue zai wan Qing zhishi fenzi zhong de yingxiang—Shanghai Gezhi shuyuan Gezhi lei keyi fenxi" (Impact of modern science on late Qing intellectuals, 1886–1894—analysis of the Classified Essay Contests held by Shanghai Polytechnic), *Qingshi yanjiu* 3 (August 2001).

———, "Nan Huairen Qiongli xue de zhuti neirong yu jiben jiegou" (Main contents and basic framework of Ferdinand Verbiest's Studies to exhaustively master principles), *Qingshi yanjiu* 3 (2003).

———, *Mingmo Qingchu (1582–1687) de gewu qiongli zhi xue* (Studies investigating things and exhaustively mastering principles in the late Ming and early Qing). Chengdu: Sichuan jiaoyu chubanshe, 2003.

Shangshu tongjian (Concordance to the Documents Classic). Beijing: Chaori wenxian Press, 1982 reprint.

Shao Yong, *Huangji jingshi shu* (Treatise on the ultimate axis for managing the world). SKQS edition.

Shen Guowei (Shin Kokui), ed., *Rokugo sōtan no gakusaiteki kenkyū* (Scholarly study of the *Shanghae Serial, 1857–1858*). Tokyo: Hakuteisha, 1999.

Sheng Zengxiu et al., *Wenbing xuepai si dajia yanjiu* (Research on the four masters of the heat therapy scholarly tradition). Beijing: Zhongguo Zhongyi yao chuban she, 2000.

Shi Yunli and Lü Lingfeng, "Lizhi, chuanjiao yu jiaoshi ceyan: Qing Qintianjian dang'an zhong de jiaoshi jilu toushi (Calendrics, missionaries, and eclipse observations: A perspective on eclipse records in the archives of the Qing dynasty Astro-calendric Bureau), *Ziran bianzheng fa tongxun* 24, 6 (2002).

Sima Qian, *Shiji* (Records of the official historian). Beijing: Zhonghua shuju, 1972.

Su Jing, *Qingji Tongwen guan ji qi shisheng* (Qing foreign language schools and their teachers and students). Taibei: Shanghai yinshua chang, 1985.

Su Yunfeng, *Cong Qinghua xuetang dao Qinghua daxue* (From Qinghua public school to Qinghua University). Beijing: Sanlian shuju, 2001.

Sun Jiji, *Kang-Yong-Qian shiqi yutu cehui yu jiangyu xingcheng yanjiu* (Study of maps and surveys during the Kangxi, Yongzheng, and Qianlong periods and the formation of territorial boundaries). Beijing: Zhongguo renmin daxue chuban she, 2003.

Taiping yulan (Materials of the Taiping Xing Guo Era [976–983] for the emperor to read). Reprint, *Sibu congkan* edition. Taibei: Zhonghua shuju, 1960.

Tan Sitong, "Yitai shuo" (On the ether), in *Tan Sitong quanji* (Collected works of Tan Sitong). 2 vols. Beijing: Zhonghua shuju, 1981.

———, "Xing suanxue yi" (Proposal to promote study of mathematics), in *Tan Sitong quanji*.

Tan Tai, "Chouren jie" (Explanation of mathematical astronomers), in CRZ.

Tang Caichang, "Zhuzi yulei yiyou Xiren gezhi zhi li tiaozheng" (Proof that passages in Conversations with Master Zhu [Xi] classified topically already contain principles of Western science), in *Tang Caichang ji* (Collected works of Tang Caichang). Beijing: Zhonghua shuju, 1980.

Tian Tao, "Zhongxi wenjian lu, Gezhi huibian yingyinben xu" (Preface to the reproduction of the *Peking Magazine* and the *Chinese Scientific Magazine*), in *Zhongxi wenjianlu*. Reprint in 4 vols. Nanjing: Guji shudian, 1992, vol. 1.

Vagnoni, Alfonso, *Kongji gezhi* (Treatise on the composition of the universe), reproduction of the 1633 version in *Tianzhu jiao dongchuan wenxian sanpian*. Taibei: Wenjin chuban she, 1979.

Wan'guo gongbao wenxuan (Selections from the Review of the Times), edited by Qian Zhongshu and Zhu Weizheng. Beijing: Sanlian shudian, 1998.

Wang Baoping, "Riben Hu Wenhuan congshu jingyan lu" (Record of seeing Hu Wenhuan's collectanea in Japan), in Lu Jian and Wang Yong, eds., *Zhongguo dianji liuchuan yu yingxiang*. Hangzhou: Hangzhou daxue chuban she, 1990.

———, "Zhongguo Hu Wenhuan congshu jingyan lu" (Record of viewing Hu Wenhuan's Collectanea in China), in *Zhong Ri wenhua luncong* (1991).

———, "Mindai no kakushoka Ko Bunkan ni kan suru kōsatsu" (An inquiry into the Ming dynasty printer Hu Wenhuan), *Kyūko* 36 (1999).

Wang Bing, "19 shiji zhongqi zhi 20 shiji chuqi Zhongguo he Riben de wulixue" (Chinese and Japanese physics from the mid-nineteenth century to the early twentieth), *Ziran kexue shi yanjiu* 13, 4 (1994).

———, "Jindai zaoqi Zhongguo he Riben zhi jian de wulixue jiaoliu" (Early modern exchanges between China and Japan in physics), *Ziran kexue shi yanjiu* 15, 3 (1996).

———, *Qinmin zhi shi: Nan Huairen* (A diligent literatus: Ferdinand Verbiest). Beijing: Kexue chuban she, 2000.

Wang Ermin, *Qingji bing gongye de xingqi* (Rise of the armaments industry in the late Qing period). Taibei: Institute of Modern History, Academia Sinica, 1972.

———, "Rujia chuantong yu jindai Zhong-Xi sichao zhi huitong" (The literati tradition and the compatibility of Chinese and Western ideas in modern times), *Xinya xueshu jikan* 2 (1979).

———, *Shanghai Gezhi Shuyuan zhilue* (An account of the Shanghai Polytechnic). Hong Kong: Zhongwen daxue chuban she, 1980.

Wang Lixin, *Meiguo chuanjiao shi yu wan Qing Zhongguo xiandai hua* (American missionaries and late Qing Chinese modernization). Tianjin: Renmin chuban she, 1997.

Wang Ping, *Xifang lisuan xue zhi shuru* (Influx of Western mathematical astronomy). Taibei: *Jindai shi yanjiu suo*, Academia Sinica, 1966.

———, "Qingchu lisuan yanjiu yu jiaoyu" (Early Qing research on mathematical astronomy and education), *Jindai shi yanjiu suo jikan* 3 (1972).

———, "Ruan Yuan yu *Chouren zhuan*" (Ruan Yuan and the Biographies of mathematical astronomers), *Jindai shi yanjiu suo jikan* 4 (1973).

Wang Qingren, *Yilin gaicuo* (Corrections of errors in the forest of medicine). Jinreng: Shuye deji, 1847.

Wang Renjun, "Lueli" (Overview), in *Gezhi guwei* (Ancient subtleties of science). Wuchang: Zhixuehui, 1897 edition.

Wang Shixiong, *Wenre jingwei* (Warp and weft of warm and hot factor diseases). Shanghai: Qianqing tang, 1863.

Wang Xianming, *Jindai xinxue: Zhongguo chuantong xueshu wenhua de shanbian yu chonggou* (New studies in the modern period: Change and restructuring in Chinese traditional intellectual culture). Beijing: Shangwu yinshu guan, 2000.

Wang Xinzhong, "Fuzhou chuanchang zhi yange" (Reform of the Fuzhou Navy Yard), *Qinghua xuebao* 8 (December 1932).

Wang Yangming, *Wang Wencheng quanshu* (Complete works of Wang Yangming). SKQS edition.

———, *Wang Yangming quanji* (Complete collection of Wang Yangming). Taipei: Kaozheng chubanshe, 1972.

Wang Yangzong, "Qingmo Yizhi shuhui tongyi keji shuyu gongzuo shuping" (Evaluation of the work of the late Qing School and Textbook Committee of the Education Society in the unification of scientific and technical terminology), *Zhongguo keji shiliao* 12 (1991.2).

———, "Gezhi huibian zhi Zhongguo bianjizhe kao" (Inquiry into a Chinese editor of the Scientific Journal], *Wenxian* 63 (January 1995).

———, "Gezhi guwei tiyao" (Outline of the Ancient subtleties of science), in ZGKX ("Zonghe juan," 1996), vol. 7.

———, "Mingmo Qingchu Xixue Zhongyuan lun xinkao" (New analysis of the late Ming and early Qing view that Western learning originated in China), in Liu Dun, ed., *Kexue xinzhuan: qingzhu Du Shiran xiansheng congshi kexue shi yanjiu sishi zhounian xueshu lunwen ji*. Shenyang: Liaoning Education Press, 1997.

———, "Liuhe congtan zhong de jindai kexue zhishi ji qi zai Qingmo de yingxiang" (Knowledge of modern science in the *Shanghae Serial* and its impact in the late Qing), *Zhongguo keji shiliao* 20 (1999.3).

———, *Fu Lanya yu jindai Zhongguo de kexue qimeng* (John Fryer and the beginnings of modern Chinese science). Beijing: Kexue chuban she, 2000.

———, "Hexuli Kexue daolun de liangge Zhong yiben" (Two Chinese translations of Huxley's Introductory for the *Science Primers*), *Zhongguo keji shiliao* 21, 3 (2000).

———, "1850 niandai zhi 1910 nian Zhongguo yu Riben zhi jian kexue shuji de jiaoliu shulue" (Summary of Sino-Chinese exchanges of science books from the 1850s to 1910), *Tōzai gakujutsu kenkyūjo kiyō* (Kansai University) 33 (March 2000).

Wang Zhenghua, "Shenghuo, zhishi yu wenhua shangpin: wan Ming Fujian ban 'riyong leishu' yu qi shuhua men" (Daily life, commercial knowledge, and cultural consumption: late Ming daily use compendia on calligraphy and painting), *Jindai shi yanjiu suo jikan* (Academia Sinica, Taiwan) 41 (2003).

Wang Zhong, "Buyi" (Appended parts), in Wang, *Shuxue* (Discourses on learning). Taibei: Guangwen shuju Reprint, 1970.

Wu Boya, *Kang Yong Qian sandi yu Xixue dongjian* (The Eastern dissemination of Western learning and the three emperors Kangxi, Yongzheng, and Qianlong). Beijing: Zongjiao wenhua chuban she, 2002.

Wu Jishou, "Xu" (Preface), in Cheng Dawei, *Suanfa tongzong*.

Wu Renshu, "Wan Ming de lüyou huodong yu xiaofei wenhua—yi Jiangnan wei taolun zhongxin" (Late Ming travels and consumption culture—with Jiangnan as the center of discussion) *Jindai shi yanjiu suo jikan* (Academia Sinica, Taiwan) 41 (2003).

Wu Weizu, compiler, *Siku caijin shumu* (Catalog of works presented to and selected for the Imperial Library). Beijing: Commercial Press, 1960.

Wylie, Alexander, and Li Shanlan, trans., *Shuxue qimeng* (Compendium of arithmetic). Shanghai: Mohai shuguan, 1853.

———, and Li Shanlan, trans., *Daishu xue* (Elements of algebra). Shanghai: Mohai shuguan, 1859.

———, and Li Shanlan, trans., *Dai weiji shiji* (Elements of analytical geometry and of the differential and integral calculus, lit., "Step by step in algebra, the differential and integral calculus"). Shanghai: Mohai shuguan, 1859.

———, and Li Shanlan, trans., *Tantian* (The outline of astronomy). Shanghai: Mohai shuguan, 1859.

Xiong Mingyu, "Zixu" (*Personal preface*) to the *Gezhi cao* (Draft for investigating things and extending knowledge). 1648 edition.

Xiong Yuezhi, "Gezhi huibian yu Xixue chuanbo" (The Chinese Science Magazine and the transmission of Western learning), *Shanghai yanjiu luncong* 1 (1989).

———, *Xixue dongjian yu wan Qing shehui* (The eastern dissemination of Western learning and late Qing society). Shanghai: Renmin chuban she, 1994.

Xu Guangqi, *Nongzheng quanshu* (Complete works on agrarian management). Reprint. Shuhai lou edition, 1843.

———, *Xu Guangqi ji* (Collection of Xu Guangqi). Taibei: Mingwen shuju, 1986.

Xu Guangqi nianpu (*Chronological* biography of Xu Guangqi). Shanghai: Guji chuban she, 1981.

Xu Guangtai, "Mingmo Qingchu xifang gezhi xue de chongji yu fanying: yi Xiong Mingyu *Gezhi cao* wei li" (The impact and reaction to Western natural studies in the late Ming and early Qing: Xiong Mingyu's Draft for Investigating Things and Extending Knowledge as an example), in *Shibien chunti yu geren*. Taibei: Taiwan National University, 1996.

Xu Haisong, *Qingchu shiren yu xixue* (Early Qing literati and Western learning). Beijing: Dongfang chubanshe, 2000.

Xu Jianyin, "Fanli" ("Conventions"), in Xu Jianyin, ed., *Gezhi congshu* (Collectanea of science). Shanghai: Zhuji shuzhuang, 1900–1901.

Xu Ke, *Qingbai leichao* (Classified jottings on Qing dynasty unofficial history). Shanghai: Commercial Press, 1920.

Xu Shiying, *Zhongguo mulu xue shi* (History of Chinese cataloging studies). Taibei: Chinese University Press, 1982.

Xu Shou, "*Gezhi huibian xu*" (Preface to *Chinese Scientific Magazine*), in the GZHB, vol. 1.

Yan Jing, "Xu" (Preface) to the *Shiwu jiyuan* (Record of the origins of things and affairs), in the GZCS.

Yan Ruoju, *Qianqiu zhaji* (Yan Ruoju's Reading Notes). Taibei: Commercial Press, 1973.

Yang Boda, "Qingdai boli gaishu" (Overview of Qing dynasty glassware), *Gugong bowuyuan yuankan*, 1983.

Yang Cuihua (Tsui-hua), "Ren Hongjun yu Zhongguo jindai de kexue sixiang yu shiye" (Ren Hongjun and modern Chinese scientific thought and achievements), *Jindai shi yanjiu suo jikan* (Academia Sinica, Taiwan) 24 (June 1995).

Yang Jian, "Kindai Chūgoku ni okeru butsuri gakusha no tanjō to sono shakai teki haikei" (Birth of physicists in modern China and their social background), *Kagakushi kenkyū* II, 35, 197 (1996).

————, "Chūgoku ni okeru butsuri gaku no senmon kyōiku no seiritsu" (Establishment of specialized education of physics in China), *Kagakushi kenkyū* II, 36, 202 (1997).

"Yangwu yundong dashi ji" (Record of important matters during the foreign studies movement), in Xu Tailai, ed., *Yangwu yundong xinlun* (New views on the foreign studies movement). Changsha: Hunan People's Press, 1986.

Yatsumimi Toshifumi, "Shizen shingaku to shizen kagaku no aide de" (Between natural theology and natural science), in Shen Guowei (Shin Kokui), ed., *Rokugo sōtan no gakusai teki kenkyū.*

Ye Fang, "Xu fangshi shi" (Preface from the dawn of facts), in GZCS.

————, "Houxu" (Afterword), in GZCS.

Yoda Yoshiie, "Jūkyūseiki kōhan ni okeru Nisshin ryōgoku no shokusan seisaku ni tsuite" (Concerning industrial development policies of Japan and the Qing in the second half of the nineteenth century), *Shakai kagaku tōkyū* 12, 3 (March 1967).

Yu Chang, "Xu" (Preface), in Yu, *Yuyi cao* (Notes of indirectly expressed ideas). *Jingyuan shushi kan* edition, 1643.

Yu Weigang, "Hu Wenhuan yu Gezhi congshu" (Hu Wenhuan and the Collectanea of works investigating things and extending knowledge), *Tushuguan zazhi* (November 1982).

Yu Yue, "Xu" (Preface), in Wang Renjun, *Gezhi guwei.*

————, "Zixu" (Preface) to *Huang Qing jingjie jianmu* (Topical index to the Qing exegesis of the classics). 1886 edition.

Yuanming yuan (Lofty Pavilion), 1 (November 1981).

Yuzhi shuli jingyun (Imperially instituted, Treasury of mathematics), in *Zhongguo kexue jishu dianji tonghui* (Compendium of classical works in Chinese science and technology), edited by Ren Jiyu et al, vol. 3. Zhengzhou: Henan jiaoyu chuban she, 1993.

Zeng Guofan, "Xu," (Preface), in Alexander Wylie's and Li Shanlan's enlarged edition of Matteo Ricci and Xu Guangqi, trans., *Jihe yuanben* (Euclid's Elements of geometry). Nanjing: Nanjing Arsenal, 1865.

Zeng Jize, "Ji" (Note), in GZQM (1886), vol. 1.

Zeng Jianli, "Gezhi guwei yu wan Qing 'Xixue Zhongyuan' lun" (The Ancient subtleties of science and the theory of the Chinese origins of Western learning), *Zhongzhou xuekan* 6 (November 2000).

Zhang Binglin, "Zhizhi gewu zhengyi" (Correct meaning of extending knowledge and investigating things), in *Zhang Taiyan quanji* (Complete collection of Zhang Binglin). Shanghai: Renmin chuban she, 1982, vol. 5.

Zhang Hailin, *Wang Tao pingzhuan* (Critical biography of Wang Tao). Nanjing: Nanjing daxue chuban she, 1993.

Zhang Huang, *Tushu bian* (Compendium of maps and materials). Reprint. Taibei: Chengwen chuban she, 1971.

Zhang Jian, "Zhongxi wenjian lu shulue" (Summary of *Peking Magazine*), *Fudan xuebao (shehui kexue ban)* 4 (1995).

Zhang Junmai, "Rensheng guan" (Outlook on life), in *Kexue yu rensheng guan* (Science and outlooks on life). 2 vols. Shanghai: Dongya tushuguan, 1923, vol. 1.

Zhang Rongmei, "Nihon ni okeru 'butsuri gaku' to iu yakugo no keisei" (The formation of the translation for physics in Japan), *Kanzawa daigaku daigakuin shakai kankyō kagaku kenkyū* 2 (1997).

Zhang Xia et al., *Qingmo haijun shiliao* (Historical materials on the late Qing navy). Beijing: Haijun chuban she, 2001.

Zhang Yongtang, *Mingmo Qingchu lixue yu kexue guanxi zailun* (Reevaluation of the relation between studies of principle and science during the late Ming and early Qing). Taibei: Xuesheng shuju, 1994.

Zhang Yuanfang, "Xixue shiliu zhong xu" (Preface to Western learning in sixteen kinds), *Xixue qimeng*. Shanghai: Zhuyitang, 1896.

Zhang Yufa, "Fuzhou chuanchang zhi kaichuang ji qi chuqi fazhan" (The founding and early development of the Fuzhou Navy Yard), *Jindai shi yanjiu suo jikan* (Taiwan) 2 (June 1971).

Zhang Zhidong, "Xu" (Preface), in Wang Renjun, *Gezhi jinghua lu* (Record of the essence of science). n.d.

Zhao Yuanyi, ed., *Gezhi shuyuan jiawu keyi* (1894 Shanghai Polytechnic prize essay competition). Shanghai lithograph, 1897.

Zhongguo taoci shi (History of Chinese porcelain). Beijing: Wenwu chuban she, 1982.

Zhonghua min'guo chuqi lishi yantaohui lunwen ji 1912–1927. 2 vols. Taibei: Academia Sinica, Taiwan, Institute of Modern History, 1985.

Zhongwai shiwu cewen leibian dacheng (Great compendium of policy questions on Chinese and foreign affairs classified topically). 1903 edition.

Zhou Zhiping and Willard Peterson, eds., *Guoshi fuhai kaixin lu: Yu Yingshi jiaoshou rongtui lunwen ji* (Record of opening new directions in Chinese history across the seas: Collected essays honoring the retirement of Professor Yu Yingshi). Taibei: Lianjing, 2002.

Zhouli jishuo (Collected sayings on the Rituals of Zhou). SKQS edition.

Zhu Kaijia, "Gezhi xinbao yuanqi" (Beginning of the *Scientific Magazine*), in *Gezhi xinbao* 1 (1898.3.13), included in *Jindai Zhongguo shiliao congkan* (Taibei: Wenhai chuban she, 1987), vol. 231.

Zhu Shijie, *Suanxue qimeng* (Primer of mathematics). Reprint. Shanghai: Jiangnan Arsenal, 1871.

Zhu Xi, *Zhouyi benyi* (Original meanings in the *Zhou Change*). Reprint of the National School edition. Taibei: Hualian chubanshe, 1971.

———, *Zhuzi yulei* (Conversations with Master Zhu [Xi] Classified Topically). 1473 edition. Reprint. Taibei: Zhongzheng Bookstore, 1973. See also the SKQS edition.

Zhuang Jifa, *Jingshi daxue tang* (The Metropolitan University of Beijing). Taibei: National Taiwan University, 1970.

Zhuangzi yinde (Concordance to the *Zhuangzi*). Taibei: Chengwen Bookstore, 1974.

Zou Zhenhuan, *Wan Qing Xifang dilixue zai Zhongguo* (Western geography in China during the late Qing). Shanghai: Guji chuban she, 2000.

Acknowledgments

Research for this project began in earnest while I was Mellon Visiting Professor in Traditional Chinese History and Civilization in the School of Historical Studies at the Institute for Advanced Study, Princeton, from 1999 to 2001. In addition to the faculty in the school, I interacted with faculty in East Asian Studies and History at Princeton University. Several members of the school participated with me in colloquia and seminars on "Culture and Science in Late Traditional China": Cynthia Brokaw (Ohio State University), Qi Han (Chinese Academic of Sciences, Beijing), Pingyi Chu (Academia Sinica, Taiwan), Roger Hart (University of Texas, Austin) and Thomas Wilson (Hamilton College).

Tony Grafton, Graham Burnett, and I organized a workshop, "Science and Philology in China and Europe, 1550–1850," sponsored by the Princeton History of Science Program in conjunction with the Shelby Cullom Davis Center for Historical Studies. I benefited from discussions on my work at the Princeton East Asian Studies Lunch Colloquium, the colloquium "Qing History (1600–1900) through 'Things,'" cosponsored by Princeton and the Institute for Advanced Study, and the Early Chinese History Seminar at Harvard. I also presented talks at Academia Sinica, Taiwan, and Tokyo University.

I thank the organizers of several important conferences I attended in Germany for allowing me to present my papers in their precincts: Michel Lackner (Erlangen) and Natascha Vittinghoff (Frankfurt), both then at Göttingen; Hans-Ulrich Vogel at Tübingen, and Axel Schneider (Leiden) and Susanne Weigelin-Schwiedrzik (Vienna), both then at Heidelberg. The title "On Their Own Terms" of my volume was inspired in part by a subheading in *New Terms for New Ideas,* edited by Lackner and others for E. J. Brill in Leiden and published in 2001. Pierre-Etienne Will invited me to present a talk for the Chinese Studies Seminar at the Collège de France. I also presented a paper at the International Sinological Center at Charles University, Prague, for a conference organized by Chün-chieh Huang and Thomas Lee, sponsored by the Himalaya Foundation and the Chiang Ching-kuo Foundation.

Norton Wise commented on my paper for a panel at the American Historical Association Annual Meeting. R. Bin Wong invited me to present some portions of the project at the University of California, Irvine. I also presented papers at the Columbia University Pre-modern China Seminar, the Columbia Modern China Seminar, the Princeton Renaissance and Early Modern Colloquium, and the workshop "Rethinking the Legacy of Nineteenth Century China," sponsored by the Center for Chinese Studies at the University of Michigan. I thank Lydia Liu and James Lee for inviting me.

Joachim Kurtz (Emory) and Hugh Shapiro (Nevada, Reno), then members of the Institute for Advanced Study, attended my seminar on "Modern Science in China, 1600–1900." Their comments, along with those of graduate students Bruce Rusk and Jonathan Seitz, helped me clarify key issues and eliminate mistakes. Graduate students who attended my science seminars at UCLA also helped this project immensely. They include Zvi Aziz Ben-Dor, Avner Ben-Zaken, Yuanling Chao, Pingyi Chu, Xiaoping Cong, Roger Hart, Sonja Kim, Eugenia Lean, Minghui Hu, Yue Meng, Min Suh Son, and Soyoung Suh. Karine Chemla and Catherine Jami, both at CNRS Paris, kindly sent me their valuable articles on Chinese mathematics.

Since 1999, I visited libraries and archives in Los Angeles (UCLA East Asian Library), Princeton (Gest East Asian Collection), Berkeley (UC, Berkeley Fryer Collection), Cambridge (Harvard-Yenching Library), Stanford (East Asian Library), Shanghai (Fudan University Library), Beijing (History of Science Library, Chinese Academy of Sciences), Taiwan (Academia Sinica, Institute of Modern History Library), Kyoto (Institute for Humanistic Studies, Oriental Library), and Tokyo (Tokyo University Oriental Library). While I was in Beijing, Wang Junyi of the Chinese Academy of Social Sciences Press and Chen Zuwu, Director of the Institute of History, Chinese Academy of Social Sciences, assisted my research.

I received a Princeton University Faculty Research Grant to spend four weeks in the summer of 2003 at the Needham Research Institute at Cambridge University. John Moffett, the librarian of the East Asian History of Science Library, and Susan Bennett at the Needham Institute were very helpful during my stay, as were other research fellows, particularly James Benn and Kim Taylor. Christopher Cullen, the director, invited me to present a summary of my work on the Protestants and Cambridge science in China. The Needham Institute gave me access to the specialized library that had informed Joseph Needham's *Science and Civilisation in China* series, which facilitated my efforts to complete the final research for the project.

The strengths of the project draw from the many fruitful interactions and generous library support summarized above. Pamela Crossley invited me to present the complete manuscript at the Dartmouth Critical Manuscript Seminar on May 3, 2004. I thank all the commentators there for their many critical observations. Michael Gordin and Liam Brockey at Princeton also read the manuscript and gave me helpful comments. Margaret Jacob of UCLA encouraged me to think of this book as a first step to prepare a more general textbook for the New Histories of Science, Technology, and Medicine Series published by Harvard University Press. Nathan Sivin and Kent Guy read the entire manuscript for Harvard University Press, and each offered vigorous comments that greatly improved the project. Soren Edgren, Martin Heijdra, and Willard Peterson of Princeton University offered frequent advice.

Credits

Maps

1. Physiocratic Macroregions of Agrarian China in Relation to Major Rivers from G. William Skinner, ed., *The City in Late Imperial China*. Copyright © 1977 by the Board of Trustees of the Leland Stanford University.

3.1 and 3.2. *Map of China and Foreign Areas (Huayi tu)* and *Map of the Tracks of Emperor Yu (Yuji tu)* from Joseph Needham, *Science and Civilization in China*, vol. 3. Copyright © 1959. Reprinted with the permission of Cambridge University Press.

3.4. *Capital Edition of a Complete Map Based on Astronomy (Jingban tianwen quantu)*. Reprinted with the permission of Richard Smith and Rice University, Fondren Library.

Figures

P.1. A Ming painting of "Broadly Examining Antiquities." National Palace Museum, Taiwan, Republic of China.

P.2. Late Ming pottery-vase in the shape of a ritual jade blue-and-white porcelain. China Institute Gallery.

4.1. The Order of the 64 hexagrams in Bouvet's studies of the *Change Classic*. Copyright © 1989. Reprinted with the permission of David E. Mungello.

5.2. Blue-and-white porcelain bowl from the Qing dynasty. China Institute Gallery.

10.2. Nakamura Shūkō, "The Great Victory of Japanese Warships off Haiyang Island, 1894." Meiji era, printed and published October 1, 1894. Photograph © 2004 Museum of Fine Arts, Boston. Anonymous gift, RES.23.288.

10.3. Kobayashi Kiyochika, "Illustration of our Naval Forces in the Yellow Sea firing at and sinking Chinese Warships, 1894." Meiji era, printed and published October 1894. Photograph © 2004 Museum of Fine Arts, Boston. William Sturgis Bigelow Collection, 21.1765.

Index

545